Autonomous Robotic Systems

Studies in Fuzziness and Soft Computing

Editor-in-chief
Prof. Janusz Kacprzyk
Systems Research Institute
Polish Academy of Sciences
ul. Newelska 6
01-447 Warsaw, Poland
E-mail: kacprzyk@ibspan.waw.pl
http://www.springer.de/cgi-bin/search_book.pl?series=2941

Further volumes of this series can be found at our homepage.

Vol. 95. T.Y. Lin, Y.Y. Yao and L.A. Zadeh (Eds.)
Data Mining, Rough Sets and Granular Computing, 2002
ISBN 3-7908-1461-X

Vol. 96. M. Schmitt, H.-N. Teodorescu, A. Jain, A. Jain, S. Jain and L.C. Jain (Eds.)
Computational Intelligence Processing in Medical Diagnosis, 2002
ISBN 3-7908-1463-6

Vol. 97. T. Calvo, G. Mayor and R. Mesiar (Eds.)
Aggregation Operators, 2002
ISBN 3-7908-1468-7

Vol. 98. L.C. Jain, Z. Chen and N. Ichalkaranje (Eds.)
Intelligent Agents and Their Applications, 2002
ISBN 3-7908-1469-5

Vol. 99. C. Huang and Y. Shi
Towards Efficient Fuzzy Information Processing, 2002
ISBN 3-7908-1475-X

Vol. 100. S.-H. Chen (Ed.)
Evolutionary Computation in Economics and Finance, 2002
ISBN 3-7908-1476-8

Vol. 101. S.J. Ovaska and L.M. Sztandera (Eds.)
Soft Computing in Industrial Electronics, 2002
ISBN 3-7908-1477-6

Vol. 102. B. Liu
Theory and Practice of Uncertain Programming, 2002
ISBN 3-7908-1490-3

Vol. 103. N. Barnes and Z.-Q. Liu
Knowledge-Based Vision-Guided Robots, 2002
ISBN 3-7908-1494-6

Vol. 104. F. Rothlauf
Representations for Genetic and Evolutionary Algorithms, 2002
ISBN 3-7908-1496-2

Vol. 105. J. Segovia, P.S. Szczepaniak and M. Niedzwiedzinski (Eds.)
E-Commerce and Intelligent Methods, 2002
ISBN 3-7908-1499-7

Vol. 106. P. Matsakis and L.M. Sztandera (Eds.)
Applying Soft Computing in Defining Spatial Relations, 2002
ISBN 3-7908-1504-7

Vol. 107. V. Dimitrov and B. Hodge
Social Fuzziology, 2002
ISBN 3-7908-1506-3

Vol. 108. L.M. Sztandera and C. Pastore (Eds.)
Soft Computing in Textile Sciences, 2003
ISBN 3-7908-1512-8

Vol. 109. R.J. Duro, J. Santos and M. Graña (Eds.)
Biologically Inspired Robot Behavior Engineering, 2003
ISBN 3-7908-1513-6

Vol. 110. E. Fink
Changes of Problem Representation, 2003
ISBN 3-7908-1523-3

Vol. 111. P.S. Szczepaniak, J. Segovia, J. Kacprzyk and L.A. Zadeh (Eds.)
Intelligent Exploration of the Web, 2003
ISBN 3-7908-1529-2

Vol. 112. Y. Jin
Advanced Fuzzy Systems Design and Applications, 2003
ISBN 3-7908-1537-3

Vol. 113. A. Abraham, L.C. Jain and J. Kacprzyk (Eds.)
Recent Advances in Intelligent Paradigms and Applications, 2003
ISBN 3-7908-1538-1

Vol. 114. M. Fitting and E. Orłowska (Eds.)
Beyond Two: Theory and Applications of Multiple Valued Logic, 2003
ISBN 3-7908-1541-1

Vol. 115. J.J. Buckley
Fuzzy Probabilities, 2003
ISBN 3-7908-1542-X

Changjiu Zhou
Darío Maravall
Da Ruan
Editors

Autonomous Robotic Systems

Soft Computing and Hard Computing Methodologies and Applications

With 263 Figures
and 21 Tables

Springer-Verlag Berlin Heidelberg GmbH

A Springer-Verlag Company

Prof. Dr. Changjiu Zhou
Singapore Polytechnic
School of Electrical and Electronic Engineering
500 Dover Road
Singapore 139651
Republic of Singapore
zhoucj@sp.edu.sg

Prof. Dr. Darío Maravall
Universidad Politécnica de Madrid
Department of Artificial Intelligence
Faculty of Computer Science
28660 Boadilla del Monte, Madrid
Spain
dmaravall@fi.upm.es

Dr. Dr. h.c. Da Ruan
Belgian Nuclear Research Centre (SCK·CEN)
Boeretang 200
2400 Mol
Belgium
druan@sckcen.be

ISSN 1434-9922

DOI 10.1007/978-3-7908-1767-6

Cataloging-in-Publication Data applied for
A catalog record for this book is available from the Library of Congress.
Bibliographic information published by Die Deutsche Bibliothek
Die Deutsche Bibliothek lists this publication in the Deutsche Nationalbibliografie; detailed bibliographic data is available in the Internet at <http://dnb.ddb.de>.

Originally published by Physica-Verlag Heidelberg in 2003.
MyCopy version of the original edition 2003

Foreword

The key words *soft computing*, *hard computing* and *autonomous robotics* used in the title of this edited volume *Autonomous Robotic Systems: Soft Computing and Hard Computing Methodologies and Applications*, reflect upon the development of robotic systems and some human-like attributes such as perception and cognition.

It was in 1965 when our friend and mentor, Professor Lotfi A. Zadeh, first introduced the notion of *fuzzy sets*. In 1991 he founded the Berkeley Initiative in Soft Computing (BISC).

The principle constituent methodologies of soft computing employ a number of emerging theoretical tools: fuzzy logic (FL), neural computing (NC), fuzzy-neural computing (FNC), genetic algorithms (GAs), fuzzy-genetic algorithms (FGAs), neuro-genetic algorithms (NGAs), fuzzy neural-genetic algorithms (FNGAs), evolutionary computing (EC), probabilistic computing (PC) and components of machine learning theory (MLT).

Traditionally, theoretical tools in science have led to a better understanding of the world in which we live. In the development of these theoretical tools, we have employed many mathematical tools and concepts inherent in the natural sciences. But as we move further into the area of autonomous robotic systems, a major area of our research is the understanding of robust tasks performed by humans. This includes the tasks of vision, perception, cognition, thinking, reasoning, speech understanding, pattern recognition, decision-making, and reasoning and control, amongst others.

We must congratulate the editors of this volume, Dr. Changjiu Zhou, Dr. Darío Maravall, and Dr. Da Ruan, for bringing together this collection of research papers in this important field of *Autonomous Robotic Systems*. This volume contains eighteen invited chapters co-authored by 45 international researchers from nine different countries.*

These eighteen chapters are divided into four parts. In Part 1 (2 chapters), the authors deal with the development of some of the basic principles and methodologies of soft computing for intelligent robotic systems. In Part 2 (6 chapters), the authors introduce some basic cognitive aspects in path planning and navigation, which lead to the design of intelligent sensory and control mechanisms. In Part 3 (7 chapters), the authors describe their work on learning, adaptation and control

* Canada (4 authors, 1 chapter); Germany (2 authors, 1 chapter); Italy (1 author, 1 chapter); Korea (1 author, 1 chapter); Singapore (8 authors, 4 chapters); Spain (16 authors, 8 chapters); Sweden (2 authors, 1 chapter); UK (6 authors, 2 chapters); and USA (5 authors, 1 chapter).

mechanisms in an uncertain mobile environment. Finally, in Part 4 (3 chapters), the authors introduce several provocative thoughts on vision and perception - the basic sensory elements in humans. Such theoretical notions and mathematical tools may contribute significantly to the further development of robust robotic systems.

With the rapidly growing research interest in the theoretical aspects of intelligent systems, and the increasing fields of applications of intelligent robots (aerospace, process control, ocean exploration, manufacturing and resource based industry, etc.), there is a need for books that deal with their theoretical foundations, implementations and applications. I am pleased to see that the editors of this volume conceived these ideas by their interactions with the intelligent systems community and invited researchers in this field.

It is gratifying to acknowledge the devotion of the many researchers that has helped the exploration of new theoretical approaches, the stimulation of exchanges of scientific information, and the reinforcement of international cooperation in this important field.

It is this vision that underlines the authoritative up-to-date and reader-friendly exposition of *Autonomous Robotic Systems* in this book. It is a *must reading* for students and researchers interested in exploring the potentials of the fascinating field that will form the basis for the design of the intelligent machines of the future.

Madan M. Gupta
Professor and Director
Intelligent Systems Research Laboratory
College of Engineering
University of Saskatchewan
Saskatoon, Saskatchewan, Canada
S7N-5A9
guptam@sask.usaask.ca

July 2002

Preface

Autonomous Robotic Systems aim at building physical systems that can accomplish useful tasks without human intervention and perform in the unmodified real-world situations that usually involve unstructured environments and large uncertainties. Hence, ideal autonomous robots should be capable of determining all the possible actions in an unpredictable dynamic environment using information from various sensors such as computer vision, tactile sensing, ultrasonic and sonar sensors, and other smart sensors. There are several methodologies that are capable of solving existing problems in the field of autonomous robotic systems. The well-developed field of conventional hard computing (HC) techniques offers efficient solutions to a wide variety of existing applications of robotics. However, HC techniques are model-based schemes that in most cases are synthesized using incomplete information and partially known or inaccurately defined parameters. They are extremely sensitive to the lack of sensor information and to unplanned events and unfamiliar situations in the working environment. It does not seem that such techniques by themselves can cope very well with uncertain and unpredictable environments all the time. On the other hand, the advent of soft computing (SC) techniques does provide us with powerful tools to solve demanding real-world problems with uncertain and unpredictable environments.

As indicated by Professor Lotfi A. Zadeh, soft computing may be viewed in two related perspectives. In one view, SC - in contrast with HC - is aimed at an accommodation with pervasive impression of the real world, exploiting the tolerance for impression, uncertainty, and partial truth to achieve tractability, robustness, low solution cost, and better rapport with reality. In another view, SC is a coalition or consortium of methodologies that share this objective. At present, the principal members of the coalition are: fuzzy logic; neurocomputing; evolutionary computing; probabilistic computing; chaotic computing; and machine learning. Thanks to their strong leaning and cognitive ability and good tolerance of uncertainty and impression, the SC techniques have shown their great potential to solve demanding problems in the field of autonomous robotic systems. The emerging SC and conventional HC should not be viewed as competing with each other but rather as complementary. From the existing literature and successful applications, it can be concluded that SC methodologies can enhance and extend traditional HC methods, and the fusion of SC and HC techniques has provided innovative solutions for autonomous robotic systems. Therefore, it is interesting to gather current trends and provide a high-quality volume for scientists and engineers working in SC, HC and autonomous robotic systems areas. As we are entering a new information technological era, the fusion of HC and SC techniques will certainly play a significant role in the field of autonomous robotic systems.

This volume is in part based on the recent invited sessions at the two international conferences. One is on "Autonomous Mobile Robotics" at the 6th International Work-Conference on Artificial and Natural Neural Networks, June 13-15, 2001, Granada, Spain. Another is on "Soft Computing and AI Methods for Autonomous Robotic systems" at the International Conference on Computational Intelligence, Robotics and Autonomous Systems, November 28-30, 2001, Singapore. We have also invited some known authors as well as announced a formal Call for Papers to several research groups related to SC and robotics to contribute the latest progress and recent trends and research results in this field. The primary aim of this volume is to provide researchers and engineers from both academic and industry with up-to-date coverage of new results and trend toward mobility, intelligence and autonomy in an unstructured world.

The volume is divided into four logical parts containing eighteen chapters written by some of the world's leading experts in the field of autonomous robotic systems in conjunction with SC and HC methodologies.

Part 1 on *Basic Principles and Methodologies* contains two chapters that contribute to a deeper understanding of the methodologies. In the first chapter, Mira and Delgado address some methodological issues on symbolic and connectionist perspectives of AI and answer a very important question to the researchers working in the area of robotics – where is knowledge in robotics? The second chapter by Oussalah explores how the fusion methodology can be decomposed into a set of primary subtasks where the elicitation and the architecture play a central role in the fusion process. A mobile robot application is given to show how the different steps of the fusion architecture have been handled.

In the second part on *Planning and Navigation*, various algorithms for planning and navigation as well as methods for mapping the environment of a robot are discussed. In the third chapter, Tunstel *et al.* address computing strategies designed to enable field mobile robots to execute tasks requiring effective autonomous traversal of natural outdoor terrain. The primary focus is on computer vision-based perception and autonomous control. HC methods are combined with applied SC strategies in the context of three case studies associated with real-world robotics tasks including planetary surface exploration and land survey/reconnaissance. The fourth chapter by De Lope and Maravall describes a hybrid autonomous navigation system for mobile robots. The control architecture proposed is highly modular and is based on the concept of behavior. In the fifth chapter, Peters *et al.* present a rough neurocomputing approach for line-crawling robot (LCR) navigation. The sixth and seventh chapters address some conventional HC techniques on navigation and planning. Urdiales *et al.* propose a hybrid layered architecture, which is used to navigate in totally or partially explored environments using sonar sensors. The main advantage of the proposed scheme is that it can operate in both known and unknown environments rapidly and efficiently. The seventh chapter by Paz *et al.* presents an analytical method for decomposing the external environment representation task for a robot with restricted sensory information. In the eighth chapter, Vadakkepat *et al.* discuss the application of the evolutionary artificial potential field (EAPF) in mobile robot path planning. The parameters of the EAPF are optimized with the multi-objective evolutionary algorithm.

Seven chapters on *Learning, Adaptation, and Control* are presented in Part 3. In the ninth chapter, Saffiotti and Wasik demonstrate the use of hierarchical fuzzy behaviors to implement a set of navigation and ball control behaviors for a Sony four-legged robot operating in the RoboCup domain. They also show that the logical structure of the rules and the hierarchical decomposition simplify the design of very complex behaviors, like the "GoalKeeper" behavior. In the tenth chapter, Maravall and De Lope present a robotic mechanism, which aims at navigating in unconventional environments. The proposed method follows the perception-reason-action paradigm and is based on a reinforcement learning process guided by perceptual feedback, which can be considered as biologically inspired at the functional level. It can be straightforwardly applied to real-time collision avoidance for articulated mechanisms, including conventional manipulator arms. In the eleventh chapter, Hagras *et al.* show how intelligent embedded agents situated in an intelligent domestic environment can perform learning and adaptation. In the twelfth chapter, Wong and Ang provide a survey on the different uses of SC methods in the different aspects of legged robotics. In the thirteenth chapter, Zhang and Rössler propose a self-valuing learning system based on continuous B-spline model, which is capable of learning how to grasp unfamiliar objects and how to generalize the learned abilities. The fourteenth chapter by Er and Gao presents a robust Adaptive Fuzzy Neural Controller that is suitable for identification and control of a class of uncertain Multi-Input-Multi-Output (MIMO) nonlinear systems. In the fifteenth chapter, Er and Sun propose a new approach towards optimal design of a hybrid fuzzy proportional-integral-derivative (PID) controller for robotics systems using the genetic algorithm.

In the last part on *Vision and Perception* for autonomous robotic systems, the sixteenth chapter by Balsi and Vilasís-Cardona shows how Cellular Neural Networks (CNNs) can provide the necessary image processing to guide an autonomous mobile robot in a maze made of black lines on a light surface. The system consists of a fuzzy controller performing the elementary navigation tasks fed by the result of processing the image only by CNN techniques. In the following chapter by Camacho *et al.*, the authors address multiresolution vision in autonomous systems. Multiresolution systems are one alternative to cover wide fields of view without involving high data volumes and, therefore, considerably reduce the constraints imposed by off-the-shelf uniresolution vision systems. In the last chapter, Buenaposada and Baumela focus on the real-time location and tracking of human faces in video sequences. The tracking is based on the cooperation of two low-level trackers from colour and template information. As a result of the coordination of these two trackers, it emerges a robust real-time tracker that accurately computes face position and orientation in varying environmental conditions.

This volume highlights the advantages of fusion of SC and HC methodologies and applications to autonomous robotic systems. Each chapter is self-contained and also indicates the future research direction on the topic of autonomous robotic systems.

We would like to thank Professor Madan Gupta, University of Saskatchewan, Canada, for his willingness to write a foreword for this volume; to Professor Janusz Kacprzyk, Editor-in-Chief of the Book Series "Studies in Fuzziness of Soft Computing", for his kind acceptance to publish this volume; to all the contributors for their kind cooperation to this book; to Hnin Wai Yin and Phyu Phyu Khing, final-year-project students of Singapore Polytechnic, for their editorial assistance, and to Katharina Wetzel-Vandai and Judith Kripp of Physica-Verlag for their advice and help during the production phases of this book.

Changjiu Zhou, Singapore Polytechnic, Singapore
Darío Maravall, Universidad Politécnica de Madrid, Spain
Da Ruan, The Belgian Nuclear Research Centre (SCK•CEN), Belgium

July 2002

Contents

Part 1

BASIC PRINCIPLES AND METHODOLOGIES

Where is Knowledge in Robotics? Some Methodological Issues on Symbolic and Connectionist Perspectives of AI

J. Mira and A.E. Delgado

Dpto. de Inteligencia Artificial,
Facultad de Ciencias y ETSI Informática, UNED, Spain
{jmira,adelgado}@dia.uned.es

Abstract. In this chapter we consider a number of methodological issues to which little importance is normally attributed in robotics but which we consider essential to the development of integrated methods of soft and hard computing and to the understanding of the artificial intelligence (AI) purpose and fundamentals.

The basic conjecture in this chapter is that knowledge always remains at the knowledge level and in the external observer's domain. To the robot only pass the formal model underlying these models of knowledge. Consequently, there are neither essential differences between symbolic and connectionist techniques nor between soft and hard computing. They are different inferences and problem-solving-methods (*PSMs*) that belong to a library and that are selected to be used in a sequential or concurrent manner according to the suitability for decomposing the task under consideration, until we arrive to the level of inferential primitives solved in terms of data and relations specific of the application domain.

The distinctive characteristics of hard and soft computing methods are related to the balance between knowledge and data available in advance, the granularity of the model or the necessity and capacity of learning in real time. Nevertheless, in all these cases the knowledge (the meaning of the entities and relations of the model) always is outside the robot, at the knowledge level, the "house" of models.

In many publications, the robotic programs are described without including any distinction between levels and domains of description of a calculus. As a result, it is generally difficult to determine what the robot actually performs, which knowledge has been represented, and which is artificially injected during the human interpretation of the robots behavior.

In order to make clear this methodological issues we consider the taxonomy of levels introduced by Marr [1] and Newell [2] (Knowledge, Symbols, and Hardware) put on the top of the two domains of description of a calculus (the *domain* proper of each level and that of the *observer* external to the computation of the level). Then, we describe the usual approach to modeling and reduction of models from the knowledge to the symbol level and finally we illustrate the analogies and differences between different models and reduction processes including the operational stage, either symbolic, connectionist, probabilistic or fuzzy. In all the cases our conviction is that most of the work must be made by modeling tasks and *PSMs* at the knowledge level, where it is crystal clear that soft and hard computing are complementary and ready to be integrated.

1 Levels of Description of a Calculus

When we are faced with the complexity of a robot, both in *analysis* tasks in which we attempt to understand how it works, and in *synthesis* tasks, in which we want to obtain the functionally equivalent system from a set of specifications, it is useful to make use of a *hierarchy of levels* which sections the complex whole when producing a description [3].

Each descriptive level is characterized by a phenomenology, a set of entities and relationships, as well as some organizational and structural principles of its own (Figure 1), including a causality law. Each level has its own environment, which is a set of signals to be understood and with which it interacts using a common language specific to the level. When the description of the level is carried out by an external observer, there is always a particular semantics (sign-significance relationship) and a set of issues pertinent to that specific level.

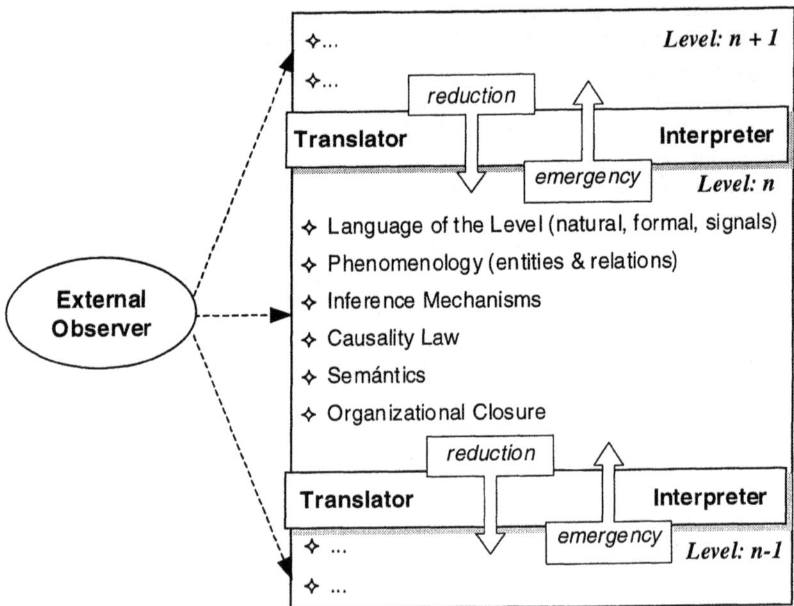

Fig. 1. Characterization of each level of description of a calculus in terms of the entities, relations, causality law and language specific of that level

A frequent source of error in Artificial Intelligence (AI) and robotics lies in the mixing of entities belonging to different levels and the attempt to explain data and processes from a high semantic level (such as the knowledge level) with data structures and logical operators characteristic of the lower levels (such as digital electronics or automata theory). Analogously, neurology tries to explain global aspects of behavior such as memory, intelligence, purpose, emotions and feelings (pleasure, anguish, anxiety) using processes at the cellular and protoplasmic level. Hence, evolution, history, culture and the social dimension of behavior are all neglected.

Every level is partially closed to organization and possesses a set of formal tools, which have proved adequate to mold its dynamics (integro-differential equations, logic, automata theory, algorithmic and heuristic descriptions, programming languages and natural language). This organizational and structural closure brings about the stability of the level and its inclusion within a hierarchy in which each level is linked to the one below by means of a process of *reduction* which follows pre-established and unequivocal laws in the case of the formal languages used in computation (translation, compilation and interpretation programs) and by means of evolutionary and genetic laws (generally unknown), in the case of nervous system. The link with the level above is carried out by means of *emergency* and *interpretation* processes which –in order to be understood– require the *injection of knowledge* from an external observer in order to add semantics to the entities at the lower level. The algorithm and data structure of the program being executed at that moment cannot be deduced from the detailed knowledge of what is happening in the inverters, registers, arithmetic units, and the rest of the architecture of a computer. Analogously, knowing the algorithm does not enable us to deduce the content at the knowledge level of the computation, which has been programmed. Nor is there one single inverse step. There are various possible algorithms for one same computation and multiple possible implementations for one same algorithm (serial, parallel, different language and environment, and finally, different physical machine). Analogously, when we study nervous systems we are not able to deduce from the detailed knowledge of what is happening in every synapse of every neuron, the "algorithms" of our neural nets, or the content at the knowledge level of the computation, which has "programmed" evolution, genetics and interaction with the environment.

The specific characteristics of the different emergency (interpretation) and reduction processes depend on the domain, on the task and on the perspective (analysis or synthesis) at which we distinguish levels of organization. Among these characteristics, robustness against syntactic and semantic disturbances is fundamental. The lower levels are much less robust that the knowledge level. Analogously, biological tissue and neural nets are much more robust than the silicon and digital electronics which supports artificial computation. Actually, the most distinctive computational characteristic is robustness against syntactic error.

In *computation*, Marr [1] and Newell [2], introduced the theory of levels, even though there were clear precedents in Chomsky [4] who introduced the concepts of *"competence"* and *"execution"* to differentiate natural languages from the arbitrary and formal systems of symbol manipulation. To have mastery of a language means to be able to understand what a person talks to us and to answer with other signals, which prove that they carry a meaningful semantic interpretation of the first message. The sentences in a language possess an intrinsic meaning determined by specific rules. We say that whoever possesses these rules has developed a specific *linguistic competence.* However, the use of that language (the *execution*) is not a mere reflection of those rules, but includes many other factors (additional knowledge of the person speaking, context, and beliefs), which are determinants in the understanding and generation of a discourse. Linguistic competence is separable from execution, which lies at a different level. The parallel between the work of Chomsky and that of Newell becomes manifest when Chomsky's concept of linguistic competence is associated to Newell's level of knowledge.

On the other hand, and as Newell himself indicates, the ideas inherent in the description of computation by levels are present in the normal methods of all field professionals. The key lies in making these ideas explicit and taking them into consideration when it comes to analyzing or synthesizing an AI program.

Every computational level can be generally represented in terms of an *input space* and an *output space*, along with the *transformation rules* which link the representations in both spaces, as shown in Figure 2. An important point is to recognize that these spaces are *representational*, with a sign-significance structure which acquires value only in the domain of the external observer (as we shall see later on) and which is characteristic of grammar at that level and of the knowledge to be modeled by the problem-solving method.

The input space is theoretically a multidimensional space containing those characteristics of the process, which are considered relevant. Thus, for example, if we are interested in visual perception, the input space is determined by the physical nature of the stimuli and their relevant properties (illumination, brightness, color, contrast), as well as time. When we consider systems which are not directly faced with an external physical world (as, for example, a decision problem, or one of heuristic search within a planning space) the definition of the coordinates in the input space has to do with our knowledge of the problem and our theoretical and practical conceptions about the structures of the level (the domain model description at this level). The important thing here is to remember that in every case, the different variables of the input space X_i are associated to a significance table S_i, such that the complete representation is the set of pairs (X_i, S_i).

The output space is once again a representation space, also generally multidimensional, in which the results of the computation are selected as a function of time, of the inputs, and of the transformation rules characteristic of the process at that level. Once more, when the level is low, the signals possess a low semantics and the output space may be physical, as in the case of a robotics manipulator. In general, however, this space will be symbolic or linguistic, with some formal variables, Y_j, which have to be associated with

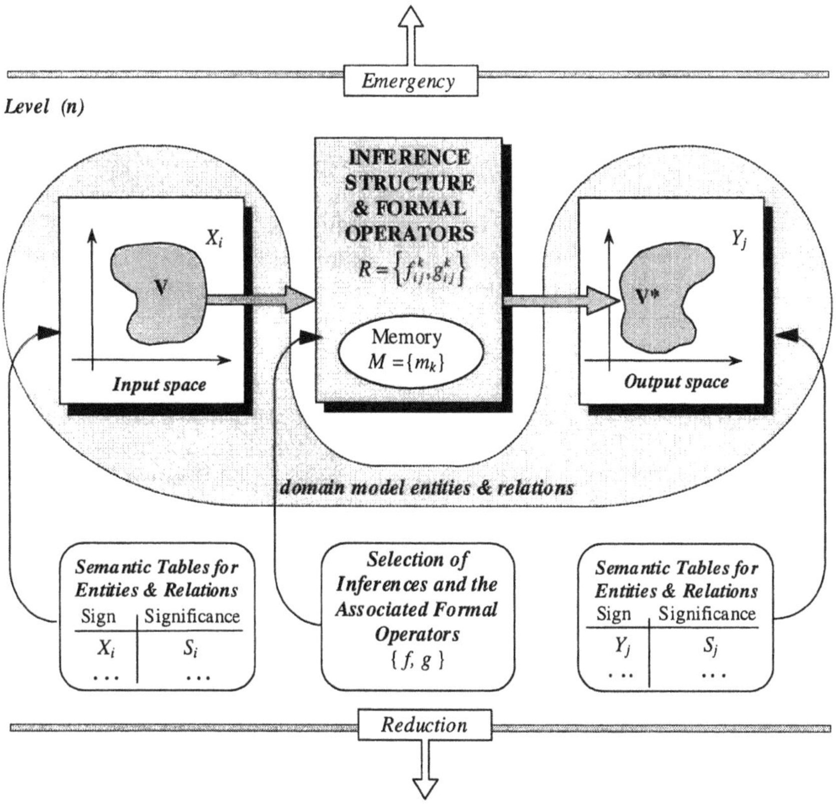

Fig. 2. Computational model at one level as transformations, which link two representational spaces using rules (inference structure) and that level's primitives (entities and relations of the domain model). The tables of semantics depend on the causality law of the level and on the inferential knowledge of the method used to solve each specific task

a semantics table S_j in order to obtain the full representation of the pairs (Y_j, S_j).

Finally, one level's computation is completed by the description of the transformations, which produce new values in the output space based on the sequence of previous values in both spaces. However, here there is a marked difference between the different levels. At the physical level there is an analytical or logical law which links X_i with Y_j, independently of the semantics (S_i, S_j), such that we can write: $Y_i = F(X_i, Y_j)$. To obtain these formal results, there is no need to make reference to the meaning of the variables. It is a process of pure manipulation of symbols. We shall see later on that these formal processes correspond to the level's *own domain* description. Assigning significance to these results (S_j) and their correlation with those at the input (S_i), generates the interpretations of the computation at that level. These interpretations only exist in the domain of the external observer.

According to Chomsky, we can say that the grammar of language $L^{(n)}$ specific to level n generates and recognizes a set of links (X,S), where X is the physical or formal representation of the signals accepted by the level, and S is the semantic interpretation assigned to X by the rules of the characteristic language at that level. These grammars are *predefined* for the formal languages characteristic of all computation, and are absolutely rigid (regular languages, independent of context and structured into sentences along with their equivalent automata).

The problem with *AI* and robotics is that the representational language which needs to produce a model of human knowledge is very close to natural language and we have still not found a universal grammar, which is sufficiently robust against the syntactic and semantic disturbances in an automatic translator. This problem is one of direct engineering: given a set of functional specifications, to find a theory, an algorithm and an implementation after the reduction to the symbol level.

Per contra the problem in Neuroscience is one of inverse engineering. Given a brain, to find the set of specifications from which it originated. For example, by looking for its representation at the equivalent intermediate levels: (1) an anatomic and physiological level where we have neurons, dendro-dendritic circuits and synaptic memory and inference. (2) A set of neurophysiological symbols "equivalent" to programs in a neurophysiological language and (3) a natural language description of tasks, methods and more global theoretical constructs such as perception, learning, control, memory, or decision-making.

2 Marr's Proposal: A Theory of Calculus

Let us briefly introduce the three computational levels proposed by Marr in 1982. Marr [1] was looking for a computational theory of visual perception but his proposals are applicable to other tasks specific to *AI* by for example, changing perception to planning, decision or learning. The starting point is the recognition that any explanation of visual perception based only on the functioning of the neural nets from the retina to the cortex is entirely insufficient. We need to have a clear understanding of what should be calculated, how it must be done, the physical assumptions on which the method is based, and some types of analysis of the algorithms necessary to carry out the calculation [1].

It is therefore not enough to know the computation at the hardware and software levels, there is a need for an additional level of understanding in which the nature of the information processing tasks carried out during the computation of the task (perception, motor control) is analyzed and understood. This additional level of description (the "theory of the calculus") is independent of the particular data structures and algorithms, which implement this model. If we change brain to computer, Marr's suggestion becomes clear (Figure 3). To analyze or synthesize a computational task, it is necessary to use three levels:

1. *Theory of calculus.*
2. *Formal representation of data and algorithms.*
3. *Implementation.*

At the first level we have the theoretical grounds of the computation, the presentation of the problem in natural language, and a possible solution scheme in terms of the knowledge of the domain. We shall see later how this Marr's "theory of calculus" is equivalent to Newell's knowledge level [2] put forward as a means of modeling the knowledge we need to inject into a system. Unfortunately, because the complexity of AI and robotics, it is not always possible to make use of a "theory of calculus." On many occasions, we only have available an imprecise and incomplete set of functional specifications.

The second level of description of a calculus is the selection of a *representational* language for the input and output spaces and an *algorithm*, which carries out the transformations, which link those representations. This Marr's second level of coincides partially with Newell's symbol level and partially with the Newell's knowledge level. In other words, once we have described everything we know about a process at the level of knowledge, that description has to be reduced to the symbol level in terms of a program.

The third level of Marr's proposal has to do with the entire implementation process, which leads us from the algorithm to the hardware. It includes the selection of a programming language and the construction of a program. From this point on there is a *translator,* which generates the code, which can be directly executed by the physical processors. Note that our worries regarding computation end here, as long as we obtain a description from which a translator program can connect us to the physical level. This takes place between the second and third level, but not between the first and the second one. In other words, there is no general and effective procedure for translating the models of tasks and problem-solving-methods at knowledge level into programs.

For a concrete task the three levels of description are related, though not in a single, causal manner. In moving down to a lower level, information is always lost, as can be easily demonstrated if we later attempt to run the inverse route (from processors to symbol level and from here to the "Theory of Calculus"). The reason is that there is no single representation at the lower level. Neither the machine, which implements an algorithm, nor the algorithm, which resolves a problem is uniquely determined. Figure 3 illustrates these losses and injections of knowledge in each level change.

The strong hypothesis in *AI* is that in spite of these semantic losses in going from the knowledge level to the level of physical processor where

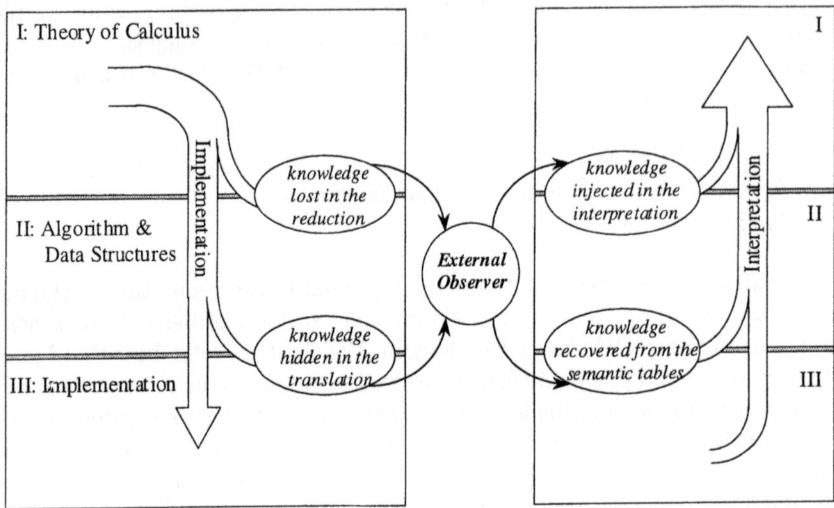

Fig. 3. Levels of description as proposed by Marr with the illustration of the knowledge lost in the reduction process and injected in the interpretation of the results of a program

semantics is intrinsic (what happens in digital electronics is determined by the architecture and local functions of its basic operators – gates and delays), it is nonetheless possible to make human intelligence computational.

3 Newell's Knowledge Level

Even though we have already mentioned that the knowledge level partially coincides with Marr's "theory of calculus" level, the coincidence is not complete since there are many processes for which we still have no theory (reasoning about any realistic domain always involves some degree of uncertainty). Furthermore, these processes without a clear, complete and unequivocal theory happen to be the most genuine ones in *AI* and robotics.

In 1981, Newell introduced the knowledge level [2] as a new level of description beyond the symbol level which characterizes the behavior of a system in terms of its goals, its knowledge, its beliefs, its inference procedures and a general "rationality" principle which guaranteed that the agents used that knowledge and those reasoning mechanisms to reach their goals (Figure 4). The key to the level is its abstract, generic, and independent nature, both in the application domain and in the languages used to represent the knowledge (frames, rules, logic or nets), and to use it (induction, deduction or abduction). It thus looks for general and reusable architectures to specify tasks, and problem-solving methods (*PSMs*) in line with the concept of *competence* introduced by Chomsky, and KADS-type methodologies.

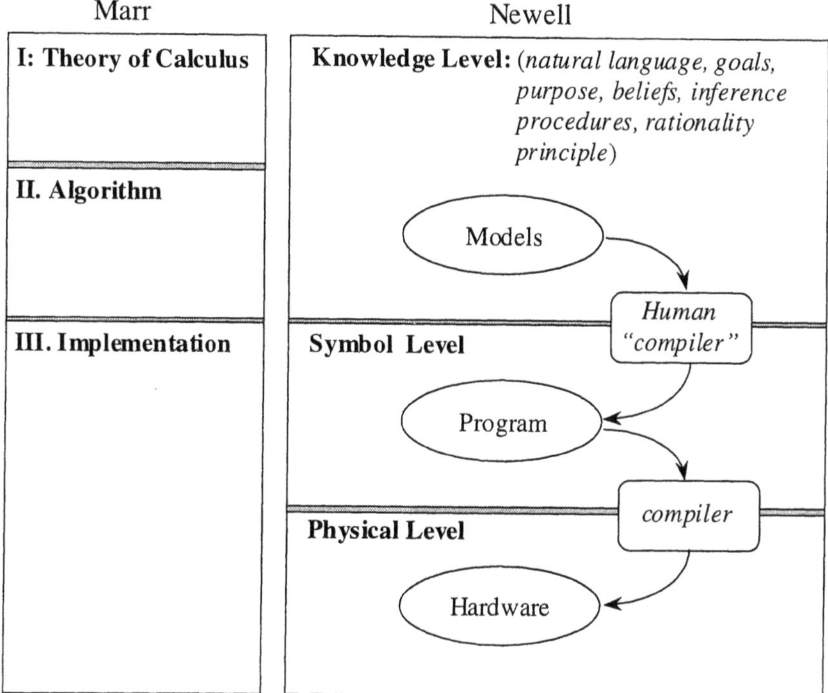

Fig. 4. Newell's proposal of the three levels of description of a calculus (knowledge, program and hardware). Mind the correspondence with the Marr's proposal (theory of calculus, algorithm, implementation)

Pylyshyn [5] has associated the knowledge level to intentional calculation which has to do with processes that could be explained by resorting to the semantic content of the representations, without going down to its syntax. Clancey [6] has indicated the relativistic nature of description at the knowledge level. They are descriptions of *an agent* in its *environment* and depend on its goals and the procedures used to reach them. These descriptions will allow us to predict the behavior of a system facing semantically analogous situations, even if the data structures and the algorithms supporting them are completely different.

An intentional computation (at the knowledge level) is *invariable* in the face of changes in the representations at the symbol level, much as a computation at the symbol level (a program) is invariable in the face of changes at the implementation level.

Newell's sets up three objectives: (1) to reason about the nature of knowledge, (2) to propose the existence of a specific level of knowledge, the basic components of which are beliefs, objectives, plans and intentions, and the rationality principle which causally connects intentions with actions and (3) the description of this level and its connections with lower levels (symbols and physical implementation).

In the first point, Newell distinguishes between knowledge and its representation. Newell's efforts here are directed at differentiating knowledge from its possible representations and at endowing the first with an identity of its own, analogous to Chomsky's distinction between *linguistic competence* and *execution* in natural language. For Newell, what was the nature of knowledge? What does a system have when we say that it possesses a specific knowledge?

> *If a system has –and uses– a data structure or procedure and the programmer accepts that, with an appropriate semantic table, these entities from the symbol level represent other specific entities of the domain (concepts, causal relations), we say that representation possesses the reduced version of the initial knowledge.*

Having accepted that knowledge is prior and totally distinguishable from its representation, Newell introduces a new level –the knowledge level– located above the symbol level and characterized by knowledge as a medium, and the rationality principle as a general law of behavior (Figure 4).

To situate this level relative to the others (symbols and implementation), five aspects are introduced to characterize them: *system*, *medium*, *components*, *composition laws* and *behavior laws* (see Table 1). At the physical level, the *system* is the computer. At the symbol level, it is the view that a programmer who is familiar with the language and the operating system has of this computer. And at the knowledge level, it is the "intelligent agent" (the human expert who possesses the knowledge) needed to build up the models.

The *medium* is the material or the way which is going to be processed in the representation typical to each level. Hence, at the physical level we have only cut-off or saturation states in the output transistors of the inverters with which the logical gates are built (logical levels 0 and 1). At the symbol level, the *material* is the symbols and their relational expressions, in accordance with the grammar of that level, which guarantee an unequivocal translation at the physical level. On the other hand, at the knowledge level the *material* is the knowledge itself that we must *reconstruct* using a representational language (frames, rules, nets) to facilitate its reduction to the symbol level.

The *components* constitute the primitive and basic processes from which all the rest can be built by repeatedly using some of the composition laws. At the physical level, we have gates, combinational circuits, counters, registers, arithmetic and logical units, RAM, EEPROM and FIFO memories and sequencing and control circuits. At the symbol level, the components are the language primitives used by the programmer, its operators and the control structures. Finally, at the knowledge level, the components are goals, actions, purposes, intentions, beliefs, wishes and meanings. This level is a purely *semantic* one. Reasoning here is carried out without descending to representation and its own particular language (extraordinarily robust) is a natural language for which there are no known compilers.

Table 1. Characteristic aspects proposed by Newell to specify the different levels of description of a computation

Levels	Knowledge	Symbols	Physical
System	- Human Expert "intelligent agent"	- Human Programmer	- Hardware
Medium	- Natural language description of Knowledge	- Programming Lang. - Symbols - Relational expressions	- Boolean vectors - Logic States in gates & registers
Components	- Goals - Actions - Intentions - Inferences - Concepts	- Prog. Lang. Primitives - Control Structures	- Gates - Registers - ALU's - RAM's Memory - Control Circuits
Composition Laws	- Natural language (verbs, nouns) - Reasoning Patterns	- Designation - Association - Composition	- Boolean Algebra
Behavior Laws	- Rationality principle - Semantic Causality	- Sequential interpretation of programs	- Automata Theory

The *composition laws* at the physical level are those characteristics of combinatorial logic (Boolean algebra), which turn into processes of designation and association at the symbol level. At the knowledge level, these composition laws are not well known. To uncover these laws is the main purpose of AI (knowledge "elicitation" and modeling). They manifest themselves through natural language descriptions of human experts and have to do with the way in which the knowledge of this expert concerning some tasks and the corresponding problem-solving-method (*PSM*) turns into actions that effectively solve the task. A large proportion of the studies on knowledge modeling, ontologies, libraries of *PSMs*, and other common patterns in the reasoning of human experts [7,8] seek to formalize these composition laws.

The *behavior laws* at the physical level are those characteristic of sequential logic (Automata Theory). At the symbol level, they correspond to the sequential interpretation of programs. Finally, at the knowledge level, there are the laws of the *rationality principle* or of semantic causality.

The link between this principle and the symbol level is established by means of reasoning by analogy:

> *When we say that "program X knows K", we mean that it contains a data structure K^* which represents K and that this structure K^* participates in the selection of formal action A^* in the same way demanded by the rationality principle (the conceptual model) to select action A at the knowledge level.*

Along with the positive fact that a single general principle can explain and guide the conduct of an agent at the semantic or intentional level, there do exist tough problems, among them those associated to *logical omniscience* (the agents know everything, so that when faced with a set of facts and rules, they infer all deducible consequences instantly), and the lack of an effective procedure for reducing the principle to the symbol level. In other words, the step from what has to be done (functional specifications), to how to do it.

The levels of description of a calculus (knowledge, symbols and electronics) are not altogether independent. They are related in terms of what it is necessary to add to one of them so as to obtain the next one. Thus the medium of one level plus an additional set of structures, define the medium of the next level up. The Boolean vectors at the physical level and its organization in registers, plus the structures to link fields define the medium at the symbol level.

Unfortunately this fails at the link between the knowledge level and the symbol level (the program). This is because we are dealing with initial entities (concepts, relations, goals, purposes inductive and abductive reasoning, beliefs and multiple nested semantic spheres) which are described using a natural language but which are not easy to reconstruct using only the entities and relations of a formal language. There are always knowledge loses in the process of reduction (re-writing) of models from one level to the next one down and, consequently, we always need to inject that knowledge in the inverse problem of interpretation of the results of a calculus. The meaning of a calculation always remains at the knowledge level and in the domain of the external observer (the house of models).

From the perspective of applied *AI*, the problem now is to find a procedure to reduce the knowledge level to the symbol level. In other words, to go from natural language descriptions of the task carried out by an agent, to a re-writing (always cut by the limitations at the symbol level) in a language of representation of knowledge which is directly accessible at the symbol level, where the rationality principle, the goals, the beliefs and the actions are projected in data structures and algorithms. Finally, everything is reduced to the primitives of a programming language and its logical representation at the physical level.

To make this reduction process easier, there is a search underway for libraries of reusable components of modeling with clear and unequivocal symbolic counterparts. Among these are found the ontologies of domain knowledge and the libraries of PSM's, roles, and primitive inferences (*select*, *adapt*, *refine*, *evaluate*, *abstract*, *match*). Additionally, different types of adapters are defined including "bridges" and "refiners". The "bridges" are adapters which explicitly model the relation between two different components with the aim of making them fit together. The "refiners" express how a component can be specialized [8,9]. Several current approaches point in this direction, such as Common KADS [10], the framework UPML [11], or Protége-II [12,13].

4 The Observer Agent and the Two Domains of Description

In order to complete our understanding of the significance of a calculus at the three levels previously described (knowledge, symbols and physical) and to fairly assess many of AI's results in general, and of robotics in particular, it is necessary to make use of a distinction between two domains of description: the level's *own domain* (OD) and the *domain of the external observer* (EOD) which interprets computation at that level and, in most cases also acts first as an analyzer and programmer and finally as an interpreter.

The introduction of the figure of the external observer and the differentiation between a phenomenology and its description comes from physics and has been re-introduced and elaborated in the field of biology by Maturana [14] and Varela [15] and in AI and neural computation by Mira and Delgado [3, 16, 17]. Trying to make evident the instrumental character of computation, Herrero and Mira [18, 19] have recently applied this distinction between two domains of description to the study of causality. The "program interpreter" (instrument-user) dyad is necessary for dealing with all causality levels underlying a computation. This is fine from the first conceptual model at a knowledge level to the final version of the program in the physical machine.

When we acknowledge the existence of an observer external to the computation in the prescription and descriptions of the functions of an AI program, we are introducing the idea of different reference systems in which magnitudes and their significances are represented (Figure 5). When we observe a computation at the physical or symbol level, the description of that which was observed should be done within two reference systems. One is that specific to the physical level (where the variables are logical vectors and the operators come from Boolean algebra), or at the symbol level where the variables are the primitives of the language used or at the formal part of the knowledge level where the variables are the data structures and the linking formal processes (deterministic, probabilistic or fuzzy rules) to be finally represented in the primitives of the programming language. We call this domain, which encompasses the right part of three levels, the level's *own* domain (OD) or self-contained. The other domain is that of the external observer (EOD), which uses natural language to describe and give significance to the OD processes.

It should be remembered that each level is characterized by a phenomenology and by properties, and causality laws, which are intrinsic to the entities, which constitute them. Hence, everything that happens in the descriptions at the level's own domain is *causal*, they are relations of *necessity*. That which "has to happen," happens because there is a coincidence of structure and function. So, the connections between observable magnitudes follow their own laws. In the physical level's OD, processes cannot be separated from the processors that realize them. Inverters invert, adders add, a D flip-flops delays

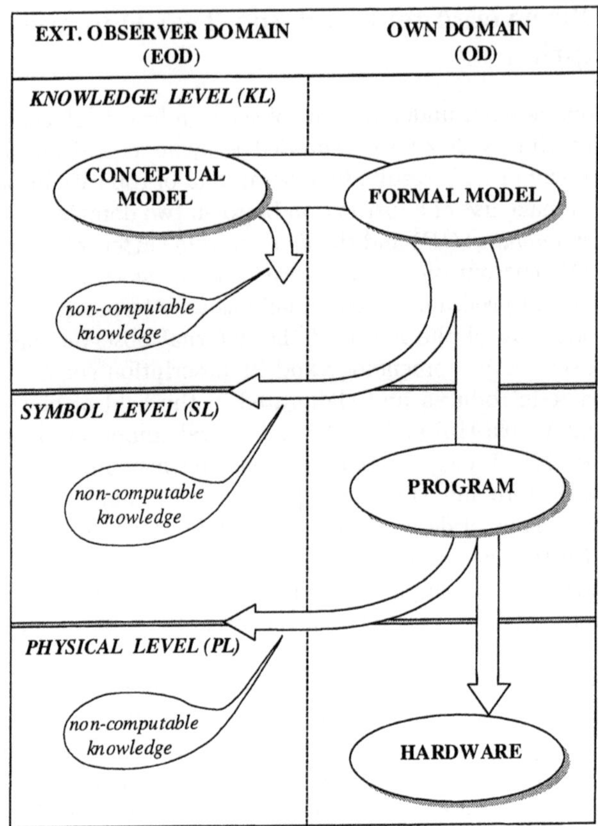

Fig. 5. Introduction of the external observer agent and the subsequent distinction between two domains of description for each level

the signal by one interval which is the function of its internal structure and ALU's carry out arithmetical and logical operations according to its truth table. In every case, the semantics is intrinsic.

Analogously, within the symbol level's own descriptions we can also find causal connections. Thus, the relationships between the primitives of a programming language are specified by its compiler and follow the laws of its grammar. And the same happens within the formal side of the knowledge level where its formal counterparts specify the primitives of a *PSM* that decomposes a task in terms of roles and inferences.

On the other hand, the observer can always act at the three levels. His or her language includes not only natural language but also formal and programming languages. His or her function is first to model the knowledge and then re-write these models taking account of the limitations of the symbol level and the knowledge loosed in the design and programming processes.

This knowledge lost in the reduction (but not other) has to be injected in the interpretation. In the knowledge level of the *EOD* there is always more knowledge than that which can be reduced to the symbol level. To the computer only pass the formal model underlying the conceptual models.

A frequent source of confusion and misunderstanding in AI and robotics is the mixing *OD* entities and relations with others in the *EOD* with which they have no causal relation. In other words, it is obvious that we can talk about inverters, registers and decodifiers (physical level), along with the primitives of a language (we are doing it now), but much care must be taken to not mix these entities operating causally in their *OD* (electronics and programming language), but not doing so at the knowledge level in the *EOD*. The important problem in *AI* emerges when semantic tables from both domains are mixed with the intention of applying significance from the characteristic entities of the level of knowledge in the *EOD* to entities at the *OD* of the other two levels (symbols and hardware).

Varela [15] clearly distinguishes between the explanations at the *EOD* and the *operational* explanations, causal in the *OD*. In both cases, the entities and processes of the level are described in two languages, which are independently consistent and subject to cross-referencing. The difference lies in the fact that in the operational descriptions (*OD*), the terms used to refer to processes can neither leave the domain in which they operate in syntax, nor in semantics (we cannot take them out). On the other hand, the terms used in descriptions at the knowledge level (*EOD*) belong to natural language and make reference to the knowledge of the domain although their referents are not obliged to follow the physical or formal laws of the *OD*. The links in *EOD* do not operate in *OD* unless we limit ourselves to the models the observer possesses at the physical or symbol level. That is to say, unless we limit ourselves to electronics, combinatorial logic or automata theory. The rest of the meanings, including the concept of *knowledge* itself, stay always in the *EOD*.

Many of the criticisms received by *AI*, robotics and computational intelligence (fuzzy systems, artificial neural networks, neuro-fuzzy networks and genetic algorithms) come from the lack of distinction between *OD* and *EOD* entities in the reduction from models of knowledge to symbol level programs and subsequent optimistic interpretations of the supposed functionalities of these programs. The reason for the error (voluntary or not) is that in descriptions with natural language, the entities proper of the *OD* and that of the *EOD* are mixed and the words that represent them coincide (purpose-natural-language versus "purpose" frozen label). A clear distinction appears and we use explicit semantic tables when we re-write the models from the *EOD* to the *OD*, both in the knowledge level and previous to the reduction from the *KL* (*OD*) to the *OD* of the symbol level.

It has to be this way because in the *OD* there are no purposes, no goals, no intelligent agents, no learning and no knowledge. There are only data structures and formal algorithms with causal laws of their own. And lower down there are only logical states in electronic circuits with new laws of their own which are absolutely causal and immutable. The most frequent error in *AI* is to start with a complex phenomenology, assign variables and operators from lower levels, and interpret the results once again at the higher level, without explicitly mentioning the rise and fall in semantics and the external knowledge which has had to be injected in these level jumps. The law, however, is clear:

Does entity X play a causal role at the symbol (or physical) level?
Yes $\Rightarrow$ *Then X belongs to OD of this level*
No $\Rightarrow$ *Then X belongs to EOD*

What characterizes descriptions in natural language in the *EOD* is that they allow us to ignore the causal links at the physical level. Consequently the knowledge and semantics associated to all the entities of a model of knowledge, which do not constitute causal elements intrinsic to the symbol level *OD*, are not computable in strict sense. Inversely, to interpret a computation in the *EOD*, you have to add to the *OD* results everything about the phenomenology of the level, which is not included in the formal counterpart of the model (from T=integer for a computer programmer, to T=temperature for a physicist, and then to T=fever for a physician).

Figure 5 summarizes the relationship between *OD* and *EOD*. In the *EOD*, the observer always describes the models of tasks and *PSMs* in a natural language. Compilers link the symbol and physical levels and, consequently, the relation between these two levels do not pose problems. Thus when we refer to the relations between the two domains, we always refer to the relations between the knowledge and the symbol level, except in same real-time applications of connectionist *AI* (artificial neural nets) in which the reduction of the knowledge level is, sought directly at the physical level as co-processor or preprocessor. We shall discuss this perspective in a later section of this chapter.

The methodological point is that in *AI* and in robotics we can and should use descriptions in both domains –provided that we do not generate confusion in mixing entities of different semantics– using the enormous integrating capacity of natural language. A syntactic analyzer would not have problems accepting the sentence "Yesterday I went out for dinner and dancing with *GoldWorks*." The error emerges when we recognize that we are talking about a "software" package.

5 Different Interpretations of the Same Formal Model

The way we proceed in the methodological building of levels and domains of description of a calculus previously introduced is outlined as follows:

I. First, we have to obtain a conceptual model of expertise at the *KL* and in the *EOD*. This model provides a detailed description of the task to be solved and the *PSM* used to decompose this task.
II. Then, we have to abstract the entities and relations of this conceptual model in terms of the entities and relations of a formal model. So we move from the *EOD* to *OD* without leaving the knowledge level.
III. Finally, we establish a correspondence between the entities and relations of the formal model (*OD* of *KL*) and the primitives of a programming language to produce the program code at the *OD* of the symbol level.
IV. Additionally, the results of the program (*OD* of the *SL*) are interpreted (*EOD*).

Let us examine the following natural language description of a decision task [18–20] to illustrate that semantic knowledge is not explicit in the formal model (*OD* of *KL*) and, consequently, lies outside the implementation (*OD* of *SL*):

> *"Two employees fit together if neither of them is a boss and if both smoke or none of both smokes"*

This description appears in Figure 6 at the *KL* in the *EOD*, where the observer knows the semantics about the decision problem this statement is supposed to describe (meaning of "employee," "boss," "fit-together" and "smoke").

Then we move a step forward and write a formal description where the labels still remain.

```
∀X ∀Y ∀Z1 ∀Z2 (X [fit_together::Y] ←
    X [smoker : Z1] ∈ employees ∧
    Y [smoker : Z2] ∈ employees ∧
    ¬(X ∈ bosses) ∧ ¬(Y ∈ bosses) ∧
    Z1 = Z2).
```

Here "employee" and "smoker" are only for the use of the external observer but are not explicit neither can be deduced from the formal model. Consequently it could never be expressed by the implementation.

To put it crystal clear let us remove all the elements that are not causal in the *OD*. Given that the causality of this formal expression does not correspond with "employees" or "bosses" we rewrite it using only terms such as "sets," "elements of a set," "formal implication," "truth" or "falsehood," "belonging to a set," and other formal entities and relations.

```
∀X ∀Y ∀Z1 ∀Z2(X [a::Y] ←
    X [b : Z1] ∈ c ∧
    Y [b : Z2] ∈ c ∧
    ¬(X ∈ d) ∧ ¬(Y ∈ d)∧
    Z1 = Z2).
```

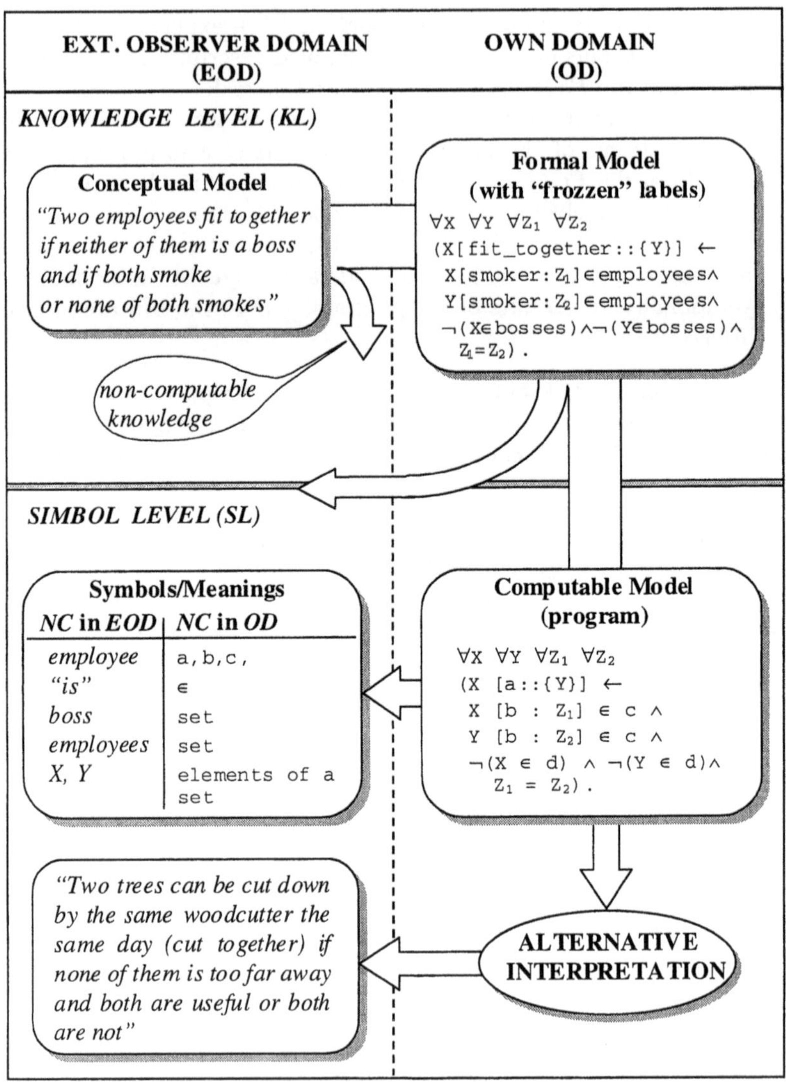

Fig. 6. Two different interpretations of the same formal model. We start with a conceptual model at the *KL* and into the *EOD* (natural language). Then move to the *OD* using "frozen" symbols, as labels, but with the causality and meaning of the formal model (sets, elements of a set). Finally, we move to the *SL*. When we try to recover the original model we find many different (alternative) interpretations of the same program [19]

There is no doubt that this expression is the only one that can be implemented (*OD* at the *SL*) and that neither can be one to one related with the conceptual model nor can be considered the computational counterpart of the semantics and causality proper of the original natural language description (*EOD* at *KL*). For example another of the many alternative interpretations of this formal model is:

> *"Two trees can be cut down by the same woodcutter the same day (cut together) if none of them is too far away and both are useful or both are not."*

In this interpretation, the knowledge level causality and meanings are completely different, «since the reasons the woodcutter considers for cutting or not cutting a tree have nothing to do with the reasons which prevent two employees from working together. We do not understand this when we read the second formal expression because we know something, which lies outside the formal expression.» [19] As commented before, many more examples can be brought up. For instance, in a typical robotic work cell we could define de following rule:

> *"Two objects placed on the conveyor belt can be pick up simultaneously by the same manipulator if none of them is an allen wrench or both are ready for use or both are defective."*

«If we admit that several expressions like the one above constitute a formalization of a model at the knowledge level, the formal model causality only establishes relationships between formal values of truth and falsehood; these "truth" and "falsehood" names coincide with the real ones of truth and falseness, but they are not the same thing: the latter is an interpretation of the former. But, even if we admit that predicate as well as propositional calculus are models of a part of human reasoning, it must be admitted that they only convey the causal relationships between truths and falsehoods, but they do not convey anything about to what these truths and falsehoods refer. Therefore the corresponding knowledge is not in the formalization, but the formalization acquires meaning when someone interprets it correctly, and this correctness criterion lies outside the formalism itself» [19].

These previous statements have deep implications in AI in general and in robotics in particular because many apparently relevant questions are in fact ill posed, such as, are there intelligent robots? Could we design emotional and social robots? [21] Can the computer think? The computer (and the robot) always needs human assistance in order to find the models and provide the right interpretations. A great part of the knowledge attributed to the robot has not been neither modeled nor implemented. Futhermore, a part of the knowledge, which has been modeled, has not been implemented because it is related with the interpretation of the formal model in terms of the specific concepts of the problem in the *EOD* of thc *KL*.

6 Knowledge Modeling at the Knowledge Level (*KL*)

The central problem of both hard and soft knowledge engineering is to construct models [22] of tasks and *PSMs* at the *KL* and in the domain of the external observer (*EOD*). Then we have to reduce these models of expertise from *KL* (*EOD*) to *SL* (*OD*). That is to say we have to go from natural language description of the *PSM* (Newell's agents, goals, purposes and intentional reasoning), to a *rewriting* of the conceptual model in terms of formal tools (graphs, automata, rules, fuzzy operators, neural nets). Finally a new re-writing of the formal model is made in terms of the primitives of a programming language to produce the program (from *KL* in the *OD* to *SL*, also into the *OD*).

The usual approach to modeling at the *KL* and to facilitating the subsequent model reduction to the *SL* has been to develop methodologies, which permit the reuse of generic aspects of this models of human knowledge by separating the *reasoning layer* (tasks, *PSMs* and control) from the *domain layer* (entities and relations of the knowledge specific of each domain of application) (Figure 7). We also seek to capture recurrent abstractions occurring in the domain knowledge by means of *domain ontology*.

The idea of component reuse in system design is implicit in all material and energy engineering and in particular in electronic engineering where we design using the same integrated circuits catalogues, the same libraries of components (adders, drivers, multiplexes, timers, CPUs, RAM memories).

These components come with a clear, complete, precise and unequivocal description of their functionality, and of the logical and electrical characteristics of each of their terminals, including their "roles" (preset, clear, clock, supply voltage), these being a direct consequence of their internal structure. The distinction between *signals* (from the application domain) and *circuits* (generic components of the task and inference structure layer) is not disputed by anyone either. For this reason, the description, in knowledge-level analysis and design models, of a global function using these components is unequivocally paralleled in the physical-level implementation. Furthermore, the underlying computational model (logic and automata theory) is clear, and the process of model reduction is reversible.

In knowledge engineering (*KE*) there is a great diversity of terms without a unique, clear, complete and unequivocal meaning, and as a consequence, there is a notorious lack of agreed-upon libraries of reusable components. As a consequence *KE* lacks robustness, completeness and stability against the changes in design personnel that are common in electronic engineering.

Considering the unavoidable, intrinsic complexity of non-analytic human knowledge, characteristic of the tasks and methods *KE* attempts to model, and the lack of a "theory of knowledge" similar to the physical theory that supports other engineering disciplines, the methodological perspective that has most contributed to bringing knowledge engineering closer to the other

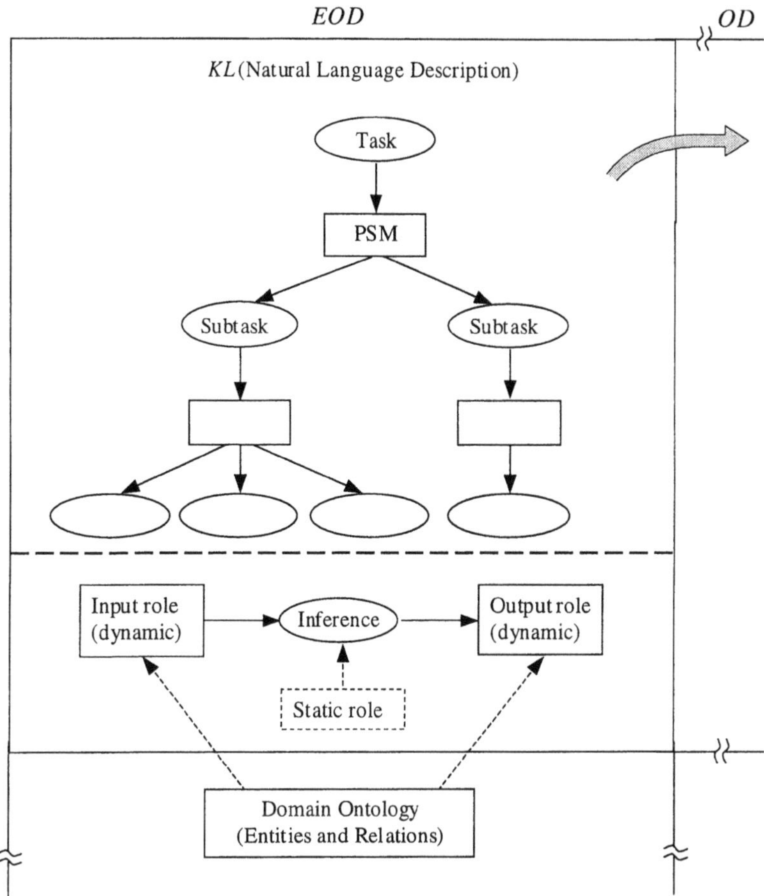

Fig. 7. Expertise modeling components at the *KL* and into the *EOD*. Here we have tasks, *PSMs*, dynamic and static roles, inferences and domain knowledge entities and relations (domain ontology)

material and energy-engineering disciplines is based on the following developments [23]:

1. The proposals of Clancey [6] and Chandrasekaran [24], which seek to capture recurrent abstractions in terms of generic tasks and *PSMs* [25].
2. The attempts to agree on libraries of *PSMs* and inferences [7, 26, 27] with different degrees of proximity to the final formal tools and program entities.
3. The introduction of intermediary entities (adapters, filters, bridges) that cope with the difficult problem of reusability: what initially seem generic ends up being dependent, to a large extent, on the method used and on the application domain [9, 28, 29].

4. The development of ontologies [30, 31] of reusable components in the domain layer (entities, roles and relations) and the subsequent libraries of these entities for each sort of tasks and *PSMs*.
5. Acceptance of the restrictions and limits that are common to any computable model. Computable knowledge is a very small part of the knowledge that the human expert uses for the same task (its discourse only includes the logic-relational components, the facts and concepts that can be defined by means of features with numerical values or labels that have a pre-defined meaning). The rest is interpretation and this makes the instrumental nature of the whole of *KE* obvious [3, 18, 19, 32, 33].

Relevant examples of the above-mentioned approaches include the CommonKADS methodology [8,12], the formal framework UPML [11], the general purpose framework Protége-II [12, 34].

In this methodological context, the knowledge modeling process that we summarized at the beginning of section 5, starts at the *KL* in the *EOD*, as illustrated in Figure 7 and follows the next steps [23, 35]:

1. *Describe* in a natural language the task you try to model and code, and disregard the terms that are not causal in the reasoning process.
2. *Identify the entities and facts of the domain knowledge.* These entities play the same role as physical magnitudes in an analytical model. They represent separate *concepts* that the human expert considers necessary and sufficient to describe his/her knowledge. Usually are associated to *names* whose referent can be physical or mental. Examples in medicine are *sign*, *finding*, *proof*, and *qualitative observation*. There are libraries of entities for specific tasks and domains structured in ontologies of reusable components for the domain layer. These entities are described by means of a set of attributes, such as name, description, and unit's possible values. The state of an *entity* is a *fact*.
3. *Identify the relations* between these entities that appear explicitly or implicitly in the expert's description. These relations are oriented and dyadic and are instantiated as *connections* between *facts*.
4. *Search for inferential components* of the reasoning, usually verbs (establish, refine, select, match, abstract), which are used by the human expert to describe his/her reasoning steps in natural language. These *inferences* are the components from which we will build the *PSMs*.
5. *Describe*, for each one of these inferential verbs, the *input* and *output roles* to be played by the domain entities. For example, "observation", "hypothesis" or "diagnostic".
6. *Try to sketch the inferential circuit* (a graph) corresponding to the knowledge flow through the dynamic roles and the different inferences according to the sequence, concurrences, and loops that more closely represent the reasoning pattern followed by the expert. This is the first draft of the customization of the *PSM* for each expert. Usually you can find that

this *PSM* can be selected from a library of methods ("*abstract-match-refine*", "*establish-and-refine*", "*propose-critique-modify*", "*generate-and-test*", "*cover-and-differentiate*"), or libraries of more specific one's ("ranking methods", or "abstract methods"). In both levels of abstraction additional knowledge is usually needed for adaptation of the *PSM* to the task (task-*PSM* bridge) and to the domain (*PSM*-domain bridge) [9].

7 Operationalization of the Inferences (from *KL* into *EOD* to the *OD*, still at *KL*)

At the end of the last step of the analysis phase we have:

1. A set of *entities* and *relations* of the domain model.
2. A set of *inferences* with the corresponding input and output roles.
3. An *inferential circuit* connecting these inferences through the dynamic roles.
4. A *control structure*.

That is to say, we have a *KL* conceptual algorithm (into the *EOD*) to solve the task. The next step in the way to build the code (*SL* into the *OD*) is to make operational each one of these inferences. That is, to rewrite them in formal terms (from the *EOD* to the *OD*, without leaving the *KL*). Having repositories of formal models underlying the different inferences (abstract, select, classify, measure) is one step further towards facilitating the integration of soft and hard computing methods, as illustrated in Figure 8. It is also necessary to provide additional knowledge for the proper selection or integration of these formal tools into hybrid architectures. This additional knowledge is related to the balance between *knowledge* and *data* available to solve the task and to the sort of *knowledge* (precise, uncertain) and *data* (labeled, unlabeled).

If we have more knowledge than data, *hard* computing methods and formal tools are usually preferred. *Per contra*, if we have more data than knowledge *soft* methods (neuro-fuzzy) are preferred. A new refinement is also of common sense. If we have labeled data, then we can use soft methods with supervised learning strategies. If we only have unlabeled data, then unsupervised or self-organizing learning tools are proper. Finally, in real world problems such as those of robotics, the more efficient approach is the hybrid one, combining both symbolic and connectionist methods of operationalization of the different inferences according to the particular balance between data and knowledge for each one of the inferences included in the *PMSs* used to solve the task.

8 Integration of Symbolic and Connectionist *PSMs* and Inferences for the Classification Task

Whether or not it is stated explicitly many of the *AI* analysis tasks (pattern recognition, diagnosis, planning and prediction) are based on the inference

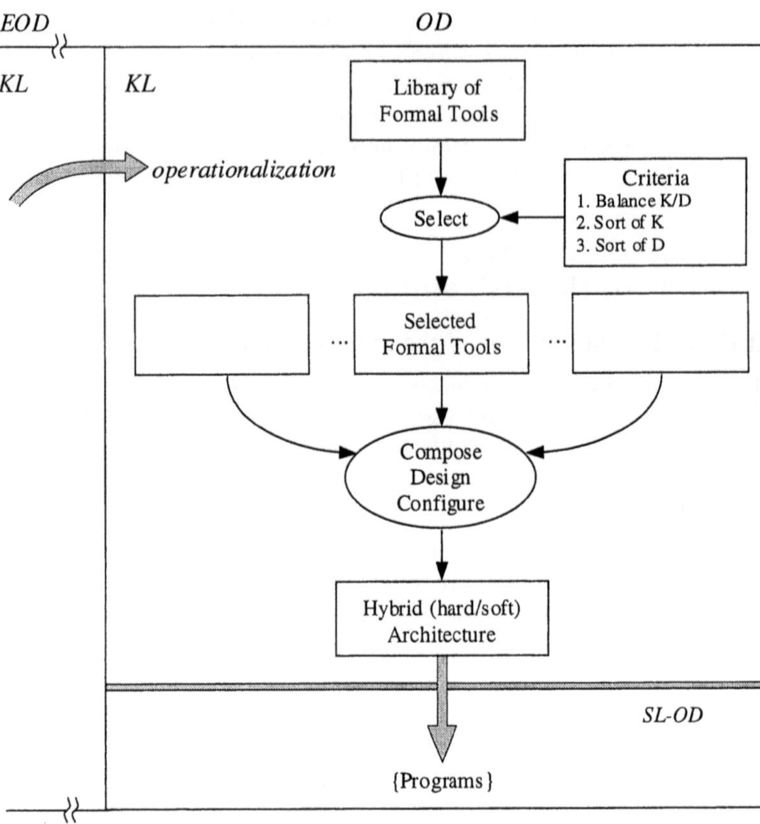

Fig. 8. The hybridization (integration, composition, configuration) of architectures is made at the *KL*, at the frontier between the *EOD* and the *OD*. For each inference a formal tool is selected according to the criteria of balance between knowledge and data. Finally the hybrid architecture selected for the whole inferential scheme is composed, assembled or configured

structure of a *classifier*, which associates input configurations, $\{X_i\}$, $i = 1, \ldots, n$, to output configurations, $\{Y_j\}$, $j = 1, \ldots, m$, with $m < n$, as shown in Figure 9.a. The input entities play the role of *observable* and the output entities play the role of categories or *classes*. Both sort of entities are described by the measuring units and the list of labels or numeric value range that each entity can take. The spaces of *observable* and *classes* should be fully specified at length before the classifier begins to infer, since what classification does is to order output categories or select one.

The *PSMs* used to decompose the classification task depend on the type of knowledge we have available. Thus if we have a clear, complete and unequivocal knowledge on the domain, there is no need for connectionist *PSMs*, since all the knowledge required by the different inferences is available. In these

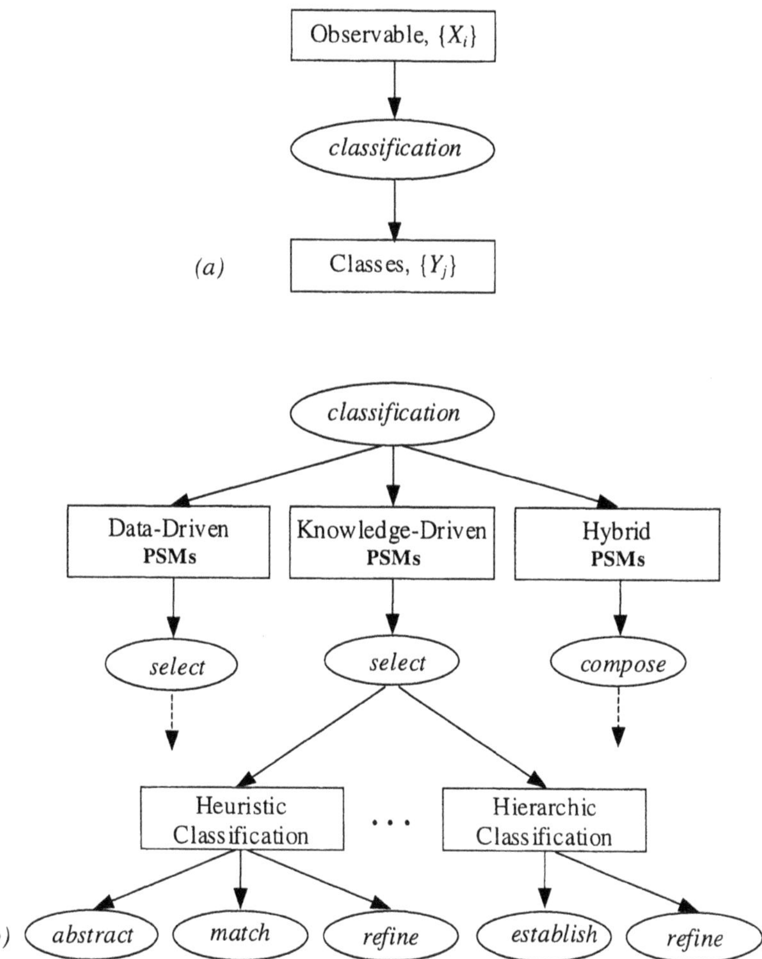

Fig. 9. **(a)** Classification task. **(b)** Different types of *PSMs* to be selected or combined according to the knowledge-data balance: knowledge-driven *PSMs* (heuristic classification, hierarchic classification), data-driven *PSMs* (neuronal, fuzzy, neuro-fuzzy) or Hybrid *PSMs*

cases we must use symbolic *PSMs* like "Hierarchical Classification" ("*establish*" and "*refine*") or Clancey's "Heuristic Classification" ("*abstract*," "*match*," and "*refine*") as shown in Figure 9.b and Figure 10.

If we have situations having to do with changing, partially know classification rules, variable environments, labeled data and with the requirements of fault tolerance and real time learning to improve the classification knowledge, then the connectionist *PSMs*, alone or combined, are probable the proper se-

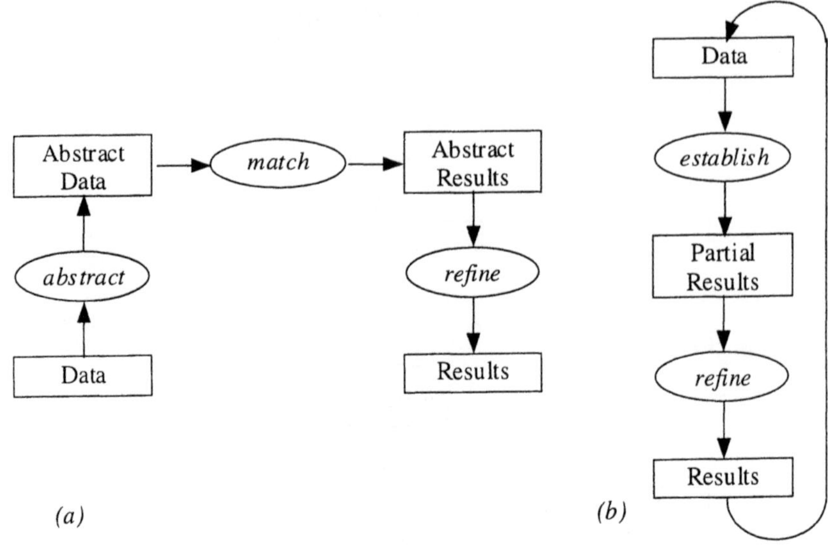

Fig. 10. Knowledge-driven *PSMs*. **(a)** Heuristic Classification. **(b)** Hierarchical Classification

lection. The following is a list of possible questions, which can guide us in the selection of a neural *PSM* [3, 36]:

C1: What is the knowledge available for the classification task?
C2: Can some or any of the inferences be formulated as an *ANN*?
C3: Does it make sense to use supervised (labeled data) or non-supervised (raw-data) learning procedures?
C4: Assuming the answers for C1 to C3 are affirmative, what is the architecture of each layer of the net?. What type of neuron and learning algorithm is needed for each layer?. What type of knowledge is available to be used to partially specify the architecture, select the most adequate type of neuron for each layer, and initialize the parameter values?
C5: Finally, does the possibility of hybrid solutions exist?. If the answer is yes, specify both subsets (connectionist inferences, symbolic inferences) and establish the corresponding integration mechanisms.

In Figure 11 we shown the inference structure associated to most of the connectionist *PSMs*: "*Property-extraction-similarity measures-ordering or maximum selection-learning*". Generally, each inference corresponds with a layer of neurons and we use all the knowledge available about the problem to improve and refine this inferential circuit and the previous data analysis.

The neural *PSMs* always work on labeled lines, which do not lose their identification during the whole process. Therefore, domain knowledge modeling involves the selection of input and output entities for each layer. Every

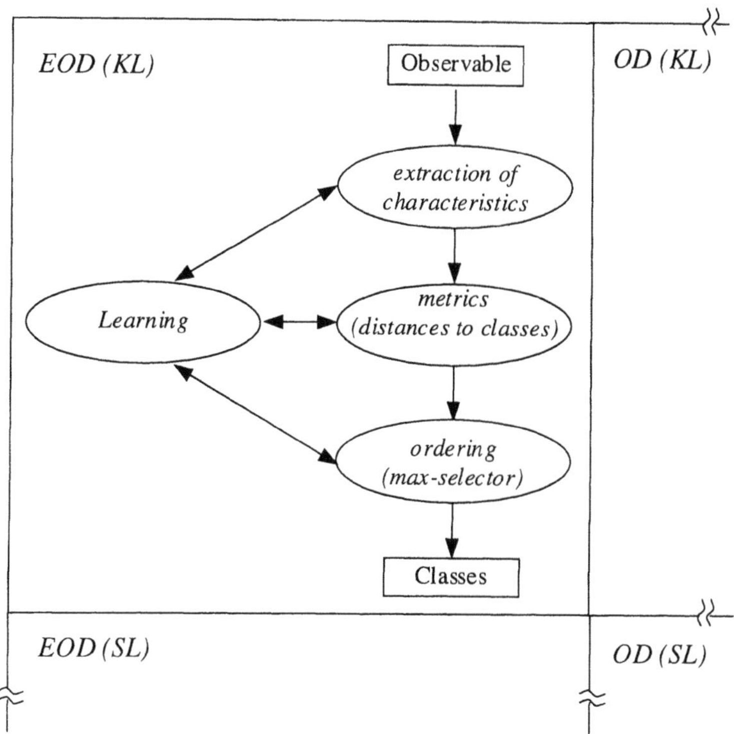

Fig. 11. Data-driven (connectionist) *PSM* for the classification task. The domain model entities (*observable* and *classes*) are labeled lines and each inference is associated to a layer

observable becomes a labeled line and the corresponding meaning always remains inside the *EOD*, at the knowledge level. The net only handles *numbers* and "doesn't know" what is being classified.

Whatever the nature of the data to be classified (voice signals, images, clinical signs) connectionist classification begins with the construction of a space of characteristic (data analysis) with a fixed part and another, which is adaptive. These characteristics play the output role of the first layer. We have also a set of intermediate input and output roles connecting the hidden layers.

The second inference (layer) incorporates the metrics associated to the *equivalence classes* (Euclidean, Manhattan, projection, probabilistic, possibilistic). Its function is to calculate the proximity of each pattern of characteristics to each one of the classes. The third inference is the *selection* of the class, which is least distance or the calculation of the membership function of that point of space of features to the different classes, in fuzzy formulations. Each output neuron is again a labeled line and that what triggers is telling us that "*this is the label of the class*" to which belong the configuration which is being classified at this moment.

Thus is we consider necessary n *classes*, we need n neurons for the last layer. Probabilistic and fuzzy neurons are included if we admit outputs range $[0, 1]$ interval and consider the responses of the neurons as a discrete "curve" shape that measures probabilities or membership functions.

Finally, the last inference in all of connectionist *PSMs* is *learning* (supervised, unsupervised, competitive or by reinforcement), according to the sort of data available. For any specification from the other inferences, a procedure for the evaluation of the quality or efficiency of the classifier will always be needed. This evaluation (the error or the mean quadratic value) is used to adjust the parameters (weights, cluster radii, threshold) of the neurons in the rest of the layers.

The next step is to refine each of the inferences of Figure 11 into a specific layer of neurons selected from two libraries of formal models and learning algorithms, as shown in Figure 12. It is worthwhile to remember that these formal models should not be restricted to simple adders followed by sigmoids. On the contrary, if we want to integrate connectionist and symbolic *PSMs* we have to use in our library of neural operators conditional "*if – then – else*" rules:

If <condition A_1> then <assignment B_1>
If <condition A_2> then <assignment B_2>
......
If <condition A_n> then <assignment B_n>

where each branching condition A_i can be any arbitrary combination of logic and relational operators. The outputs B_i are labeled lines and the assignment (A_i, B_i) is carried out by means of a look up table (*LUT*), for example. This allows us to include fuzzy and probabilistic models as previously mentioned [35].

The refinement process can go on by zooming in the *LUT* evaluation field again or in the conditional action field in order to include learning, from a parametric description of the neuron.

Usually, most real problems solving methods are neither only symbolic (knowledge driven), nor only connectionist (data-driven), but hybrid, in the middle of these extreme situations. For all these cases the proper solution is a hybrid architecture that combine inferences of both types, integrating the available knowledge to improve the connectionist inferences and the available data to improve the knowledge-based inferences.

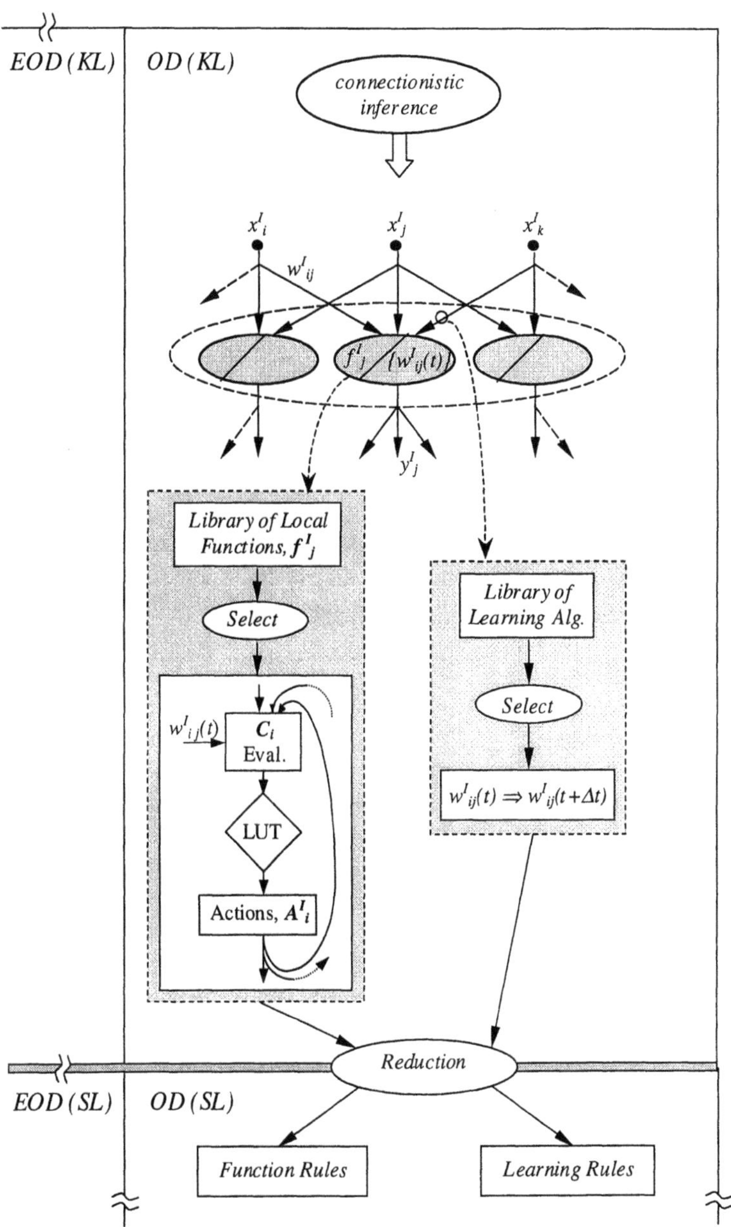

Fig. 12. The refinement of each layer develops the connectionist inferences into a parallel graph (the neural network) where the nodes are operationalized as *if-then* rules and the arcs are interpreted as learning rules for the adaptation of the weights

9 Final Comments on the Interplay between Symbolic and Connectionist Inferences

In this chapter we have presented the taxonomy of levels and domains of description of a calculus in order to support two conjectures: (1) the knowledge always remains at the knowledge level and in the domain of the external observer. To the robot only pass the formal model underlying these models of knowledge. (2) Integration of symbolic and connectionist inferences and *PSMs* has to be made at the *KL* where it is crystal clear that soft and hard computing are complementary and ready to be integrated, according to the data and knowledge available to configure the method.

This final section of the chapter will be devoted to discussing some similarities and differences between the two *AI* perspectives (symbolic and connectionist), attempting to emphasize their complementary nature and their significance in the processes of knowledge modeling and reduction [3,35,37,38].

AI had connectionist origins, then became predominantly symbolic, and after the rebirth of neural computation in around 1980, we have reached a stage of cooperation between both alternatives, dependant on the nature of the problem. They are two complementary, and in some cases symbiotic, techniques.

Having studied the different computational levels, and the difference between the inferences and the formal operators which support them, along with the introduction of the figure of the observer who handles descriptions in the two domains, it is relatively simple to understand the analogies and differences between symbolic and connectionist techniques.

Actually, all computation ends up being connectionist in the physical level's own domain in the sense we have given those terms. Symbolism is born in the domain of the external observer and at the *KL* during the process of models building and reduction from the knowledge level to the symbol level, and in the inverse interpretation process. When reduction is done through the primitives of a programming language, then *AI* is symbolic. By contrast, when the translation is done through the primitives of a neurosimulator or through a net of special purpose neural processors, then the *AI* is connectionist. The distinction at the processor level lies in the programming necessity of the symbolic case, which, in the connectionist case is substituted with the training of the neural net

The term "*neural inference*" refers to all computation which is modular, distributed, "small grain" and self-programmable. Its architecture in general is organized into layers without feedback or with it (recurrent nets), and each processor's local function is usually the weighted sum of its inputs, followed by a non-linear decision function, However, if we extend the neural model to substitute the weighted sum by a conditional (if-then-else rule), then this extension *eliminates* the frontier between symbolic and connectionist inferences, including fuzzy and probabilistic nets.

A comprehensive vision of hybrid *PSMs*, with both symbolic and connectionist inferences, can use neural techniques to refine symbolic inferences and knowledge driven techniques to improve neural architectures.

References

1. Marr D (1982) Vision. Freeman, New York
2. Newell A (1981) The knowledge level. AI Magazine, summer 1-20
3. Mira J, Delgado AE, Boticario JG, Díez FJ (1995) Aspectos básicos de la inteligencia artificial. Sanz y Torres, SL Madrid
4. Chomsky N.(1968) Language and Mind. Harcourt Brace & World Inc., N. York
5. Pylyshyn ZW (1986) Computation and cognition. Towards a foundation for cognitive science. The MIT Press. Cambridge, Mass
6. Clancey WJ (1985) Heuristic classification. Int J Artif Intell Neural Netw, Complex Problem-Solving Technol 27: 289-350
7. Motta E (1999) Reusable components for knowledge modelling. IOS Press, Amsterdam
8. Breuker J, Van de Velde W (eds) (1994) Common KADS library for expertise modelling. IOS Press, Amsterdam
9. Taboada M, Des J, Mira J, Marín R (2001) Diagnosis systems in medicine with reusable knowledge components. IEEE Intelligent Systems vol 16 no 6: 68-73
10. Schreiber et al. (1999), Engineering and managing knowledge: The CommonKADS methodology. The MIT Press
11. Fensel D et al., (1999) UPML: A framework for knowledge system reuse. Proc 17th Int'l Joint Conf Artificial Intelligence (IJCAI99), Morgan Kaufmann, San Francisco
12. Eriksson H et al. (1995) Task modeling with reusable problem-solving methods. Artificial Intelligence vol 79 no 2: 239-326
13. Gennarí JH, Grosso W, Musen M (2000) A method-description language: An initial ontology with examples. In: Proc Europ Knowledge Acquisition Conference (EKAW-2000). LN in Artificial Intelligence. Springer, Berlin
14. Maturana H (1975) The organization of the living: a theory of the living organization. Int J Man-Machine Studies; 7: 313
15. Varela FJ (1979) Principles of biological autonomy. The North Holland Series in General Systems Research. North Holland
16. Mira J (1995) Reverse neurophysiology: The 'Embodiments of Mind Revisited'. In: Moreno R, Mira J (eds) Brain processes, theories and models. The MIT Press.Massachusetts, pp 37-49
17. Mira J, Delgado AE (1987) Some comments on the antropocentric viewpoint in the neurocybernetic methodology. Proc of the Seventh International Congress of Cybernetics and Systems; 2: 891-95
18. Herrero JC, Mira J (1999) SCHEMA: A knowledge edition interface for obtaining program code from structured descriptions of PSM's. Two cases study". Applied Intelligence 10:2/3, pp 139-153
19. Herrero JC, Mira J (2000) Causality levels in SCHEMA: A knowledge edition interface. IEE Proceeding-Software Engineering, vol 147 No 5 pp 191-200
20. Fensel D, Anciele J, Studer R (1994) The specification language KARL and its declarative semantics. IS-CORE94 Workshop, pp 27-30

21. Mira J (2000) Computación y antropomorfismo en robótica emocional. In: Mora F (ed) El cerebro sintiente. Ariel, Barcelona, pp 153-185
22. Ford K, Bradshaw JM, Adams-Webber JR, Agnew NM (1993) Knowledge acquisition as a constructive modeling activity. Int J of Intelligent Systems 8: 9-32
23. Mira J, Álvarez JR, Martínez R (2000) Knowledge edition and reuse in DIAGEN: A relational approach. IEE Proceedings-Software, vol 147 no 5 pp 151-162
24. Chandrasekaran B (1986) Generic tasks in knowledge based reasoning: High-level building blocks for expert system design. IEEE Expert fall: 23-29
25. Puerta AR, Tu SW, Musen MA (1993) Modelling tasks with mechanism. Int J of Intelligent Systems 8: 129-52
26. Benjamins R (1995) Problem-solving methods for diagnosis and their role in knowledge acquisition. Int J of Expert Systems: Research and Applications, vol 2 no 8: 93-120
27. Benjamins R, Fensel D (1998) Editorial: Problem-Solving Methods. Int J Human-Computer Studies vol 49 no 4: 305-313
28. Fensel D (1997) The tower-of-adapter method for developing and reusing problem-solving methods. In: Benjamins R, Plaza E et al. (eds) Knowledge acquisition, modeling and management. LNAI 1319. Springer-Verlag, Berlin, pp 97-112
29. Wielinga BJ, Akkermans JM, Schreiber A (1998) A competence theory approach to problem-solving method construction. Int'l J Human-Computer Studies vol 49 no 4: 315-338
30. Chandrasekaran B, Josephson JR, Benjamins R (1999) What are ontologies, and why do we need them?. IEEE Intelligent Systems vol 14 No 1: 20-26
31. Guarino N (1997) Understanding building and using ontologies. Int J Human-Comput Studies 46, (2-3): 293-310
32. Herrero JC, Mira J (1998) In search of a common structure underlying a representative set of generic tasks and methods: The hierarchical classification and therapy planning cases study. In: Mira J, del Pobil AP, Moonis A. (eds) Methodology and Tools in Knowledge-Based Systems. LNAI 1415. Springer-Verlag, Berlin, pp 21-36
33. Mira J, del Pobil AP (2000) Knowledge modelling for software component reuse. IEE Proc-Softw, vol 147 no 4 pp 149-150
34. Eriksson H, Shahar Y, Tu SW, Puerta AR, Musen MA (1996) Task modeling with reusable problem-solving methods. Artificial Intelligence 79 2:293-326
35. Mira J, Herrero JC, Delgado AE (1998) Where is knowledge in computational intelligence?: On the reduction of the knowledge level to the level below. Proc 24th EUROMICRO Conference, IEEE vol II pp 723-732
36. Mira J, Delgado AE (1995) Computación neuronal avanzada: fundamentos biológicos y aspectos metodológicos. In Barro S, Mira J (eds) Computación neuronal. Universidad de Santiago de Compostela, Cap. VI, 125-178
37. Mira J, Martínez R, Álvarez JR, Delgado AE (2001) DIAGEN-WebDB: A connectionist approach to medical knowledge representation and inference. In: Mira J, Prieto A (eds.) Connectionist models of neurons, learning processes, and artificial intelligence. LNCS 2084, Springer Verlag, Berlin, pp 772-782
38. Moreno R, Mira J (1995) Logic and neural nets: variations on themes by WS McCulloch. In: Moreno R, Mira J (eds) Brain processes, theories and models. The MIT Press. Massachusetts, pp 24-36

Introduction to Fusion Based Systems – Contributions of Soft Computing Techniques and Application to Robotics

M. Oussalah

City University, CSR
10 Northampton Square, EC1V 0HB, London, UK
M.Oussalah@city.ac.uk

Abstract. Data/information fusion, as a methodology to integrate information stemming from different sources to get a more refined and meaningful knowledge, has gained a lot of interest within several communities as it sounds from the number of publications and successful applications in this area. This chapter is aimed to explore how the fusion methodology is decomposed into a set of primary subtasks where the elicitation and the architecture play a central role in the fusion process. Particularly the contributions of soft computing techniques at various levels of the fusion architecture are laid bare. Some exemplifications, through the use of serial and parallel architectures, employing both probabilistic and possibilistic approaches, have been carried out. Finally a robotics application consisting in a localization of a mobile robot has been performed and shows how the different steps of the fusion architecture have been handled.

1 Introduction

Data/Information fusion is an expanding area of research that deals with the integration and the combination of information issuing from multiple sources to obtain a better information in terms of confidence and efficiency than that supplied by the individual sources taken separately[1]. A source here refers to any root or cause that can generate any sort of information for the decision-maker from its interaction with the environment or its own record. This includes physical devices like, sensors, measuring the intensity of the signal-input, expert opinion which relies on his own experience and expertise to provide an output, and database which rests on its own registered records. More generally, a source represents any entity, either physical or not, that can translate the observation into

[1] We shall adopt here a unified definition of data fusion pointed out by EARSel –SEE-EMP working group [1] which states "data fusion is a formal framework in which are expressed means and tools for the alliance of data originated from different sources. It aims at obtaining information of a better quality; the exact meaning of 'greater quality' will depend upon the application"

some *internal interpretation* that will be used later on by the decision-maker or other devices.

Motivated by opportunities in areas like commerce and geographical research and due to a large support of the military institutions, several information/data fusion approaches have been developed during the last years as it can be noticed from the very long extensive paper of Luo and Kay [2]. As well as from the increasing number of scientific research associations that work in that area. Among these approaches which include conventional statistical and probabilistic methods [3], it is worth mentioning the new emerging area of soft computing. This embraces fuzzy logic [4], possibility theory [5, 6], evidence theory [7, 8], fuzzy measures [9], capacity [10, 11] among a wide range of upper and lower probabilities [11-13], neural networks and genetic algorithms [14, 15], etc. Further, the emergence of new sensors, advanced fusion algorithms, together with improved hardware and software technologies, make real-time data fusion a practical option for automated target recognition systems, applied robotics systems, air traffic control and weather forecast, etc. This stresses further motivations to the information fusion field. Indeed, for instance, in recent book titled STAR 21 *Strategic Technologies for the Army of the Twenty First Century*, published on behalf of the National Research Council by the National Academic Press in 1992, p.278, listed *data fusion* as one of the defense-critical technologies. Example of applications using the concept of information fusion includes pooling of expert opinions [16], multi-target tracking [17], image classification [18], localization of enemy position [19, 20], autonomous robotics systems with multiple sensors [19, 21, 22], transportation systems [23], sensor technology [24], radar [17, 20, 25] etc. On the hand several projects have been successfully carried out in this area. We may mention, for instance, ADVANCE (Advanced Driver Vehicle Advisory Navigation Concept) [26] in United State, AGVs (Autonomous Guided Vehicles) [22] in United Kingdom, PROMETHEUS (Program for European Traffic with Highest Efficiency and Unprecendented Safety) [27], etc.

Strictly speaking, the advantage of a multi-sensor architecture, when appropriately designed, relies on its redundancy, timeliness, complementary and cost aspects. Redundancy increases the confidence in the sense that any information perceived by two different sensors is likely to be more trusted than that perceived by a single sensor. Further redundancy enables the system to overcome the failure of a single component. Complementary aspect permits the system to extend its capability over the intrinsic limits of individuals. For instance, if each sensor is focused on one portion of the whole region then the overall picture is drawn only by the integration of the different sensors. From economical considerations, building a reasonable multiple unit system is often cheaper and more efficient in terms of time requirement than building a single unit with very high dependability requirement. Basically, as pointed out by Dasarathy [28, 29], data fusion can be implemented at different levels of complexity. The low level deals with data supplied by the different sources and produces another more refined data, that is, "data in – data out" fusion architecture. While more elaborated levels involve feature-based, or decision-based fusion. Feature-based corresponds to a more abstract representation and supports the synthesis of more meaningful information for guiding human decision-making like images, classes, categories. While the decision level allows, usually on the basis of lower fusion levels, the system to make hard or soft decisions concerning the context of interest like absence or presence of a given target, etc.

Even if most multisensor data fusion systems described in the literature follow an application specific approach, there are always some elementary bases that govern the construction of any fusion-based system. Crucial in that system is the architecture used to handle the problem at hand. Such architecture, which is mainly motivated by the aim of the fusion process, type of knowledge in each source, environmental constraints, among others, is of paramount importance in determining the performance of the fusion process. Basically, the architecture amounts for the type of communication between the different units, the handling of uncertainty pervading the system at different stages of the fusion process, the degree of decision required at each level, etc. Further, uncertainty analysis and data modelling are highly related to the underlying architecture. For instance, we may be more demanding in terms of precision and certainty when dealing with homogenous data, while this requirement should be relaxed in case of complementary information. A simple example of fusion, induced by a special architecture, is a temporal fusion when a single sensor is fired several times over time, provided that the state of the system is static, and the outcomes are merged accordingly. More importantly the question of whether the result of the fusion process is better than that of a single source is strongly a matter of the used architecture. In this respect, the fusion architecture is more related to the concept of integration[2], which indicates the way by which the different information are handled before being combined into a single unified entity. Consequently, any uncertainty theory, whatever are its soundness and sensibility, cannot improve the performances of the fusion process if the fusion architecture is inadequate. Further, this also addresses the problem of the limit of information fusion; that is, does the fusion indefinitely increase the performances of the overall system, or is there an upper limit on the number of individual components beyond which the overall performance tends to deteriorate [30]. This stresses some interest on the area of fusion architecture. On the other hand the soft computing field, which includes all the wide range of nonadditive measures like upper and lower probabilities, evidence theory, fuzzy sets, possibility theory, fuzzy measures, among others, neural network and genetic algorithms, has grown rapidly since the sixties. Several appealing properties in this framework have been pointed out, and their interests for the information fusion field are certainly not negligible

This chapter is mainly dedicated to this purpose of fusion-based systems. Meanwhile, we review and explore the possible contributions of the soft-computing framework to this area. Section 2 of this chapter addresses some preliminary relevant issues, which constitute the basis for any fusion-based system. Special focus on the architecture of such system is highlighted. This includes sensor configuration, various fusion-types, *homogenization* of

[2] We notice that other definitions of integration are provided in Luo and Kay [2]. Some other authors do not make any distinction between the concept of integration and fusion. We shall advocate here that integration is only a part of the whole fusion process and is more related to the architecture of the fusion process.

information and dealing with complementary information. We then survey how the different soft computing techniques may contribute to the fusion process at various stages. Section 3 presents an example of treatment of serial and parallel systems based on both probability and possibility theories and shows the feasibility and advantages of one of the soft based approach. Section 4 describes how a fusion-based approach can be applied for absolute localization of a mobile robot in a structured environment while reasoning in the framework of a possibility theory.

2 How to Handle the Fusion Problem?

2.1 Introduction

When dealing with any fusion problem, several fundamental questions are of primary importance for building the appropriate architecture. This includes:

i) What is the goal of the fusion for the problem purpose?

ii) What are our expectations about the outcome of the fusion process?

iii) What are the major difficulties to be solved?

iv) What are the characteristics of the data (e.g. where is located uncertainty, what are the relations between data, are they generic or factual, etc.)?

v) How the different pieces of knowledge should be represented, and will all be represented in the same basis or in different basis?

vi) At what extent we can trust the information supplied by each source, i.e., reliability or dependability of each source?

vii) Which methodology should be chosen to reach the goal and cover, at least partly, our expectations while agreeing with the available knowledge?

viii) How will such methodology be evaluated and validated before and/or after real time implementation?

ix) What are the limits of the foreseen methodology?

x) Etc.

These questions among others are necessary before deciding on appropriate architecture for the fusion process. Indeed, the model attached to the information issued from the source[3] is strongly related to the general context of the knowledge constraining the environment. For instance representations of cardinal, ordinal and logical inputs are very different from each other, which, obviously, entails different models. Also, if one expects an interval-valued outcome, the single inputs are unlikely to be modeled in terms of single valued entities. Expectations may as well include a set of desirable properties either algebraical of behavioral in agreement with the result of the fusion operation. For instance, we may require that the result of the fusion process is independent of the order by which the inputs

[3] We shall use the world "source" and "sensor" indifferently throughout this chapter unless stated otherwise.

are combined, presents some absorption property, agrees with the idempotency property that ensures the combination of similar inputs provides no change, etc. Central in the representation of individual knowledge is the associated temporal, spatial and attribute frame. The level and the type of uncertainty pervading the data provide further insights on the appropriateness of a special modelling issue. That is, when dealing with a random process, deterministic representations are likely to be avoided. Rather, a *distributional* representation based on probability measure is more appropriate. However, in the absence of randomness and the presence of imprecision and vagueness as main part of uncertainty, the theories of upper and lower probabilities including fuzzy sets, possibility measures, belief functions among others, sound to have certain favor. The fusion methodology, which encompasses the fusion architecture, should, therefore, deal with the purpose of achieving the expectations under the modelling and architectural constraints. Obviously the simplest methodology consists in taking the standard average or weighted average of these data. While more elaborated methodologies involve transformation of the input into a more abstract level defining some algebra and then using some decision support systems or logical systems to decide on the type of aggregation of the transformed data. Importantly, it matters the degree to which the decision level is involved in the fusion process. That is, can each source provide an individual decision by its own, or is the central unit the only one that performs the decision.

Evaluation and validation tools play a central role in determining the feasibility of the proposed methodology. Examples of evaluation tools cover simulation platforms in which real data are roughly reproduced. In some special cases of estimation problems, where the fusion aims at assessing the value of a given control parameter, evaluation and validation can be performed considering the true value of that parameter. Among the techniques suitable for such purpose we may mention the Monte Carlo simulations [31], and covariance error prediction and analysis [17].

2.2 Type of Knowledge and Information Fusion Level

Depending on the nature of the information stemming from the different sources, Dasarathy [28] distinguished between three levels of inputs: data, feature and decision. The decision-input corresponds to the highest level in this hierarchy. Data-input, which consists of the manipulation of the raw signal data, is the lowest level. The two other types of inputs subsumes that the source is endowed with some processing ability which enables the source to transform the signal input into a more abstract level of representation. Feature corresponds to any categorization of initial data as an abstract description like image, symbolic description, etc. While in decision level, we rather deal with some logical, probabilistic or any kind of measure-based description of the context of interest. For instance, in the case of the target detection problem, the decision level corresponds to the yes/no response whether the target is detected or not, or to a confidence value that the target has been detected. While the feature level might be any geometrical representation of the target generated through successive analysis of a sequence of raw data over a short period of time. Viewed in terms of black-box system, considering only the input and the output of the system, we may distinguish: data-in data-out fusion, data-in feature-out fusion, feature-in feature-out fusion, feature-in decision-out fusion and decision-in decision-out fusion. Basically, in each of these descriptions there are basic methods that can apply. For instance, estimation techniques, including averaging, least squares [17], Kalman filtering [17, 32], are appropriate for data-in data-out fusion. While classification techniques [33, 18] are more suited for data-in feature-out fusion, and Bayesian [17], multi-criteria decision making [34], fuzzy logic based approach [4, 35] among others are more suited to handle decision-in decision-out fusion. Strictly speaking, such decomposition (data, feature and decision levels) is mainly context dependent, while we may find more intermediate levels between the data and the decision levels. For instance, Hall and Llinas [3] pointed out in the context of military type application five levels of hierarchy: existence of entity, position and/or velocity, identity of emitter or platform, behavior of an entity, situation assessment, threat analysis (decision).

On the other hand, we may also find more abstract forms of description of the data, which includes preference, or ordinal information in which estimation techniques can not apply. The reader may consult [34, 36, 37] for preference manipulation.

Elicitation of the knowledge stemming from the different sources plays a central role in any fusion process. This corresponds to the modelling issue that involves a (mathematical) model through which the output of the source can be recovered, or at least approximated, from its input. Typically, such model should implicitly amounts for the nature of *uncertainty*, which is basically inherent to any measurement, pervading the information issued from that source. For instance, an ultrasonic sensor, which measures a distance to an object using the "time-of-flight" of the pulse, is uncertain about the angle from the sensor at which this distance is measured. This is because the sound-pulse, which is reflected by that object lies somewhere within the cone generated by the ultrasonic beam. Basically, in general context, uncertainty occurs here as a consequence of

i) Physical limitations of the sensors like the limited range, possible failure in the physical system, which leads to erroneous output.

ii) Environmental constraint, which induces for instance occultation phenomenon, e.g., when some objects are occluding the view of the sensor.

iii) Reliability of the sensor in the sense that it often occurs that the manufacturer itself supplies a reliability degree or function. That is, it is more likely the sensor repeating the same measurement under the same conditions would not supply the same result.

iv) Lack of specificity in the sense that the sensor provides several distinct values (no steady state) and there is no further evidence to trust one of them more than others. This occurs for instance in situation where the analysis of the sensor output shows a multimodal distribution.

v) Huge abundance of information in such a way that looking for a relevant information is, if not impossible, almost very expensive. Take for instance a source, which consists of a web search engine whose output for a given demand (input) contains more than one million of responses!

vi) Calibration issues which adjust the sensor readings in the way that they more likely agree with the model.

vii) Model simplification in the sense that the full modelling of a sensor is too complex, for instance, due to a large amount of electronic devices. So, a simplified model, which is computationally more attractive is used. In turn, uncertainty here occurs as a gap between the simplified model and the non-simplified one (but very complex).

2.3 Sensors' Configuration and Architecture

Basically the architecture characterizes the type of communication between the different sources, which form a set of nodes. It also determines the degree of autonomy of each source, which, endowed with a processing capability, acts as an intelligent agent that may communicate with other agents or with a central unit, or, at some extent, take some local decision (s). From the viewpoint of the disposition of the sources with respect to the decision processor, we can distinguish [28] a parallel and serial suite (cf. Fig. 1).

- Parallel suite: the sources are interrogated in parallel and more or less simultaneously, where each source acts independently of another one. The outcomes of the sources, which might be data, features or decisions, are then combined accordingly.
- Serial suite: the outcome of a given sensor serves as input for another source, the outputs of the different sources are then combined sequentially.

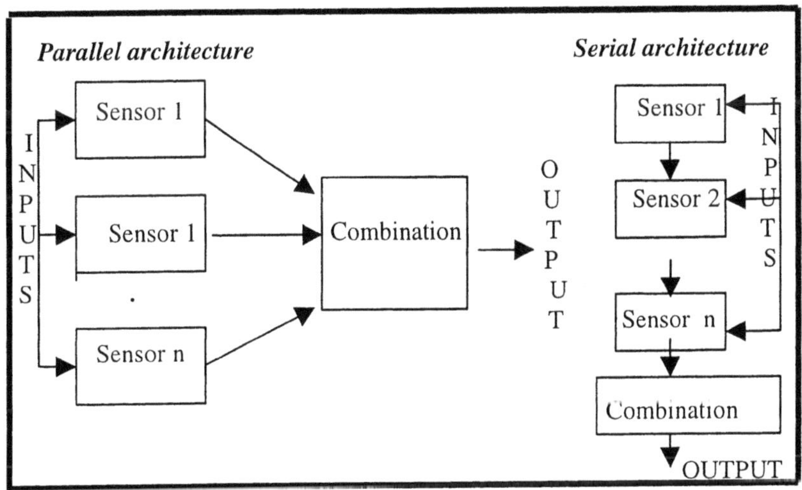

Fig 1.. Parallel versus Serial architecture

Other types of architectures corresponding to mixtures of serial and parallel systems can also be envisaged when groups of parallel (serial) are combined in serial (parallel). More complex mixtures can similarly be constructed. However, in practice, the highly redundant systems should be balanced by the cost of highly reliable switches that allow the system to pass from one branch to another in appropriate time. Indeed, many failure in communication systems, for instance, are due to the failure of the switch, which induces lack of synchronization.

Further, the architectures differ in the degree to which the single sources are able and allowed to make decisions for the overall system. In this respect, we may distinguish:

- Centralized architecture: no decision is performed at the source level. Rather, all the data-input are transferred to a central unit who will then perform the overall decision. Consequently, any change in one source will be immediately propagated to the global decision. Usually, many of the existing fusion systems use a centralized architecture. Basically, the centralized architecture is theoretically optimal as no loss of information is allowed, but very expensive in terms of time requirements and less robust to conflictual information and erroneous data. Further recent advances in computing and communications have made other non-centralized architectures more attractive and allow the system to execute some intelligent functionalities which are not allowed in centralized architecture, like cooperation.
- Decentralized architecture. In this case, the central unit is not necessary linked to all sources, rather some possible links between sources (or nodes), acting as single agents, are permitted. Basically, the whole set of sources is partitioned into multiple sets with a fusion node for each set. The different fusion nodes can then communicate with each other and exchange relevant information. This produces a set of local decisions, which are then, if required, combined to yield an overall decision. Consequently, the failure of some of these sources is not systematically propagated to others. However, a key issue in such architecture is how to combine the results issuing from two fusion nodes. Particularly, in contrast to centralized architecture where it can rationally be assumed that errors in data sources are independent, this assumption is no longer valid in decentralized architecture since the link between different nodes implicitly introduces dependency. Another important issue is the partitioning of the sources into multiple sets; that is, with whom each node is going to communicate to? This issue is mainly context dependent even if some feasibility study can be carried out beforehand. However, in a fully distributed architecture, which is unfortunately less mature area of research even if it appears to be very promising for future autonomous systems, each node can communicate to any other nodes depending on the information content, the needs of individual (intelligent) agents and subject to connectivity constraints [38,39]. From the modelling perspective, basic elements from graph theory [40] are well suited to model the communication and the interactions between the different nodes.
- Hierarchical architecture is a kind of decentralized architecture where the final decision is reached after several stages. Each stage involves a combination of a certain number of nodes. The result is then communicated to

the higher level in the hierarchy. This allows the structure to refine the output accordingly. Hierarchical architecture sounds to be natural for some type of applications. For instance, in radar type application, we may have fusion node for radar data and another fusion node for infrared type data, and then another node combines the result of the two nodes. Further, we may also distinguish hierarchy with feedback or without feedback depending on whether the low level is allowed to revise its knowledge in the light of the result at higher levels. The hierarchical architecture without feedback can easily be converted into a singly connected information graph by using tracklets or restarting at high level. This provides a simple approach for de-correlation and fusion even if it requires the synchronization between the different nodes which is quite costly [20, 25].

From the viewpoint of the degree of interaction between the different sources, Durrant-White [38, 39] distinguished between cooperative fusion, competitive fusion and complementary fusion. Cooperative fusion occurs when one sensor relies on the observations of another sensor to make its own observations. For instance, a touch sensor may refine the estimated curvature of an object previously sensed by range sensors. Typically, cooperative fusion operates in either centralized or decentralized architecture, and allows the system to reduce the overall uncertainty. In complementary fusion, each sensor supplies only a partial information of the environment, such that their integration leads to a more complete view of the environment. This type of fusion is not systematically aimed at reducing the overall uncertainty, but rather resolves the incompleteness of the sensor data. Remark that, at some extent, cooperative fusion can also be viewed as a kind of complementary fusion. Competitive fusion is similar to a fusion of redundant information, in the vocabulary of Luo and Kay [2], Dasarathy [28, 29], Hall and Llinas [3], among others, where all the sources are sensing the same object and execute the same task simultaneously. This allows the system to reduce the effect of uncertain and erroneous measurements and compensate failure of one sensor by another one. A simple example of such fusion is when repeating the sensor measurement several times, under the assumption of static system, yielding a temporal fusion. This sort of fusion is primary predominant in the literature as noticed by the extensive review of Luo and Kay [2].

2.4 Selection and Voting Schemes

Central in any fusion architecture is the logic underlying the way by which the sources are handled. Is it better to perform the combination of the information stemming from the different sources, or to restrict the output to a single source, which sounds better according to some criteria? Will all the sources be taken into account or only few of them according to some logic? This type of questions has often been raised in applications involving fault-tolerant systems where it matters for instance, whether a single version software is better than N-version software [41]. Such logic is often generated by the voting mechanism underlying the fusion process. In this respect, a certain number of votes are assigned to each source, and the combination process is carried out over those sources whose total number of votes exceeds certain threshold. The latter is basically determined by the voting strategy employed by the decision-maker. This includes, among others, see [42] for a more detailed review:

- Majority voting which requires that the sources that have to be combined must have in total more than the half of the total number of votes.
- Unanimity voting where the total number of votes must be reached before the combination is carried out. This means that the only way to discard sources is to attach them zero votes.
- M-out of V, in which the total votes attached to the output must be greater than M, where V is the total number of votes. If M is less than the majority (V/2) then the result at the output may not be unique.

Central in the voting algorithm is the weight assignment procedure where each source is attached a finite number of votes. Several algorithms have been developed for such purpose. These algorithms differ in the criteria that constraint the goodness of the assignment. For instance, maximization of the overall system reliability is one of the rational criteria for this purpose. Further, as mentioned by Parhami [42], the voting problems are getting complicated in some applications where there is a range of values, which are considered as correct. For instance, such effect often occurs in numerical applications, due to round-off errors, and in clock synchronization due to clock drift and variation in message transmission times. At some extent this makes non-additive models (like possibility or evidence theories [4, 8, 11]) that exhibit nonspecificiy as a main facet of uncertainty very promising for such purpose.

2.5 Homogenization of the Data

When dealing with redundant sources of information, which ensure uncertainty reduction and increase the overall reliability, it is required that the information stemming from the initial sources are homogeneous, or, at least, can be transformed through some deterministic relations into homogenous information. Roughly speaking, if the fusion process involves redundant information, then the fusion operation can be accomplished in three main steps.

i) *Alignment step*, where the data are converted into the same spatial, temporal and attribute frame, e. g, initial information are converted into a common representation. This permits the system to get homogeneous information, which will then be merged to yield a refined result about the same entity. Typically, in target-tracking type application, the alignment deals with the spatial registration, which induces converting the data issuing from each sensor to a common coordinate system, or a temporal prediction of the target track based on the input of sensor suite. Temporal alignment involves extrapolation of the tracks to the same time origin. This step typically involves coordinate transform, unit adjustment and utility, handling of environmental constraints, etc. So, issue of understanding environment constraints, sensor modelling and measurement theory [38,39,36] are very important for this purpose.

ii) *Association and correlation*, where the obtained data are filtered and the relationships between each datum with the associated sensor (s) are established. Basically, the association and the correlation are responsible for partitioning the data into sets of measurements that could have been originated from the same targets. Usually, statistical tests [17, 28, 43], similarity based techniques [44-46], classification methods [33, 45], multiple hypothesis tree [17, 47, 48], among others are possible candidates to ensure the coherence between the obtained classes of measurements and the underlying targets. This also allows the system to handle conflicting situations, where some measurements are simply discarded as being erroneous. In some literature, this step is expressed as situation assessment, as there is a sort of pattern matching between model outputs and measurements, as pointed out by Hall and Llinas [3].

iii) *Combination* which consists of aggregating the established homogenous results, after being filtered and re-arranged in association / correlation stage, in order to obtain the best single representation in terms of confidence, faithfulness, etc. Basically two types of aggregation can be distinguished, depending on whether the process involves conditioning or not. For instance, Kalman filtering [17, 32] approach belongs to the former while averaging or weighted average belong to the latter. Oxenham et al. [49] divided this step again into two other steps: a) attribute data fusion, or the merging of the features from two or more representations of the same object with different information sources into new features for the object; b) analysis data fusion, or the aggregation of two or more representations into a new representation and the generation of an interpretation of the object for further use. Strictly speaking, depending on the architecture employed by the fusion process, the combination operation may occur either at low level, i.e., data-input, or feature level, or decision level. So, the step a) corresponds to the performance of the feature input fusion type. Further, according to the context under consideration, each level requires special types of combination operations. For instance, averaging like operation is more suited for low level, high-level image processing or graph reduction techniques [40] for feature level, and rules simplification [47, 50] or measure aggregation [11] for the higher level. Similarly, the alignment and the association stages may occur at different levels depending on the architecture and the type of the inputs (data, feature or decision).

These steps are expressed differently in [49, 51, 52] even if some redundancy between authors' claims can be highlighted. Thus, each fusion methodology has to address its own interpretations about the previous steps. Sometimes it is refereed to the first two steps as *pre-treatment* of data, which, of course, play a central role in the whole process. Crowley [51, 53] has pointed out some very closed notions as general principles for fusion. They can be summarized into five main points: i) observations and model should be expressed in a common co-ordinate frame; ii) observations and model should be expressed in a common vocabulary; iii) primitives should include explicit representation of uncertainty, where primitive refer to the elements, which is composed of the different properties that may be observed or inferred from different sources; iv) primitives should be expressed as a vector of parameters; v) primitives should be accompanied by a confidence factor.

Examples of applications of such decomposition (alignment, association and combination) are described in [54] for robotics-type application using soft computing approach, and radar-type application in [3, 43] using standard estimation theory.

2.6 Dealing with Complementary Data

In some fusion problems, the information issued from the different sources can not be transformed into homogeneous information. In this situation, either the information are conflictual, due, for instance, to human errors or lack of consistency, or complementary. The latter aspect (complementary) as pointed in the introduction of this chapter attempts to provide another facet of the context under consideration. For instance, if a sensor 1 provides a distance from the target T to a point A and sensor 2 gives a distance from the same target to another point B, then the two sensors are complementary and allow us to represent the distance into two-dimensions space. In other words, the feature space or the representation space becomes augmented. Another example of complementary consists of synthesizing the global view of some geographical relief when sensed by different images. Consequently, processing regular fusion operations like averaging or median or so, in this situation, has no meaning. Therefore, it turns out, if the system is constituted only of complementary sources, then the system is unlikely to reduce the uncertainty of the overall system. Complementary sources can also be understood even in the case where the sensors pertain to the same data attribute, spatial and temporal frame in situations where, at least, for some attributes, one sensor is more accurate than others. For instance, if the context of interest is to estimate the x-y position of some unknown target, we may have situations in which sensor 1 is more accurate in direction x and less in y while sensor 2 is more accurate in y and less in x direction. In this respect, both sensors complement each other.

2.7 Architecture and Reliability

Data fusion and reliability are ultimately linked in the sense that data fusion architecture may serve as a tool to generate reliable inputs. On the other hand, reliability analysis may serve as an aid to induce appropriate fusion architecture. See Fig. 2 for a summary of the general methodology of the fusion process.
Intuitively, one may be more confident when a given target is confirmed by two or several sources than when only one single source supports the hypothesis. Primary, it is quite common that the use of parallel architecture (redundancy) leads

to an increase in reliability of the overall system in the sense that the lifetime of the overall system will be extended. So any defect or disfunctioning of one sensor is compensated by another one, which, in turn, results in an increasing reliability. Besides, multisensor system may be even motivated from economical ground when the construction of a single highly reliable sensor is very expensive, while its performance can be almost achieved through a suite of less reliable and, therefore, cheaper sensors. In a broader way, improvement of reliability can also be achieved by appropriate choice of maintenance strategy [31]. In this case, the fusion architecture is crucial to determine the optimal cycle of the maintenance phase that achieves the goals. In this course, when the components have different single reliability, the determination of optimum maintenance period becomes an optimization problem.

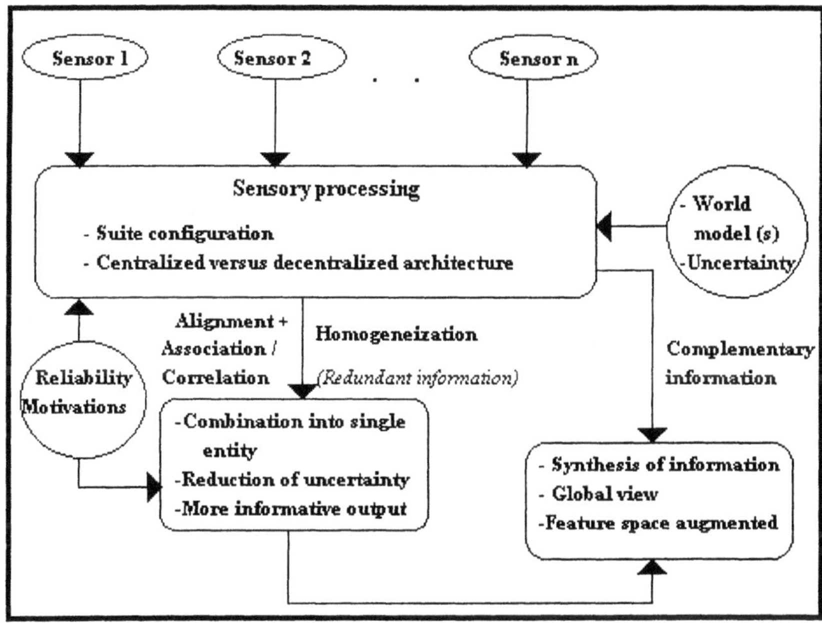

Fig. 2. Synoptic of general fusion process

A simple example in robotics consists of a sensor configuration; namely, what is the optimal location of the different sensors that achieves the reliability requirements, given that a maintenance phase, which primary includes a re-calibration of the sensors, is required. In other words, letting a larger maintenance cycle may lead to a degradation of the performances of the sensors while too short cycles often induce a decreasing in real time performances of the robot task. So, a compromize situation is required between these two extreme situations. Unfortunately, this area of research still is premature due to actual technological limits of autonomous systems, while it is gaining high importance in safety-critical systems.

2.8 Application of Soft Approaches in Data Fusion

Strictly speaking the application of soft techniques, including fuzzy logic, possibility theory, evidence theory, general upper and lower probabilities, neural network, genetic algorithms, is becoming very popular in several communities including the information fusion community. It suffices to analyze the very respectable fusion projects mentioned in the introduction of this chapter, to see that soft techniques have been employed at different levels. For instance, expert systems, neural network, Kalman filtering and fuzzy logic have been used for ADVANCE [26]. The AGV [22] involves mainly a fuzzy logic framework, while Kalman filter, neural network and expert system have all been used in PROMETHEUS [27]. The successful and popular Pathfinder project [55] on autonomous vehicle for space applications sponsored by NASA also involves fuzzy logic approach in some of its modules. Basically, the applications of soft computing techniques are rather performed in the third or second level fusion dealing respectively with decision or feature-like inputs. Indeed, the powerfulness of soft techniques in classification and manipulation of more abstract representations of entities make these approaches more suited to guide human decision-making. Fuzzy logic and other non-additive measure theories are well suited for handling imprecise and ill-defined entities. While endowed with approximate reasoning framework [56], which includes for instance "if then" rules, enables the structure to handle linguistic valued entities and perform high level data processing. Further, fuzzy and non additive measure theories can also be accommodated to reason in low level fusion when dealing with pure imprecise entities or ordinal information. For instance the use of fuzzy numbers [57] endowed with Zadeh's extension principle [58] permits the manipulation of imprecise real valued entities, like "approximately 40cm" or "not far from 20°", etc. Random set theory [59] and approximate reasoning tools [60] might also be alternative candidates for handling imprecise real-valued entities. Neural network can be applied as robust classification tools in high fusion levels, and, because of their ability to process several operations in parallel, offer as well a powerful tool for the first level fusion. Moreover, considering the relationships between the different nodes in the fusion architecture, neural network offers appealing properties to model the communication between the different nodes, particularly when the neural network is endowed with some learning capabilities.

Basically, we may summarize the main possible contributions of the soft-based approaches to the fusion problem into i) flexibility, ii) understandability, iii) generalization, iii) fast computation and, iv) intuition in handling uncertainty.

From uncertainty viewpoint, *Flexibility* is a straightforward result of the relaxation of the additivity axiom in probability. This yields, sometimes, outcome with multiple interpretations as a result of the presence of ambiguity and fuzziness. From the architecture viewpoint, flexibility results in more robustness against small changes that may occur in the disposition of the different sources as it will be pointed out in Section 3 when dealing with possibility theory for instance.

Understandability is mainly justified by approaches like approximate reasoning involving a set of rules "if... then..." such that, at any stage, the system is able to provide explanations for its findings and actions by relying on the causes underlying a given result. While such property is not feasible in classical algorithmic approach. From the viewpoint of multiple-interpretations paradigm, understandability allows the decision-maker to justify the joint occurrence of several candidates and also provides means to make suitable selection by relying on the confidence value attached to each candidate.

Generalization arises from the fact that fuzzy controllers as well as neural network are universal approximators [4, 14]. Consequently, when dealing with estimation-like problems this allows us to handle non-linear systems in quite approximate but reasonable fashion.

Fast computation is mainly due to neural network property as a result of its ability to perform several subtasks in parallel. This makes certain classes of neural networks as well as genetic algorithms very suited to real time applications and allow us to handle the first level fusion very efficiently. Strictly speaking, neural network becomes very fast only once the training part, which is usually very expensive in terms of time requirement, is ended. So, it matters to know whether the training samples are sufficiently enough and representative of the possible state of unknown situations.

The intuition in handling uncertainty results from the multiple facets of uncertainty conveyed by the non-additive framework. Indeed, the latter allows the system to handle specificity, conflict, certainty, randomness, etc. While standard probability theory only relies on randomness nature of uncertainty. Klir and his colleagues (see, for, instance [61, 62]) have written interesting papers in this topic. For example, possibility theory provides more powerfulness in representing the nonspecificity, by letting more than two alternatives to be fully possible. Conflict is represented in evidence theory, with respect to some focal set A, by allowing a nonzero mass to other focal sets which are completely disjoint with A. Certainty can represent any necessity measure in possibility theory or a degree of belief in evidence theory. While randomness can be embedded either in a random set representation of fuzzy set or by allowing a more generalized framework of fuzzy random variables [59, 63, 64].

Now let us summarize some possible soft methodologies that may be applied at different stages of the fusion process. Table 1 and Table 2 provide a range of tools issued from soft computing approaches to handle the different stages and types of fusion process. Primary, one may retain the following:

Neural networks (and at some extent genetic algorithms) [14, 15, 47] can be used as tools to compute the common operations performed in standard (non-soft) theory of estimation, image processing or decision making. From this perspective, the soft-computing approach contributes through the aforementioned fast computation property to improve the computational aspect of the fusion process.

Through the property of fuzzy controller and neural network as universal approximator [4], this allows the system to circumvent the modelling issue when nonlinearity and quantification of randomness are often delicate problems in standard data fusion problems. This offers more flexibility in modelling either the sensor output or the models governing each stages of the fusion process (alignment, association and combination).

Table 1. Summary of soft computing tools in fusion process

Modelling of source informat-ion	- Linear or nonlinear models - Black box model through neural network - Fuzzy controller whose inputs are the inputs of the sources and its output the output of the source - Use of linear or nonlinear model but with fuzzy parameters - Use linear or nonlinear models with nonrandom error (non-probabilistic uncertainty), etc.		
	Level 1 fusion - Data input -	Level 2 fusion - Feature input-	Level 3 fusion -Decision input-
Alignment	-Extension principle applied to environ. constraints -Translation of frame of discernment in evidencial reasoning. -Allocate the task performed by stand. methods to neurons. -Uncertainty minim. criterion that guides the search for estimate parameter in a common frame.	- Robust image filtering tools. -Use extension principle over vector and matrix entities. - Allocate standard tasks to neural netowk. -Classification tools like cluster analysis, fuzzy c-means, etc. - Orthographic projections from projective geometry.	-Extension principle based approach where the universe lies in [0,1] interval (measure scale) -Translation of frame of discernment in evidence theory. -Structural matching of strings - Transformation of rules via modus ponens or modus tolens .
Associat-ion	- Similarity measures -Fuzzy minimization tools. -Allocate statistical validation techniques to neurons in neural network -Re-calibration of the source output	- Pattern matching techniques using like certainty factors - fuzzy graph tools - Validation tests using measures. -Dependency analysis - Evaluation measures -Nearest neighbor tool	-Utility based approach. - Allocate statistical validation techniques to neurons -Evaluation measures brought from nonadditive measures -Logical templating
Combinat-ion	- Fuzzy estimation tools like fuzzy linear regression, variance minimization, etc -Fuzzy controller, neural network (universal approx.)	-Object aggregation - Elimination of redundant information - Merging the list of feature nodes and the corresponding graphs. - Knowledge-based models -Neural network used as a classification tool	-Fuzzy knowledge based systems -General aggregation of fuzzy measures -, Dempster rule , fuzzy or Choquet integral -Application of insertion, deletion, truncation and merging rules.

Table 2. Summary of soft computing tools in fusion process (Continuation)

	Level 1 fusion - Data input -	Level 2 fusion - Feature input-	Level 3 fusion -Decision input-
Reliability	- Assign a single value to observation - Assign distribution to the observation	- Assign a single value to the observation, - Assign a distribution to the observation	- Assign single value to observation - Assign a distribution to the observation
Complementary sources	- State vector augmentation - Composition of universal sets - Composition of uncertainties like in fuzzy random variable framework -Use reliability evaluation to restrict to the most reliable one (s).	- Image insertion and deletion tools - (fuzzy) graph extension tools - Composition of uncertainty - Allocate different neural network for different sets of redundant sources	- State vector augmentation - Composition of universal sets -Allocate different neural network for different sets of redundant sources

Soft computing techniques extend the standard approaches in case where other types of uncertainty occur. For instance, fuzzy set or possibility distribution can be used to quantify the imprecision pervading some ill-known parameter of the model. In this situation, we may mention fuzzy linear regression tools [35, 65, 66] as counterpart of standard linear squares at first level fusion. As well in feature based systems, fuzzy classification tools offer more flexibility for handling imprecision and ambiguity inherent in data structure [44, 45]. Further, soft computing tools and standard approaches can be used in cooperative fashion, like allocating some tasks in the fusion process to the soft tool and another to the standard approach. This is motivated by the fact that some of operations are more efficient and/or more attractive when performed by one formalism rather than other. This kind of reasoning has been carried out for data tracking applications [67]. Besides, the various connections between the different nonadditive models, including probability [9, 11, 62] go in this direction. Issues from cooperative games are also helpful for this purpose [68].

Nonadditive measures offer a variety of choices in choosing a proper combination mode ranging from a pure conjunctive mode to a disjunctive mode passing through different compromise, adaptive and prioritized situations [11, 56, 69]. This sounds for instance natural when comparing the consistency in nonadditive models and additive model. Clearly, in the former, an inequality is enough to restore consistency, thereby, leading to wide classes of consistent results (see, for instance, [11, 21, 69]). This offers more flexibility for the decision-maker and a broader framework for interpreting multiple (non-unique) results that appear at first glance conflicting. Also, the revision of information in the light of a new information becomes more flexible and allows the system to model a wide variety of intuitive meanings [70-73], while Bayes' rule is unique in probability theory.

In rule based systems, the use of soft computing tools offers more richness in expressing the expert knowledge. For instance, the use of rules "if x then y", modeled, for example, as joint possibility distributions, allows the decision-maker, through the use of the whole set of fuzzy connectives like projection and composition operations [4, 56, 60], to generate a wide class of behaviours. Further the qualification of the α-certainty statements provides the user the ability to express in formal way the reliability of the different assertions [56].

3 Analysis of Some Examples of Fusion Architectures

Throughout this section, we shall examine in more detail an example of the previous architectures when dealing with probabilistic and possibility theories. We shall consider parallel and serial architecture. Moreover, we will restrict to the third level fusion where the source inputs and outputs are decision based.

3.1 Parallel Architecture

Consider the case where the sensors provide some confidence measure μ about the state x given evidence Z_i (for source i), see Fig. 3. It is as well assumed that the information are homogeneous and not conflicting, so there is no need to carry out alignment and association steps

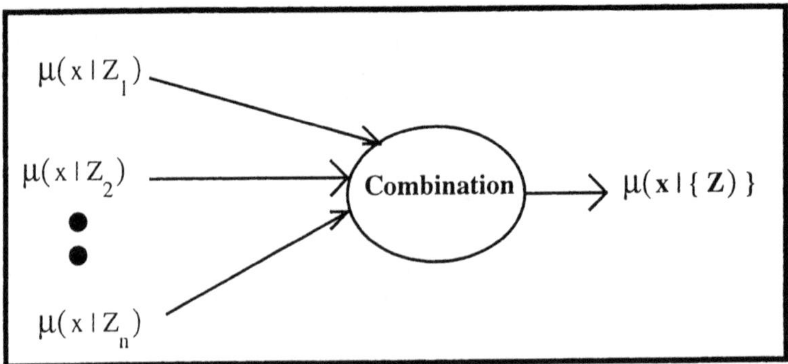

Fig. 3. Decision fusion in parallel architecture

- ***Case of probability measures***

Suppose that μ is a probability measure. Then we may distinguish at least two main issues for generating the result $P(x \mid \{Z\})$, with $Z = \{Z_1, ..., Z_n\}$.

- ***Linear pool***

In this case, the conditioning aspect of the inputs is forgotten, and assuming the existence of a set of positive weights λ_i summing up to one, i.e., $\sum_{i=1}^{n}\lambda_i = 1$, the result is given by:

$$P(x \mid \{Z\}) = \sum_{i=1}^{n}\lambda_i P(x \mid Z_i)\,. \tag{1}$$

The weight λ_i may model the reliability of the source i. In the case where all the sources are equally reliable, then w_i are assigned 1/n value (uniform probability).

More generally, the linear pool has been justified by McConway [74] as being the unique combination mode F when we require marginalization[4] and weak zero probability (e.g., if all input probabilities are zero valued then the resulting probability is as well zero valued).

- *Bayesian pool*

Here the result mainly lies in the application of Bayes' rule and its variants. That is, the probability of the estimate x given a measurement z is given by

$$P(x \mid z) = \frac{P(x,z)}{P(z)} = \frac{P(z \mid x)P(x)}{P(z)}, \tag{2}$$

where P(x) corresponds to the prior probability, i.e., evidence about the estimate x before observing the measurement (s) z, and P(z|x) corresponds to the likelihood of the source. While P(z) is a normalization term.

Now consider the sources as in Fig. 3, and assuming the conditional independence of the data supplied by the different sources, the application of Bayes' rule leads in case of two-sources to

$$P(x \mid Z_1, Z_2) = \frac{P(x \mid Z_1)P(Z_2 \mid x, Z_1)}{P(Z_1 \mid Z_2)}. \tag{3}$$

The independence assumption stipulates that: $P(Z_2 \mid x, Z_1) = P(Z_2 \mid x)$ and $P(Z_1 \mid Z_2) = P(Z_1)$. Consequently, (3) becomes

$$P(x \mid Z_1, Z_2) = \frac{P(x \mid Z_1)P(Z_1 \mid x)}{P(Z_1)} = \frac{P(x \mid Z_2).P(x \mid Z_1)P(Z_1)}{P(x)P(Z_1)} = \frac{P(x \mid Z_1).P(x \mid Z_2)}{P(x)}. \tag{4}$$

The reader may consult [17] for a quick refresh on these notions. Extension of (4) to n sources yields

---

[4] The marginalization stipulates that combining probabilities P_1, .., P_n defined on probability space (Ω, S) into a new probability $F^S(P_1, .., P_n)$ such that for all sub-σ-algebra S' of S and for all A of S': $[F^S(P_1, .., P_n)]^{S'}(A) = F^{S'}(P_1^{S'}, .., P_n^{S'})(A)$.

$$P(x \mid \{Z\}) = \frac{1}{[P(x)]^{n-1}} \prod_{i=1}^{n} P(x \mid Z_i) . \tag{5}$$

Remark that (4-5) assumes that all the sources have the same prior. Extensions to sources with different priors can also be envisaged.
Interestingly, (5) can be expressed in iterative form, assuming a vector of measurements, say $Z^{n-1} = \{Z_1, Z_2, ..., Z_{n-1}\}$, and denoting P_{n-1} the probability obtained at n-1 iteration, then it is easy to check that

$$P(x \mid \{Z\}) = P_n(x \mid Z^n) = \frac{1}{P(x)} P_{n-1}(x \mid Z^{n-1}) P(x \mid Z_n) . \tag{6}$$

Particularly, the underlying iterative process enables the fusion operation to possess some modularity property that reduces the complexity of the operation.

- ***Case of possibility measures***

In possibility theory [5, 6] the confidence measure on the estimate x can be expressed in terms of possibility measure, or equivalently, its dual necessity measures, $\Pi(x \mid \{Z^k\})$ (where the source i is assigned $\Pi_i(x) = \Pi(x \mid \{Z_i^k\})$).
It has been shown by Dubois and Prade [75] that the only eventwise manner of combining possibility measures into a new possibility measure, i.e., $\forall A, \Pi(A) = f(\Pi_1(A),..,\Pi_n(A))$, is given by $\Pi(A) = \max(f_1(\Pi_1(A)),.., f_n(\Pi_n(A)))$, where f_i are monotone increasing with $\forall i$, $f_i(0)=0$ and $\exists j$ $f_j(1)=1$. Consequently, a a trivial consensus in the course of the preceding would be

$$\Pi(x \mid \{Z\}) = \max(\Pi(x \mid Z_1),..., \Pi(x \mid Z_n)) . \tag{7}$$

Reliability of the individual sources can be incorporated, for instance, as an index, say λ_i, attached to the information supplied by each source, such that using a weighted min combination for example, with $\max_i [\lambda_1] = 1$, leads to

$$\Pi(x \mid \{Z\}) = \max(\min(\lambda_1, \Pi(x \mid Z_1)),..., \min(\lambda_2, \Pi(x \mid Z_n)) . \tag{8}$$

Other forms of expressing reliability of individual sources can also be applied provided that the underlying functions f_i satisfy the aforementioned requirements. For instance, we may use exponential function of x, say $1- e^{-\lambda x}$, $\lambda > 0$:

$$\Pi(x \mid \{Z\}) = \max((1 - e^{-\lambda_1 x}).\Pi(x \mid Z_1),...,(1 - e^{-\lambda_n x})\,\Pi(x \mid Z_n)) .$$

Clearly, (8) sounds to be a direct counterpart of the linear pool in probability theory. However, if the sources supply distribution functions instead of single valued measure (at particular value of the estimate x); that is, a source i generates a possibility distribution $\pi(x \mid \{Z_i\})$ [5], then a natural consensus can be obtained by allowing any min-combination:

$$\pi(x \mid \{Z\}) = \min(\pi(x \mid Z_1, ..., \pi(x \mid Z_n)) . \tag{9}$$

In this case, (9) can be generalized by substituting any triangular norm operator T [76] to the min operator. Remark that (7-9) also fulfills the zero possibility property; that is, the result takes zero value when all sources supply zero value possibility. Consequently, in contrast to probability theory, a non-conditional combination in possibility theory offers a variety of pooling methods to accomplish the fusion operation.

Now in order to consider the counterpart of the conditional rule (5), we use the following definition of conditional rule in possibility theory [73]

$$\Pi(A \mid B) = \frac{\Pi(A \cap B)}{\Pi(B)}, \text{ which is equivalent to } \pi(u \mid v) = \frac{\pi(u,v)}{\sup\limits_t \pi(t,v)}.$$

On the other hand, conditional independence of ordinary event A and B given C is expressed as $\Pi(A \cap B \mid C) = \Pi(A \mid C).\Pi(B \mid C)$, or, equivalently, $\pi(a,b \mid c) = \pi(a \mid c).\pi(b \mid c)$.

Under such requirements, it can be checked that the counterpart of (6), using similar notations is

$$\Pi(x \mid \{Z\}) = \Pi_n(x \mid Z^n) = \frac{1}{\Pi(x)} \Pi_{n-1}(x \mid Z^{n-1}) \Pi(x \mid Z_n) . \tag{10}$$

It is worth noticing that the previous result is mainly due to the definition of conditional possibility and the associated conditional independence utilized here. While the use of others conditioning like Hisdal's conditional rule [77] no longer supplies the same result. From (10) it is clear that possibility theory in this context (conditioning) does not perform worse than the standard approach of probability theory. While the uniqueness property of conditioning in probability seems to offer certain advantages over possibility.

3.2 Serial Architecture

In this case each sensor is communicating with another sensor. As shown in Fig. 4, except for the first sensor in the sequence, each sensor i will receive two inputs: the input from the demand space $\mu(x \mid Z_i)$ and the output from the previous sensor i-1.

---

[5] We recall that, for ordinary subsets A of some universal set U, a possibility distribution π is linked to possibility measure Π by Π (A)=sup{ π (u), u$\in$ A}. While possibility measure is characterized by Π (U) =1, Π ($\varnothing$) =0 and $\Pi(A \cup B) = \max(\Pi(A), \Pi(B))$. More details on this topic can be found in [4, 56].

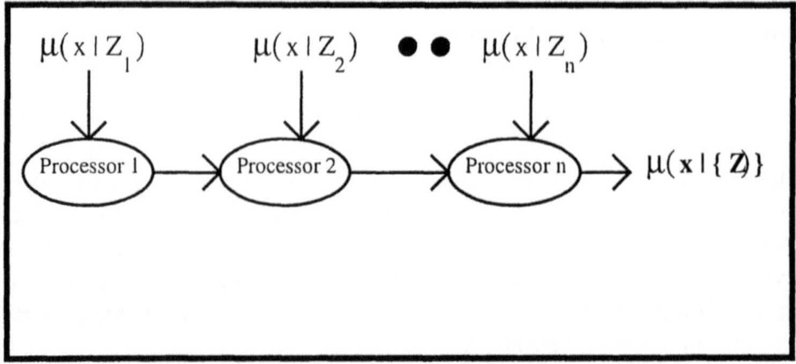

Fig. 4. Example of Serial decision combination

- **Use of probability theory**

Let us assume that the first input $P(x \mid Z_1)$ undergoes no transformation from the sensor 1, whose output will then be $P(x \mid Z_1)$. From this perspective, the result is mainly dependent on the nature of the operation performed at each processor (c.f. Fig. 4).

- Linear pool

Let us use consider the following further assumption:

A*ssumption 0*: there is no preference in terms of confidence degree between the two inputs: output of processor (i-1) and the observation $P(x \mid Z_i)$.

This assumption is very reasonable and very intuitive in the sense that the decision-maker at each stage believes the obtained result is fully reliable as the reliability of sources has already been taken into account in previous steps. As a result, due to the normalization constraint, both the output of the processor (i-1) and $P(x \mid Z_i)$ are assigned a weight of $\lambda_i/2$ (sharing equally the initial weight λ_i). Then using the notations employed for the definition of iterative process and applying the normalization constraint on the weights λ_i, it holds that

$$P_2(x \mid Z^2) = \lambda_2 P(x \mid Z_1) + (1-\lambda_2) P(x \mid Z_2)\text{, with } \lambda_2 \in [0,1]$$

.

$$P_k(x \mid Z^k) = \frac{\lambda_k}{2} P_{k-1}(x \mid Z^{k-1}) + \frac{\lambda_k}{2} P(x \mid Z_k)\text{, } \lambda_k \in [0,1]\text{.} \quad (11)$$

For instance, in case of a sequence of three sources, we have

$$P_3(x \mid Z^3) = \frac{\lambda_3\lambda_2}{2} P(x \mid Z_1) + \frac{\lambda_3(1-\lambda_2)}{2} P(x \mid Z_2) + \frac{\lambda_3}{2} P(x \mid Z_3). \qquad (12)$$

It can be seen that the weights λ_1 and λ_2 tend often to decrease the confidence attached to the sensor 1 and 2 as the multiplication of λ_1 and λ_2 leads to a smaller weight. For instance, if we require all the P(x | Z$_i$) will be assigned similar weights, then we should have $\lambda_1\lambda_2 = (1-\lambda_1)\lambda_2 = 1-\lambda_2$, which ensures $\lambda_1 = 0.5$ and $\lambda_2 = 2/3$. In other words, there are no equal weights $\lambda_i = \lambda_j, \forall i, j$ that ensure consistency in view of (12).

- Bayesian pooling

However, in case where the processors perform a Bayesian updating then the fusion process looks differently. Fusing $P(x \mid Z_1)$ and $P(x \mid Z_2)$ leads to $P(x \mid Z_1, Z_2)$, which combined again with $P(x \mid Z_3)$ yields

$$P(x \mid Z_1, Z_2, Z_3) = \frac{P(x \mid Z_1)P(x \mid Z_2)P(x \mid Z_3)}{[P(x)]^2}. \qquad (13)$$

The latter expression is exactly equivalent to that pointed out in the case of parallel architecture. This is basically justified through the iterative relation (6), which enhances the modularity of the conditioning combination process.

- **Use of possibility theory**

Extension of (7) in the case of serial architecture is straightforward, and due to the associativity of the max-combination, the result remains unchanged. That is, (7) is not sensitive to any change of the architecture (so is for the combination via (9)). The use of a weighted combination in the spirit of (10) leads to a different result. Indeed, in case of two-sources combination, we have

$$\Pi_2(x \mid \{Z^2\}) = \max(\min(\lambda_1, \Pi(x \mid Z_1)), \min(\lambda_2, \Pi(x \mid Z_2)), \qquad (14)$$

with a normalization condition $\max(\lambda_1, \lambda_2) = 1$

Now in order to generalize this result to more than two sources, we may need a proper interpretation of Assumption 0. Clearly, from this perspective, supposing the decision-maker is confident in the output of the subsystem leads to a reliability value one. Further, in contrast to a probabilistic pooling, the consistency still is held even if we assign a reliability one to the output of the subsystem and reliability λ_i ($\lambda_i \in [0,1]$) to the input $\Pi(x \mid Z_i)$. Consequently, for three sources:

$$\begin{aligned}\Pi_3(x \mid \{Z^3\}) &= \max(\min(1, \max(\min(\lambda_1, \Pi(x \mid Z_1)), \min(\lambda_2, \Pi(x \mid Z_2))), \min(\lambda_3, \Pi(x \mid Z_3))) \\ &= \max(\min(\lambda_1, \Pi(x \mid Z_1)), \min(\lambda_2, \Pi(x \mid Z_2)), \min(\lambda_3, \Pi(x \mid Z_3)). \qquad (15)\end{aligned}$$

For n sources, (15) is extended as

$$\Pi_n(x \mid Z^n) = \max(\lambda_1, \Pi(x \mid Z_1), ..., \min(\lambda_n, \Pi(x \mid Z_n))) \,,\quad \max(\lambda_1, \lambda_2) = 1. \tag{16}$$

Clearly, the results (15-16) are very similar to (8) in the case of parallel architecture, except that the weights are normalized differently: $\max(\lambda_1, \lambda_2) = 1$ versus $\max_{i=1,n} \lambda_i = 1$.

However, it deserves mentioning that the former normalization condition entails the latter one, while the inverse implication does not hold. In this respect, if we set up the weights such that either λ_1 or λ_2 are equal to one, then both the serial and the parallel architecture under the aforementioned assumptions will provide the same result. Consequently, in this context, possibility offers more flexibility in dealing with serial and parallel architecture in the sense that the result is less sensitive to the choice of the architecture than it is in the case of probabilistic-based-reasoning under the same assumptions.

While the use of conditioning based reasoning would lead to the same result as in probability, in terms of modularity, due to the iterative result pointed out in (10). That is, even in this respect, possibilistic-based approach is not worthless than its probabilistic counterpart approach.

4 Robotics Oriented Application

In this section, we shall consider the application of the aforementioned three-step fusion decomposition (alignment, association & correlation and combination) to determine an accurate position of a mobile robot in a structured environment, possibly containing a non-modeled obstacle.

4.1 Introduction: Localization From Information Fusion Perspective

Loosely speaking, most of localization techniques found in the literature (see, for, instance, [38, 51, 78] for an overview) use multiple sensors paradigm. So, basically, the concept of data fusion is implicitly involved even if it is not explicitly mentioned. Therefore the link to a fusion methodology is implicitly established. Interestingly is to look to the extent to which the established three-step decomposition (alignment, association and correlation, combination) holds for well founded localization techniques. In this course, the review of these techniques shown that such decomposition can be rationally demonstrated for most of these localization techniques. To see this, we may review the following four classes of localization techniques, determined according to the way by which the given method uses proprioceptive sensors (sensors without interaction with external environment like odometer), and the exteroceptive sensor (sensors related to the environment like ultrasonor (US), camera). We shall say the fusion is dynamic when the time is an active parameter of the fusion process, and static otherwise.

i). D*ead-reckoning* like method (dynamic fusion of proprioceptive sensors) is concerned by fusing data collected over time. The alignment step consists in the relationship between X_k and X_{k+1} (where X_k stands for the vector position, i.e., $X_k = (x_k, y_k, \theta_k)$ corresponding to the x-y position of some reference point in a robot platform and its orientation with respect to robot axis at time k-1). This permits the datum pertaining to the sample "k+1" to be brought down to the same frame as the one of k-th sample. The association step can be reduced to some noise analysis where the covariance matrix, for instance, allows for the noise propagation over time. Also some additional procedures can be added to detect erroneous data resulting from the slide or the skid effect. The combination step corresponds to the recursive formulation underlying the state and its associated variance-covariance estimation from the time origin (t=0).

ii). *Triangulation* like method (Static fusion of exteroceptive sensors) consists in using the multiplicity of measurements (exteroceptive sensor readings) referring to the different targets in the environment to solve the relative equation of the robot positioning parameters. Typical example of this class involves telemetric data measured from a set of beacons at known locations in the environment. So, at least three distance measurements, or one distance and two angles, or two distances and one angle, are necessary for 2D localization. The alignment step is carried out by the geometrical relationship between the measurements and the position vector. The correlation and association step is usually handled through the use of different codes for sensors so that each beacon's signal can be recognized and will not be confused with other signals. Moreover, data tracking procedures like statistical validation tests may be added for noise analysis. The combination step consists of the analytic resolution of the system when the number of measurements is optimal, or using like least squares approach when there is redundancy.

iii). Predictive fusion of exteroceptive sensors (Dynamic fusion of exteroceptive sensors) makes use of the sensor models in addition to the sensor readings, which provides the system with predictive information. Usually, some representation of the environment is needed in order to get the measurement model (s). Crowley [51], Cox [78] used a geometric model of the environment, which will be updated at each time increment using Kalman filter [17] and some stochastic relation. The alignment step is carried out using both the state and the observation equations. The second step can be done by data tracking techniques and Jacobian matrix (in case of non-linear models) for error propagation. The third one is represented by the analytic resolution in terms of state estimation and variance-covariance. In the case of a cellular representation of the environment, the problem comes down to finding the state of each cell (the robot itself corresponds to some cells) in terms of the occupancy or the freedom. The first step consists here in finding the degree of occupancy (and freedom) of each cell using a probabilistic model of the sensor and Bayes' theorem. The second one boils down to overcoming some ambiguity situations induced by the previous step when, for instance, one cell is found to be occupied and free with both at high probabilities. The third one makes use of some decision-making procedures like Bayesian inference to induce an appropriate choice among available cells.

iv). Predictive fusion of a proprioceptive and exteroceptive sensors (d*ynamic fusion of* proprioceptive and exteroceptive sensors). The most common approach in this category is Kalman filter where the odometer readings are updated by the ultrasonic or vision information. The alignment corresponds to the state and the observation equations that bring all the measurements down to the same state frame. In the second one, in addition to Jacobian matrix, some selection procedure based on distance threshold has been added. The refined data are combined in the third step according to Kalman's expressions.

Consequently most localization methods agree with data fusion point of view in the sense of the above mentioned steps.

4.2 Experiment Setup

The robot is equipped with eight Polaroid ultrasonic sensors, seven of which are situated in the front side spaced out with 30° (cf. Fig. 5).

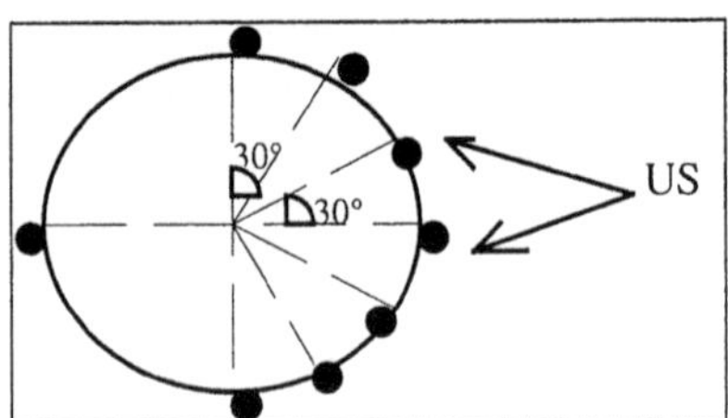

Fig. 5. Robot's ultrasonic sensors

Two DC motors control the wheels with an incremental encoder. The goal is to determine accurately the position of the mobile robot that navigates in a structured environment, while possibly non-modeled obstacle may be encountered.

4.3 Outline of the Solution

The approach advocated here consists in constructing a set of local solutions, each one makes use of odometric reading and one ultrasonic measurement using the structure of the environment. The obtained local solutions are then combined to get a global solution. The following algorithm (cf. Fig. 6) summarizes this approach.

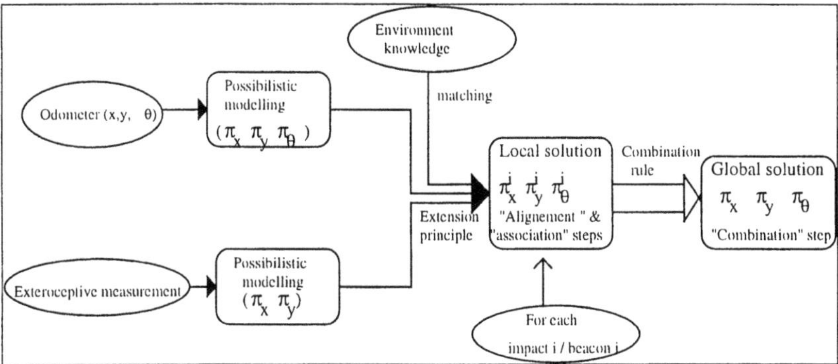

Fig. 6. General scheme

The following assumptions are put forward.

- Each datum is modeled in terms of a possibility distribution that takes into account its precision and certainty.
- There is no uncertainty attached to the geometrical locations of the natural beacons of the environment like walls and corners.
- The ultrasonic measurements are independent.
- Using odometric readings, each ultrasonic measurement induces a proper choice for the localization problem referring to as a *local solution*, which corresponds to alignment and association step of the combination process.
- These local solutions are then combined to generate a more representative result called *global solution*, which fits the combination step.

Now, let us investigate each of the previous parts.

4.4 Sensor Modelling

To each measurement, we assign a possibility distribution that roughly captures the uncertainty pervading the data. We assume for both sensors (ultrasonic and odometric) a trapezoidal possibility distribution. This shape is usually well used in the literature because of its intuitive interpretation and computational purpose. For the ultrasonic sensor, if ρ_m^i and α_m^i denote respectively the measurement and the orientation (with respect to the horizontal axis) of the sensor i, the corresponding possibility distributions π_ρ and π_α, centered around ρ_m^i and α_m^i, are represented in Fig. 7a and Fig. 7b respectively.

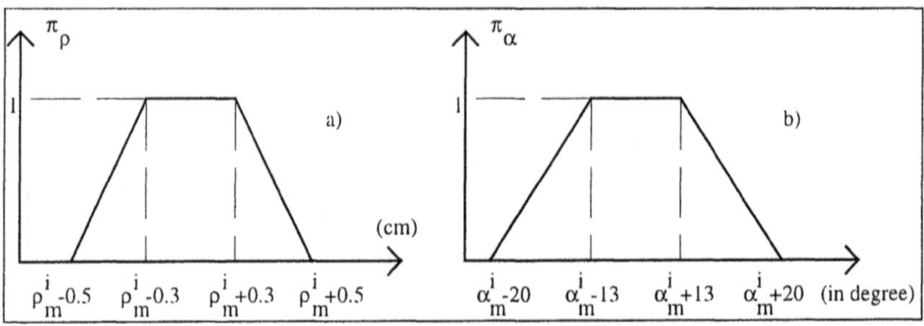

Fig. 7. Ultrasonic sensor modelling
a) distance measurement, b) orientation

Fig. 7a has the following intuitive interpretation: if the sensor "i" provides ρ^i_m as a possible value for the distance from the robot to a target, then the true distance robot-target cannot exceed a range of 0.5 cm around the measurement value ρ^i_m. This range forms a region where the true distance lies with different degrees of certainty levels. The range 0.3 around the central value corresponds to the imprecision of the reliable sensor output. That is, every value lying within this interval is considered as a completely faithful measurement. In other words, if we repeat the same experiment under the same conditions, then it is most likely that the output ranges from ρ^i_m-0.3 to ρ^i_m+0.3. The values 0.3 and 0.5 have been checked through some statistical validation tests, and agree with those provided by Patrioux et al. [79] in close circumstances. Fig. 7b models uncertainty pervading the beamwith of the ultrasonic sensor. Strictly speaking, it is well known that due to uncertainty caused by the beamwidth of the sensor the orientation of the target with respect to the axis of the US beam is very uncertain. In many applications, it is only assumed that the target is located on that axis, while it is, in reality, within the beam. That core of the distribution stands for the beamwidth (26^0) of the sensor, while the support delimits the worst case situations.

The odometric data are also modeled through a trapezoidal possibility distribution where the parameters are similar to those used in [79] (cf. Fig. 8). The parameter r stands here for the elementary moving either l^k_L or l^k_R pertaining respectively to the left and the right wheel after deleting some consistent bias, between sample k-1 and k, which justifies the symmetry in distribution of Fig. 8.

The distribution in Fig. 8 asserts that the parameters pertaining to the left and the right spreads are proportional to the modal value r. This is motivated by the fact that in dead-reckoning the imprecision increases with the traveled distance. The support and the core are chosen respectively around 5% and 2.5% of the modal value "r", which agree in regular ways with the odometric errors obtained in a structured environment without sliding.

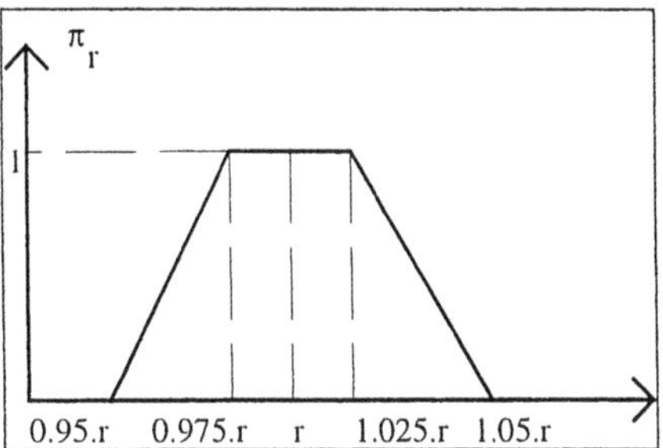

Fig. 8. Odometric sensor modelling

To get the possibility distributions attached to the x-y components and the orientation θ , it suffices to apply the extension principle [57, 58] to the odometric equations (17), where E represents the distance between the two wheels. l_R^k and l_L^k are respectively the incremental moving of the right and the left wheel of the robot between time increment k and k-1.

$$\begin{bmatrix} x_{k+1} \\ y_{k+1} \\ \theta_{k+1} \end{bmatrix} = \begin{bmatrix} x_k + \frac{l_R^k + l_L^k}{2}.\cos(\theta_k + \frac{l_R^k - l_L^k}{2.E}) \\ y_k + \frac{l_R^k + l_L^k}{2}.\sin(\theta_k + \frac{l_R^k - l_L^k}{2.E}) \\ \theta_k + \frac{l_R^k - l_L^k}{E} \end{bmatrix}. \qquad (17)$$

4.5 Alignment Step

This step consists in deriving a 2D location C_i of the *impact* of the ultrasonic measurement ρ_m^i pertaining to the sensor "i" whose orientation with respect to the horizontal axis is α_m^i , using the odometric readings in direction of x, y, θ . It is then easy to check that, provided that r_0 is the distance between the referential point of the robot and the location of the sensor i in the robot configuration.

$$\begin{cases} X_{C_i} = X_k + (r_0 + \rho_m^i).\cos(\theta_k + \alpha_m^i), \\ Y_{C_i} = Y_k + (r_0 + \rho_m^i).\sin(\theta_k + \alpha_m^i). \end{cases} \qquad (18)$$

Due to the measurement errors and the possible existence of a non-modeled obstacle, C_i may not be the true target of the ultrasonic sensor, since it may not be situated in any walls of the piece (environment). Similarly to (17), the use of the extension principle in (18) assigns possibility distributions to the x-y components of C_i.

The next phase involves the use of the environment knowledge, which consists of the location of the six walls (cf. Fig. 9), in order to correct the impact C_i such that it will be rather situated on one of the existing walls. At the same time, we update the odometric knowledge, leading to a *fictitious impact location* I_i , which, in turn, induces a *fictitious robot position* Ω_i (cf. Fig. 9). Naturally, in the absence of further evidence on the presence of an obstacle, the point I_i should be situated in one of the six walls. The rational is to choose the orthogonal projection. Namely, for the impact C_i, we seek the orthogonal projection of C_i to each wall, and the one inducing the minimal distance (C_i I_i) is considered. Moreover, some threshold test using the dimensions of the room permits elimination of inconsistent data due to multiple reflections or odometric errors.

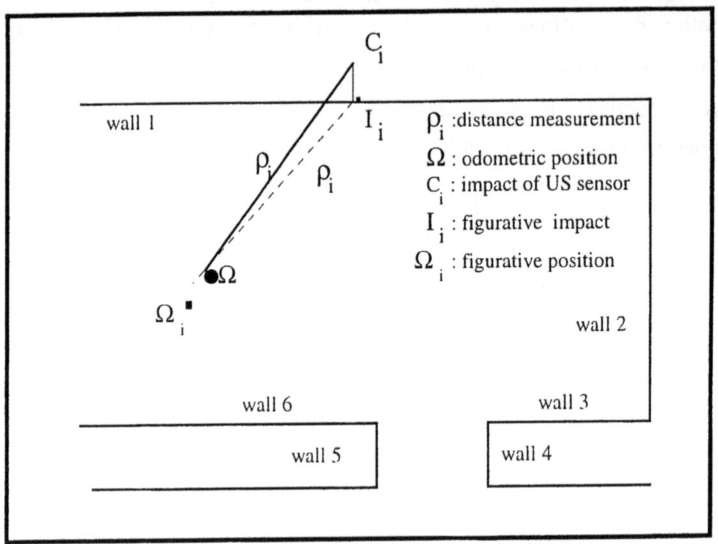

Fig. 9. Illustration of alignment step

The alignment point Ω_i corresponds to the correction of the odometric position Ω when changing the US impact from C_i to I_i. It is obtained such that the distance from I_i to Ω_i is ρ_i while Ω_i, Ω and I_i are lined up. Further, this choice is optimal when we restrict the available information to the odometric readings, the sensor "i" and the environment knowledge. Indeed, it corresponds to the nearest point to the odometry position Ω considering the above knowledge. That is, Ω_i stands for the solution involving a minimal change in the position of Ω. So, the principle of insufficient reason well known in probability may serve as an essence for a justification of such procedure. Ω_i is called a "local solution" since only the i-th ultrasonic measurement is involved.

Refinement of the procedure

In the case where large deviations (of the sensor) are fully possible, then the above procedure even optimal is not the most rational one, despite the use of possibility theory, which allows a range of admissible candidates. This is, basically, due to the effect of the orientation θ_k, which induces large geometrical errors that makes the orthogonal projection criterion debatable. So, the appropriate impact may be situated somewhere around I_i along the given wall. The procedure can then be significantly improved by noticing that larger the distance ρ_i the more likely the true impact is different from I_i. The general algorithm is shown in Fig. 10.

```
FOR each sensor measurement ρi
        test with respect to the threshold ε
          use odometric data Ω and 6 to obtain the impact C,
        rank the distances ρi in increasing order
    FOR ρl (the smallest distance)
        find the closed wall to the impact Cl,
        find the appropriate impact Il,
        find the position Ωl such that
        Il, Ω and Ωl are lined up and d(Il, Ωl)= ρl
    END (P4)
  FOR each other distances ρj .
        Change the orientation with ±15° around the current value.
      FOR each angular subdivision k
        find the impact Ck
        find the closest wall
        find the appropriate impact Ik
        deduce the position estimation Ωk
      END (P3)
    Choose the closest point Ωj (thus Ij and Cj) to Ωl
  END (P2)
END (P1)
```

Fig. 10. Algorithmic illustration of the alignment improvement

Basically, we consider the set of all US measurements (ρ_i, i=1 to 7) issued from the robot platform, let Ω_l be the local solution obtained using the smallest distance ρ_l among these sensors. Ω_l is then considered as the most reliable solution. We consider some *discretization* points around C_j (situated at ρ_i distance from Ω). This is equivalent to take into account the beam of the US sensor. So, for each point within the beam (of course situated at the same distance

from Ω), the previous optimal procedure is repeated, leading to Ω_j^k (k=1 to N (number of discrete points). Next, we only keep the point providing Ω_j^m, which is the closest one to Ω_1 (the most reliable one).
This refined procedure can be justified in the setting of reliability, where smaller distance entails higher reliability and the requirement of taking into account the beam of the ultrasonic sensor.

On the other hand, it is worth mentioning that the use of smallest distance to guide the pattern-matching scheme should be employed with prudence in the case where further evidence shows the presence of possibly non-modeled obstacle. In this situation, the smallest distance still is a good candidate for further exploration. Particularly, if it is found that the local solution provided by that distance is far away from all others generated by others US for more than one iteration, then a hypothesis of a presence of non-modeled is generated.
Notice that these different results are also expressed in terms of possibility distributions obtained using extension principle [57] applied to the different involved formulations. The following algorithm summarizes the preceding.

4.6 Association Step

The association step that helps in elevating some confusion and conflict among the data is performed at two stages using thresholding. First the sensor measurements are pruned with respect to a threshold ε' corresponding to the largest distance in the piece (environment). This discards echoes due to multiple reflections for instance. However, this test is still insufficient, which justifies the use of a second threshold ε introduced in the alignment step. The association sensor-measurement, here ultrasonic sensor – target, is handled by a proper use of odometric reading. That is, assuming that there is no important odometric degradation between two successive increments, the target pertaining to each sensor (US) can easily be determined through the orientation component. Further, the improved method previously mentioned in the alignment step also contributes to this purpose.

4.7 Combination Step

Given the set of local solutions provided in terms of π_x^i, π_y^i and π_θ^i evaluations, where i stands for the i-th valid ultrasonic measurement, the final step consists in combining these entities to obtain the resulting π_x, π_y and π_θ distributions. That is, all π_x^i, for instance, will be combined together to yield π_x. So are the distributions π_y^i and π_θ^i. The combination step is carried out by a progressive combination rule pointed out in [54]. Basically, this combination turns to a conjunctive combination via the minimum operator when the input possibility distributions are in full agreement as indicated by a non-zero intersection of their cores. In this case, we may have, for instance for π_x, $\forall s \in IR^6$, $\pi_x(s) = \min_i \pi_x^i(s)$. While as soon as the sources π_x^i are less in agreement then, rather, a trade-off combination between conjunctive and disjunctive modes is employed. In this case, the peak of the underlying distribution is dictated by the consensus among the initial distribution, while its support explicitly accounts for the distance from the consensus zone to the remaining data. The complete formulation of the rule is developed in [54].

4.8 Results

Fig. 11 shows some global results displaying in the same plot odometric readings, the estimated and the true robot positions along the whole trajectory. These results are obtained considering the case of the presence or absence of an unexpected obstacle. In each case, a defuzzification like method [5] is performed over the obtained distributions π_x and π_y in order to get a geometrical representation shown in Fig. 11. Notice that in both cases, the results are quite similar, except for one or two points pertaining to the unexpected obstacle and because of the lack of knowledge induced by other sensors. While at the end of the trajectory, the estimation seems less accurate because outside the piece, the environment is not modeled, but the estimation does not yield worse results than the odometry.

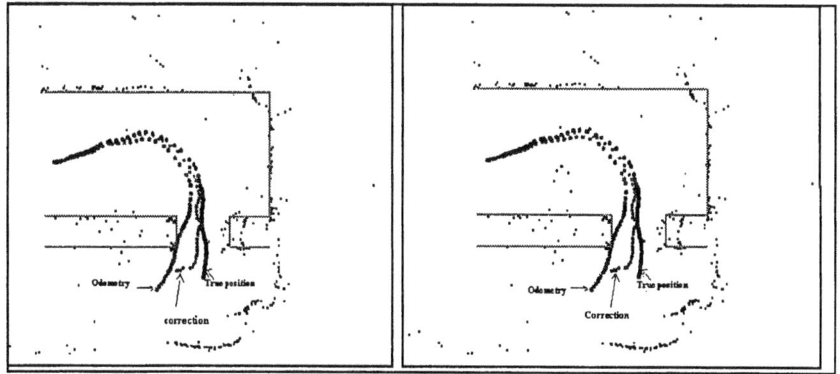

Fig. 11. Example of correction phase over the whole robot trajectory Without unexpected obstacle (on the left) and With an obstacle (on the right)

[6] IR stands for the set of real values. It indicates that the components x, y and θ of the positioning vector are real valued.

Furthermore, in terms of time computation aspect, the possibilistic approach, though slower than some traditional method like Kalman filter, can be applied on-line as soon as the speed of the robot is relatively low.

The preceding shows that the estimated robot position based on the new elaborated fusion-based approach always provides a better result than the standard dead-reckoning approach. Further, compared to other more elaborated approaches like Kalman filtering, the new approach offers the advantageous of not requiring any statistical support. Indeed, modeling through possibility distributions allows the decision-maker to express more freely his knowledge according to his intuition, without recourse to extensive and non trivial statistical justification tests, and circumventing like Gaussian assumption used Kalman filtering like approach.

6. Conclusions

Throughout this chapter, we have attempted to provide an overview of the fusion-based system. Particularly different types and stages of a fusion operation have been highlighted as well as the architecture governing the fusion process. This particularly matters when the question of whether the integration of several sources would lead to a better result than a single input. Special interest has been focused on the contribution of soft computing techniques to this area. It has been mainly shown that, by means of the variety of tools and approaches driven from such techniques, the contributions cover the whole range of fusion types at their various stages. An exemplification through the standard use of probability and possibility theories provides an order of magnitude of the feasibility of one of these soft techniques. Particularly, considering the architecture of the system, it has been shown that possibilistic result provides a more robust result when considering possible unexpected changes that may occur in that architecture. Finally, a robotics-oriented application consisting of localizing a mobile robot in a structured environment, with possibly the presence of non-modeled obstacles, has been investigated. The application demonstrates the feasibility of the fusion-based methodology in the framework of a possibility theory.

Acknowledgement

This work is partly supported by the interdisciplinary research collaboration DIRC funded by British EPSRC, which is gratefully acknowledged.

References

1. Data Fusion Lexicon, Data Fusion Subpanel of the Joint Directors of Laboratories, Technical Panel for C^3, F.E. White Code 4202, NOSC, San Diego, CA, 1991
2. Luo R. C., Kay M. K. (1992) Multisensor integration and fusion in intelligent systems, In: M. A., Abidi and R. C., Gonzalez (eds), Data Fusion in Robotics and Machine Intelligence, Academic Press 1992, 1-135.
3. Hall D. L., Llinas J. (2001), An introduction to multisensor data fusion, Proceedings of the IEEE **85**(1), 6-21

4. Gupta, M. M. (1992) Fuzzy Sets and Systems, McGraw-Hill Encyclopedia of Science & Technology. McGraw-Hill: New York
5. Dubois D, Prade H. (1980) Fuzzy Sets and Systems, Theory and Applications, Kluwer Academic, New York
6. Zadeh L. A. (1978) Fuzzy sets as a basis for a theory of possibility theory, Fuzzy Sets and Systems **1**, 3-28
7. Shafer G. (1976) A Mathematical Theory of Evidence, Princeton University Press Princeton, New Jersey
8. Smets P. (1988) Belief Functions, in: Non Standard Logics for Automated Reasoning. In: P. Smets, A. Mamdani, D. Dubois and H. Prade (eds) Academic Press, London, 253-286
9. Wang Z., Klir G. J. (1995) Fuzzy Measures, Plenum Press, New York
10. Choquet G. (1953) Theory of Capacities, Annals of Institute of Fourrier **5**, 131-295
11. Grabish M., Nguyen H. T., Walker A. E. (1994) Fundamentals of Uncertainty Calculi with Applications to Fuzzy Inference, Kluwer Academic Publishers
12. Dempster P. (1967) Upper and Lower probabilities induced by a multivalued mapping, Annals of Mathematical Statistics **38**, 325-339
13. Walley P. (1991) Statistical Reasoning With Imprecise Probabilities, Chapman and Hall
14. DeClaris N. (1992) Neural Network, McGraw-Hill Encyclopedia of Science & Technology, McGraw-Hill: New York
15. Masters T. (1993) Practical Neural Network Recipes in C++, Academic Press, Inc, Harcourt Brace and Company Publishers, Boston
16. Genest C., Zidek J. V. (1986) Combining probability distributions: A critique and an annotated bibliography (with discussions). Statistical Science, **1**, 114-148
17. Bar-Shalom Y., Foortmann T. E. (1988) Tracking and Data Association, New York Academic Press
18. Tou J. T., Gonzalez R. C. (1974), Pattern Recognition Principles, Addison-Welly Publishing Company
19. Hughes T. J. (1989) Sensor Fusion in a Military Avionics Environment, Measurement and Control, 203-205
20. IEEE, Special issue: data fusion, February 1997, Proceedings IEEE
21. Abidi M. A., and Gonzalez R. C. (1992) Data Fusion in Robotics and Machine Intelligence, Academic Press
22. Harris C. J., Read A. B. (1988) Knowledge-based Fuzzy Motion Control of Autonomous Vehicles. In: Proceedings of the IFAC Workshop Artificial Intelligence in Real-Time Control, Swansea, UK, 139-44
23. Dailey D. J.,Harn P., Lin P. J. (1996) ITS Data Fusion, Research, Technical Report T9903, Washington State Transportation Center (TRAC)
24. Pau L. F. (1988) Sensor Data Fusion, Journal of Robotics Intelligent Systems, **1**, 103-116
25. Blackman S. S. (1986) Multiple Target Tracking with Radar Applications, Artech House Inc Norwood, MA
26. Kirson, A., Smith, B.C., Boyce, D., Shofer J. (1992) The Evolution of ADVANCE, In: Proceedings of the Third International Conference on Vehicle Navigation & Information Systems. IEEE, 516-523
27. Martinez, D., Esteve, D., Demmou H. (1990) Evaluation of a modular multilayer architecture for recognizing dangerous situations in car driving, In: Proccedings of Neuro-Nimes '90. Third International Workshop, Nimes, France, 71-80
28. Dasarathy V. B. (1994) Decision Fusion, IEEE Computer Society Press
29. Dasarathy V. B. (1997) Sensor fusion potentiel exploitation -innovative architecture and illustrative applications, Proceedings of IEEE **85**, 24-39
30. Ashton R. H. (1986) Combining the judgement of experts: how many and which ones? Organizational Behavior and Human Decision Process, **38**, 405-414

31. Hoyland A., Rausand M. (1994) System Reliability Theory, Models and Statistical Methods, John Wiley & Sons
32. Kalman R. E. (1960) A new approach to linear filtering and prediction problems, Trans. ASME, J. Basic Engineering, **82**, 34-45
33. Kittler J., Fu K. S., Pau L. (1981) Pattern Recognition Theory and Applications, NATO ASI series, Dordrecht, The Netherlands: D. Reidel
34. Roubens M., Vincke P. (1989) Preference Modeling, Springer-Verlag, Berlin
35. Tanaka H., Uejima S., Asai K. (1982) Fuzzy linear regression model, IEEE Trans. Systems Man Cybernetics **12**, 903-907
36. Orlovsky S. A. (1978) Decision making with a fuzzy preference relation, Fuzzy Sets and Systems, **1**, 155-167
37. Roberts F. S. (1979) Measurement Theory, Encyclopedia of Mathematics and its Applications Vol. 7, Addison-Wesley Publishing Company
38. Durrant Whyte H. F. (1988) Integration, Coordination and Control Multi-Sensor Robot Systems, Kluwer Academic Publishers
39. Durrant-Whyte H. F., Rao, B.Y.S., Hu H. (1990) Toward a Fully Decentralized Architecture for Multi-sensor Data Fusion. In: Proceedings of IEEE International Conference on Robotics and Automation. Los Alamitos, Vol. 2, 1331-1336
40. Zhan C. T. (1971) Graph theoretical methods for detecting and describing Gestalt cluster, IEEE Trans. on Computer C.20
41. Eckhardt D. E., Lee L. D. (1985) A theoretical basis for the analysis of multiversion software subject to coincident errors, IEEE Trans. Software Engineering, **11**, 1511-1517
42. Parhami B. (1994) Voting algorithms, IEEE Trans. On Reliability, **43**(4), 617-629
43. Hall D. I. (1992) Mathematical Techniques in Multisensor Data Fusion, Artech House
44. Bezdek J. C. (1981) Pattern Recognition with Fuzzy Objective Function: Algorithms, Plenum press
45. Bezdek J. C., Pal S. K. (1991) Fuzzy Models For Pattern Recognition, IEEE Press
46. Zwick R., Carlstein E., Budescu D. V. (1987) Measures of similarities, fuzzy concepts: a comparative analysis. Int. J. of Approximate Reasoning, **1**, 221-242
47. Aggarwal J. (1989) Multisensor Fusion, ASI series, Heidelberg, FRG: Springer-Verlag.
48. Waltz E., Llinas J. (1990) Multisensor Data Fusion, Artech House Inc. Norwood MA
49. Oxenham M. G., Kewley D. J., Nelson M. J. (1996) Measures of information for multi-level data fusion, SPIE Vol. 2755, 271-282
50. Liebowitz J. (1993) Rule-based Expert Systems, In: R.C. Dorf (ed.), The Electrical Engineering Handbook. CRC Press: Boca Raton, FL
51. Crowley J. L. (1993) Principles and Techniques for Sensor Data Fusion, in: Data Fusion and Computer Vision, edited by J.K. Aggarwal, Springer-Verlag
52. Linn R. J., Hall D. L. (1991) A Survey of Multi-sensor Data Fusion Systems, In: Proceedings of the SPIE - The International Society for Optical Engineering SPI, Orlando, Vol. 1470, 13-29
53. Crowley J. L., Demazeau Y. (1993) Principles and Techniques for Sensor Data fusion, Signal processing **32**, 5-27
54. Oussalah M., Maaref H., Barret C. (2001) New Fusion Methodology Approach and Application to Mobile Robotics: Investigation in the Framework of Possibility Theory. International Journal of Information Fusion, **2**(1), 31-48
55. Sumner R. (1991) Data Fusion in Pathfinder and TravTek, In: Proceedings of VNIS '91. Vehicle Navigation and Information Systems Conference, Dearborn, Vol. 1, 71-75
56. Dubois D., Prade H. (1991) Fuzzy sets in approximate reasoning: Part 1: Inference with possibility distributions, Fuzzy Sets and Systems 25th anniversary memorial volume, **40**, 143-202

57. Dubois D., Prade H. (1987) Fuzzy numbers: An overview, In: J. Bezdek (ed), Analysis of Fuzzy Information, CRS Press, 112-148
58. Zadeh L. A. (1975) Calculus of fuzzy restrictions, in: L.A Zadeh, K.S. Fu, M. Shimura and K. Tanaka (eds), Fuzzy Sets and Their Applications to Cognitive and Decision Processes, Academic Press, 1-39
59. Goodman I. R., Nguyen H. T. (1985) Uncertainty Models for Knowledge Based Systems, North-Holland, Amsterdam
60. Zadeh L. A. (1979) A theory of approximate reasoning, In: Machine intelligence 9, J.E. Hayes, D. Michie and L.I. Kuich (eds), John Wiley & Sons, New York, 149-194
61. Klir G. J. (1990) A principle of uncertainty and information invariance, Int. Journal of General Systems **17**, 249-275
62. Klir G. J., Wierman M. J. (1988) Uncertainty-Based Information, Physica-Verlag
63. Goodman I. R. (1982) Fuzzy sets as equivalence classes of random sets, In: R.R. Yager (ed.), Recent Advances in Fuzzy Sets and Possibility Theory, Pergamon Press, New York, 327-432
64. Kwakernaak H. (1978) Fuzzy random variables, I, Definition and theorems, Information Sciences, **15**, 1-29
65. Chang Y. H. O. (2001) Hybrid fuzzy least-squares regression analysis and its reliability measures, Fuzzy Sets and Systems, **119**, 225-246
66. Miyazaki A., Kwon K., Ishibuchi H., Tanaka H. (1994) Fuzzy regression analysis by fuzzy neural networks and its application, In: Proceedings of IEEE International Conference on Fuzzy Systems, 52-57
67. Oussalah M., De Schutter J. (2002) Hybrid fuzzy probabilistic data association filter and joint probabilistic data association filter, to appear in Information Sciences
68. Aumann R. J., Hart S. (1994) Handbook of Game Theory with Economic Applications Vol. I and Vol. II, Elsevier
69. Yager R. R. (1980) A general class of fuzzy connectives, Fuzzy Sets and Systems, 4, 235-242
70. Bobrow D. G. (1984) Qualitative Reasoning about Physical Systems, Artificial Intelligence, 24, North Holland, Amsterdam
71. Boutilier C. (1994) Conditional logics of normality: a modal approach. Artificial Intelligence, **68**, 87-154
72. Boutilier C. (1994) Unifying default reasoning and belief revision in a modal framework, Artificial Intelligence, **68**, 33-85
73. De Campos L. M., Huete J. F. (1999) Independence concepts in possibility theory: Part I, Part II, Fuzzy Sets and Systems, **103**, 127-152
74. McConway K. J. (1981) Marginalization and linear opinion pool, J. American Statistical Association, **76**, 410-414
75. Dubois D., Prade H. (1990) Aggregation of possibility measures, In: J. Kacprzyk and M. Fedrizzi (eds), Multiperson Decision Making Using Fuzzy Sets and Possibility Theories, Kluwer Academic Publishers, Netherlands, 55-63
76. Klement E. P., Mesiar R., Pap E. (2000) Triangular Norms, Kluwer Academic Publishers Dordrecht
77. Hisdal E. (1987) Conditional possibilities independence and non-interaction, Fuzzy Sets and Systems, **1**, 283-297
78. Cox I. J (1991) Blanche, An experiment in guidance and navigation of an autonomous robot vehicle, IEEE Transaction on Robotics and Automation, **7**, 193-204
79. Patrouix O., Jouvencel B. (1993) Range information extraction using U_BAT an ultrasonic based aerial telemeter, In: Proceedings of Int. Conf. on Robotics and Automation, 1460-1465

Part 2

PLANNING AND NAVIGATION

Applied Soft Computing Strategies for Autonomous Field Robotics

Edward Tunstel[1], Ayanna Howard[1], Terry Huntsberger[1], Ashitey Trebi-Ollennu[1], and John M. Dolan[2]

[1] NASA Jet Propulsion Laboratory, Caltech, Pasadena, CA 91109, USA
[2] Carnegie Mellon University, Pittsburgh, PA 15213, USA
tunstel@robotics.jpl.nasa.gov

Abstract. This chapter addresses computing strategies designed to enable field mobile robots to execute tasks requiring effective autonomous traversal of natural outdoor terrain. The primary focus is on computer vision-based perception and autonomous control. Hard computing methods are combined with applied soft computing strategies in the context of three case studies associated with real-world robotics tasks including planetary surface exploration and land survey or reconnaissance. Each case study covers strategies implemented on wheeled robot research prototypes designed for field operations.

1 Introduction

Hard computing methods which address robot perception and control issues rely upon strong mathematical modeling and analysis [1,2]. The various approaches proposed to date are suitable for control of industrial robots and automatic guided vehicles that operate in structured environments and perform relatively well-defined repetitive tasks, such as manipulator positioning or tracking fixed/pre-programmed trajectories. Operations in unstructured environments, on the other hand, require robots to perform more complex tasks for which sufficient analytical models for control are often difficult to develop. This is typically the case in field robotics, which is concerned with development and application of robotic systems as tools for performing tasks or missions in unstructured, demanding, and/or hazardous natural environments (e.g., land surfaces and subsurfaces, sea, air, space, etc.). For applications in which analytical models are available, it is questionable whether or not the models are complete, or whether uncertainty and imprecision are sufficiently accounted for [3,4]. These issues become very significant in applications of autonomous field and space robotics [5], where target environments are not always known in sufficient detail to enable robust robotic system performance using hard computing techniques alone.

Field robotic vehicles intended for operation on land surfaces are the focus of this chapter. The target environments consist of natural rugged terrain, as opposed to relatively flat ground surfaces such as paved roadways. For the latter case (as with indoor environments), the motion controls for robot mobility systems can often be designed based on linear system models, or simpified nonlinear models of vehicle kinematics or dynamics assuming motion on a 2-D plane [2]. Practical mobility control systems for use in outdoor rugged terrain must account for a wider array of real-world complexities [6], the most fundamental of which is the fact that complex motions in the third dimension occur quite frequently. The situation is similar for rovers that interact with complex planet surfaces by roving across or burrowing through terrain [7,8]. For the robotic spacecraft that delivers them to the surface of remote planets, conventional estimation and control techniques have proven quite useful [9]. This is due to the fact that the physical laws of orbital mechanics and planetary atmospheric aerodynamics are reasonably well understood and well-behaved in space. Unfortunately, the complexities of terrestrial surface mobility and navigation recur once rovers are deployed on remote planet surfaces (and are sometimes compounded by reduced-gravity effects).

Autonomous mobile field robots must be designed to handle complex, and often uncharted, terrain encountered through the course of navigation, while remotely situated relative to human supervisors. Robust mission execution requires systems that can autonomously, and with minimal supervisory communication from humans, operate in natural environments and perform goal-directed tasks. Robotics researchers at the NASA Jet Propulsion Laboratory (JPL) have integrated soft computing techniques with more conventional hard computing methods to address some of the problems related to surface exploration of planet Mars using autonomous rovers, as well as control problems specific to terrestrial land vehicles. Soft computing strategies for rover perception, navigation decision making, and control have been developed and applied to complement conventional methods in an effort to achieve required levels of robustness. These include applications of fuzzy logic and neural network computing methods that facilitate reasoning and action selection in unfamiliar environments in order to enable reliable and safe mission execution. Using three case studies, several approaches are presented in this regard that have been integrated on field robot research prototypes and validated through operation on natural terrain. Each case study is presented following a brief overview in the context of general field mobile robot problems.

2 Case Study Overview

Autonomous mobile robot computational systems include functional and/or behavioral components that process sensory data, and maintain internal state information in order to perform repetitive cycles of processing designed to achieve specified tasks. Repetitive activities typically include the following: generation of perceptions (and sometimes models) of the target environment, utilization of sensing and perceptions to compute intelligent navigation or motion planning decisions, and execution of motion decisions via automatic control of actuators. This is the so-called *sense-plan-act* cycle common to many robotics tasks. Three isolated case studies are presented below as representative examples of how soft computing strategies have been applied to address some important problems, associated with this cycle, that arise in field robotics applications. Each covers a specific field robot implementation of the strategies applied. The implementation target in each case is a different wheeled mobile robot, and although the cases are isolated, they share the following common thread. Each case study addresses a function typically included in the repetitive cycle of activities associated with the overall task of effectively traversing natural terrain for purposes such as planetary surface exploration, land survey, and reconnaissance. The techniques are also relevant to other real-world field applications involving military and search/rescue operations, as well as agricultural, construction, and mining automation. The overall focus here is on visual perception and intelligent control. In particular, the case studies cover: (1) stereo vision-based perception for autonomous navigation, (2) monocular vision-based perception for effective mobility, and (3) control synthesis for locomotion systems complicated by nonlinearity and/or lack of suitable mathematical models. A thorough coverage of complementary soft computing-based navigation techniques can be found in [3].

Case studies 1, 2, and 3 center around a prototype planetary rover, a commercially available mobile robot research platform, and an unmanned ground vehicle, respectively. The intended application environment for case studies 1 and 2 is a barren landscape with soils of various consistencies and cluttered with rocks, such as the surface of planet Mars. For case study 3, the application environment may consist of unpaved roads, foliage, and/or grass. In all cases, the application environments are neither engineered nor structured in any way to accommodate the field robots. To overcome some of the practical problem constraints associated with task complexity and environmental challenges, it is sometimes advantageous to augment conventional hard computing methods with soft computing techniques. The case studies detail three instances for which this was achieved in the respective computing implementations for robot perception and control. Prior to field deployment, applied soft computing strategies must undergo extensive testing to ensure that the host robotic systems function properly, and are reliable enough to survive mission duration and achieve mission success. Advanced technologies are therefore developed in stages leading to increased readiness for field application; special concerns are measured against specific requirements of particular tasks/missions. Case studies presented herein reflect the present state of development of several soft computing strategies, and the scope of each is limited to overviews of ongoing technical research. Sections 3 and 4 present the case studies covering practical approaches to intelligent vision-based perception of natural terrain for autonomous navigation and mobility. This is followed by coverage of an adaptive speed control solution for field locomotion in Section 5. Each section describes the problem(s) addressed and identifies the hard computing and soft computing aspects of the applied solution. Motivation for adopting particular soft computing techniques is discussed, followed by descriptions of the implementation.

3 Case Study 1: Stereo Processing for Navigation

In 1997, the successful NASA Mars Pathfinder mission demonstrated that geological science exploration tasks can be performed within short range of a lander using small lightweight rovers with modest computational resources [10]. On future missions, more complete coverage of the planetary surface will require the development of higher level autonomy onboard the rovers. However, it is likely that advanced autonomy capabilities will also have to be realized using modest computational resources. This is due to practical limitations on mass and power typical of space flight avionics, as well as limited processing capabilities characteristic of radiation-hardened CPUs. A key component of required advanced autonomy is long-range navigation with obstacle detection and avoidance capabilities. Traditional methods used for achieving such autonomy often rely on range maps generated from a stereo image pair. Depending on the implementation, stereo image processing can consume significant portions of the time required to complete the sense-plan-act cycle, and/or require relatively large computational resources. For long-range rover missions, fast and efficient algorithms are desirable for reducing the amount of time required for stereo processing. A prototype technology rover called Sample Return Rover (SRR) is used for development and testing of such algorithms at JPL (Fig. 1). The sensor suite for navigation consists of a stereo pair of hazard cameras body-mounted to the front of the rover, and a single goal and stereo pair of cameras mounted to an articulated mast. This case study addresses algorithm designs for the generation of fast obstacle detection and avoidance using the front hazard cameras.

A variety of computational methods apply to stereo processing including maximum likelihood [11], dynamic programming [12], and biologically-inspired methods based on neural networks [13] and wavelets [14]. This case study employs a fuzzy self organizing feature map (FSOFM) algorithm combined with wavelet image processing to generate navigation commands for a behavior-based control system called BISMARC (Biologically Inspired System for Map-based Autonomous Rover Control). The system architecture for BISMARC is shown in Fig. 2. The stereo processing system does not build range maps as is done traditionally, but instead relies on a raw encoding of the image disparity information for reactive action generation.

3.1 Motivation and Approach

Striate cortical cells in cats have been found to respond to both monocular and binocular inputs [15]. In addition, cells in the retinal mammalian visual system respond to small lines and edges [16]. The behavior of these cells can

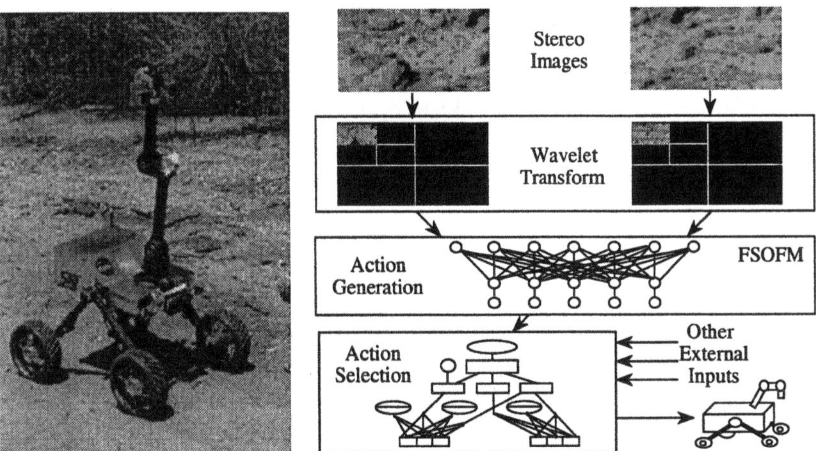

Fig. 1. SRR, a JPL rover.

Fig. 2. BISMARC system architecture.

be modeled using Gabor wavelets by determining the maximal response of filters that are tuned to specific frequency and orientation [14]. This method of analysis detects salient features that correspond to points of significant curvature change in the image. The vision preprocessing level for BISMARC uses such wavelet-based algorithms to decompose the images generated by each of the stereo cameras. The wavelet decomposition provides information about the scale, location, and orientation of features in an image. Because the wavelet decomposition contains information about the local frequency content of an image, it can represent visually important features (such as edges) more compactly than many of the other transforms commonly used in image processing [17].

After the wavelet transform is performed on each of the stereo images, a vector is formed using the multiresolution information from the two highest levels of the wavelet horizontal and vertical detail channels. This vector of length $\frac{5n^2}{4}$ elements, where n is the image width, is the input to the FSOFM — the output being any of six action states: go forward, backup, turn right, turn left, stop at goal, or pick either direction to turn. Raw stereo visual information is being encoded without any attempt to label individual features or objects beyond the desired action associated with the input pattern. Although the FSOFM is an unsupervised network, it can be trained by presenting it with samples and labeling the output nodes that correspond to each of the generated actions. The weights are then clamped for all subsequent runs.

The process of depth analysis can be functionally simulated using a combination of wavelet-based preprocessing coupled with a neural network that responds to binocular inputs. However, computation of Gabor wavelet coefficients for the full range of frequencies and orientations necessary to completely encode an image is computationally expensive, which limits their usefulness for fast image processing. There is another class of wavelet transforms that is more suited for constrained rover computing environments.

3.2 Wavelet and Neural Network Implementation Strategy

The standard wavelet decomposition of a general signal (or image) f is a representation in terms of linear combinations of translated dilates of a single function ψ:

$$f(x) = \sum_{\nu} \sum_{k} c_{\nu,k} \psi_{\nu,k}(x), \tag{1}$$

with $\psi_{\nu,k}(x) = 2^{\nu/2}\psi(2^{\nu}x - k)$ for integers ν and k. The mapping that takes f into the corresponding coefficients $c_{\nu,k}$ is called the wavelet transform of f. Since the transform is linear, all of the information content of the image or signal is contained in the $c_{\nu,k}$s. The detail channels of the coefficient space are defined as the orientation-specific difference between average channels at resolution ν and $\nu - 1$. A hierarchical representation of f is obtained by varying the scale from coarse (small ν) to fine (large ν).

There are a number of limitations to standard wavelets, including border artifacts and a dyadic (power of 2) restriction on image sizes. A method that can be used to avoid these problems involves the use of biorthogonal wavelets [18,19]. A biorthogonal wavelet decomposition uses two families of basic building blocks $\psi^i_{\nu,k}$ and $\tilde{\psi}^i_{\nu,k}$ to decompose the image as

$$f(x) = \sum_{i finite} \sum_{\nu} \sum_{k} \left\langle f, \tilde{\psi}^i_{\nu,k} \right\rangle \psi^i_{\nu,k}, \tag{2}$$

where $\left\langle f, \tilde{\psi}^i_{\nu,k} \right\rangle$ is defined as the inner product of f and $\tilde{\psi}^i_{\nu,k}$. This type of wavelet can be defined to exist within an arbitrarily shaped region [18,19]. We have previously used this form of the transform for texture analysis in adaptive, stereo range map generation for local rover navigation [20]. An example of the biorthogonal wavelet transform applied to a stereo pair of images from SRR's hazard cameras is shown in Fig. 3, where the hierarchical display format of Mallat is used [17]. The set of coefficients that correspond to the average channel at the coarsest resolution ν is in the upper left hand corner of the figure. The coefficients corresponding to the vertical, diagonal, and horizontal detail channels are shown in clockwise order from the upper right hand corner of the figure, and at different levels of resolution while traveling from the lower right to the upper left corner.

Fuzzy Self-Organizing Feature Map The neural network model [21] used for the action generation process is a modification of the self-organizing feature maps of Kohonen [22]. It is also a member of the generalized class of clustering networks developed by Pal, Bezdek, and Tsao [23] (hereafter referred to as GLVQ). The FSOFM, shown in Fig. 4, consists of three layers.

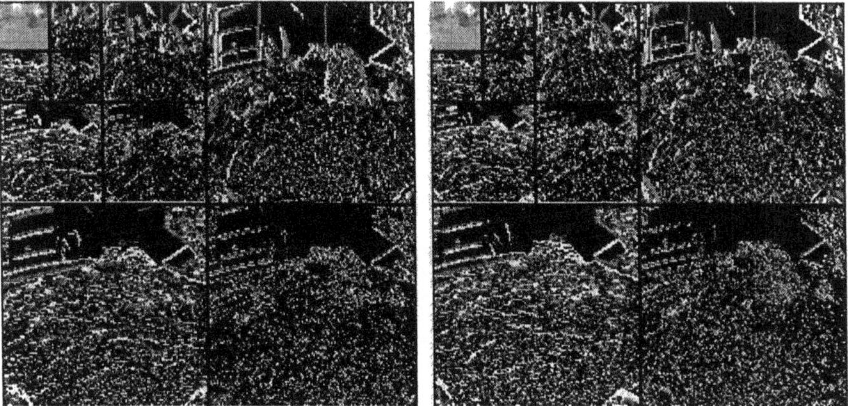

Fig. 3. Wavelet processed stereo pair used as FSOFM preprocessing step.

The input layer feeds forward into a distance layer which determines the distance between the input vector and the current weights using a predefined metric. The distance layer then feeds forward into a membership layer which calculates the membership values $u_{ji} \in [0.0, 1.0]$ of the input vector to the set of all output vectors. This is done using the following form derived from the fuzzy c-means algorithm [24] and used in a previously developed computer vision system [25]:

$$u_{ji} = \frac{1}{\left[\sum_{l=0}^{j-1} \left(\frac{d_{ji}}{d_{li}}\right)^{\frac{2}{(m-1)}}\right]} \tag{3}$$

where j is the number of output nodes, d_{ji} is the distance between the input vector ξ_i and weight vector W_j, and $m \in [2.0, \infty)$ is a weighting exponent. These membership values u_{ji} are then fed back into the network and participate in the weight update rule as:

$$W_j = W_j + u_{ji} * dw_j \tag{4}$$

where $dw_j = (\xi_i - W_j)$. The addition of a feedback loop allows the network to respond to the localized patterns of activity in the distance layer. The sum of membership values for any given input vector to all of the j sets is normalized to 1.0. This means that an input vector that is not close to any of the previously defined sets (an outlier) will have an equal membership value to all sets approximately equal to $1/j$. Such a vector would be classified as unfamiliar. The complete algorithm is given as [26]:

P1. Randomly initialize weights W_j $\forall$ j to values between 0 and 1. Set the *limit* value to be sufficiently small, e.g., 0.001, and set *total_diff* to 0.

P2. For each vector input ξ_i, $i = 1, 2, ..., n$ where n is the number of inputs:
   1. Calculate d_{ji} and determine the feedback membership value u_{ji} for each input vector using Eq. (3).

2. Update the weight change factor dw_j such that $dw_j = (\xi_i - W_j)$.
3. Save the current weight W_j as W_old_j before updating weight.
4. Update weight W_j using Eq. (4).
5. Return the value $diff$, where $diff = \sum(W_j - W\_old_j)^2$. Update $total\_diff$ using $total\_diff = total\_diff + diff$.

P3. If $total\_diff > limit$ then go to P4, else go to P5.
P4. Reset $total\_diff$ to 0 and go to P2.
P5. Write out weight file and determine the fuzzy membership value u_{ji} for each input vector ξ_i by using Eq. (3).

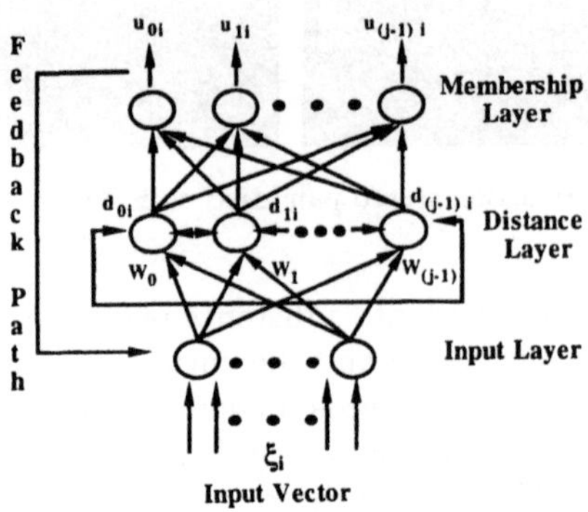

Fig. 4. Fuzzy self-organizing feature map neural network.

The FSOFM algorithm has certain advantages over GLVQ [23] in that there is no need to determine a winning output node. This property of FSOFM leads to a uniform weight update rule for all nodes. Details and performance of sequential and parallel versions of the algorithm can be found in the original paper [26]. Experimental studies indicated that the network reproduced the segmentation results from an earlier standard implementation of the fuzzy c-means algorithm [25] with close to two orders of magnitude increase in performance.

The network is trained with a set of wavelet processed images that are a representative sample of the types of obstacles that a rover would encounter. The right image from some of the stereo pairs (taken by hazard cameras on SRR) that were used to train the navigation FSOFM are shown in Fig. 5. The actions generated by the FSOFM for each stereo image pair are clockwise from top left: turn right, turn left, backup, go either way, stop at goal, and go forward. A total of five hundred stereo pairs were used for the training session, which took 637 epochs to converge. This should be compared to a backpropagation implementation, which typically takes hundreds of thousands of epochs to converge. Recall of the trained images was 100%. An advantage of using the FSOFM for the action generation level lies in the membership values that are generated at the output nodes. The sum of these values is normalized to one, and the relative size of the membership values gives a ranking

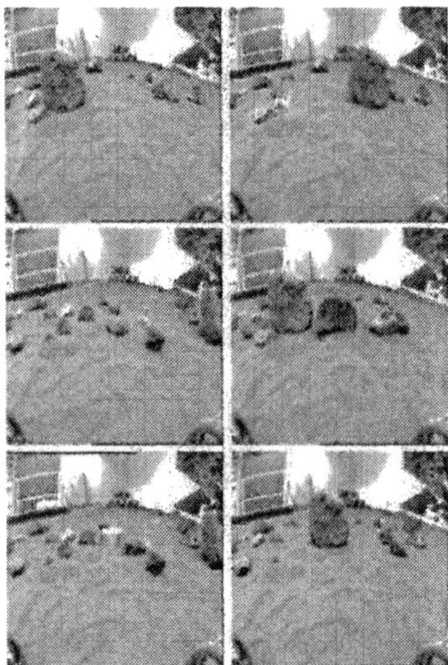

Fig. 5. Right stereo images from FSOFM training samples.

of the actions that are possible. The response to a unknown ambiguous input would be 0.16 from every output node, which means that any output greater than this is the favored action, with the largest membership value being favored. The response of the network to inputs that it was not trained with are shown in Fig. 6. Actions (membership values) that were generated for the images from left to right in the figure are: turn left (0.52), go forward (0.41), and turn right (0.37). The middle image also returned a membership value of 0.27 for turn left, which indicates that movement towards the left will be necessary to avoid the obstacle in the long run. Movement far from an obstacle is automatically generated, and the rover will only turn when its field of view has the vector of wavelet coefficients that indicate looming of an obstacle large enough to require avoidance behavior.

Fig. 6. Actions returned from FSOFM with unknown inputs. Although the system was not trained with these images (right image from stereo pair shown), it was able to generalize to the appropriate actions.

In this application, the FSOFM is employed to provide useful information for collision-free navigation. In particular, the FSOFM returns membership values that give a strong indication of the obstacle content of the local environment. To date, the wavelet and neural network strategies have been used to navigate the SRR platform in laboratory testbeds comprised of sand and rocks. Additional phases of testing and validation in uncontrolled outdoor terrain are planned for the near future. This case study focuses on a high-level vision-based strategy for terrain perception using efficient stereo image processing to facilitate rover navigation. In the following section, Case Study 2 describes a terrain perception strategy that employs low-level monocular vision for maintaining mobility performance during navigation via supervisory-level control of vehicle speed and terrain-aided position estimation on varied terrain surfaces.

4 Case Study 2: Terrain-Aided Mobility

To achieve effective mobility in natural terrain, it is desirable for motion controllers to accommodate physical interactions between the mobility system and rugged terrain while maintaining vehicle safety and reasonably accurate position estimation. While a rover traverses outdoor terrain, perception of the type and condition of the terrain surface provides clues for safe mobility assessment. Human automobile drivers are able to perceive certain road conditions (e.g., oil slicks, pot-holes, and ice patches) as measures of safety, and react to them in order to reduce the risk of potential accidents. In a similar manner, rover potential safety can be inferred and reacted on the basis of knowledge about the terrain type or surface condition. Two prevalent effects of wheel-soil interactions are wheel slippage on low tractive surfaces and wheel sinkage on soft surfaces. On dry paved roads, traction performance is maximal for most wheeled vehicles due to the high coefficient of friction/adhesion between the road and the tread. On off-road terrain, however, a variety of surface types are encountered on which rover wheels are susceptible to slippage. Excessive wheel slippage reduces the effective traction that a rover can achieve and, therefore, its ability to make significant forward progress. On soft soils, such as fine-grained sand, excessive wheel slippage can often lead to wheel sinkage and eventual entrapment of the vehicle. Frequent loss of traction during a traverse from one place to another will also detract significantly from the ability to maintain good position estimates for rover localization. Non-systematic localization errors due to wheel slip are compounded by errors due to wheel sinkage. As the load-bearing strength of the terrain/soil varies so does the amount of wheel sinkage. This has the effect of varying the *effective wheel radius*, which is an important parameter in the kinematic models used for position estimation. Unfortunately, wheel slippage and sinkage are often difficult to measure and estimate in a straightforward manner.

Some progress has been made, however, in developing statistical estimation approaches for planetary rovers [27].

This case study addresses the problem of mitigating the effects of wheel slippage and sinkage through active management of traction via speed control, and effective wheel radius estimation. The target platform is the Pioneer-AT (All-Terrain), a commercially available mobile robot designed for rough terrain mobility. At JPL, it is utilized for research and development of perception and navigation concepts for eventual integration on more capable field robots. The factory configuration of the robot includes a low-profile chassis with a PID-controlled locomotion system of 4 wheels (driven by DC motors) and an array of ultrasonic range sensors, managed by an embedded 16 MHz MC68HC11-based microcontroller. The system was enhanced at JPL with additional onboard computing (Pentium II laptop) and a vision system for real-time terrain assessment (see Fig. 7). The laptop computer hosts all high-level intelligence and commands the low-level motor control system. The ultrasonic sensors are not utilized for this case study.

Fig. 7. Pioneer-AT rover with enhancements.

4.1 Motivation and Approach

Traction control solutions are often derived from analyses based on the following equation for wheel slip ratio, λ_s, which is defined non-dimensionally as a percentage of vehicle forward speed, v [28]:

$$\lambda_s = \left(1 - \frac{v}{r_w \omega_w}\right) \times 100. \tag{5}$$

Here, r_w is the nominal radius of the vehicle wheel and ω_w is the wheel rotational speed. Eq. 5 expresses the normalized difference between vehicle and wheel speeds. When this difference is non-zero, wheel slip occurs. The objective of traction control is to regulate λ_s to maximize traction. This is a relatively straightforward regulation task if v *and* ω_w are both observable. The wheel rotational speed ω_w is typically available from shaft encoders or tachometers. However, it is often difficult to measure the actual over-the-ground speed v for off-road wheeled vehicles. This is due to a lack of adequate practical sensing solutions as well as nonlinearities and time-varying uncertainties caused by wheel-soil interactions [29]. Effective solutions have been found for automotive applications, but in many of these cases, measurement of v is facilitated by an even surface on which the vehicle travels, or by special sensing arrangements engineered for the operating environment. In large part, the available automotive solutions are not directly transferable to off-road vehicle applications. The traction management problem for off-road robotic vehicles can be addressed using a soft computing approach that does not rely on accurate sensing of over-the-ground vehicle speed to compute wheel slip. Instead, the strategy relies on visual perception of terrain surface texture to infer appropriate tractive speed controls.

Distinct terrain surfaces reflect different textures in visual images. The ability to associate image textures to terrain surface properties such as traction, hardness, or bearing strength is directly useful for autonomous traction management. To provide this capability, we make use of an onboard monocular camera pointed such that its field-of-view (FOV) covers the ground area in front of the rover as illustrated in Fig. 8. On the Pioneer-AT shown in Fig. 7,

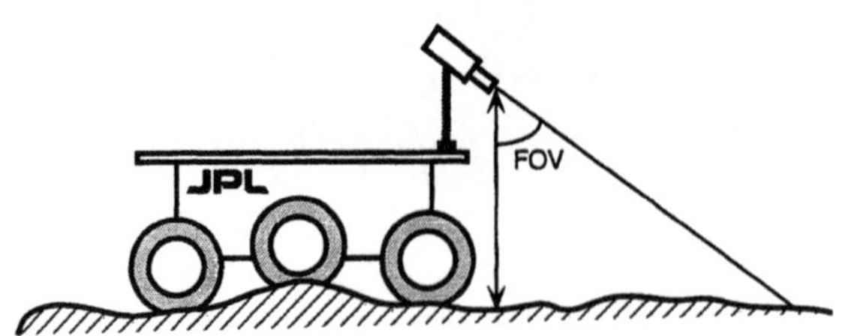

Fig. 8. Camera mounted on rover.

the ground-facing camera on the front of the rover is mounted $0.3\,m$ above the ground, tilted downward 45° with a 45° FOV. This camera enables surface traction classification using soft computing-based image analysis. (Cameras shown mounted on the raised platform are used for strategic navigation.) In particular, hard computing methods of computer vision are combined with an image texture classification approach using an artificial neural network (ANN). This provides an automated method of classifying the terrain surface just ahead of the rover with respect to tractive quality. The ANN is adopted to realize this capability due to its effectiveness for representing arbitrary input-output relationships. This qualitative knowledge serves as input to a supervisory fuzzy logic speed controller, which produces speed recommendations associated with the current perception. The fuzzy logic implementation also serves to absorb the effects of potential uncertainties in quantities representing the traction classification and inferred speeds. Simple fuzzy speed control rules are formulated based on off-road driving heuristics for facilitating maintenance of wheel traction, thereby mitigating excessive wheel slip. For a variety of potential surface types, the maximum speeds achieved before the onset of wheel slippage (tractive speeds) are determined from empirical traction tests performed with the actual rover. Given this information, commanded vehicle speed can be modulated during traversal based on visual classification of the terrain surface type in front of the rover. This is analogous to the perception-action process that takes place when a human driver notices an icy road surface ahead and decelerates to maintain traction. The perception of the surface type is also used to roughly estimate the effective wheel radius used for dead-reckoning on classified terrain surface types. Details of the implementation follow.

4.2 Vision-Based Strategy and Implementation

Considering the typical surfaces that the rover may encounter, three different texture prototypes are selected to train the ANN: sand, gravel, and compacted soil. Classification of the different textures is achieved using the following strategy:

- Extract a set of 40×40 image blocks from image data.
- Reduce image data dimensionality using orthogonal sub-space projection.
- Train the ANN on a set of texture prototypes projected on the eigenvector set.
- During run-time, feed projected texture images to the trained ANN.
- Extract texture prototype output from network & classify ground surface type.

Assuming the section of the ground image just ahead of the front wheels is free of obstacles, a set of 40×40 pixel image blocks is randomly selected from a camera image of size 320×280 pixels. The assumption of an obstacle-free area immediately ahead is plausible since a separate strategic navigation module handles the more forward-looking function of guiding the vehicle towards traversable terrain; navigation details are reported in [30]. To reduce the large data dimensionality inherent in typical computer vision-based applications, a filtering step is performed using a standard technique called Principal Component Analysis (PCA) [31]. PCA is a linear optimal method for reducing data dimensionality by identifying the axis about which the desired feature set varies the most. This orthogonal sub-space projection of the image subset permits effective extraction of features embedded in the surface image data set in real time. This technique reduces the dimensionality of the image set while preserving as much of the signal as possible. PCA computes a set of orthonormal eigenvectors (filters) of a data set that captures the greatest correlation between features. The filters associated with a given feature set are derived from the distribution of potential dynamic features embedded in the images. To characterize the distribution of these features, the covariance matrix, R, is found for image subsets containing the desired dynamic features. The eigenvector problem, $R\mathbf{w} = \lambda\mathbf{w}$, is solved to derive the set of filters, $\mathbf{w}$, used in the algorithm outlined above to maximize the greatest correlation between features. A total of 30 eigenvectors is used to reduce the 40×40 image block (1600 pixel values) to a pattern set of 30 values (Fig. 9) by projecting the image data onto the most significant filters. This reduced data set is then used to train the ANN (Fig. 10) to associate texture with several surface types.

The ANN uses 30 input nodes corresponding to the projected image block and 1 output node representing the surface type classification; its hidden layer has 20 processing elements. Initially, the ANN is trained using backpropagation by finding a set of weights that produce the desired surface type classification for given training data input. After training the network on typical image data representing different texture prototypes imaged in different illumination conditions, it is utilized to classify surface types during run-time. For the algorithm, the network output provides the qualitative information needed to make any necessary adjustments to wheel speed in order to maintain traction on the classified surface. Fig. 11 shows several images of real terrain data properly classified by the trained ANN; these images were not included in the data set used to train the ANN.

Fig. 9. Texture eigenvectors.

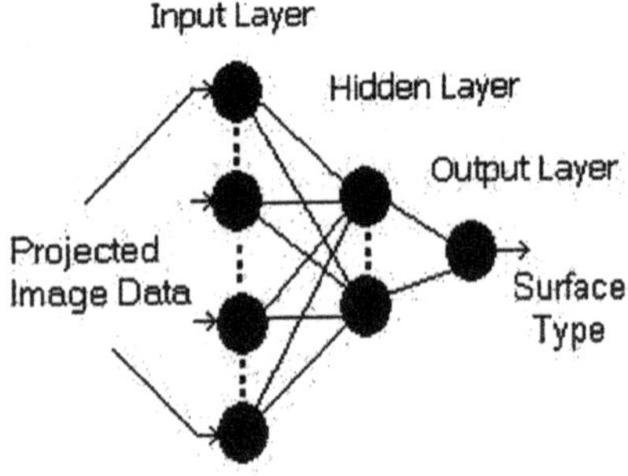

Fig. 10. ANN for surface classification.

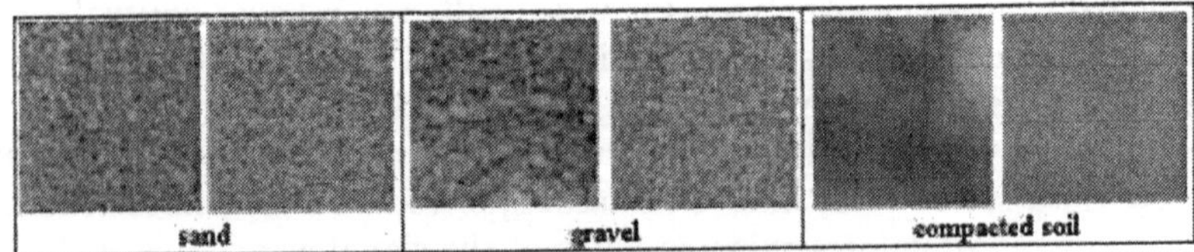

Fig. 11. Terrain surface images classified by the ANN.

Tractive Speed Modulation The ANN is trained to classify ground textures and produce surface type outputs represented as numerical values in the unit interval [0, 1], with 0 corresponding to surfaces of very low traction (e.g., ice) and 1 corresponding to surfaces of very high traction (e.g., dry cement). This is a design decision motivated by a desire to establish some *intuitive* correlation to actual wheel-terrain coefficients of friction. In this way, we can make a qualitative association between output of neural networks and expected terrain traction in front of the rover. We will refer to the texture prototype output as the *traction coefficient*, denoted by C_t. The range of traction coefficients, [0,1], obtained from the ANN is partitioned using three fuzzy sets with linguistic labels of LOW, $MEDIUM$, and $HIGH$ as shown in Fig. 12. Based on these definitions, a single-input-single-output fuzzy logic

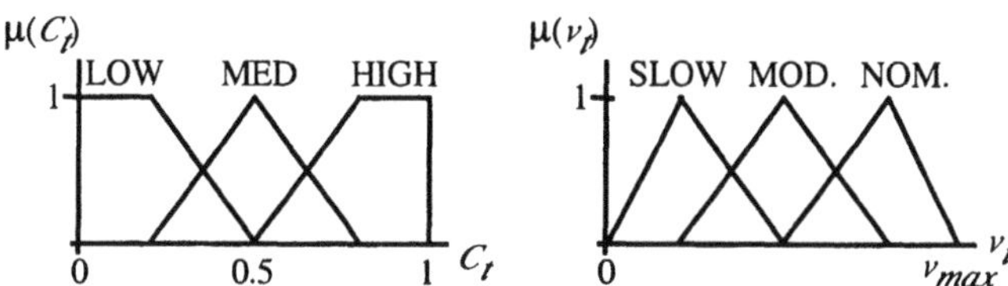

Fig. 12. Fuzzy sets for traction management.

controller infers tractive speed set-points, v_t, for input to the low-level PID wheel motor controller. The range of allowable speeds is partitioned using fuzzy sets labeled $SLOW$, $MODERATE$, and $NOMINAL$ as shown in Fig. 12, with v_{max} specified according to the application. The following fuzzy logic rules are applied to modulate tractive speeds in proportion to expected terrain traction in front of the rover:

- IF C_t is LOW, THEN v_t is SLOW.
- IF C_t is MEDIUM, THEN v_t is MODERATE.
- IF C_t is HIGH, THEN v_t is NOMINAL.

Complete definition of membership functions for tractive speeds v_t is based on results of prior traction tests as mentioned earlier. As such, these membership function definitions are vehicle-dependent and reflect knowledge derived about non-slip speeds achieved when the vehicle was tested on various terrain surfaces. For best results, traction tests should be performed on surfaces that represent the expected roughness, hardness, and slope variations of the rover operating environment. Once membership functions are defined, the robot is equipped to perceive and react to learned terrain conditions in a similar manner as a human driver. A more sophisticated approach to perception and reaction on very challenging terrain is proposed in [7], and is better suited for articulated rover mobility mechanisms that are capable of kinematic reconfiguration. That strategy uses look-ahead stereo visual perception as well as onboard kinematic analysis of potential tip-over and expected traction to plan subsequent stable and tractive configurations.

Effective Wheel Radius Modulation The sophistication of localization methods for outdoor mobile robots depends on the available sensor suite onboard the vehicle. A variety of viable hard computing solutions exist, many of which employ several sensor types (e.g., wheel odometry, inertial measurement units, GPS, visual odometry, sun sensing, etc.) and perform sensor fusion for position and attitude estimation, often based on Kalman filter formulations [32–35]. The vehicle employed in this case study uses odometry based on wheel encoder data as the primary means of position estimation. Such dead-reckoning is prone to accumulated error during traversal. As such, it is not recommended as a sole means of position estimation for field mobile robots over significant distances. In lieu of more sophisticated estimation methods due, for example, to limitations of onboard sensing, it is highly desirable to adopt techniques for improving dead-reckoning accuracy. One such technique is a wheel velocity synchronization approach described in [32], which is designed for improved odometry and power efficiency of 6-wheeled rovers with independently-driven wheels. As each wheel experiences different loading profiles, the wheel velocity synchronization serves to minimize motion interactions between them which may cause excessive wheel slippage and the tendency to side-slip as the suspension system traverses over obstacles. The Pioneer-AT platform used in this case study is kinematically simpler. The two wheels on each side of the platform are physically coupled for synchronized drive motion, and the platform is steered differentially due to relative action of the paired wheels on either side.

For the simpler Pioneer-AT, the traction coefficient described above can be used to address a more fundamental aspect of the position estimation problem, namely, position error reduction via intelligent estimation of effective wheel radius. As mentioned earlier, the accuracy of kinematic models used to compute rover localization updates depends on the specification of a nominal wheel radius, r_w. This kinematic parameter is used to compute the equivalent linear distance $r_w\theta_w$ traveled by a wheel after any rotational wheel displacement θ_w on the terrain. For the Pioneer-AT, these linear distances are used to compute robot position and heading updates in a reference coordinate frame according to the following kinematic equations:

$$x_{k+1} = x_k - r_w \left(\frac{\theta_{w,r} + \theta_{w,l}}{2} \right) \sin \phi_{k+1}, \tag{6a}$$

$$y_{k+1} = y_k + r_w \left(\frac{\theta_{w,r} + \theta_{w,l}}{2} \right) \cos \phi_{k+1}, \tag{6b}$$

$$\phi_{k+1} = \phi_k + r_w \left(\frac{\theta_{w,r} - \theta_{w,l}}{d} \right). \tag{6c}$$

Here, x and y are Cartesian coordinates of the robot position on the terrain surface and ϕ is its heading; $\theta_{w,r}$ and $\theta_{w,l}$ are the rotational displacements of the right and left wheels (measured by encoders in the two front wheels), and d is the lateral distance between them. The displacements of wheels on the same side are assumed to be equal due to the physical coupling of the Pioneer-AT locomotion mechanism. The influence of r_w on the localization update is apparent in these equations. For compliant wheels/tires such as those on the Pioneer-AT, r_w is reduced as the tire compresses in normal reaction with the terrain to an effective wheel radius, r_{eff}, which should be used instead of r_w in Eqs. (6a)-(6c) to yield more accurate updates. Non-compliant wheels produce a similar effect. For example, as a vehicle with non-compliant wheels traverses terrain with both hard-packed soil and soft sand, use of r_w in the kinematic model is valid only over the hard-packed terrain. As the terrain load-bearing strength decreases (over softer soil), so does the effective wheel radius, r_{eff}, and the accuracy of the model. To reduce the effect of propagating this non-systematic error during rover traverses on varied terrain, we make further use of C_t to estimate the varying r_{eff} according to the following linear regression relationship,

$$r_{eff} = r_w C_t + c_{reg}, \tag{7}$$

where c_{reg} is a regression constant, which can be evaluated for a given vehicle via regression analysis of empirical data produced by multiple runs over varied terrain surfaces. Here, linear regression is assumed to produce a sufficient estimation model. However, an effective wheel radius estimation model for a given vehicle and terrain could require nonlinear regression to achieve desired improvements. It should be noted that this may still be insufficient depending upon the localization accuracy required in a given application. With the luxury of additional onboard sensing, such as an inertial measurement unit (including rate gyroscope and accelerometers), one could improve the position estimate further by employing the well-known hard computing techniques of Kalman filtering [2]. A good example of this is provided in [33] where a Kalman filter is formulated to improve localization for the Pioneer-AT robot.

This case study describes a low-level vision-based strategy for terrain perception with supervisory-level fuzzy control of tractive vehicle speed and terrain-aided position estimation. Statistical techniques [36] and color-based methods [37] for terrain classification have also been developed at JPL for navigation and mobility in vegetated terrain. These applications employ laser rangefinders as well as color and infrared stereo cameras for day and night vision, and the target platforms are cross-country military unmanned ground vehicles. In the next section, Case Study 3 covers an application of fuzzy logic speed control for a military all-terrain autonomous vehicle that differs significantly in size and propulsion from the field robots discussed thus far. The soft computing strategy, employed at a lower level, serves to overcome the lack of a complete analytical model.

5 Case Study 3: Adaptive Fuzzy ATV Speed Control

In this case study, an adaptive fuzzy speed control design for a throttle-regulated internal combustion engine on an All Terrain Vehicle (ATV) is presented. The ATV, shown in Fig. 13 is one of several mobile platforms used in the *CyberScout* project at Carnegie Mellon University, which aims to develop distributed mobile robotics technologies that will extend the sphere of awareness and mobility of small military units. The low-level ATV speed controller is one of the subsystems crucial to achieving complete autonomy.

Various automatic speed controllers based on hard computing techniques alone (e.g., adaptive, robust, and sliding model) have been implemented on automobiles. However, the proposed control techniques are not directly applicable to the ATV throttle control problem for the following reasons. First, the ATV engine is mechanically controlled via a carburetor, unlike most automobiles, which have microprocessor-based engine management systems that guarantee maximum engine efficiency and horsepower. Second, the ATV carburetor clearance makes it difficult to incorporate a sensor to measure the throttle plate angle, which is required in virtually all of the automotive speed controllers reported in the literature. Third, and most importantly, automobile cruise control does not work well at speeds below 30 miles per hour (mi/h) due to engine nonlinear torque and speed fluctuations. Finally, the ATV throttle is actuated via an R/C servo, with no explicit position feedback, instead of a pneumatic actuator which is the preferred actuator in most automobiles. The adaptive fuzzy throttle control strategy is applied to address these constraints for the ATV. Fuzzy logic rules are formulated from extensive experiments conducted by human operators and based on quantitative data. An adaptive control law is applied to augment the soft computing based control.

Fig. 13. Autonomous CyberATV "Lewis".

5.1 Motivation and Approach

The intended military operations task for the ATV requires that the speed control system be designed to provide smooth throttle movement, zero steady-state speed error, constant vehicle speed over varying road slopes, and robustness to system variations and operating conditions for a 2-30 mi/h speed range. The two main challenges in designing an effective speed controller for the ATV are: 1) the lack of a complete mathematical model for the engine, and 2) the highly nonlinear nature of the engine dynamics, especially for the targeted low speed range of 3-30 mi/h. Both of these factors make the use of classical control strategies such as PID control ineffective. To elaborate, initial open-loop experiments conducted on level terrain revealed that humans could not easily drive the ATV at speeds below 10 mi/h, shedding light on the nature of the nonlinear relationship between throttle valve openings and speed. Also, the throttle valve-opening threshold for initiating vehicle movement varied from one trial to the next, indicating a shifting operating point. Attempts to apply conventional PID control revealed that a PI controller could be used for higher speeds where the carburetor operation is fairly linear, i.e., speeds above 15 mi/h, thus covering the upper portion of the target speed range. (The measured speed signal is very noisy, so it was not feasible to implement a derivative component for a PID controller.) This result indicates that a possible approach is to use more than one control strategy via lookup tables, depending on the speed range. However, it is very difficult to apply conventional control techniques for the ATV since a complete mathematical model of the engine is not available, and developing this model requires information about the engine, which the manufacturer was unwilling to provide. An alternative approach is fuzzy logic control (FLC), since human experience combined with the extensive quantitative and qualitative results can be employed effectively in fuzzy systems. It was found that while fuzzy control provided very smooth throttle movement, it was insufficient for achieving the required steady-state accuracy, particularly for substantial changes in the terrain (up and down hill). Several attempts to tune the fuzzy membership functions did not significantly reduce the steady state error, suggesting the need for adaptivity. Since the FLC strategy worked fairly well, an adaptive control law based on that strategy was considered. Details of the fuzzy control strategy and adaptive control design are discussed below.

5.2 Fuzzy Speed Control Design and Implementation

The design goal for the ATV speed control is to minimize the magnitude of the speed error, ε, defined as the difference between desired and actual speed. Human operators can control the speed of the ATV via a throttle lever, which opens and closes the throttle valve to increase or reduce the speed of the ATV. Based on familiarity with the speed response to this action, fuzzy rules were formulated using speed error and change in control input to the throttle actuator. Automatic actuation of the ATV throttle is achieved via a Futaba R/C servomotor (added to the system) which is controlled by pulse width modulation (PWM) such that pulse widths of $1\,ms$, $1.5\,ms$, and $2\,ms$ correspond to idle, half, and full throttle, respectively. Speed feedback is obtained via a tachometer mounted in the gearbox. The change in throttle control is defined using the two past values of the control input, u, and can be expressed as $\Delta u = u_{(k-1)T} - u_{(k-2)T}$ where discrete time $t = kT$, $k = 0, 1, 2, \ldots, n$ and T is the sampling and control update period.

Fuzzy input and output membership functions are shown in Fig. 14. The five linguistic labels for speed error input (ε) are: Negative Large (NL), Negative Small (NS), Zero, Positive Small (PS), and Positive Large (PL). Three linguistic labels are similarly defined for the input change in throttle opening (Δu). The five linguistic labels for the output throttle opening (u) are: Zero, Small (SM), Medium (MED), Large (LG) and Very Large (VLG). The ranges of these linguistic variables were determined by experimentation and the physical constraints of the sensors employed, e.g., the R/C servomotor input pulse width command range of 1-2 ms. The Zero membership function center for throttle opening is defined to be slightly above idle engine speed. The FLC was implemented using product inference and a center-average defuzzifier [38]. Fig. 15a shows the complete set of fuzzy rules for u in tabular form along with a block diagram of the control structure (Fig. 15b), in which the derivative block represents Δu. Rules in the table of Fig. 15a can be interpreted in linguistic form. For example, the rule specifying "Very Large" for the throttle opening may be written as: $IF\, \varepsilon\, is\, PL\, and\, \Delta u\, is\, Zero, THEN\, u\, is\, VLG$. The fuzzy linguistic label names used here give an intuitive sense of how the rules apply. Experimentation and tuning of the membership functions, revealed that this rule set was sufficient to encompass all realistic combinations of inputs and outputs. Recall, however, that this fuzzy control was not sufficient to achieve the required steady-state accuracy alone. The theoretical formulation of the adaptive fuzzy control is outlined below.

5.3 Adaptive Fuzzy Control Solution

Assume that the rule base consists of multiple-input single-output rules of the form $R^{(j)} : IF\, x_1\, is\, A_1^j\, and \ldots and\, x_n\, is\, A_n^j, THEN\, y\, is\, C^j$, where $\mathbf{x} = (x_1 \ldots x_n) \in N$, $y_j \in S$ denotes the linguistic variables associated with inputs and outputs of the fuzzy system. A_i^j and C^j are linguistic values of linguistic variables $\mathbf{x}$ and y in the universes of discourse N and S respectively; $j = 1, 2, \ldots, Q_R$ (number of rules). A fuzzy system consisting of a singleton fuzzifier, product inference, center-average defuzzifier and triangular membership functions can be written as [38]

$$f(\mathbf{x}) = \frac{\sum_{j=1}^{Q_R} \bar{y}^j \left[\prod_{i=1}^{n} \mu_{A_i^j}(x_i)\right]}{\sum_{j=1}^{Q_R} \left[\prod_{i=1}^{n} \mu_{A_i^j}(x_i)\right]}, \tag{8}$$

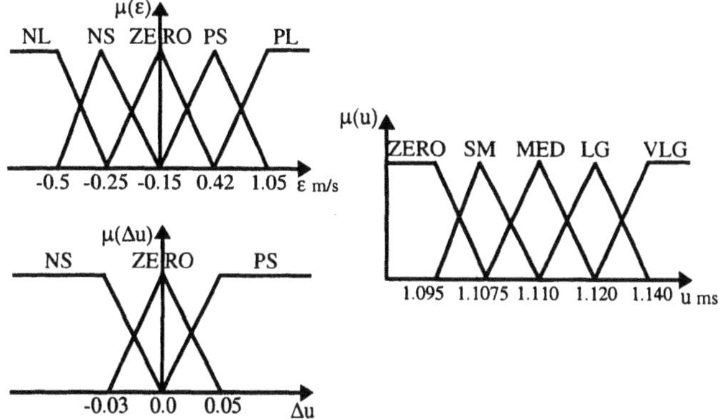

Fig. 14. ATV speed control input and output membership functions.

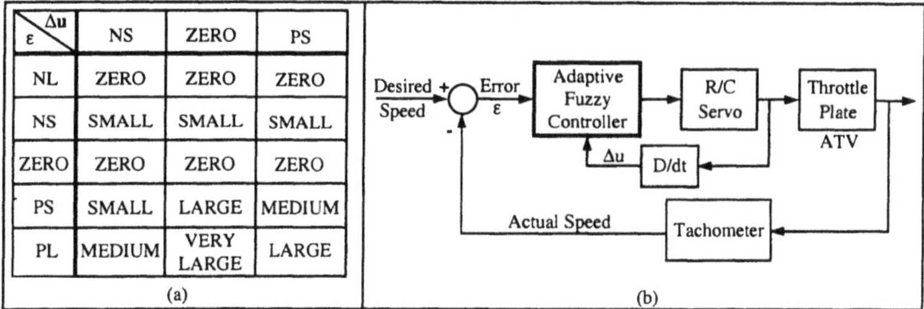

ε \ Δu	NS	ZERO	PS
NL	ZERO	ZERO	ZERO
NS	SMALL	SMALL	SMALL
ZERO	ZERO	ZERO	ZERO
PS	SMALL	LARGE	MEDIUM
PL	MEDIUM	VERY LARGE	LARGE

(a)

Fig. 15. (a) Fuzzy rule table for ATV speed control; (b) Controller block diagram.

where $f : N \subset \Re^n \to \Re$, $\mathbf{x} = (x_1 \dots x_n)^T \in N$, $\mu_{A_i^j}(x_i)$ is a triangular membership function and $\bar{y}^j$ is the point in S where μ_{C^j} is maximum or equal to 1. If $\mu_{A_i^j}(x_i)$ and $\bar{y}^j$ are free (adjustable) parameters, then (8) can be written as

$$f(\mathbf{x}) = \vartheta^T \mathbf{\Psi}(\mathbf{x}), \tag{9}$$

where $\vartheta = (\bar{y}^1 \dots \bar{y}^{Q_R})$ is a parameter vector and $\mathbf{\Psi}(\mathbf{x}) = (\psi^1(\mathbf{x}) \dots \psi^{Q_R}(\mathbf{x}))^T$ is a regression vector with the regressor given by

$$\psi_i(\mathbf{x}) = \frac{\prod_i^n \mu_{A_i^j}(x_i)}{\sum_{j=1}^{Q_R}(\prod \mu_{A_i^j}(x_i))}. \tag{10}$$

Eq. (9) is referred to as an adaptive fuzzy system [38]. There are two main reasons for using adaptive fuzzy systems as building blocks for adaptive fuzzy controllers. Firstly, they have been proven to be universal function approximators [38]. Secondly, all the parameters in $\mathbf{\Psi}(\mathbf{x})$ can be fixed at the beginning of the adaptive fuzzy system expansion design procedure, so that the only free design parameters are ϑ. In this case $f(\mathbf{x})$ is linear in the parameters.

The advantage of this approach is that very simple linear parameter estimation methods can be used to analyze and synthesize the performance and robustness of adaptive fuzzy systems. If no linguistic rules are available, the adaptive fuzzy system reduces to a standard nonlinear adaptive controller. This approach is adopted to synthesize the adaptive control law.

Adaptive Law Synthesis and Implementation From theoretical considerations and quantitative data, the following model was developed for the ATV engine [40]:

$$\begin{bmatrix} \dot{N} \\ \ddot{N} \end{bmatrix} = \begin{bmatrix} \frac{-d_1}{J_e} & 0 \\ \frac{-c_t c_e^2 \eta_{vol}^2 m_a}{J_e} & \frac{-d_1}{J_e} \end{bmatrix} \begin{bmatrix} N \\ \dot{N} \end{bmatrix} + \begin{bmatrix} 0 \\ \frac{-c_t c_e \eta_{vol} D_1}{J_e} \end{bmatrix} \alpha + \mathbf{E} \tag{11}$$

where N is the engine speed (RPM); d_1 and D_1 are respectively engine friction and airflow constants; J_e is the engine moment of inertia; c_e is derived from the intake manifold and engine displacement; c_t is derived from the engine efficiency, combustion heat, and air-to-fuel ratio; η_{vol} is the engine volume efficiency; m_a is the net manifold air-mass flow rate; α is the throttle plate angle; and E lumps together higher-order cross-coupling terms and load torque. This engine model (11) represents a primitive mathematical model of internal combustion engine dynamics and by no means captures all the nonlinear parameters of the engine. The model can be expressed as

$$\dot{\mathbf{z}} = \mathbf{Az} + \mathbf{Bu} + \mathbf{E(z)} \tag{12}$$

where $\mathbf{A}$ is Hurwitz and $\mathbf{E(z)}$ is the uncertainty in the model expressed as a function of the state $\mathbf{z}$. Therefore, there exists a unique positive definite matrix $\mathbf{P}$ that satisfies the Lyapunov equation

$$\mathbf{A}^T\mathbf{P} + \mathbf{PA} = -\mathbf{Q} \tag{13}$$

If the control input, $\mathbf{u}$, is expressed as an adaptive fuzzy system then (12) becomes,

$$\dot{\mathbf{z}} = \mathbf{Az} + \mathbf{B}\vartheta^T\psi(\mathbf{z}) + \mathbf{E(z)} \tag{14}$$

Let [39],

$$\dot{\hat{\mathbf{z}}} = \mathbf{A}\hat{\mathbf{z}} + \mathbf{B}\vartheta^{*T}\psi(\hat{\mathbf{z}}) \tag{15}$$

be the ideal engine model with no uncertainty with $\varepsilon = \mathbf{z} - \hat{\mathbf{z}}$, where ϑ^* denotes the optimal parameter vector. Therefore,

$$\dot{\varepsilon} = \mathbf{A}\varepsilon + \mathbf{B}\phi^T\psi(\varepsilon) + \hat{\mathbf{E}} \tag{16}$$

where $\phi = \vartheta - \vartheta^*$, and $\hat{\mathbf{E}}$ is an estimate of the upper bounds of the uncertainties. The following Lyapunov function (with $\gamma > 0$ as a design parameter) is used to derive the adaptive control law given by Eq. (18) below, which ensures that $\varepsilon \to \mathbf{0}$ as $t \to \infty$ (see [40] for details on this formulation).

$$V = \frac{1}{2}\left(\varepsilon^T \mathbf{P} \varepsilon + \frac{\phi^T \phi}{\gamma ||\hat{\mathbf{E}}||}\right) \tag{17}$$

$$\dot{\phi} = -\gamma ||\hat{\mathbf{E}}|| \, ||\varepsilon^T \mathbf{P} \mathbf{B}|| \, \psi(\varepsilon) \tag{18}$$

The fuzzy rule table of Fig. 15a was used to implement (18) for adaptive fuzzy speed control on the ATV. The insight gained from the non-adaptive FLC was used to select the ϑ values to lie within the interval [1, 2]. The remaining control parameters were set as follows: $\mathbf{Q} = diag(3,3)$, $\hat{\mathbf{E}} = [120 \;\; 0]^T$, $\gamma = 0.00025$; and $\mathbf{\Psi}(\varepsilon)$ was formulated using the IF part of fuzzy rules in Fig. 15a. Finally, $\mathbf{P}$ is obtained from the iterative solution of (13), and $\hat{\mathbf{E}}$ and γ are obtained empirically.

Typical ATV performance with the adaptive fuzzy control law is shown in Figs. 16 and 17, which depict representative ATV responses to selected speed commands ($2.97\,mi/h$ and $3.4\,mi/h$). Acceptable steady-state error performance is achieved; in addition, considerable improvement over the non-adaptive FLC was observed with respect to disturbance rejection (load and terrain). The adaptive algorithm responds to varying terrain by continuously minimizing the speed error by tuning the center of the input membership functions.

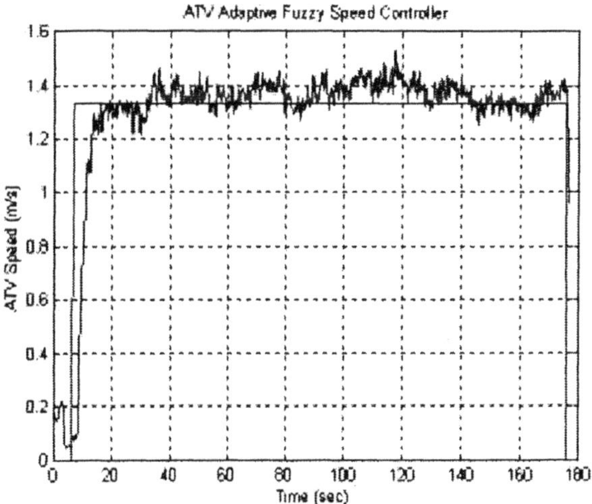

Fig. 16. ATV speed response to $1.3\,m/s$ command.

At very slow speeds for the ATV (e.g., $1.0\,m/s$), the speed response overshoot and settling times tend to increase. Since the ATV is back-heavy, considerable momentum is required to initiate motion on slight inclines, and brief

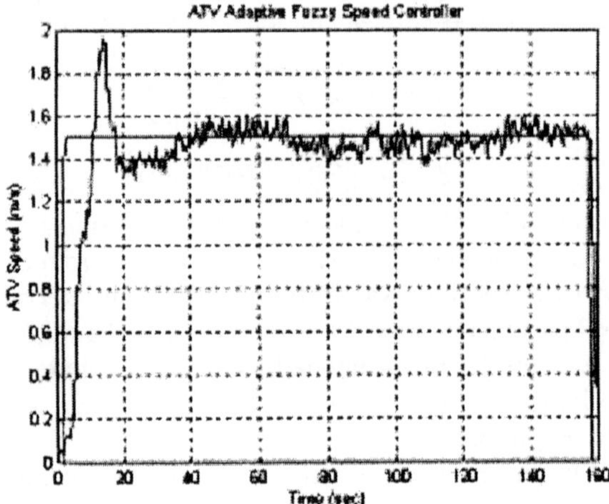

Fig. 17. ATV speed response to 1.5 m/s command.

slippage in its automatic transmission causes considerable drops in speed to well below the commanded speed. In such cases, the adaptive fuzzy control responds with small increases in throttle input (rather than the large response that a PI controller would typically generate) and regains the commanded velocity without significant overshoot once the slipping ceases. For this case study, the combined soft and hard computing-based control solution successfully achieves the desired properties of smooth and accurate control at speeds within the desired range, as well as robustness to varying terrain.

6 Discussion and Conclusion

Autonomous robotic operations in natural outdoor environments require robots to perform complex tasks for which analytical models of sufficient accuracy or completeness are not always available. To address such issues while producing reliable and safe field robotic systems it is often advantageous to supplement conventional hard computing methods with particular soft computing techniques. This chapter presented several case studies in this regard describing soft computing techniques that have been integrated on physical field mobile robot systems and validated in natural terrain environments.

Case study 1 focused on a high-level vision-based strategy for terrain perception using efficient stereo image processing to facilitate rover navigation. A hybrid wavelet processing and neural network system was described which generates appropriate navigation actions based on input of a stereo pair of images. The system reacts to a pattern of wavelet coefficients in much the same way as binocular neurons in the mammalian visual pathways. Case Study 2 described a terrain perception strategy that employs low-level monocular vision to facilitate maintaining mobility performance during navigation using supervisory-level control of tractive vehicle speed. This case also presented a fundamental approach to improving position estimation enabled by the same perception system. The techniques applied in Case Study 2 may be preferred for systems with minimal onboard sensing and computing resources. Case Study 3 presented an adaptive fuzzy throttle control design for controlling speed of an autonomous ATV driven by an internal combustion engine. The control algorithm has been implemented on two other ATVs with very little recalibration of the control parameters. The formulation of the fuzzy rules for the system may be relevant to other practical applications where a complete mathematical model is not available.

Soft computing has been successfully applied to address perception using neural networks based on both fuzzy self-organizing feature maps and backpropagation, as well as supervisory and low-level speed control using rule-based fuzzy logic and adaptive fuzzy systems, respectively. However, implementation experiences associated with the case studies have revealed some disadvantages of applying these soft computing methods for perception and control. A key disadvantage, from the point of view of the perception applications, relates to specification of the initial knowledge that is embedded within the system and ultimately used as the basis for motion decisions. For example, the generalization capability (and therefore, success) of the vision-based neural networks depends on the richness of a set of training images and how well they represent the target environments. If the training set is insufficient, there may be visual scenes for which the system has not learned appropriate motion responses. It is not always obvious when this is the case, and extensive testing may not reveal such insufficiencies. The strategies in this chapter make use of empirical tests and adaptive learning techniques in order to ensure that once initial knowledge is established, it can be improved based on actual system behavior in subsequent tests. By allowing adaptive capability, the initial establishment of a thorough knowledge base is not as critical if significant testing is performed.

Regarding control applications, soft computing techniques have been subject to debate and skepticism, since they do not have the formal rigor of conventional control techniques. This can be attributed to the fact that most successful applications of soft computing are based on simulation and experimental results, rather than the theoretical proofs relied upon by classical control theorists and practitioners. Additionally, most soft computing applications have been in mass non-critical consumer goods with less stringent performance criteria than the traditional control areas of aerospace and industrial systems. Nevertheless, soft computing has been demonstrated to solve some practical problems that have been beyond the reach of conventional control techniques. The rapid growth of research in this area establishes the fact that there is a genuine need for soft computing-based control strategies. While technological tools exist for implementing them, tools for synthesis of performance criteria, robustness, and design methodology are lacking, however. Hence, there is a need for further research in this direction. Until techniques and tools can be developed that can subject soft computing control systems to such factors, the debate and skepticism are likely to remain.

The computing strategies described herein for perception and control were developed for application on wheeled robots to be potentially deployed for space and military missions. The case studies serve only as existence proofs of the concepts presented, as well as practical implementation examples validated on research prototype vehicles. For real missions, various requirements and operability or safety restrictions are specified on a task by task basis. Compliance of combined hard and soft computing strategies with real mission constraints must be verified and validated through extensive testing prior to field deployments. Note that the utility of the strategies presented is not limited to the specific field robots described in this chapter. Indeed, the soft computing strategies may be adapted and integrated with hard computing solutions on a wider variety of autonomous field robots to address common problems of perception, navigation, and control. In practice, it is typically the *solutions* to the common problems that must be tailored for specific robotic systems. Particular design and implementation decisions must be governed by the real-world task constraints, mission objectives and requirements, and/or vehicle capabilities.

Acknowledgment

The rover research described in this chapter was performed at the Jet Propulsion Laboratory, California Institute of Technology, under a contract with the National Aeronautics and Space Administration. The ATV research described in this chapter was performed at Carnegie Mellon University, sponsored by DARPA contract F04701-97-C-0022, "An Autonomous, Distributed Tactical Surveillance System." Publication support was provided by the Jet Propulsion Laboratory, California Institute of Technology, under a contract with NASA.

References

1. Fu, K. S., Gonzalez, R. C., Lee, C. S. G. (1987): Robotics: Control, Sensing, Vision, and Intelligence. McGraw-Hill, New York
2. Dudek, G., Jenkin, M. (2000): Computational Principles of Mobile Robotics. Cambridge University Press, Cambridge, UK
3. Driankov, D., Saffiotti, A., (Eds.) (2001): Fuzzy Logic Techniques for Autonomous Vehicle Navigation. Fuzziness and Soft Computing Series **61**, Physica-Verlag, Heidelberg
4. Thrun, S. (2000): Probabilistic Algorithms in Robotics. Tech. Report CMU-CS-00-126, Computer Science Department, Carnegie Mellon University, Pittsburgh, PA
5. Dubowsky, S., Farritor, S., et al. (1996): Research on Field and Space Robotic Systems. In: Proc. ASME Computers in Engineering Conf., Irvine, CA

6. Hagras, H., Callaghan, V., Colley, M. (2001): Outdoor Mobile Robot Learning and Adaptation. IEEE Robotics & Automation Magazine **8**, 3, 53–69
7. Schenker, P. S., Pirjanian, P., et al. (2000): Reconfigurable Robots for All Terrain Exploration. In: McKee, G. T., Schenker, P. S. (Eds.): Proc. SPIE **4196**, Sensor Fusion and Decentralized Control in Robotic Systems III. Boston, MA
8. Gauss, V. A., Bay, J. S. (1998): A Fuzzy Logic Solution for Navigation of an Autonomous Subsurface Planetary Exploration Robot. In: Proc. IEEE Intl. Symp. on Intelligent Control, Gaithersburg, MD, 559–564
9. Kaplan, M. H. (1976): Modern Spacecraft Dynamics and Control. John Wiley & Sons, New York
10. Mishkin, A., Morrison, J., Nguyen, T., et al. (1998): Experiences with Operations and Autonomy of the Mars Pathfinder Microrover. In: Proc. IEEE Aerospace Conf., Aspen, CO
11. Matthies, L. (1992): Stereo Vision for Planetary Rovers: Stochastic modeling to near real-time implementation. Intl. J. Computer Vision **8**, 71–91
12. Belhumeur, P. N. (1993): A Binocular Stereo Algorithm for Reconstructing Sloping, Creased, and Broken Surfaces in the Presence of Half-occlusion. In: Proc. Intl. Conf. on Computer Vision. New York, NY, 431–438
13. Churchland, P. S., Sejnowski, T. J. (1993): The Computational Brain. MIT Press, Cambridge, MA
14. Daugman, J. G. (1985): Uncertainty Relation for Resolution in Space, Spatial Frequency, and Orientation Optimized by Two-dimensional Visual Cortical Filters. J. Optical Society of America A **2**, 1160–1169
15. Hubel, D. H., Wiesel, T. N. (1963): Shape and Arrangement of Columns in the Cat's Striate Cortex. J. of Physiology **165**, 559–568
16. Dobbins, A., Zucker, S. S., Cynader, M. S. (1987): Endstrapped Neurons in the Visual Cortex as a Substrate for Calculating Curvature. Nature **329**, 438–441
17. Mallat, S., Zhong, S. (1992): Characterization of Signals from Multiscale Edges. IEEE Trans. PAMI **14**, 710–732
18. Andersson, L., Hall N., Jawerth, B. et al (1994): Wavelets on Closed Subsets of the Real Line. In: Schumacher, L. L., Webb, G. (Eds.): Topics in the Theory and Applications of Wavelets. Academic Press, New York, NY
19. Cohen, A., Daubechies, I., Jawerth, B., Vial, P. (1993): Multiresolution Analysis, Wavelets and Fast Algorithms on an Interval. Comptes Rendes **316, I**, 417–421
20. Hoffman, B. D., Baumgartner, E. T., Huntsberger, T., Schenker, P. S. (1999): Improved Rover State Estimation in Challenging Terrain. Autonomous Robots **6**, 2, 113–130
21. Huntsberger, T. L. (1990): Comparison of Techniques for Disparate Sensor Fusion. In: Proc. SPIE Symp. Sensor Fusion III: 3-D Perception and Recognition **1383**, Boston, MA, 589–595
22. Kohonen, T. (1989): Self-Organization and Associative Memory, 3rd Edition. Springer-Verlag, Berlin
23. Pal, N. R., Bezdek, J. C., Tsao, E. C.-K. (1993): Generalized Clustering Networks and Kohonen's Self-organizing Scheme. IEEE Trans. Neural Networks **4**, 549–557
24. Bezdek, J. C. (1981): Pattern Recognition with Fuzzy Objective Function Algorithms. Plenum Press, New York
25. Huntsberger, T. L., Jacobs, C. L., Cannon, R. L. (1985): Iterative Fuzzy Image Segmentation. Pattern Recognition **18**, 131–138

26. Huntsberger, T. L., Ajjimarangsee, P. (1990): Parallel Self-organizing Feature Maps for Unsupervised Pattern Recognition. Intl. J. of General Systems **16**, 357–372
27. Wilcox, B. H. (1994): Non-geometric Hazard Detection for a Mars Microrover. In: NASA/AIAA Conference on Intelligent Robotics in Field, Factory, Service, and Space. Houston, TX, 675–684
28. Gillespie, T. D. (1992): Fundamentals of Vehicle Dynamics. Society of Automotive Engineers, Inc.
29. Tunstel, E., Howard, A. (2002): Approximate Reasoning for Safety and Survivability of Planetary Rovers. Fuzzy Sets and Syst. Elsevier Science, In press.
30. Howard, A., Seraji, H. Tunstel, E. (2001): A Rule-based Fuzzy Traversability Index for Mobile Robot Navigation. In: IEEE Intl. Conf. Robotics and Automation. Seoul, Korea, 3067–3071
31. Diamantaras, K., Kung, S. Y. (1993): Principal Component Neural Networks: Theory and applications. John Wiley & Sons, New York
32. Baumgartner, E. T., Aghazarian, H., Trebi-Ollennu, A. (2001): Rover Localization Results for the FIDO Rover. In: Proc. SPIE Conf. Sensor Fusion and Decentralized Control in Autonomous Robotic Systems IV **4571**. Newton, MA
33. Goel P., Roumeliotis S. I., Sukhatme G. S. (1999): Robust Localization using Relative and Absolute Position Estimates. In: Proc. IEEE/RSJ Intl. Conf. on Intelligent Robots and Systems. Kyongju, Korea, 1134–1140
34. Barshan, B., Durrant-Whyte, H. F. (1995): Inertial Navigation Systems for Mobile Robots. IEEE Trans. Robotics and Automation **11**, 328–342
35. Olson, C. F., Matthies, L. H. et al (2001): Stereo Ego-motion Improvements for Robust Rover Navigation. In: Proc. IEEE Intl. Conf. on Robotics and Automation. Seoul, Korea, 1099–1104
36. Bellutta, P., Manduchi, R., Matthies, L., Rankin, A. (2000): Terrain Perception for DEMO III. In: Proc. Intelligent Vehicles Conf. Dearborn, MI
37. Macedo, J., Manduchi, R., Matthies, L. (2000): Ladar-based Discrimination of Grass from Obstacles for Autonomous Navigation. In: 7th Intl. Symp. on Experimental Robotics. Waikiki Beach, Hawaii
38. Wang, L-X. (1994): Adaptive Fuzzy Systems and Control Design and Stability Analysis. Prentice Hall, Englewood Cliffs, NJ
39. Trebi-Ollennu, A. (1996): Robust Nonlinear Control Designs using Adaptive Fuzzy Systems. Ph.D. thesis, Dept. Aerospace and Guidance Systems, School of Engrg, and Applied Science, Royal Military College of Science, Shrivenham, Cranfield University
40. Trebi-Ollennu, A., Dolan, J. M., Khosla, P. K. (2001): Adaptive Fuzzy Throttle Control for an All-terrain Vehicle. In: Proc. Instn. Mech. Engrs., Part I: J Syst. and Control Engrg. **215**, 189–198

Integration of Reactive Utilitarian Navigation and Topological Modeling

Javier de Lope and Darío Maravall

Department of Artificial Intelligence
Faculty of Computer Science, Universidad Politécnica de Madrid
Campus de Montegancedo, 28660 Madrid, Spain
{jdlope,dmaravall}@fi.upm.es

Abstract. This chapter describes a hybrid autonomous navigation system for mobile robots. The control architecture proposed is highly modular and is based on the concept of behavior, which is a generalization of the usual reactive interpretation of this term. The proposed navigation system involves a straightforward integration of reactive and deliberative modules, enabling global, model-based navigation and local, adaptive navigation. At the local navigation level, we introduce the concept of *utilitarian* navigation, which models low-level robot navigation as a functional optimization process. Thanks to this innovative perspective, we have been able to implement low-level tasks, like collision avoidance and sensory source search and evasion, which have been integrated into the hybrid navigation system. At the global navigation level, two fundamental problems are considered: (1) map or model building and (2) route planning. Fuzzy Petri nets (FPN) are used to construct topological maps. A minimum cost algorithm of the FPN propagation has been implemented for route planning and execution. This chapter also discusses the experimental work carried out with realistic simulations, as well as with a holonomic prototype built by the authors and a NOMAD-200 mobile platform.

1 Introduction

The role played by mathematical models or formal representations of the environment, or even the absence of such models, is now one of the key topics in the development of advanced robotic systems and, more specifically, in the design of what are know as autonomous navigation systems for mobile robots. As a matter of fact, most navigation methods in the field of autonomous mobile robots are based on the manipulation of models of the environment – with varying levels of formal representation – in which the robot has to exist [1].

This dominant paradigm is extremely demanding as regards the *reasoning* abilities of such robots, which are designed upon the principle of using, maintaining and even, sometimes, creating formal representations of their worlds. Furthermore, the requirements as regards the perceptual abilities of these robots are even more demanding. These demands on artificial reasoning and, especially, artificial perception are far beyond the current state of human knowledge and the technological state of the art. It is no wonder that no physical robot has ever been able to navigate in truly real environments.

The question then is how to progress toward higher levels of robot autonomy – without falling in the easy temptation of somehow structuring and externally controlling the environment or using external information –. An understandable reaction to such a formidable challenge has emerged in the shape of a tendency to reduce the complexity of both the reasoning and the perceptual tasks to be accomplished by the robots – without any kind of human intervention in the environment –, which has finally converged into a new paradigm. The chief difference between this and the dominant paradigm lies in a radical shift in the role played by the formal representations of the world, which, in the new paradigm, have lost their traditional importance or, to be more exact, their strong apriority and usual rigidity. Consequently, adaptation and learning have become the central issues concerning world representations. According to these realistic assumptions – simpler robot reasoning and perception – the approach to autonomous navigation has switched from complex, symbolic models of the world to minimalist, flexible models, where as discussed before, adaptation and learning play a central role in the construction of the robot's internal representation of its world, mainly through a direct coupling between perception and action.

An extreme embodiment of this novel paradigm is what is known as the reactive approach, [2,3], which has done away with almost any type of formal representation of the environment. Robots designed in this way are practically restricted to the interaction between perception – usually the inputs of the robot – and action – normally the robot outputs –. Although it is sometimes advisable, or even compulsory, to build robots without any model of the environment at all, the fact is that they are limited to a purely reactive existence, as the close coupling between perception and action extremely reduces the activities and the possibilities of such robots. Therefore, even pure reactive navigation systems do use some kind of intermediate processing of the perceptual information before passing to action, which is in fact, a primitive reasoning ability. Despite their minimal world modeling, reasoning and planning capacities, the reactive navigation schemes have distinctive features that make it possible to design robots able to adapt to very dynamic, uncertain and even completely unknown environments, that are prohibitive for model-based navigation schemes. Some remarkable success-cases like the Mars Sojourner, [4], speak to the practical interest – apart from the indisputable theoretical interest – of the reactive paradigm. Therefore, it is almost compulsory to integrate reactive schemes in any autonomous navigation system, and a hybrid navigation system should combine the low-level navigation ability of reactive schemes with the high-level efficiency of deliberative, model-based, schemes.

An important distinction is made in deliberative or model-based navigation schemes, depending on whether the environment models are introduced *a priori* by the designer or dynamically built after exploratory missions by the actual robot. Therefore, the respective world models, either pre-designed or learned, can be classed as topological or metric models. In a metric model, the information about the position and orientation of the robot is numerically specified in relation to a fixed reference system. In topological models, the information about the world is based on a group of reference places that can be recognized by the robot and on a group of links between accessible reference places.

The choice of a particular model depends on the problem to be solved. Metric models are suitable for situations where highly accurate robot navigation is required. Topological models are more suitable for situations in which a very precise knowledge of the robot's position is not vital and, therefore, it is sufficient for navigational purposes for the robot to recognize the reference places rather than to compute its exact position. Moravec and Elfes [5] and Cox [6] discuss examples of metric models. Topological models are proposed by Kuipers [7], Mataric [8], Kurz [9] and Serradilla and Maravall [10].

In the sequel, we present a hybrid navigation system that integrates purely reactive modules – intended for copying with dynamic and unknown elements of the environment – and deliberative, model-based schemes, necessary to exploit knowledge about the navigation world – either introduced by the designer or acquired by the actual robot –. The organization of the remainder of this chapter is as follows. First, we present a brief overview of the system architecture, which is based on a cyclical execution policy and on the general concept of *behavior*, ranging from the purely reactive, low-level behaviors to the high-level, deliberative behaviors, like map building or route planning. Within the low-level navigation category, we introduce a novel method based on the formalization of autonomous robot navigation as a functional optimization problem. From this viewpoint, a unified framework for low-level autonomous navigation emerges, where the robot is just an agent that maximizes or minimizes different performance functions. The definition of such performance functions – i.e. the robot's goals, or more exactly, purposes – and their quantitative measurement by the robot's sensors are obviously the two strategic issues related to this novel approach. This approach can be called *utilitarian*, in the sense that it converts the robot into a physical agent aiming to maximize utility functions. As an illustration of the practical interest of this utilitarian navigation based on functional optimization, we obtain a novel version of the well-known Artificial Potential Field Theory (APF), with interesting advantages, like simplicity, robustness and efficiency for dynamic obstacle avoidance. The navigation task of searching – or evading – sensory sources is also modeled using the utilitarian approach. As a result, real-time obstacle avoidance and sensory-based search can be simultaneously solved by a single gradient operator.

On the high-level, deliberative side, our hybrid navigation system is based on topological and adaptive models of the world. This chapter includes the description of two basic deliberative modules: (1) map building and (2) route planning. With regard to the first module, i.e. for the robot's internal representation of the environment, we propose a topological model based on fuzzy Petri nets (FPN). The models are automatically built by the robot during exploratory missions, in which the robot detects the different landmarks and tests different navigation strategies. Whenever one of the trials leads to a new landmark, the model of the navigation world is updated. Once the environment model has been built, the mobile robot can perform any type of navigation mission. There are basically two phases in route planning and execution: model search and model tracking. During model search, the robot detects and identifies some of the already created landmarks, so that its position is updated. During the model tracking phase, the robot must test the strategies suggested by the navigation model until it reaches a new landmark. For route planning, a minimum cost algorithm of the FPN propagation has been implemented. The chapter ends with the discussion of experimental work carried out by means of realistic simulations and with a NOMAD-200 mobile platform.

2 A Cyclical Hybrid Architecture for Autonomous Navigation

Our navigation system is based on the cyclical execution policy typical of many real-time operating systems, in which what is known as the scheduler periodically activates the basic processes of the operating system, depending on internal and external events [11,12].

In the proposed architecture [13], the concept of process is associated with the basic elements that, under the Perception-Planning-Action (P-P-A) cycle execution, are responsible for processing external information in order to activate appropriate actions. Thus, processes are the least abstract, i.e. the most physical, elements of the navigation system. A particular set of processes conforms what we call *behaviors.* Behaviors are formed by processes in such a way that they produce observable, purposeful robot activities and tasks. The appropriate sequential combination of the existing behaviors generates what we call *navigation strategies*, which can be designed *a priori* or learned by robot experience.

A distinctive feature of the proposed system is the absence of a central element in charge of coordinating the different behaviors, as there is a mechanism within each behavior for detecting whether each specific behavior is competent for dealing with the current state of affairs in the navigation system. Therefore, each behavior is responsible for activating the appropriate behavior, other than itself, for events outside its scope. The basic processes of each behavior are divided into two categories: *supervisory processes*, which supervise the objective conditions that must be present to activate the appropriate behavior, and *actuator processes*, in charge of processing the sensory information and the actions to be applied in each possible event. Although radically different as regards their intrinsic objectives, it must be emphasized that both kinds of processes are based on the same P-P-A cycle execution.

Each behavior includes at least one actuator process responsible for the respective actions, and one or more supervisory processes in charge of monitoring the suitable conditions for behavior activation. All the behaviors are organized on a priority basis, so that they are activated in accordance with a given hierarchy. Obviously, critical behaviors for robot navigation, like collision avoidance, have the highest priority. Another essential concept is behavior *inhibition*, which is aimed at raising the flexibility of the navigation system without increasing the number of behaviors. The general lay-out of the control architecture is shown in Fig. 1. The behaviors that produce inhibition-induced changes in the navigation strategies have been highlighted. Behaviors at the top of the control structure have the highest priority. Generally, the behaviors associated with strategy changes are normally inhibited, unless the robot is executing the respective navigation strategy, like, let us say, *wall following*, *sensory-based search*, *obstacle avoidance*, etc. Furthermore, only one such *strategic* behavior is activated at any given time, so that the robot is able to harmoniously perform its navigation activities: *wander without colliding*, *world modeling or map building*, *route execution*, etc.

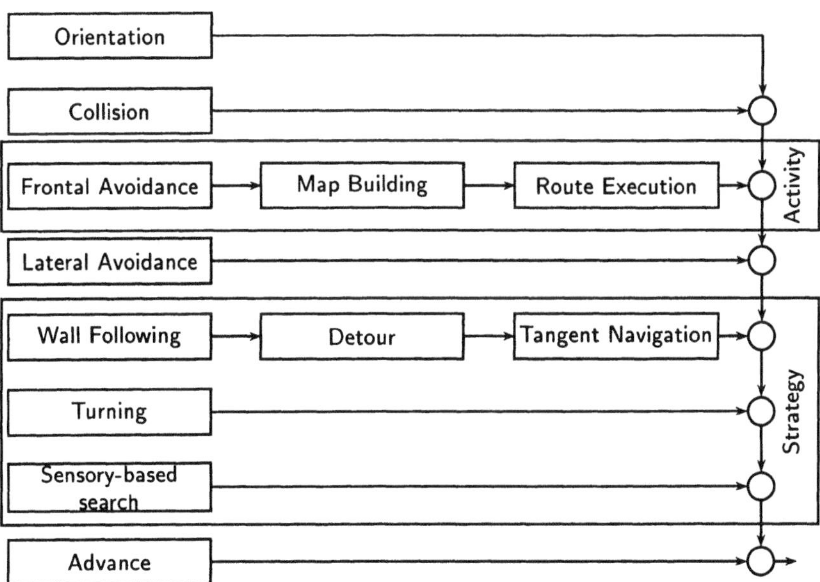

Fig. 1. General lay-out of the control architecture

3 The Organization of Robot Behaviors

Under the cyclical execution policy, robot behaviors are executed cyclically, so that one or more behaviors are activated depending on the particular state of the environment. As mentioned above, all the robot activities and tasks – which include high-level tasks like map building and route planning, as well as the local or low-level tasks necessary to guarantee real, physical navigation – are founded on the basic concept of behavior. Now, let us proceed with a brief description of some of the behaviors that enable autonomous robot navigation in complex environments.

First, we define three different groups of navigation behaviors. The first group, or *basic behaviors* group, allows the robot to perform basic navigation operations like *moving forward* and *obstacle avoidance*. Three behaviors patterns are used for this purspose: (a) *advance*, which makes the robot move in a straight line and which is the basic behavior needed by the robot in order to come back after performing a particular navigation maneuver; (b) *lateral avoidance*, which corrects the robot trajectory when a side collision is foreseen and (c) *frontal avoidance*, that produces a turn whenever an obstacle is detected in front of the robot. These three simple behavior patterns make it possible to implement collision-free trajectories. Two trajectories are shown in Fig. 2. Only the right lateral avoidance behavior is activated in the first one. In the second one, some activations of the frontal avoidance behavior contribute to a collision-free trajectory.

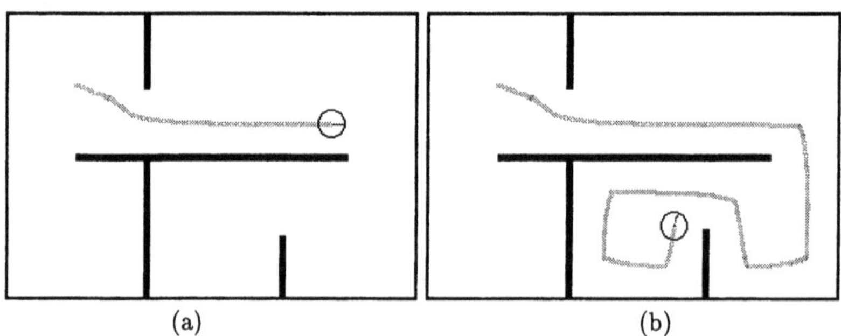

Fig. 2. Trajectories obtained with (a) lateral avoidance and (b) frontal avoidance

The second group of behaviors, or *navigation strategies group*, is formed by the behaviors that make the robot activate the basic behaviors properly, depending on the specific structure of the environment. In particular, we define four different navigation strategies: (a) *wall and obstacle following*, which generates a trajectory along the wall or the obstacle contour; (b) *detour*, which uses the obstacle as a reference for trajectory modifications and permits high-level commands like "take the first right turning"; (c) *turning*, which prompts a change of to 180 degrees in the direction of the movement and (d) *free-movement* or *wandering*, in which the robot moves along without colliding with any obstacle. Two different trajectories obtained with (a) the wall following strategy and (b) the detour strategy are depicted in Fig. 3.

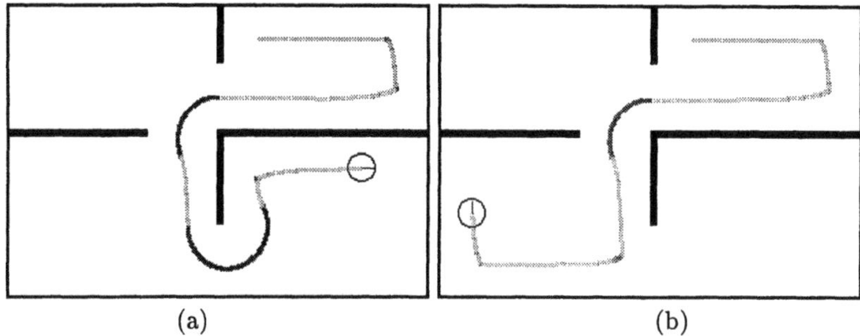

Fig. 3. Trajectories obtained with (a) wall following and (b) detour strategies

The third and last group of behaviors, or *operating behaviors group*, includes behaviors needed for high-level navigation tasks. They include the behaviors responsible for selecting the appropriate robot task – for instance, activation of an exploration mission to build environment maps or activation of an execution mission to reach a specific goal – and behaviors in charge of monitoring the internal robot states.

All the behaviors follow a cyclical-execution architecture based on the P-P-A paradigm and organized in two levels: supervision and performance. At the supervision level, all possible events for the activation of any behavior are under continuous control and surveillance. The performance level is responsible for deciding on the activation and coordination of the behaviors. During the perception cycle the system inputs can stem either from the physical sensors or from internal events generated by the navigation system itself. Similarly, the action cycle can actuate either physical elements of the robot – for instance, a modification in the advance movement – or over the internal states of the robot navigation system – for instance, update the environment model –.

As a practical demonstration of the flexibility and modularity of the proposed control architecture, the next section introduces the novel approach of utilitarian navigation, which can be integrated directly into the navigation system as an additional module – or behavior, for that matter –.

4 Utilitarian Navigation

The new paradigm for autonomous navigation, which has emerged as a consequence of the practical failures of the deliberative, model-centered paradigm, is based, as we know, on a tendency to emphasize the direct coupling between perception and action, in detriment of the powerful reasoning and deliberative facet of the traditional navigation schemes. Using a well-known distinction between local and global levels, deliberative schemes exploit the global knowledge available about the navigation world, whereas reactive schemes enable the robot to adapt to the local, unpredictable characteristics of that world. Thus, for the design of truly autonomous robots, hybrid navigation systems able to integrate both facets are the logical choice.

In this section, we introduce a novel approach to the low-level or local-navigation problem by formalizing the navigation of a mobile robot as a functional optimization problem, which makes it possible to unify two basic skills of any physical autonomous agent like a mobile robot: obstacle avoidance and sensory source search and evasion. We are proposing a somewhat bio-inspired approach to robot navigation, in the sense that the robot navigates under sensory-based utilitarian schemes, which bear a strong resemblance to the navigation of many living beings. From this utilitarian perspective, the basic navigation activities of collision avoidance and sensory source search, or in some cases the evasion, are performed as a consequence of the optimization of some utility functions or performance indices. Summarizing, at the local level, the robot's basic movements are driven by the maximization or minimization of utility functions computed by the robot sensors. This is the reactive paradigm in its quintessential state: actions, i.e. movements, fired by perceptions.

Incidentally, we shall demonstrate that the widely used APF theory [14–16] can be derived from our utilitarian concept. *De facto*, we obtain a much simpler and efficient mechanism for obstacle avoidance [17]. This novel approach to autonomous navigation apparently departs to some extent from the strictly behavior-based perspective discussed earlier, as the robot now performs its navigation tasks according to the optimization of some performance or utility functions, instead of navigating as a result of the sequential and cyclic activation of certain pre-designed behaviors and elementary processes under the P-P-A cycle execution. However, thanks to the modularity of the proposed control architecture, the utilitarian navigation modules can be integrated straightforwardly as additional behaviors, one for collision avoidance and the other for sensory search or evasion. Let us start with the latter topic.

4.1 Sensory Search and Functional Optimization

Consider the situation depicted in Fig. 4, where there is a sensory source S and a robot R. The objective for robot R is to reach the goal S as efficiently as possible – minimum energy and minimum time – supposing the goal to be attractive. The opposite situation should be a repulsive goal, which the robot had to evade. Obviously, the robot must avoid all the existing obstacles while searching for or escaping from the goal. The curves appearing in the figure represent hypothetical *isosensory* curves.

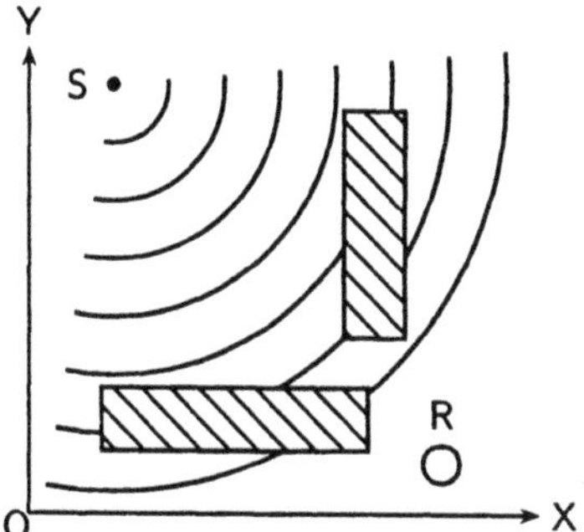

Fig. 4. Sensory source and a robot with obstacles and isosensory curves

The sensory source can be mathematically defined by a function $F(x,y,z,t)$ that determines the space-time distribution of the sensory information induced by the source S. Apart from its specific influence on the sensory function throughout its particular topography, a particular *ecological* environment introduces additional space-time effects that can perturb the sensory function as a, generally multiplicative or additive, noise function $N(x,y,z,t)$:

$$\begin{aligned} I(x,y,z,t) &= F(x,y,z,t) \cdot N(x,y,z,t) \\ I(x,y,z,t) &= F(x,y,z,t) + N(x,y,z,t) \end{aligned} \tag{1}$$

Another interesting case is the potential mobility of the sensory sources that makes autonomous navigation a much more complex task. Without loss of generality and in order to simplify the formal development of our model, we shall consider the simple case of a static source without any environmental disturbance and independent of time and coordinate z; i.e. the sensory function is given by $I(x,y)$. Going back to Fig. 4, our objective is for robot R to efficiently reach the sensory source S by simply using sensors able to measure the function $I(x,y)$. Generally, $I(x,y)$ is *a priori* unknown by the robot, which only needs to measure it at any given point. In this situation, sensory navigation aimed at searching and finding a sensory source – think about an autonomous vehicle that has to detect gas leakages, burning spots or the like – can be modeled as a functional optimization problem. In this particular case, the objective is to identify the optimum or, at least, an efficient trajectory in the plane XOY by means of which the robot can reach the goal S. Therefore, the robot's dynamic equations are as follows:

$$\dot{x} = \frac{\partial I(x,y)}{\partial x} \quad ; \quad \dot{y} = \frac{\partial I(x,y)}{\partial y} \tag{2}$$

and for the evasion maneuver:

$$\dot{x} = -\frac{\partial I(x,y)}{\partial x} \quad ; \quad \dot{y} = -\frac{\partial I(x,y)}{\partial y} \tag{3}$$

These partial-differential equations should be discretized, as the robot control and sensory measurements are based on digital processors. Then, we can re-write (2) or (3) as:

$$\begin{aligned} \dot{x}_k &\approx x_{k+1} - x_k = \mu_k \left.\frac{\partial I(x,y)}{\partial x}\right|_{(x_k,y_k)} \\ \dot{y}_k &\approx y_{k+1} - y_k = \mu_k \left.\frac{\partial I(x,y)}{\partial y}\right|_{(x_k,y_k)} \end{aligned} \tag{4}$$

in which we have introduced an additional parameter to control the optimization trajectory, μ_k, that plays an important role in simultaneously controlling the speed and reliability of the maximum or minimum search. Notice that (4) provides the new robot position (x_{k+1}, y_{k+1}), which is currently placed at point (x_k, y_k). As $\boldsymbol{r} = x_k \hat{a}_x + y_k \hat{a}_y$, a more compact expression is as follows:

$$\boldsymbol{r}_{k+1} = \boldsymbol{r}_k + \mu_k \nabla_{\boldsymbol{r}} I(\boldsymbol{r})|_{\boldsymbol{r}_k} \tag{5}$$

where $\nabla_{\boldsymbol{r}} I(\boldsymbol{r})$ is the gradient vector of the sensory function with respect to the vector position of the robot, $\boldsymbol{r}$. A remarkable and positive characteristic of expressions (4) and (5) is that the *a priori* knowledge of the sensory function is unnecessary with the gradient-based optimum search. Consequently, the gradient operator provides the robot with a very effective device for navigating without explicity knowledge of the global sensory information. Paraphrasing Elias Cannetti: "nobody knows what is good, but everybody knows what is better or worse". In summary, our robot does not need to know the precise form of the sensory objective function and it only has to reason about whether its actions – movements – are improving sensory source search. Expression (5) can be rewritten more conveniently as a robot orientation control signal:

$$\phi_{k+1} = \tan^{-1} \frac{\nabla y|_{\boldsymbol{r}_k}}{\nabla x|_{\boldsymbol{r}_k}} \tag{6}$$

In summary, (6) gives the control action, interpreted as a wheel orientation, to be applied to the robot at each discrete instant and exclusively derived from sensory-based search navigation. Let us now proceed with a similar idea for obstacle avoidance navigation.

4.2 Formalization of Artificial Potential Field Theory for Obstacle Avoidance as a Functional Optimization Problem

The interest of the APF paradigm in autonomous navigation of mobile robots, as well as in adaptive trajectory generation and execution of mechanical manipulators is clear [18]. In the following, we are going to formalize this well-known method as an optimization problem. Roughly speaking, the basic idea of the APF is to create imaginary potential curves $U(x, y, z)$ around the objects that are present in the robot's environment, so that the forces acting on the robot are generated as $\boldsymbol{F} = -\nabla U$. If the obstacles produce positive potential functions inversely proportional to distance and the goals produce negative potential functions, then the resultant forces tend to drive the robot towards the goal without it colliding with any obstacle. In spite of its conceptual elegance and practical appeal, this method is beset by many problems: difficulties in generating the potential field curves, particularly in unknown or in dynamic environments; appropriate tuning of the control forces; chaotic trajectories; local minima trapping, etc. We believe that one main reasons why APF malfunctions in many situations is because, by definition, the negative forces produced by the obstacles are normal to the potential field curves and, although these forces are offset by the attractive force of the goal, they tend to generate far from optimal trajectories. Instead, we propose using forces tangent to the potential curves that tend to generate near-optimal trajectories.

Apart from avoiding the local minima produced by equal opposite normal forces – a situation extremely frequent in indoor environments –, the main advantage of tangent forces is that they follow the obstacle contours; i.e. the optimal trajectory for navigating through obstacles is to go around, rather than just to evade them. The only problem with navigation based on tangent forces is that obstacles with a closed contour tend to trap the robot, in which case an evasion maneuver must be performed. We shall introduce a general solution for this trapping problem later on.

Another important drawback of the APF method is that the generated forces are directly applied to the robot control system, making it considerably troublesome to appropriately tune the control actions involved, like the speed and acceleration magnitudes [19]. To overcome this problem, apart from introducing the optimization approach, we propose control only the direction of the robot, which will move at a constant speed or using a speed profile that is not directly computed from the potential field curves.

This novel interpretation of APF theory means that robot dynamics is not longer based on the usual expressions:

$$F_x = m\ddot{x} + \frac{\partial U(x,y)}{\partial x} \quad ; \quad F_y = m\ddot{y} + \frac{\partial U(x,y)}{\partial y} \tag{7}$$

obtained from the application of the Lagrange equation to the mobile robot. Instead, robot dynamics, as far as obstacle avoidance is concerned, is now expressed as follows:

$$\dot{x} = -\frac{\partial U(x,y)}{\partial x} \quad ; \quad \dot{y} = -\frac{\partial U(x,y)}{\partial y} \tag{8}$$

which can be integrated directly into the same functional optimization framework introduced above for sensory-based search navigation. Note that no external forces are now present. Furthermore, it is straightforward to interpret

(8) under the utilitarian navigation umbrella, as $U(x,y)$ plays the role of a negative utility function that the robot must minimize.

In order to formalize the APF method – with the proposed two variations: (a) using the tangential instead of the normal component and (b) employing direction control rather than forces – as an optimization problem, we first define a potential field function as follows:

$$U(x,y) \equiv f(x,y) = f(1/d_o^r) \tag{9}$$

where $1/d_o^r$ means that this function is inversely proportional to the distance between the robot, placed at point (x,y), and the obstacle. If there are, let's say, N obstacles then the global potential function is:

$$U(x,y) = \sum_{i=1}^{N} f_i(x,y) \tag{10}$$

Unlike conventional APF-based algorithms that compute the forces acting upon the robot, we are going to use the potential curves to get the robot's next position, which, ultimately, will make it possible to drive the robot by its direction only. Taking into account that the objective now is to minimize the function $U(x,y)$, the robot's future position is immediately obtained from the discretization of (8):

$$\begin{aligned} x_{k+1} &= x_k - \mu_k \left.\frac{\partial U(x,y)}{\partial x}\right|_{(x_k,y_k)} \\ y_{k+1} &= y_k - \mu_k \left.\frac{\partial U(x,y)}{\partial y}\right|_{(x_k,y_k)} \end{aligned} \tag{11}$$

Note that (4) and (11) are equivalent, the only remarkable distinction being that, unlike the sensory-based navigation case, expression (4), we do not have a straightforward and easily available function to be minimized now. In other words, we still face the tricky problem of selecting a proper distance function, which plays the role of virtual sensory information. A method for obtaining such a distance function is described Section 4.4.

By composing the movements generated by the sensory information function, $I(x,y)$, and the potential function, $U(x,y)$, the robot's final control coordinates are given by the compact expression:

$$\begin{aligned} \boldsymbol{r}_{k+1} &= \boldsymbol{r}_k + \mu_k \left\lfloor \nabla_{\boldsymbol{r}} I(\boldsymbol{r})|_{\boldsymbol{r}_k} - \nabla_{\boldsymbol{r}} U(\boldsymbol{r})|_{\boldsymbol{r}_k} \right\rfloor = \\ &= \boldsymbol{r}_k + \mu_k \nabla_{\boldsymbol{r}} \left\lfloor I(\boldsymbol{r}) - U(\boldsymbol{r}) \right\rfloor_{\boldsymbol{r}_k} \end{aligned} \tag{12}$$

that actually unifies the navigation strategy of the robot by simultaneously using potential field information and sensory function information. However, instead of the normal direction computed from the gradient of the artificial potential field, we shall use the tangent direction, which is directly derived from the normal one. Indeed, the dynamics of the robot when following the tangent instead of the normal direction turns out to be

$$\dot{x} = \pm\frac{\partial U(x,y)}{\partial y} \quad ; \quad \dot{y} = \mp\frac{\partial U(x,y)}{\partial x} \tag{13}$$

which determines the robot's updated coordinates as follows:

$$\begin{aligned} x_{k+1} &= x_k \pm \mu_k \left.\frac{\partial U(x,y)}{\partial y}\right|_{(x_k,y_k)} \\ y_{k+1} &= y_k \mp \mu_k \left.\frac{\partial U(x,y)}{\partial x}\right|_{(x_k,y_k)} \end{aligned} \tag{14}$$

the double, antisymmetrical signs of the partial derivatives give the two possible tangent orientations. Once the global navigation strategy has been established, the next step is to obtain the robot's physical trajectory, which depends on its particular kinematics.

4.3 Optimal Kinematics for Utilitarian Robots

As we already know from the previous section, an utilitarian robot is controlled by the steering angle, a control case which can be modeled with a conventional feedback loop as shown in Fig. 5.

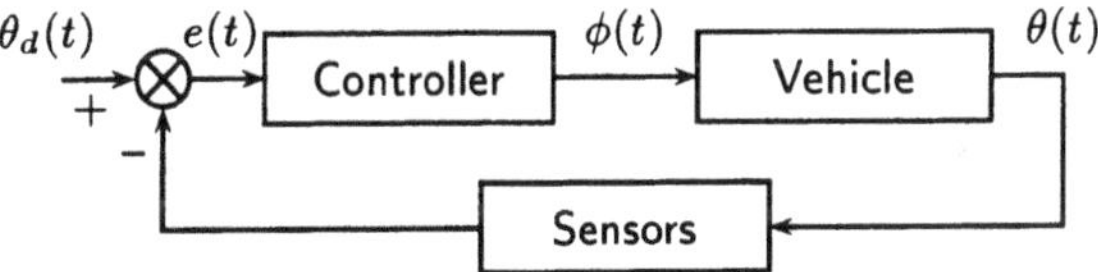

Fig. 5. Feedback loop of a mobile robot controlled by the steering angle

In this case, $\theta_d(t)$ is the robot's desired direction that is compared to its current orientation, $\theta(t)$, and as a result of the difference error, a control law generates the steering angle $\phi(t)$ that eventually drives $\theta(t)$ towards $\theta_d(t)$.

An optimal kinematics for a utilitarian autonomous robot should allow it to almost instantaneously turn around its vertical axis – centered at coordinates (x, y) – in such a way that its dynamic equations should be:

$$\begin{aligned} \dot{x}_R &= |v_R| \cos\theta \\ \dot{y}_R &= |v_R| \sin\theta \\ \dot{\phi} &= \dot{\theta} \end{aligned} \tag{15}$$

Trajectory generation and execution is extremely simple for such kinematics, as the desired control variable, $\phi(t)$, coincides with the robot's steering angle, $\theta(t)$, or, to be more correct, as the steering angle can be made almost instantaneously equal to the desired orientation. As the next step is to compute the desired orientation in real-time, by looking at the dynamic equation of the optimal robot – expression (15) – and by considering the basic equations of a purely "optimum searching" robot – expression (2) – we can finally write:

$$\phi^s(t) = \tan^{-1}\frac{\dot{y}}{\dot{x}} = \tan^{-1}\frac{\partial I/\partial y}{\partial I/\partial x} = \theta^s(t) \tag{16}$$

where superindex s indicates that these angles are generated by the search module. For the angles produced by the APF module for obstacle avoidance, we will use superindex o. Obviously, for a computer-controlled autonomous vehicle, this continuous trajectory has to be discretized, so that the desired orientation should be:

$$\phi^s_{k+1} \equiv \theta^s_{k+1} = \tan^{-1}\frac{\Delta y_k}{\Delta x_k} \tag{17}$$

where $\Delta x_k = x_{k-1} - x_k$ and $\Delta y_k = y_{k-1} - y_k$ are given by (4).

For obstacle avoidance navigation, the basic idea is exactly the same, as the orientation for the robot should be derived from the normal or the tangent directions given, respectively, by:

$$\phi^o_{k+1} = \hat{n} = \tan^{-1}\frac{\left.\frac{\partial U(x,y)}{\partial y}\right|_{x_k,y_k}}{\left.\frac{\partial U(x,y)}{\partial x}\right|_{x_k,y_k}} \quad ; \quad \phi^o_{k+1} = \hat{\tau} = \hat{n} \pm \pi/2 \tag{18}$$

The ambiguity due to the two tangent components is removed by choosing the one closer to the current robot orientation θ_k. Its implementation, however, may be quite involved, as the potential field function $U(x,y)$ must be properly established, which is not a straightforward problem as mentioned above. In the next section, we introduce an algorithmic approach to the computation of suitable potential functions using a radial structure of sonar sensors. Before proceeding with the technical details, let us give just a brief description of this procedure for computing the steering angle using the illustration in Fig. 6.

Fig. 6 shows a wheeled robot with the optimal kinematics given by (15) for two different obstacles. This vehicle has a ring of N radial sonar sensors, with readings $d_0, d_1, \ldots, d_{N-1}$. The angle θ_k stands for the robot's current orientation and θ_{k+1} for the next one, as computed by our algorithmic procedure and which coincides with the normal orientation to the imaginary radius defined by the sensor with the minimum distance reading. This algorithm tends to align the vehicle's steering angle with the tangents to the potential field curves, which is the optimal trajectory in obstacle avoidance navigation. The number of sensors and the distance between the mobile robot and the obstacles are the key elements in the precision achieved in the alignment of the steering angle with the optimal trajectory.

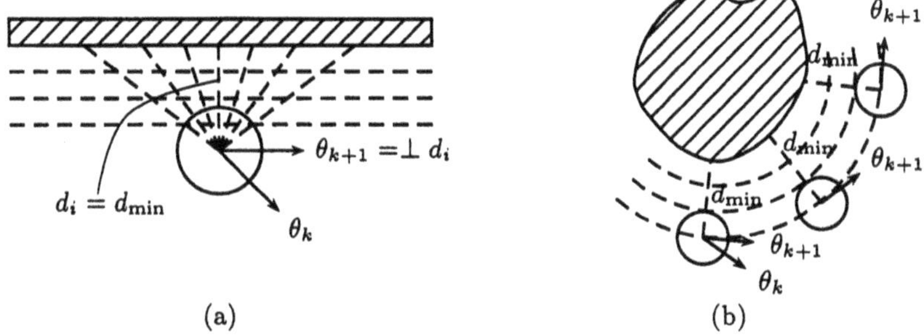

Fig. 6. Two different situations in which the vehicle navigates following tangential component of the potential field. Discontinuous lines represent the potential field

As previously stated, the final steering angle to be applied to the vehicle is the combination of both angles, i.e. the one provided by the sensory source search module, expression (17), and the angle obtained from the obstacle avoidance module, expression (18). The combination of both control angles is discussed in Section 4.5.

4.4 Sensor-based Method for Obtaining Collision Avoidance Functions

In utilitarian navigation, as we already know, the robot is under a continuous drive to optimize certain performance or utility functions, instead of being subject to forces created by artificial potential fields. In the same manner as in APF-based navigation, the key point for a utilitarian robot is to establish its utility functions. As a matter of fact, we have shown that the artificial potential fields do actually conform the performance or utility functions for collision avoidance. Thus, it is a strategic endeavor for a utilitarian robot not to collide with the physical objects existing in its navigation world and, as a consequence, to construct the appropriate collision avoidance functions. Furthermore, these functions must be computed in real-time, as in most cases of practical interest the obstacles, and sometimes the regular, permanent elements of the environment, are dynamic or *a priori* unknown.

Let us state a method for efficient real-time computation of collision avoidance functions, which can be used, indistinctly, for APF-based and utilitarian navigation. The method is based on a ring of range sensors as shown in Fig. 7.

The proposed sensory system has N circularily distributed range or distance sensors. The robot's reference orientation, θ, coincides with the position of sensor s_0. The angular separation between consecutive sensors is $\Delta\theta = 2\pi/N$ rads. The respective distance readings are d_0, d_1, ..., d_{N-1}. By way of an example, let us suppose that the robot is placed in front of a polygonal object, as depicted in Fig. 7. As the generalized robot's coordinates, (x, y, θ), in the navigation plane are known, we can immediately obtain

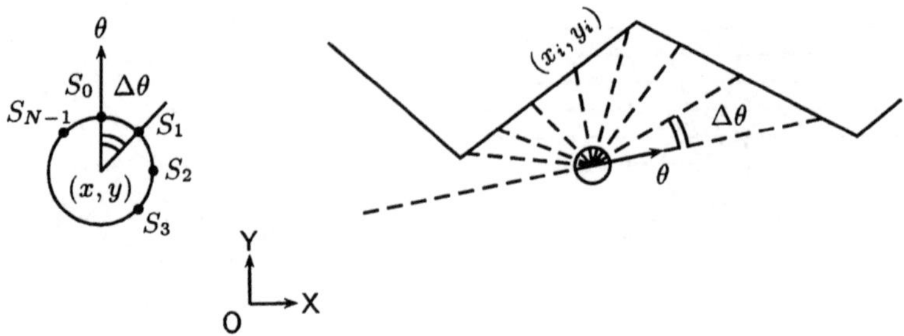

Fig. 7. The sensory system and its application to collision avoidance for a polygonal environment

the coordinates of the points $\{(x_i, y_i)\}$ belonging to the function defined by the environment contour $f(x,y) = 0$ using the sensors readings. Thus, the contour function could be reconstructed from the set of points $\{(x_i, y_i)\}$ by means of any method for function approximation. However, the measurement errors of commercial range sensors, which in the case of sonar and infrared sensors are considerable, as well as the sampling errors produced by the angular resolution of $\Delta\theta$ radians, preclude the use of the restored contour function as a practical navigation function. Fortunately, the part of the physical environment nearest to the robot, which is the most critical one, can be restored with a higher precision, as the resolution produces errors directly dependent on the distance between the robot and the obstacles. Of course, the restored contour function can be improved by increasing the number N of sensors. Although we shall later propose a much more reliable and simpler algorithm that does not need the contour function, let us suppose, for the time being, that the contour function has been properly computed, as shown in Fig. 8, where the contour has a generic curved shape.

Fig. 8. A generic curved environment with a known contour function. Discontinuous lines represent the navigation functions

Fig. 8 also depicts the virtual navigation functions that conserve and reproduce the shape of the environment contour. If the contour function $f(x,y) = 0$ is known, it is straightforward to compute, at every generic point of the navigation plane XOY, the respective navigation function $U(x,y)$ that reproduces the contour shape. Depending on whether the robot is navigating with a normal or tangent strategy, the normal and the tangent steering orientations can be computed directly:

$$\hat{n} = \tan^{-1} \frac{\pm \frac{\partial U(x,y)}{\partial y}}{\pm \frac{\partial U(x,y)}{\partial x}} \quad ; \quad \hat{\tau} = \tan^{-1} \frac{\pm \frac{\partial U(x,y)}{\partial x}}{\mp \frac{\partial U(x,y)}{\partial y}} \tag{19}$$

The main drawback, however, for obtaining these steering angles is the need to know the navigation function $U(x, y)$, which depends on the accuracy of the range sensors and on the angular resolution of the sensor ring, as mentioned above. Thus, we are going to introduce an alternative method that efficiently computes the normal and the tangent steering angles without explicit knowledge of the navigation function $U(x, y)$. Indeed, looking at Fig. 8, we find that at any generic point (x, y) that gives the current position of the robot, the normal direction of the navigation function coincides with the line defined by the shortest distance between the point (x, y) and the contour function $f(x, y) = 0$. As a result of this geometrical property, the normal and the tangent orientations can be obtained straightforwardly, as illustrated in the two examples shown in Fig. 9.

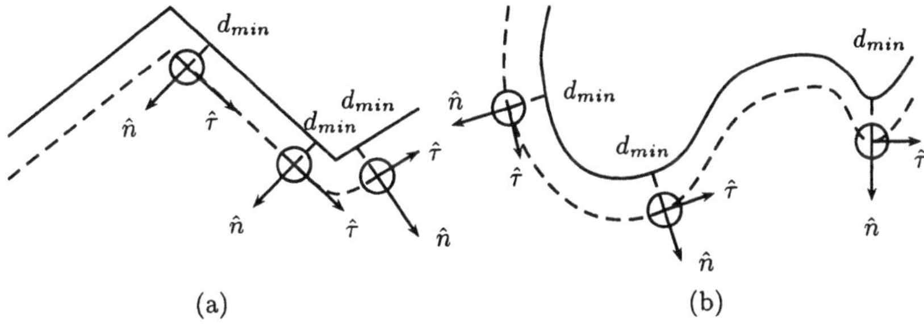

Fig. 9. Two different cases for contour shapes. In (a) a polygonal obstacle is shown. In (b) a curved obstacle is perfectly navigated, provided that the robot continuously measures its distance to the environment

The main practical problem of the method outlined above arises from the finite resolution of the proposed sensory system. Thus, let us consider a polygonal environment like the one shown in Fig. 10. Given the smallest two distances provided by the sensor ring, d_1 and d_2, the true normal and tangent orientation vectors of the contour function can be computed directly as:

$$\hat{n} \triangleq \tan^{-1} \frac{d_1 - d_2 \cos\gamma}{d_2 \sin\gamma} \quad ; \quad \hat{\tau} = \hat{n} \pm \pi/2 \tag{20}$$

where γ is the angle formed by the normal and the radial d_2.

For curved obstacles, the computation of the normal and the tangent steering angles is much more complex and, *de facto*, requires knowledge of

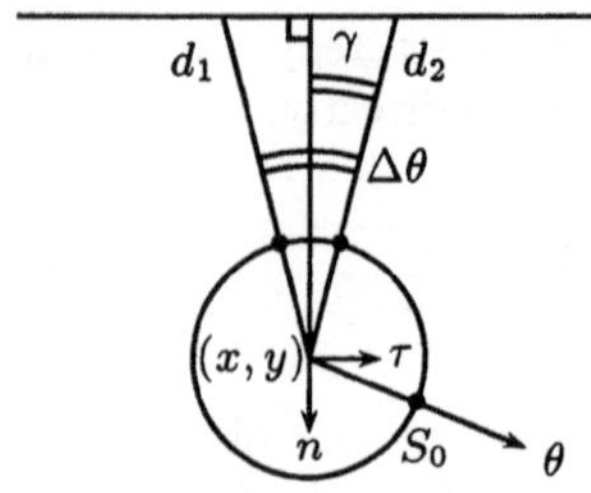

Fig. 10. Computation of the true normal and tangent vectors

the exact expression of the environment contour function $f(x, y) = 0$. Fig. 11 shows the two possible generic cases of concave and convex curves, which illustrate the difficulties in estimating the true normal and tangent steering angles for non-polygonal environments.

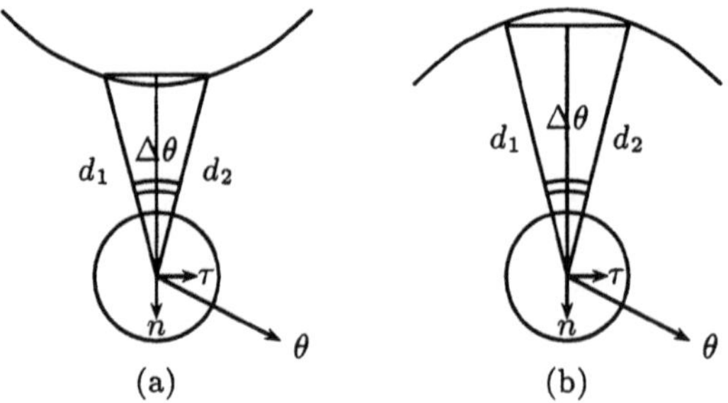

Fig. 11. The two possible cases for curved obstacles. In (a) the convex part of the environment produces a minimum between the smallest two readings, whereas for a concave obstacle such as in (b), there is a maximum

Two steps can be taken to deal with non polygonal environments. The first and simpler one is to apply a polygonal approximation between the smallest two readings of the range sensors. If not efficient enough, the number of sensors should be increased or, even better, a motorized range sensor, scanning as many environment points as needed, could be used.

4.5 Combination of Navigation Actions

The combination of both collision avoidance and searching steering angles is a crucial design issue in the utilitarian navigation, as it determines the relative importance of each control action. Let us suppose that we establish the following linear combination for both navigation actions:

$$\begin{aligned} \theta_{k+1} &= \omega_1 \phi^s_{k+1} + \omega_2 \phi^o_{k+1} \\ \omega_1 + \omega_2 &= 1 \\ 0 \le \omega_1, \omega_2 &\le 1 \end{aligned} \tag{21}$$

where ϕ^o_{k+1} is either the normal component $\hat{n}$ or the tangent component $\hat{\tau}$. Values of ω_1 close to 1 mean that the robot is almost unconcerned about avoiding obstacles, whereas the opposite case of ω_2 approaching 1 implies that the robot's main goal is to avoid obstacles without searching sensory sources. For complex environments, it can be hard to set the right balance between both navigation goals to achieve efficient global navigation. Unknown and unpredictable environments make things worse.

One of the most serious problems for a utilitarian robot is the above-mentioned situation in which the robot is trapped by an obstacle. An *ad hoc* solution would be to apply the normal direction to the robot in order to escape from the obstacle orbit, whenever a complete 360° maneuver has been detected. A more general and efficient navigation strategy is based on the idea of building a "security zone" around the obstacle by applying a distance threshold, beyond which the robot enters a state in which obstacle avoidance navigation is of maximum priority (i.e. $\omega_2 = 1$). This state is activated whenever the minimum distance value provided by the sensor readings is below the threshold. Fig. 12 depicts the two-state automaton controlling the combination of the orientation angles. State S_1 stands for the obstacle avoidance priority state and S_2 the sensory search priority state.

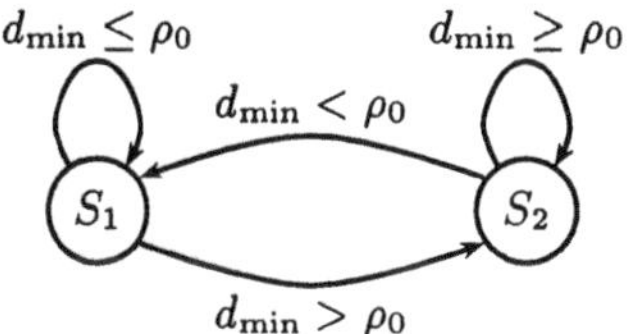

Fig. 12. The automaton changes its current state according to the simple rule explained in the text

Fig. 13 shows several simulated navigation examples, in which we have used a kinematic model for the simulated robot based on our holonomic prototype shown in Fig. 14. The simulated distance sensory system onboard the robot is a ring of 16 ideal ultrasonic sensors with a range of 5 to 110 inches. The sensory source search has been simulated by a simple point source. After thorough testing, the optimum value for the security zone happens to be $\rho_0 = 40$ inches. Note that tangent-based navigation is clearly superior to normal-based orientations.

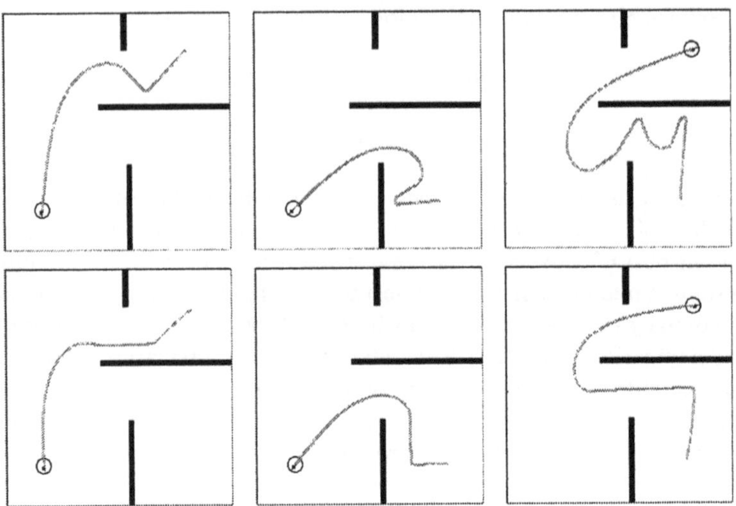

Fig. 13. Several simulated navigation examples using normal-based navigation (above) and tangent-based navigation (below)

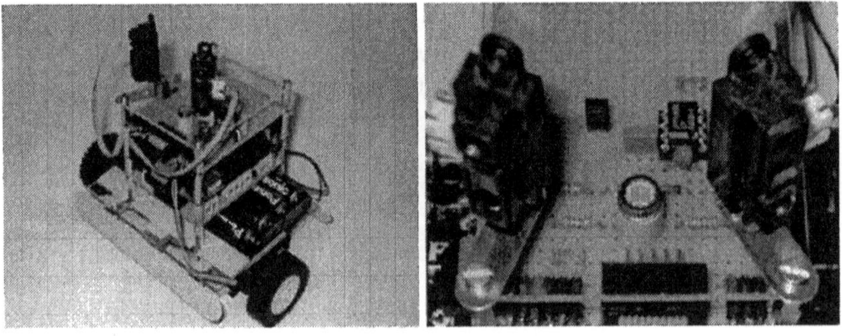

Fig. 14. A view of the holonomic vehicle (left) and a detail of the light source sensor and two range sensors (right)

5 Topological Model Building

Having discussed the low-level navigation problems, we proceed now with the high-level aspects of our autonomous navigation system. In the first place, we approach the problems of detecting and recognizing landmarks to build topological maps and, afterwards, we introduce a method for topological modeling based on fuzzy Petri nets. Finally, the route planning topic is considered.

5.1 Detection of Reference Places

Most of the mobile robot navigation systems based on topological models use sensory information as the main source for detecting reference places. For instance, Kuipers [7] employs a detection method of what are termed *distinctive places* (i.e. reference places) based on production rules about the sensory information coming from the environment. Mataric [8] defines a procedure for detecting reference places (landmarks) using the continuous monitoring of the sensory measurements, which is closely related to navigation based on wall following. Kurz [9] proposes a very similar, although more general detection approach. Serradilla and Maravall [10] introduce a novel concept, the sensory gradient, for detecting reference places. When there are big changes in the sensory gradient operator, the robot detects a *relevant sensory place* – i.e. a reference place –. Nehmzow and Smithers [20] propose a rather different approach. Instead of using sensory information to detect reference places, the robot's internal behavior produced by its control subsystem is used to detecting the reference places.

Our method for detecting reference places is based not on sensory information but on the information provided by the robot's control subsystem. The changes in the behavior modes of the control subsystem generated by the presence of obstacles allow reference places to be created. Unlike Nehmzow and Smithers' work, our method uses several navigation strategies, including wall following, which enriches the world model built by the robot. Other strategies, apart from wall following, are free movement or wandering and detour, where the robot can make decisions about alternative trajectories.

As mentioned above, the changes in the behaviors – some of which are previously defined – of the navigation system that are generated by the structure and physical characteristics of the environment allow the detection of the reference places or landmarks. Fig. 15 shows three reference places. In all of them the robot abandons its current navigation strategy in order to avoid collision with the obstacle in front of it. These situations are recognized as reference places because the robot must abruptly change its trajectory and they can be used to plan future navigation missions.

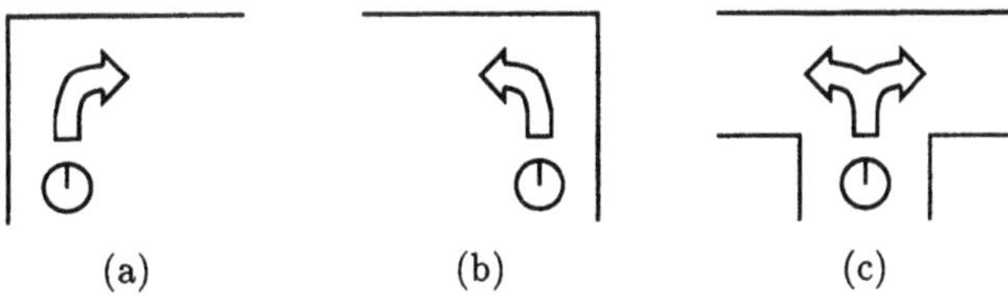

Fig. 15. Detection of reference places

In the three above-mentioned situations the robot must suddenly change its current trajectory turning to the obstacle-free side. In cases (a) and (b), there is only one alternative: to turn right and left, respectively. In case (c), the robot can turn in either directions.

Once the collision situation has been overcome, the robot must continue with a navigation strategy. For the two first cases, any of the aforesaid strategies may be used. However, the wall-following strategy cannot be activated in case (c) because it could lead the robot to an infinite loop state.

When a reference place is detected, some relevant information must be retained for later use. Basically, this information concerns the state of the robot and some features of the environment of the reference place in question. From the standpoint of topological model building, the information about each reference place is divided into two groups: (a) information about the physical place and (b) information about the navigation strategies activated since the last reference place. These two different pieces of information about a reference place are shown in Table 1. The information associated with a reference place and the number of times it has been visited are important for making decisions about the direction and the navigation strategy for future visits to a particular reference place.

Table 1. Information about places and transitions

Reference Places	Transitions
Robot Cartesian coordinates	Places of origin and destination
Robot angular coordinates	Duration
Type of reference place	Exit turn direction
Concavity for calibration	Activated strategy

In summary, the detection of the reference places serves to extract key environment points or landmarks that will be used in both the exploratory missions undertaken for map building and the actual navigation missions. Furthermore, in both exploratory and the navigation missions, it is vital to determine whether or not the detected reference place has been previously visited. During the exploratory phase, this distinction serves to correctly update the environment model or map. During the navigation phase, this distinction is critical for an optimal choice of the navigation alternatives at the robot's current position.

We have tested two methods for classifying any newly detected landmark as either already existing or completely new. The first one uses the distance between the estimated position of the robot obtained through its sensors – usually odometric information – and the position of the reference place under scrutiny in the environment model as only discriminant feature.

On many occasions, the estimation of the robot's position using standard, commercially available sensors can be quite inaccurate, and, except when using simulation, it is not very advisable to rely exclusively on this method.

Therefore, we have also implemented a more robust pattern recognizer based on artificial neural networks and enlarged the discriminant feature vector by adding the sixteen ultrasound readings of our mobile platform. By combining the distance-based recognition method and the neural network classifier, a near one hundred per cent success has been achieved in reference place recognition.

5.2 Recognition of Reference Places

The recognition of reference places by means of distance-based information works appropriately for highly structured environments and when the exact position of the robot is available, which, unfortunately, is not very often the case. Therefore, more robust methods for landmark recognition are necessary [21].

The algorithm described in this chapter employs a network of perceptrons for the landmark recognition task. The only information injected into the perceptrons – i.e. the basic neural units – is the readings of the sonar sensors onboard the robot. This information is extremely simple, making the method very attractive from the computational standpoint. A raw estimation of the robot's position can also be used for the final recognition in the event of a draw among several perceptrons.

Let $p_i \in P$ be a particular, existing reference place. We define s_i as the sensory register of landmark p_i, with s_i being a vector of dimension M and formed by the M robot sonar sensor measurements at place p_i.

The training set $S = \{s_1, \ldots, s_n\}$ is formed by the sensory registers taken during different visits to each individual landmark. Every reference place has at least one sensory register in the training set. We also define a function $f : P \times S \rightarrow [0,1]$ that associates each landmark $p_i \in P$ with the sensory registers $s_i \in S$.

A perceptron with M inputs and a single output is associated with each landmark. Graphically, we can represent the network of perceptrons as shown in Fig. 16. Each landmark holds information related to its position in the environment, links to other connected landmarks and the perceptron.

Every perceptron is trained with the complete training set S in order to be properly activated whenever the input corresponds to a sensory register of the associated landmark. Formally stated, for each perceptron Π_i ascribed to landmark p_i:

$$\Pi_i = f(p_i, s_j) \; \forall s_j \in S \tag{22}$$

Therefore, when a new reference place p_n is detected during the recognition process, the perceptrons activated by the current sensory register are associated with existing landmarks bearing the closest resemblance:

$$P_c = \{p_i\} \; \forall \, \Pi_i = 1 \tag{23}$$

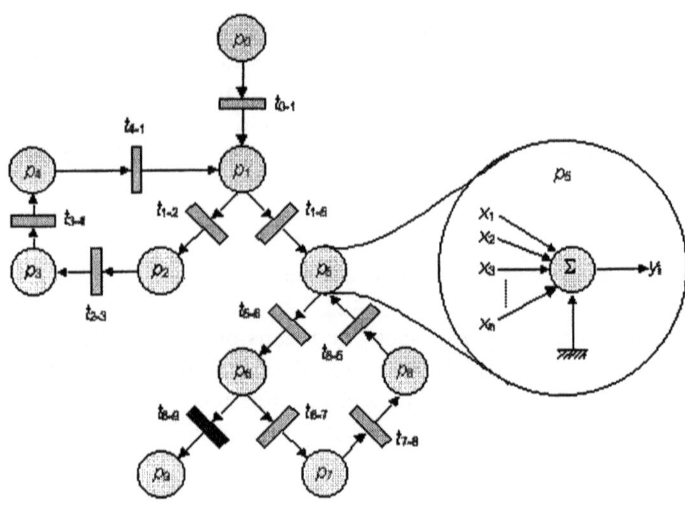

Fig. 16. Net of landmarks with detailed information about a node

Afterwards, the landmark $p_c \in P_c$ nearest to the newly detected place p_n is selected:

$$p_c \mid d_c = min[d(\boldsymbol{p}_i, \boldsymbol{p}_n)] \; \forall p_i \in P_c \tag{24}$$

Finally, a test is carried out to decide whether or not the newly detected and the existing landmarks are the same. This test is performed by applying a distance threshold t_d on d_c.

A positive test indicates that the current landmark is actually one of the existing landmarks, i.e. p_c, and a negative test means that the current landmark p_n is a new reference place. Formally:

$$p = \begin{cases} p_c \text{ iff } d_c < t_d \\ p_n \text{ otherwise} \end{cases} \tag{25}$$

The t_d threshold can be made significantly greater than the threshold used in the distance-based recognition phase, as the sensory information used by the networks of perceptrons has been able to dramatically refine landmark recognition by sharply reducing the search region.

The exact position of the robot is obviously available in our simulation and development environment – unless we introduce simulated errors – and distance-based recognition guarantees a perfect recognition of the reference places. Such accuracy permits an excellent validation of the model-building process and a reliable quality test of the neural network-based recognition method. The thresholds used in distance-based recognition are $t_d = 15$ inches, which gives a slightly higher tolerance than the robot radius of 10 inches, and $t_\delta = \pi/4$, which guarantees a suitable range of orientations for the type of landmarks occurring in our experiments – indoor environments –.

Fig. 17 shows several reference places that have been detected and recognized by just applying the distance-based recognizer. For this example, the threshold relative to the robot orientation has not been applied.

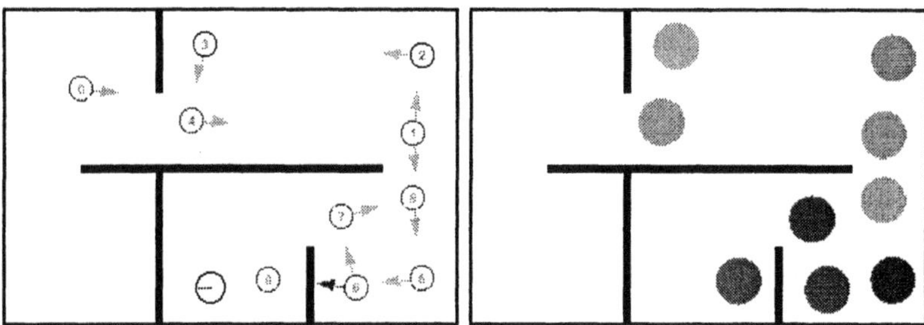

Fig. 17. An example of landmarks detection and recognition employing only the distance-based method

In principle, the recognition of landmarks using the network of perceptrons has the important advantage of being almost independent of the errors in the estimated robot position, as it is based on the sensor readings. However, the results of our first experiments were unsatisfactory, even for simulated environments. The reason for this is quite obvious: the resemblance of the sensors readings for landmarks of similar shape: walls, corners, doors and so on. As regards this problem, just observe landmarks p_2 and p_3 in Fig. 18. Both reference places are of the same type – i.e. corners – although they are different physical landmarks. The sensory registers for both landmarks are extremely similar: small magnitudes on the front and right side and big magnitudes for the rest. Similarly, landmarks p_1 and p_4 are of the same kind – i.e. bifurcations – and, therefore, have the same problem. In such situations, different physical landmarks of the same type cannot be correctly discriminated even with a sophisticated recognition algorithm.

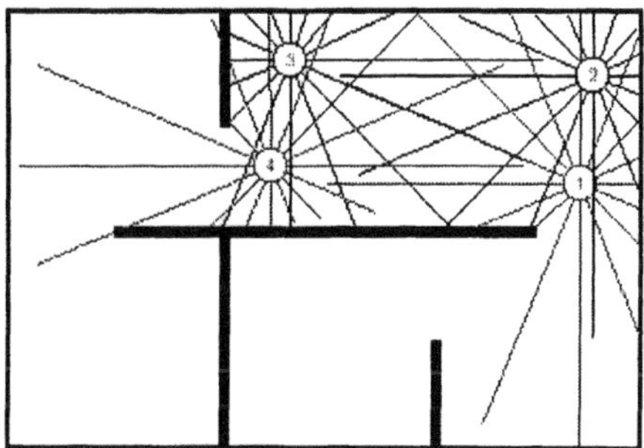

Fig. 18. Average sensory register for two similar landmarks

In order to overcome this serious problem and to increase the difference among the measurements produced by the set of landmarks, the coordinates of the robot's sensors are rotated to coincide with the robot's orientation θ. The formal expression of this base change is:

$$s_j = s'_i \mid j = \left(\frac{\theta}{2\pi/M} + i \right) \bmod M \tag{26}$$

where s_j is the sensor readings after transforming the original readings s'_i; M is the number of sensors – which in our case have been placed in a totally symmetrical radial configuration – and $i, j \in [0, M)$ are indexes for each particular sensor.

Fig. 19 shows the landmarks detected and recognized by a network of perceptrons with thresholds u_d = 30 inches – i.e. three times the robot radius – for the environment of Fig. 17. During the exploration of this particular environment, a 100% success ratio was obtained, for both new and already existing landmarks – i.e. revisited – landmarks: reference places p_1, p_5 and p_6.

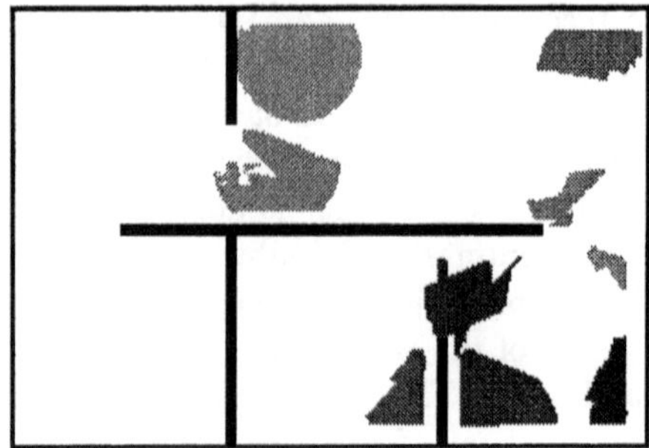

Fig. 19. Regions corresponding to the landmarks detected and recognized using the perceptron network

Whenever a landmark cannot be recognized with absolute accuracy using the perceptron network, it is advisable to simultaneously employ the distance-based method in order to get an estimation of the robot's physical position and its distance to the nearest existing landmark. In spite of positional errors, the distance-based recognizer provides a valuable second vote for landmark detection and recognition. Eventually, in the worst case of the robot not being able to recognize a particular landmark, the only consequence is that the environment model will create a new landmark instead of employing the existing landmark associated with the region in question.

If a successful recognition has been performed using the distance-based method as the background recognition process, then the respective sensory register has to be incorporated to the perceptron network training set. This new training element is incorporated immediately after the distance-based landmark recognition has been accomplished in order (a) to avoid future errors in landmark recognition by the perceptron network and (b) to update all the perceptrons with the new element. As the number of both the training instances and perceptrons is relatively small, training time is almost insignificant.

The formal description of the algorithm for the combined recognition process – i.e. the perceptron network and the distance-based recognizer – for both (1) recognition of a new landmark p_n and (2) the introduction of new instances in the training set S is as follows.

1. If the reference place or landmark p_n can be recognized by means of (25), then the recognized landmark corresponds to landmark p_c, i.e. $p = p_c$.
2. Else, determine whether or not the reference place p_n can be recognized using a distance-based method.
   (a) If it is recognized, then the reference place is $p = p_c$ and $f(p_c, s_n) = 1$ and $f(p_i, s_n) = 0 \; \forall p_i \neq p_c$ are consequently added to the training set $S = S \cup s_n$ and the set of perceptrons Π_i are trained according to (22).
   (b) Else, the new landmark is $p = p_n$ and $f(p_n, s_n) = 1$ and $f(p_i, s_n) = 0$ $\forall p_i \neq p_n$ are consequently added to the training set $S = S \cup s_n$ and the set of perceptrons are trained according to (22).

The justification of a rather simple neuron prototype as the perceptron is twofold. In the first place, we have found through experimentation that more complex neural structures, for instance the multilayer perceptron trained with a well-tested learning algorithm, perform quite similarly to the network of perceptrons used in our work. The advantages in implementation, mainly storage requirements and computation time, for the simplest neural network are evident. In the second place, the computational burden in the training phase is almost insignificant. For environments with, let us say, 50 reference places and an average of two samples for each landmark, the total computation time is always lower than 50 ms. Such fast performance is absolutely crucial for real-time navigation, as the training phase can be executed by means of a low priority process, which is almost compulsory for our control architecture [28,29]. Last but not least, the reconfiguration process in the number of neuron units due to the incorporation of a new landmark is immediate, as a new perceptron is added to the network whenever a new landmark is detected and the enlarged neural network is easily re-trained. For any other neural network architecture other than the one proposed, a maximum number of outputs should be defined at the beginning of an exploratory mission – i.e. the phase in which the robot creates the map of the environment by detecting landmarks – or it would be necessary to dynamically modify the outputs of the neural network. For multilayer perceptron networks the process could be even more involved, because the hidden layers are much more difficult to update for each new neuron unit added to the output layer. An additional advantage of the perceptron network is that the network parameters are easily updated whenever a new landmark is introduced into the environment model.

5.3 Fuzzy Petri Nets for Topological Modeling

Unlike other topological modeling methods, our approach admits two possible interpretations. On the one hand, the environment model can be viewed as a set of places accessible by the robot by means of navigation strategies. On the other hand, the inverse interpretation is also correct: the world model is formed by a set of robot states determined by the navigation strategies related to each of the reference places detected in the environment.

This double interpretation and the fact that the links between reference places are unidirectional enable the use of Petri nets [22] for model representation. More specifically we are going to introduce fuzzy Petri nets [23–25], because the application of inference tools is very useful for route planning.

The reference places correspond to net nodes or places and the robot states – generated by the navigation strategies – to net transitions. A Petri net can be interpreted either as a sequence of places or as a sequence of transitions. Regarding our navigation system, the second interpretation is more interesting, because the quantitative information about the exact position of the reference places is not necessary and, at the same time, the robot can reach these places using navigation strategies based on world features rather than on physical positions. Accordingly, the routes can be interpreted as *navigation suggestions* as in the papers by Agre and Chapman [26] and Payton [27], and the route towards a goal place can be established through navigation strategies activated at each reference place. A possible example of route planning is as follows: "go ahead towards the wall, then turn right and take the first turning on the left".

A Petri net can be defined as the tuple $PN = \{P, T, F, W, M_0\}$, where $P = \{p_0, \ldots, p_m\}$ is the finite set of places; $T = \{t_0, \ldots, t_n\}$ is the finite set of transitions; $F \subseteq (P \times T) \cup (T \times P)$ is the set of arcs; $W : F \rightarrow \{1, 2, 3, \ldots\}$ is the weighting function for arcs and $M_0 : P \rightarrow \{0, 1, 2, \ldots\}$ is the initial marking of the net that may or may not exist. As an additional restriction, we include $P \cap T = \emptyset$ and $P \cup T \neq \emptyset$.

For our navigation system, the set of places P is formed by robot positions and orientations and the set of transitions T consists of the navigation strategies such that a transition $t_q = t_{i-j}$ drives the robot from place p_i to place p_j. The weighting function W assigns value 1 to each arc linking places and transitions, as each transition determines the movement of the robot between two places. The net marking depends on the place where the robot is situated. Therefore, the firing of transition t_{i-j} depends on the position of the robot.

Fuzzy Petri nets are derived from Petri nets and can be described as the tuple $FPN = \{P, T, D, I, O, f, \lambda, \alpha, \beta\}$, where P and T stands for the same as in a PN; $D = \{d_0, \ldots, d_m\}$ is a finite set of propositions such that $P \cap T \cap D = \emptyset$ and $|P| = |D|$; $I : T \rightarrow P^{\infty}$ is an input function that defines a correspondence between the transitions and their input places; $O : T \rightarrow P^{\infty}$ is an output function that stands for the correspondence between the transitions and their output places; $f : T \to [0,1]$ is an association function that establishes a correspondence between transitions and the real-valued segment $[0,1]$ and determines the certainty of each transition; $\lambda : T \to [0,1]$ is an association function establishing a correspondence between the transitions and the interval $[0,1]$ and sets the activation threshold for each transition; $\alpha : P \to [0,1]$ is a function that establishes a correspondence between the places and the interval $[0,1]$ and $\beta : P \to D$ is a bijective correspondence between places and propositions.

When fuzzy Petri nets are used for knowledge representation, the set of propositions D defines the antecedent and the consequent of the rules, i.e. the transitions. In our navigation system, the propositions d_i are statements such as "the robot is at place p_i". The truth-value of this proposition, as in a general-purpose fuzzy Petri net, is given by $\alpha(p_i)$. The input function I applied to a transition t_{i-j} will provide the set of places p_i from which it can be fired. The output function O applied to the same transition t_{i-j} will be result of the set of places p_j accessible from p_i, which, in our case is constrained to a single place. The value given by the function f over the transition t_{i-j} is interpreted as the cost associated with this transition, and it is determined by factors like the execution time of the respective navigation movement and by navigation strategy complexity. The value given by the function λ applied to the transition indicates the activation threshold of this transition and its value is estimated from experience.

5.4 Route Planning Algorithm

After building a topological model of the environment using the fuzzy Petri net concept, the next step is to devise mechanisms for route planning. Any fuzzy Petri net-based algorithm for route planning must propagate the certainty value of places and transitions over the net. The initial choice of the place values guarantees that, after propagation, the least-valued transitions are the ones that lead faster to the goal.

It should be relevant emphasized that thanks to the concurrent nature of Petri nets, the paths that lead to several goal places can be directly planned by the algorithm merely by appropriately initializing the net places. Furthermore, by updating the transition values as is described later, the planned routes are ordered in such a way that the first one leads to the nearest goal place. Obviously, path planning for a single goal place is a particular case.

The values of places and transitions propagate from the goal places towards the robot's position. Therefore, the respective value is stored for each transition and it is possible to explicitly establish the longer paths produced by either more expensive transitions or by a higher number of transitions. Therefore, the right decision at each place is to choose the transition with the minimum value. Note that if the place and transition values were propagated in the opposite direction – i.e. from the starting place towards the goal places – then it would not be possible to evaluate the optimum transition at each place of the net.

The transition values do not necessarily have to be initialized, as they are updated at each propagation step. On the contrary, the goal places must have lower values than the other places in order to guarantee that the least-valued places lead to the goal more rapidly.

Each propagation step consists of three phases. The first one corresponds to what in fuzzy reasoning is known as *composition*. In this phase, the transition values from each node are updated by accumulating the estimated cost in crossing each transition. Each transition of the net necessarily leads to a single place because the net is an environment model. Therefore, the updated value of each transition is obtained by adding the value of the place where the transition ends and the estimated cost of the transition. The transitions ending in a goal place have a value equal to their estimated costs. For the other transitions the estimated costs are added to the previous path. Obviously, the transition values cannot be higher than 1. The transition values $f(t_i)$ are updated as follows:

$$f(t_i) = \min\left[1, C(t_i) + \alpha(p_j)\right] \tag{27}$$

where p_j is the goal place for transition t_i and $C(t_i)$ is the weighted cost associated with firing transition t_i. This cost is estimated as:

$$C(t_i) = \frac{c_i}{\sum_{j=0}^{i} c_j} \tag{28}$$

where c_i is the real cost for transition t_i, measured as a combination of the time consumed in physically performing this transition and the complexity of the behaviors emerging during the transition.

The second stage – *thresholding* – is very common in fuzzy reasoning. It assigns value 1 to the transition t_i, such that its computed value does not exceed the activation threshold, so that the number of rejected transitions for navigation planning can be easily controlled:

$$f(t_i) = 1;\ \text{iif } f(t_i) > \lambda(t_i) \tag{29}$$

Finally, the third phase, usually known as *inference* in fuzzy reasoning, updates the values of the places depending on their transitions. The final value determines the optimum place. If the value of this path has been previously computed then the place value does not change. However, if the minimum value of all the transitions leaving the place is less than the value of the actual place then the latter value must be updated with the value of the transition. The values $\alpha(p_i)$ of the net are updated as follows:

$$\alpha(p_i) = \min\left[\alpha(p_i), f(t_i)\right] \tag{30}$$

where the transitions t_j are all those leaving the place p_i – i.e. $p_i \in I(t_j)$ –.

The values of all places and transitions must continue to be propagated until the initial place is no longer equal to 1. This means that a path linking the initial place with its nearest goal place has been obtained. Furthermore, by actually building the paths linking the initial place to the goal places, the path found is of minimum cost.

After propagation ends, the transitions and the places store the information needed to make decisions about the control strategies for the navigation of the robot towards the goal. The lease-valued transition at each place is taken. Furthermore, the transitions with value 1 are rejected.

The pseudo-code of the above propagation algorithm is as follows:

1. Initialize the goal places p_g so that $\alpha(p_g) = 0$ and the other places p_i give $\alpha(p_i) = 1$.
2. As long as $\alpha(p_s) = 1$ holds for the initial or starting place p_s:
   (a) Update the transitions t_i by means of (27).
   (b) Update the transitions t_i as in (29).
   (c) Update the places p_i by applying (30).

6 Experimental Results

6.1 Simulation Results

First, we discuss several examples of model building developed using our own mobile robot NOMAD-200 simulation environment. Special care has been taken to work with realistic simulated situations, both regarding the sensors of the robot – 16 ultrasound ring and odometric information – and the environment. In these examples, the simulated world is an office-like environment measuring about 50 square meters.

Fig. 20 shows a Petri net built by the system on the simulated environment. The reference places detected by our system, which correspond to the net places, are represented by circles with their reference number. Place p_0 is the initial place of the net, from which the robot starts its exploration of the environment and it has been marked with the number 0 at the top left-hand side. The lines linking the places stand for the transitions.

Fig. 21 shows two robot trajectories, starting at place p_5 and ending in the bottom left-hand corner. In (a) the goal place is p_{20} and the places crossed by the robot are: p_6 with a free-movement strategy; p_{17} with a left-hand wall-following strategy and p_{18}, p_{19} and p_{20} with a free-movement strategy.

The $f(t_{i-j})$ values of Table 2 explain the robot trajectory. From place p_5, the robot's only alternative is to go to p_6. From this place, it is possible either to go to place p_7 with $f(t_{6-7}) = 1.0$ and a free-movement strategy or to go to place p_7 with $f(t_{6-7}) = 0.106055$ and a left-hand wall-following strategy. As the second is the minimum cost alternative, the robot uses this transition. Afterwards there is a single alternative at each place, always by means of a free-movement strategy.

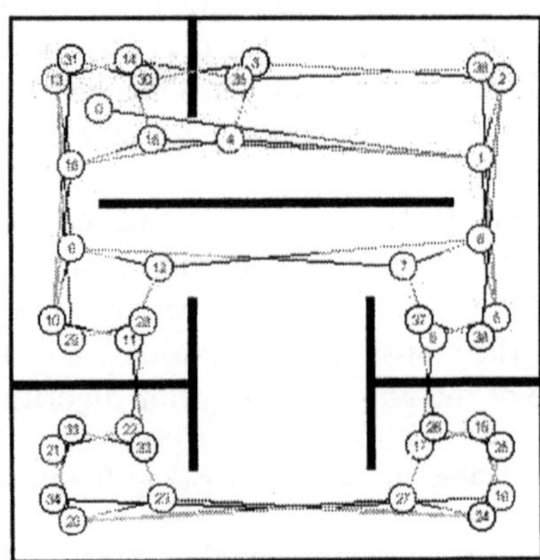

Fig. 20. Model of the environment obtained with the free-movement and wall following strategies

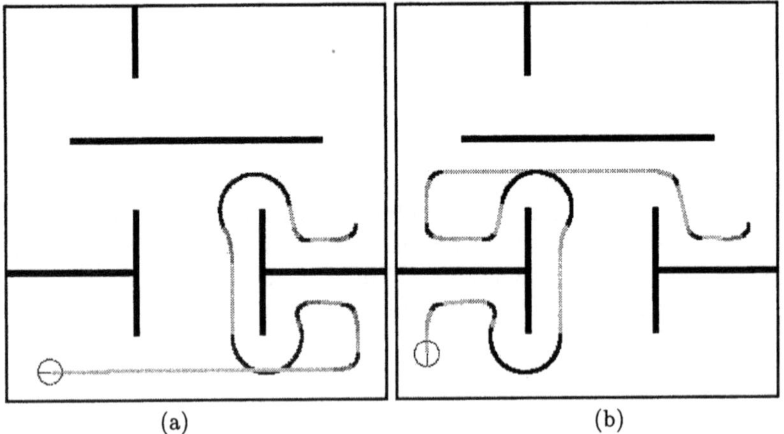

Fig. 21. Robot trajectories: (a) from p_5 to p_{20} and (b) from p_5 to p_{34}

In Fig. 21 (b) the goal place is p_{34} and the places crossed by the robot are: p_6, p_7, p_9, p_{10} and p_{11} with a free-movement strategy; p_{32} with right-hand wall-following strategy and p_{33} and p_{34} with a free-movement strategy.

Table 3 gives the values of $f(t_{i-j})$ obtained by our algorithm. For this case, place p_6 is chosen instead of p_7, because $f(t_{6-7}) = 0.122264$ and $f(t_{6-17}) = 1.0$. Another important decision occurs at place p_9, which, in this case, is to follow the path to p_{10}, as $f(t_{9-10}) = 0.085341$ and $f(t_{9-13}) = 0.102383$. The last important decision occurs at p_{11} and, because $f(t_{11-32}) = 0.069075$ and $f(t_{11-12}) = 0.103076$, the wall on the right of the robot is followed to reach place p_{32}. Following the only alternatives for the subsequent places, the robot finally arrives at the goal place p_{34}.

Table 2. The $f(t_{i-j})$ values for the trajectory from p_5 to p_{20}

Transition	$f(t_{i-j})$	Transition	$f(t_{i-j})$	Transition	$f(t_{i-j})$
$5 \rightarrow 6$	0.113172	$27 \rightarrow 24$	0.062213	$20 \rightarrow 21$	0.033987
$6 \rightarrow 17$	0.106055	$34 \rightarrow 24$	0.090440	$21 \rightarrow 22$	0.026306
$17 \rightarrow 18$	0.053048	$24 \rightarrow 25$	0.052837	$22 \rightarrow 23$	0.017729
$18 \rightarrow 19$	0.045602	$25 \rightarrow 26$	0.045269	$23 \rightarrow 20$	0.010053
$19 \rightarrow 20$	0.038038	$26 \rightarrow 27$	0.037715	Others	1.0
$23 \rightarrow 24$	0.082065	$27 \rightarrow 20$	0.029695		

Table 3. The $f(t_{i-j})$ values for the trajectory from p_5 to p_{34}

Transition	$f(t_{i-j})$	Transition	$f(t_{i-j})$	Transition	$f(t_{i-j})$
$5 \rightarrow 6$	0.129381	$12 \rightarrow 9$	0.095279	$30 \rightarrow 31$	0.109952
$6 \rightarrow 7$	0.122264	$9 \rightarrow 10$	0.085341	$31 \rightarrow 10$	0.101367
$7 \rightarrow 9$	0.114471	$2 \rightarrow 3$	0.139749	$10 \rightarrow 11$	0.077547
$9 \rightarrow 13$	0.102383	$3 \rightarrow 4$	0.117285	$11 \rightarrow 32$	0.069075
$13 \rightarrow 14$	0.119200	$4 \rightarrow 16$	0.109152	$32 \rightarrow 33$	0.016011
$14 \rightarrow 15$	0.111066	$16 \rightarrow 10$	0.093346	$33 \rightarrow 34$	0.007337
$11 \rightarrow 12$	0.103076	$3 \rightarrow 30$	0.135354	Others	1.0

6.2 Real Navigation Experiments

The navigation experiments reported below have been carried out on the premises of the Faculty of Computer Science, at the Universidad Politécnica de Madrid. Fig 22 shows two pictures of some of the Faculty premises where the experiments took place. Note the variety of obstacles and objects that are present: chairs, tables, furniture, people, etc.

Fig. 22. Two pictures of the testing environment

Fig. 23 shows two trajectories performed by the robot in the office that appears correspondly the first picture. The physical objects have been represented using the sonar readings. Thc first trajectory was obtained by activating just three basic robot behaviors: *advance*, and *lateral* and *frontal obstacle avoidance.* The second trajectory was performed with a *right-hand wall following* navigation strategy. In this trajectory, the robot has mantained a constant distance to any lateral object or obstacle placed on its right-hand side, until it reaches the door.

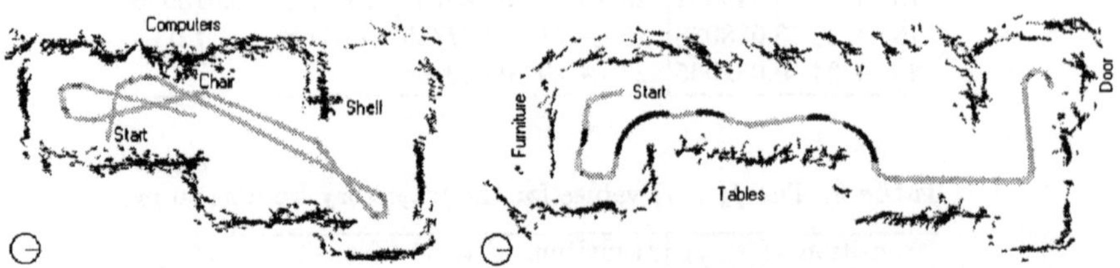

Fig. 23. Two instances of trajectories performed by the NOMAD-200 platform

As for the map building and route planning modules, a topological model created by the NOMAD-200 is shown Fig. 24.

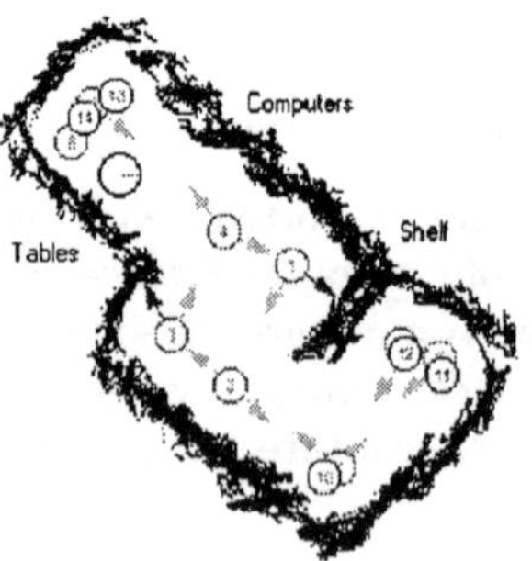

Fig. 24. Topological model built by the navigation system

Fig. 25 shows the execution of two route planning instances. Tables 4 and 5 represent the respective transitions between the consecutive landmarks visited by the robot during the execution of each route.

Table 4. The $f(t_{i-j})$ values for the trajectory to p_8 or p_{14}

Transition	$f(t_{i-j})$	Transition	$f(t_{i-j})$	Transition	$f(t_{i-j})$
$1 \rightarrow 2$	0.190164	$2 \rightarrow 3$	0.139649	$3 \rightarrow 4$	0.140599
$1 \rightarrow 5$	0.377530	$2 \rightarrow 10$	1.000000	$3 \rightarrow 8$	0.103507

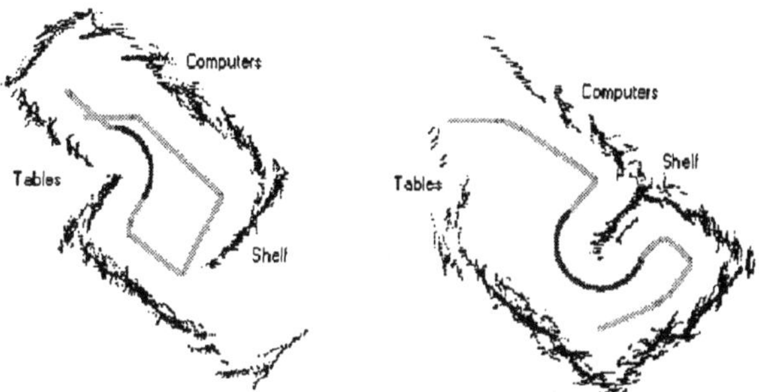

Fig. 25. Left: Path to landmark p_8 or to landmark p_{14}. Right: Path to to landmark p_7

Table 5. The $f(t_{i-j})$ values for the trajectory to p_7

ransition	$f(t_{i-j})$	Transition	$f(t_{i-j})$
$1 \rightarrow 2$	0.248690	$5 \rightarrow 6$	0.096124
$1 \rightarrow 5$	0.207966	$6 \rightarrow 7$	0.060639

7 Conclusions

A hybrid navigation system has been described. The modularity of the control architecture proposed and the generalized concept of behavior have made it feasible to integrate low-level and high-level navigation modules. The novel idea of *utilitarian* navigation has also been introduced. Utilitarian navigation transforms the Artificial Potential Field paradigm for collision avoidance into a problem of functional optimization and allows the implementation of local, adaptive navigation for obstacle avoidance and sensory source search or evasion by means of gradient operators. Another innovation is the generation of steering angles tangent to the APF or, for that matter, to the utility functions. A holonomic vehicle has been built and tested to illustrate the theoretical and practical interest of utilitarian navigation.

At the global navigation level, two basic problems have been considered: (1) map building and (2) route planning. For autonomous map building, a landmark detection method based on a network of perceptrons has been introduced. Afterwards, a dynamic learning-based method for topological modeling of environments, based on fuzzy Petri nets, has been proposed. This novel method for topological map building can be interpreted in two ways: (1) as a set of successive reference places of the environment and (2) as a sequence of robot control strategies or behaviors. The main advantage of the proposed method lies in the second interpretation; i.e. using the changes in the robot control strategies or internal behaviors – states –, rather than sensory information leads to independence from the particular sensory equipment of the robot, which is very interesting for map building with unreliable or noisy sensory information.

For route planning, an algorithm based on fuzzy Petri nets has been introduced. This algorithm is very efficent from the computational standpoint and guarantees an optimum path from the starting position to the final goal. In particular, the important issue of the propagation of the values associated with the topological map of the environment in the fuzzy Petri net has been discussed.

Experimentation with complex, realistic simulated environments was the first step before implementating the hybrid navigation system on a NOMAD-200 mobile platform. The proposed system has been fully implemented on the NOMAD-200 and succesfully tested in real-life environments.

References

1. Kortenkamp, D., Bonasso, R.P., Murphy, R. (1998) Artificial Intelligence and Mobile Robots: Case Studies of Successful Robot Systems. AAAI Press / MIT Press, Menlo Park, California
2. Brooks, R.A. (1986) A robust layered control system for a mobile robot. IEEE J. of Robotics and Automation, **2**(1), 230–237
3. Arkin, R. A. (1998) Behaviour-based Robotics. MIT Press, Cambridge, Massachusetts
4. Tunstel, E. (2001) Ethology as an inspiration for adaptive behavior synthesis in autonomous planetary rovers. Autonomous Robots, **11**, 333–339
5. Moravec, H. P., Elfes, A. (1985) High resolution maps from wide angle sonar. Proc. of IEEE Int. Conf. on Robotics and Automation, 116–121
6. Cox, I. J. (1991) Blanche — An experiment in guidance and navigation of an autonomous robot vehicle. IEEE Trans. on Robots and Automation, **7**(2), 193–204
7. Kuipers, B. J. (2000) The spatial semantic hierarchy. Artificial Intelligence, **119**, 191–233
8. Mataric, M. J. (1992) Integration of representation into goal-driven behavior-based robots. IEEE Trans. on Robotics and Automation, **8**(3), 304–312
9. Kurz, A. (1996) Constructing maps for mobile robot navigation based on ultrasonic range data. IEEE Trans. on Systems, Man and Cybernetics—Part B: Cybernetics, **26**(2), 233-242
10. Serradilla, F., Maravall, D. (1997) Cognitive modelling for navigation of mobile robots using the sensory gradient concept. In F. Pichler and R. Moreno-Díaz, (eds.), Computer Aided Systems Theory, LNCS 1333. Springer-Verlag, Berlin, 273-284
11. Baker, T.P., Shaw, A. (1989) The Cyclic Executive Model and Ada. J. of Real-Time Systems, **1**(1), 7–25
12. Burns, A., Hayes, N., Richardson, M.F. (1995) Generating feasible cyclic schedules. Control Eng. Practice, **3**(2), 152-162
13. De Lope, J. (1998) Modelado de entornos con técnicas basadas en Redes de Petri Borrosas para la exploración y planificación de robot autónomos. Ph.D. Thesis Dissertation. Department of Artificial Intelligence, Universidad Politécnica de Madrid

14. Khatib, O. (1986) Real-time obstacle avoidance for manipulators and mobile robots. Int. J. of Robotics Research, **5**(1), 90–98
15. Krogh, B.H., Thorpe, C.E. (1986) Integrated path planning and dynamic steering control for autonomous vehicles. Proc. of IEEE Int. Conf. on Robotics and Automation, 1664–1669
16. Koren, Y., Bornestein, J. (1991) Potential field methods and their inherent limitations for mobile robot navigation Proc. of IEEE Int. Conf. on Robotics and Automation, 1398–1404
17. Maravall, D., De Lope, J. (2002) Integration of Artifical Potential Field Theory and Sensory-based Search in Autonomous Navigation. Proc. of the IFAC'2002
18. Latombe, J-C. (1991) Robot Motion Planning. Kluwer Academic, Boston, Massachusetts
19. Adams, M.D. (1999) High speed target pursuit and asymptotic stability in mobile robotics. IEEE Trans. on Robotics and Automation, **15**, 230–237
20. Nehmzow, U., Smithers, T. (1991) Using motor actions for location recognition. Proc. of the First European Conf. on Artificial Life, 96–104
21. De Lope, J., Maravall, D. (2001) Landmark recognition for autonomous navigation using odometric information and a network of perceptrons. In J. Mira and A. Prieto, (eds.), Bio-Inspired Applications of Connectionism, LNCS-2085. Springer-Verlag, Berlin, 451–458
22. Murata, T. (1989) Petri Nets: Properties, analysis and applications. Proc. of the IEEE, **77**(4), 541–580
23. Looney, C.L. (1988) Fuzzy Petri Nets for rule-based decisionmaking. IEEE Trans. on Systems, Man, and Cybernetics, **18**(1), 178–183
24. Chen, S. M., Ke, J. S., Chang, J. F. (1990) Knowledge representation using Fuzzy Petri Nets. IEEE Trans. on Knowledge and Data Engineering, **2**(3), 311–319
25. Yu, S.K. (1995) Comments on "Knowledge representation using Fuzzy Petri Nets". IEEE Trans. on Knowledge and Data Engineering, **7**(1), 190–191
26. Agre, P. E., Chapman, D. (1991) What are plans for? In P. Maes (ed.), Designing Autonomous Robots. MIT Press, Cambridge, Massachusetts, 17–34
27. Payton, D. W. (1991) Internalized Plans: A representation for action resources. In P. Maes (ed.), Designing Autonomous Robots. MIT Press, Cambridge, Massachusetts, 89–103
28. De Lope, J., Maravall, D., Zato, J. G. (1998) Topological modeling with Fuzzy Petri Nets for autonomous mobile robots. In A.P. del Pobil, J. Mira and M. Ali, (eds.), Task and Methods in Applied Artificial Intelligence, LNCS-1416. Springer-Verlag, Berlin, 290–299
29. Maravall, D., De Lope, J., Serradilla, F. (2000) Combination of model-based and reactive methods in autonomous navigation. Proc. of the IEEE Int. Conf. on Robotics and Automation, 2328–2333

14. Khatib, O. (1986). Real-time obstacle avoidance for manipulators and mobile robots. Int. J. of Robotics Research, 5(1), 90–98.
15. Krogh, B.H., Thorpe, C.E. (1986) Integrated path planning and dynamic steering control for autonomous vehicles. Proc. of IEEE Int. Conf. on Robotics and Automation, 1664–[illegible]
16. Koren, Y., Borenstein, J. (1991) Potential field methods and their inherent limitations for mobile robot navigation. Proc. of IEEE Int. Conf. on Robotics and Automation, 1398–1404.
17. [illegible] and Sensory-based [illegible] navigation. Proc. of the [illegible]
18. Latombe, J-C. (1991) Robot Motion Planning. Kluwer Academic, Boston, Massachusetts.
19. [illegible] High speed target pursuit and asymptotic stability in mobile robotics. IEEE Trans. on Robotics and Automation, [illegible], 256–2[illegible]
20. [illegible] Proc. of [illegible]
21. [illegible]

Line-Crawling Robot Navigation: A Rough Neurocomputing Approach

J.F. Peters[1], T.C. Ahn[2], M. Borkowski[1], V. Degtyaryov[1], S. Ramanna[1]

[1]Computational Intelligence Laboratory, Department of Electrical and Computer Engineering, University of Manitoba, Winnipeg, Manitoba R3T 5V6, Canada,
[2]Intelligent Information Control & System Lab, School of Electrical & Electronic Engineering, Won-Kwang University, 344-2 Shinyong-Dong, Iksan, Chon-Buk, 570-749, Korea
jfpeters@ee.umanitoba.ca

Abstract. This chapter considers a rough neurocomputing approach to the design of the classify layer of a Brooks architecture for a robot control system. This paradigm for neurocomputing that has its roots in rough set theory, works well in cases where there is uncertainty about the values of measurements used to make decisions. In the case of the line-crawling robot (LCR) described in this chapter, rough neurocomputing works very well in classifying noisy signals from sensors. The LCR is a robot designed to crawl along high-voltage transmission lines where noisy sensor signals are common because of the electromagnetic field surrounding conductors. In rough neurocomputing, training a network of neurons is defined by algorithms for adjusting parameters in the approximation space of each neuron. Learning in a rough neural network is defined relative to local parameter adjustments. Input to a sensor signal classifier is in the form of clusters extracted from convex hulls that "enclose" similar sensor signal values. This chapter gives a fairly complete description of a LCR that has been developed over the past three years as part of a Manitoba Hydro research project. This robot is useful in solving maintenance problems in power systems. A description of the locomotion features of a line-crawling robot and the basic architecture of a rough neurocomputing system for robot navigation are given. A brief description of the fundamental features of rough set theory used in the design of a rough neural network is included in this chapter. A sample sensor signal classification experiment using a recent implementation of rough neural networks is also given.

1 Introduction

Studies of neural networks in the context of rough sets [1-24] and granular computing [13-14, 24-30] are extensive. An intuitive formulation of information granulation was introduced by Zadeh [31]. Practical applications of rough neurocomputing have recently been found in predicting urban highway traffic volume [6], speech analysis [13], classifying the waveforms of power system faults [4], signal analysis [21], assessing software quality [15], control of autonomous vehicles [16-17], EEG analysis [32], and handwriting recognition [33]. In its most general form, rough neurocomputing provides a basis for granular computing. A rough mereological approach to rough neural network springs from an interest in knowledge synthesized (induced) from successive granule approximations performed by neurons [24].

Various forms of rough neural networks work quite well in solving the problem of classifying noisy sensor signals. This problem is intense in the case of a line-crawling robot (LCR) designed to navigate along high-voltage power lines because of the electromagnetic field surrounding the conductors. The set approximation paradigm from rough set theory [34-36] provides a basis for the design of rough neural networks (RNNs). This chapter gives a fairly detailed overview of the design of a LCR. The LCR control system is patterned after the Brooks' subsumption architecture for a robot control system. This architecture is organized into separate layers, each with its own control function and with the ability to subsume the control functions of lower layers. This chapter focuses on the use of a RNN in the design of the classify layer of the LCR as a step towards the solution of the LCR navigation problem. This chapter is organized as follows. An overview of the LCR is presented in Section 2. A description of the architecture of the LCR control system and LCR navigation problem is given in Section 3. A brief presentation of rough set methods underlying the design of a rough neural network for the LCR is given in Section 4. The architecture of the rough neural network built into the classify layer of the LCR is described in Section 5. Experimental results using a rough neural network to classify sample LCR proximity sensor values is presented in Section 6.

2 Overview of Line-Crawling Robot

A brief introduction to two features of one form of line-crawling robot is given in this section, namely, architecture of control system and method of locomotion of robot designed to navigate along transmission lines for a power system. In this description, we do not deal with such issues as types of servos and micros or shielding required to permit operation of the robot in the presence of a high electromagnetic field associated with high-voltage (e.g., 350 KV) lines carrying up high current (e.g., 1200) amps of current. This robot has been designed to operate on transmission lines with many types of obstacles (e.g., insulators, commutators) like the ones used by Manitoba Hydro.

2.1 Line-Crawling Robot Architecture

The line-walking robot has the following functional components.

1. *Base.* A base is the robot's frame for supporting other components, electronic equipment and a payload.
2. *Gripper.* A gripper provides two essential functions:

Holding the wire. There are two possible states: "Opened" and "Closed". In the state "Closed", the gripper holds the wire and in the state "Opened", the gripper stays free. During the transition from "Opened" to "Closed", the gripper catches the wire. Design of the gripper allows catching wires and objects with varying diameter, which is very important in case of icy wires or sudden obstacles. The transition from "Closed" to "Opened" releases the wire.

Moving the robot along the wire. Being equipped with wheels and a driving unit, the gripper provides translational motion of the robot. This function is available in the "Closed" state.

3. Leg. A leg can be represented by a lever connected to the base and to the gripper with the following functions:

Delivering the gripper to the required location on the wire. The gripper can be delivered within certain limits in both horizontal and vertical directions (see Fig. 1).

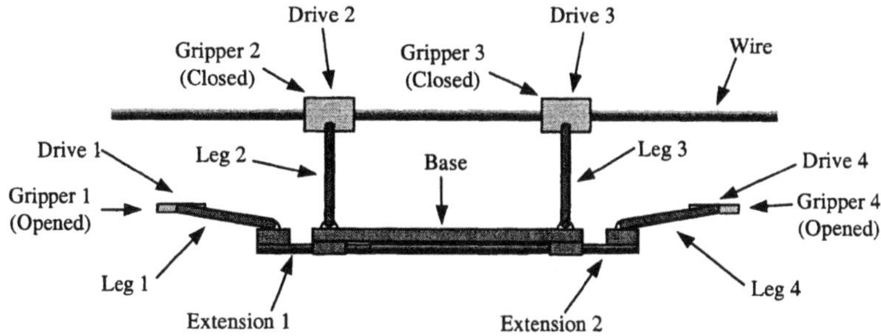

Fig. 1. LRC on a wire

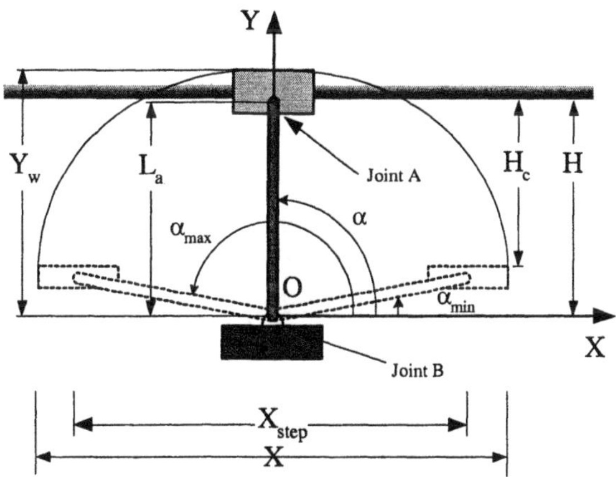

Fig. 2. Inverted pendulum model of LCR leg

The leg inclines about joint B on angle α (inclination), delivering the gripper within the arc of radius L_a (length of robot leg). The physical parameters of the leg determine its usage limitations, such as a length of a step X_{step}, distance from the gripper to the wire (clearance) H_c, horizontal and vertical sizes of "leg plus gripper" workspace X_w and Y_w, distance from the base to the wire H.

Moving the base in horizontal and vertical directions. Once the gripper catches the wire, the leg can move the base in the vertical direction and reduce the clearance between the base and wire. In this case the leg inclines about joint A on angle α, reducing clearance H_c and advancing the base in the horizontal direction.

A model of the inverted pendulum with the mass M_a and the weightless link of length L_a can represent the leg (see Fig. 2). The leg performs rotation about the joint O at angle α, within the range $\alpha_{min} < \alpha < \alpha_{max}$.

Base Extension. Extensions are the links connected to the foundations of leg1 and leg4, and which slide along supports connected to the base (see Fig. 3).

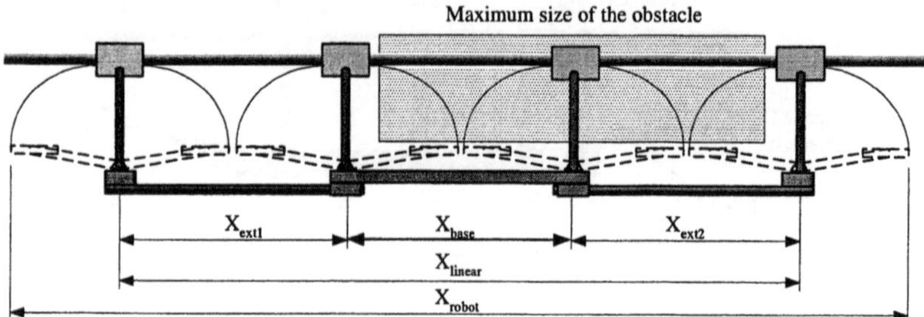

Fig. 3. LCR Movements Relative to Largest Obstacle

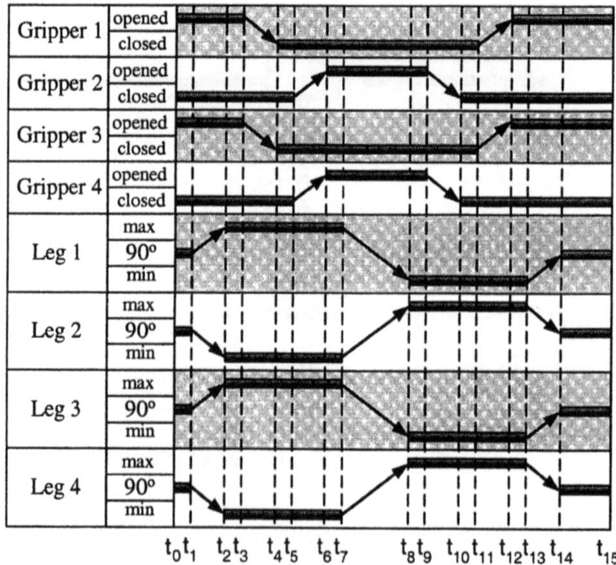

Fig. 4. Sample LCR Locomotion States

Base extensions allow the robot to change the shape of the robot by increasing the linear size of the base from X_{base} to X_{linear}. This considerably increases the workspace of the robot and allows it to avoid bigger obstacles. The base extension makes the following actions possible.

Translation of leg1 and leg4 in the horizontal direction. This motion is restricted to the length of the base extensions X_{ext} and is possible when gripper 1 and gripper 4 are in "Opened" state, and gripper 2 and 3 in "Closed" state.

Translation of the base in the horizontal direction. This motion is performed when grippers 1 & 4 in "Closed", and grippers 2 & 3 in "Opened" state (see Fig. 4).

Legged locomotion utilizes the property of a leg to move the robot's body in both vertical and horizontal directions (curvilinear translation). This happens when grippers of acting legs hold the wire and legs simultaneously deflect on the same angle with the same angular speed. Drives of the robot, at this motion, are in stopped state and the base extensions are at maximum distance, allowing the second and the third legs deflect in full range of angles. Applying different repetitive activation of patterns for legs and grippers the robot has the ability to stride along the wire.

2.2 Model of Robot Driving Unit

In this section, a model of the robot driving unit (i.e., a wheel attached to a gripper at the end of each leg) responsible for robot locomotion along a wire is given (see Figures 5 and 6).

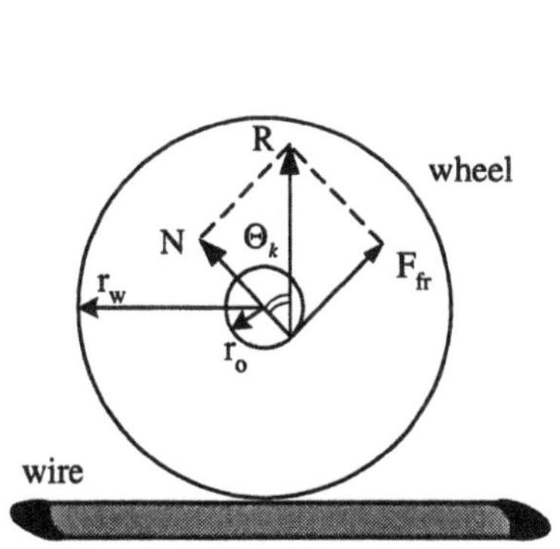

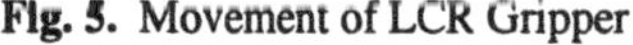

Fig. 5. Movement of LCR Gripper

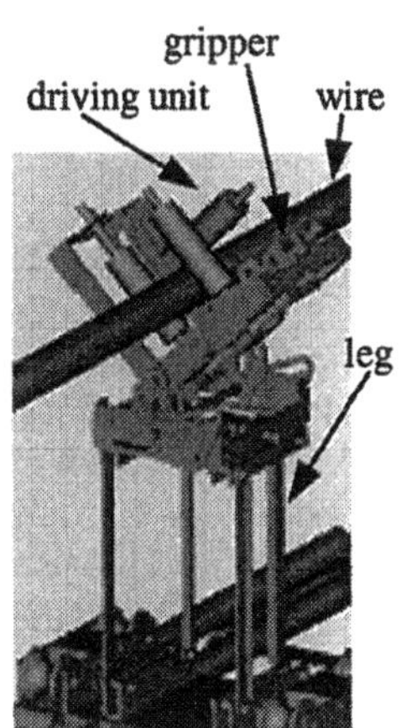

Fig. 6. LCR Driving Unit

This model will give the reader some insight into how a line-crawling robot moves along a transmission line. Mathematical models of the robot leg and gripper are considered outside the scope of this chapter. Every gripper has one driving wheel on the grasping levers. Applying the torque to this wheel, the robot performs the translational motion along the wire. The wheel is supported by two journal bearings. For maintaining the motion, the actuator torque should overcome the frictional resistance in the bearings and the moment of inertia of the robot. Let T_o N.m, J_o kg·m^2, r_o m, r_w m, ω_o rad/sec^2 denote actuator torque, moment of inertia, radius of the axle, radius of the wheel, and angular speed of the wheel. Then a model for the drive unit is given in Eq. (1).

$$T_o = J_o \cdot \frac{d\omega_o}{dt} + m_o \cdot g \cdot r_o \cdot \sin(\tan^{-1} \mu_k) \cdot sign(\omega_o) \tag{1}$$

2.3 Obstacle Avoidance Method

In this section, a brief description of obstacle avoidance for the line-crawling robot is given. To perform obstacle avoidance, the robot must be able to place at least two legs behind (on the far side of) and obstacle (see Fig. 7).

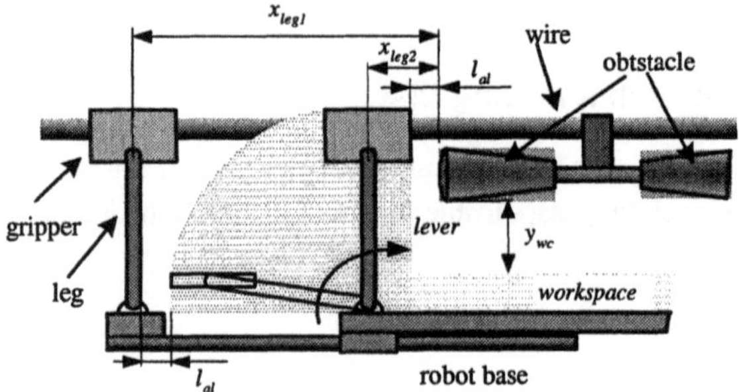

Fig. 7. LCR Copes with Obstacle

This is a safety-critical procedure where there is the greatest risk of collisions: the second leg into the first leg which is already placed behind the obstacle, or colliding the second leg into the obstacle. Possible way of placing the second leg behind the obstacle is shown in Fig. 7. Obstacle avoidance maneuver starts with placing the first leg behind the obstacle at a distance not less then x_{leg1}. This will provide enough space for workspace of the second leg and safety allowance l_{al}. Next, the base of the robot moves forward until it reaches x_{leg2}. Rotational motion of the second leg to the leg angle $\alpha=90°$ finishes the procedure. This method has advantage in its simplicity. But disadvantage is much greater – the work zone for performing this procedure is considerable and requires relatively big linear sizes of the robot, both for base and base extensions, increasing total weight of the robot (and consequently requiring greater powers for servos and amplifiers). The alternative way of placing the second leg is to combine simultaneously a rotational motion of the leg and a linear motion of the base (see Fig. 8).

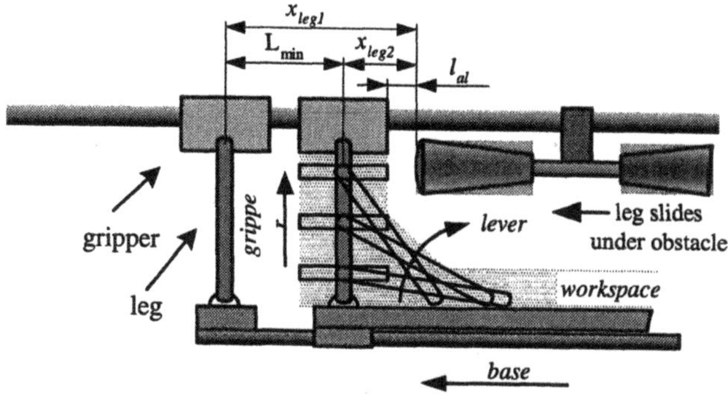

Fig. 8. Rotational Motion of LCR Leg

In this case, the first leg is placed at a distance x_{leg1}, which is equal to two safety allowances l_{al} and minimal distance between the base and the first leg L_{min}. When the first leg grasps the wire, the base of the robot starts moving forward. As soon as the second leg is at distance x_{leg2}, the second leg starts rotational motion. Control system stabilizes the second gripper at the distance x_{leg2}, providing necessary angular speed for the rotation. When the base of the robot reaches x_{leg2}, it stops moving. At the same time the second leg finishes the rotational motion, placing the gripper between the obstacle and the first leg. The advantage of this method is in significant decreasing the size of the work zone and reducing requirements to the linear sizes of the robot. The disadvantage of this method is the complexity in synchronization of two motions.

3 LCR Control System

A partial description of the LCR control system is given in this section.

3.1 Subsumption Architecture for Robot Control System

A layered architecture developed by Brooks in organizing control systems for intelligent autonomous mobile robots [37] has been incorporated into the design a line-crawling robot. The line-crawling robot is designed with the following features:

- Multiple goals: controller must be responsive to high-priority (e.g., avoiding danger like falling, navigating around obstacles hanging from wire) as well as low-priority goals (e.g., analyzing a density of wood in a wooden tower for power system transmission lines). The robot has been designed for inspection of electrical power system structures such as conductors, insultators, and towers.

- Multiple sensors: e.g., infra-red, camera, impact, ultrasonic, electromagnetic field (bolometer).
- Multiple actuators: e.g., servos, grippers, drive units.
- Robustness: robot should continue to function even if some sensors fail
- Extensibility: it is possible add more sensors, and add capabilities to robot

The requirements for the control system are expressed informally as follows:

Req0: Robot controller will consist of a set of independent processors that send messages to each other.
Req1: Each processor implements a competence module.
Req2: Competences are hierarchical (one competence per level):
(lowest) level 0: avoid contact with other objects
level 1: wander aimlessly without hitting things
level 2: explore, seeing places in the distance that look reachable, heading for them.
level 3: build map of environment, plan routes from one place to another
level 4: monitor changes in static environment
level 5: classify_{NN} objects
level 6: plan_{NN} changes to the world
(highest) level 7: reason_{NN} about behavior of objects.

Robot actions are grouped together in terms of desired classes of behaviors called competences. A *competence* is a sensor-driven process. Competences are represented by tuples (S, T, A) where S specifies sensory input from one or more sensors, T performs transformations on inputs and generates output that affect the behavior one or more actuators in the set A.

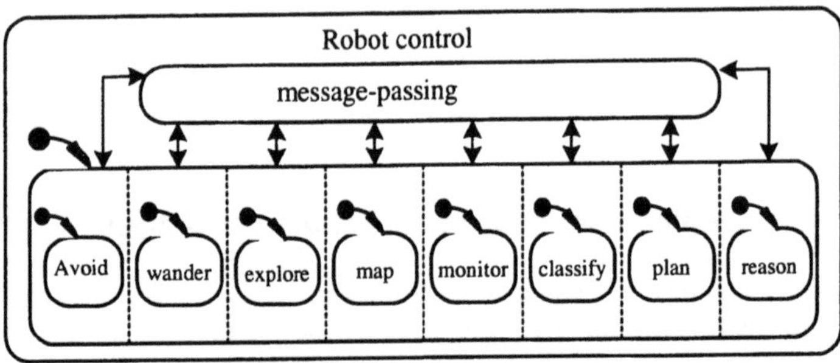

Fig. 9. Statechart for robot control system

The execution of a competence triggers some form of response to stimuli. The behaviors of a mobile robot are identified with the execution of competence modules (software which implements a particular competence) [38]. A *level of competence* is a desired class of behaviors over all environments that a robot will encounter resulting from the execution of competence modules. A partial representation of the robot control satisfying the informal requirements is given in the form of a statechart in Fig. 9. The competences are represented by orthogonal statecharts in Fig. 9 to satisfy the requirement that each competence is carried out by a separate processor that runs asynchronously (there is no direct communication between processors (only message-passing), no shared global memory, no central control). These requirements can be satisfied by separating task-achieving behaviors into layers, each with different tasks to perform relative to inputs from sensors, by viewing control in terms of layers of competence (see Fig. 10).

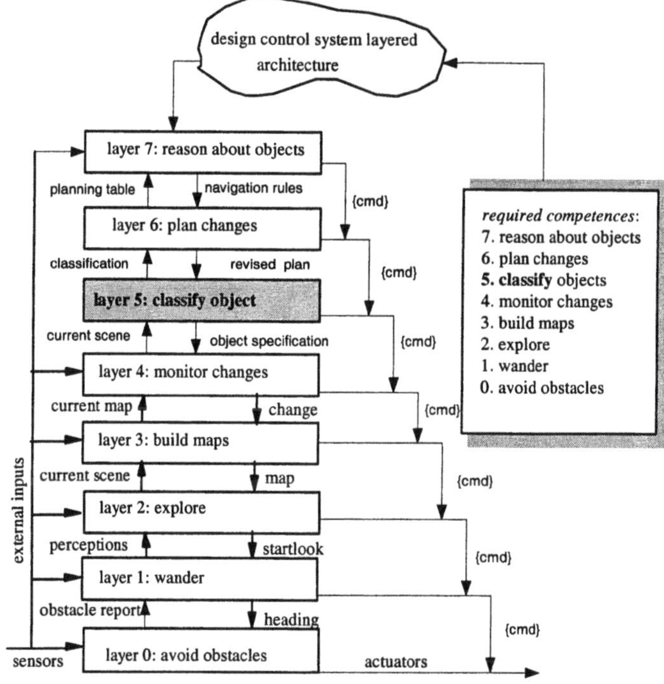

Fig. 10. Control Layers for a Mobile Robot

In the Brooks control system, higher-level layers can subsume the roles of lower level layers whenever they wish to take control. All layers have access to inputs from sensors. The layers below a particular layer (e.g., level 3) form a complete operational control system (levels 2, 1, and 0 control movements of the robot). Each layer is run by its own processor, and processors send messages asynchronously to each other. All processors are created equal (there is no central control within a layer). Neural networks are included in the design of the competences of layers 5 (classify), 6 (plan), and 7 (reason) of the control system for the line-crawling chapter described in this paper. Layer 5 communicates an object specification to layer 5, which monitors changes and updates the current scene being encountered by the robot. Layer 7 uses a planning table (updated condition vectors with corresponding responses ("decisions") that it receives from layer 6. The planning table reflects knowledge gained from each classification made by layer 5. Only the neural network design for layer 5 (classifying sensor signals) is considered. A description of the remaining layers of the line-crawling robot control system is outside the scope of this paper.

3.2 LCR Navigation Problem

Basic features of the line-crawling robot (LCR) navigation problem are described in this section. A principal task of the LCR control system is to guide the movements of the robot so that it maintains a safe distance from overhead conductors and any objects such as insulators attached to conductor wires or towers used to hold conductors above the ground. To move along a conductor, the LCR must continuously measure distances between itself and other objects around it, detect and maneuver to avoid collision with obstacles. Only two types of obstacles are considered in the description of the navigation problem in this section: (1) obstacles that are attached to the wire and protrude above the wire (e.g., insulators), and (2) obstacles that are attached to the wire and hang down from the wire. In both cases, the robot must "crawl" around these obstacles, and continue its forward or backward movement along the wire. If one were to plot the trace of typical LCR movements, the hashed line in the plot of robot positions as it moves forward relative to the wire would look like the one in Fig. 11.

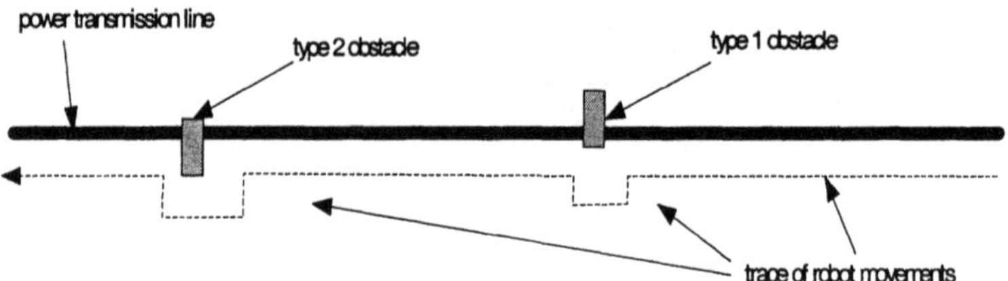

Fig. 11. Trace of Typical LCR Movements

The LCR body is equipped with 6 proximity sensor arrays, each containing 3 sensors for a total of 18 proximity sensors. Each LCR leg is equipped with an array of 4 proximity sensors (a total of 16 sensors on the legs) because legs slide independently along a track in the robot base and must avoid collision with each other as well as with obstacles in the neighborhood of the robot. Hence, LCR navigation decisions are based on the evaluation of signals from a total of 34 sensors. For simplicity, we consider only 4 of these sensors (see Fig. 12). Let a1, a2, a3, a4 denote proximity sensors (e.g., ultrasonic sensors). The following assumptions are made about the movements and sensors (see Fig. 12) on a line-crawling robot:

- The robot moves only in one direction ("forward"). The end of the robot that moves in the forward direction is called the "front" of the robot.

- A single proximity sensor (namely, a4) is positioned in the center of the robot to measure the distance between the robot and wire or the bottom of an obstacle. Sensor a4 measures objects in the [0.1, 1] meter range. If a4 < 0.4 m, then the robot lowers the robot to obtain better overhead clearance for the robot body.
- An array of three proximity sensors (namely, a1, a2, a3) is connected to the front of the robot (see Fig. 12). Each sensor detects objects in the [0.1, 3] meter range:
  - a1 is positioned to detect objects connected to and projecting above the wire. If a1 < 0.5 m, the robot executes a small obstacle avoidance routine.
  - a2 is positioned to detect objects in front of the robot. If a2 < 0.5 m, the robot executes a small obstacle avoidance routine.
  - a3 is positioned to detect objects connected to and hanging down from the wire. If a3 < 0.5 m, the robot executes a large obstacle avoidance routine.

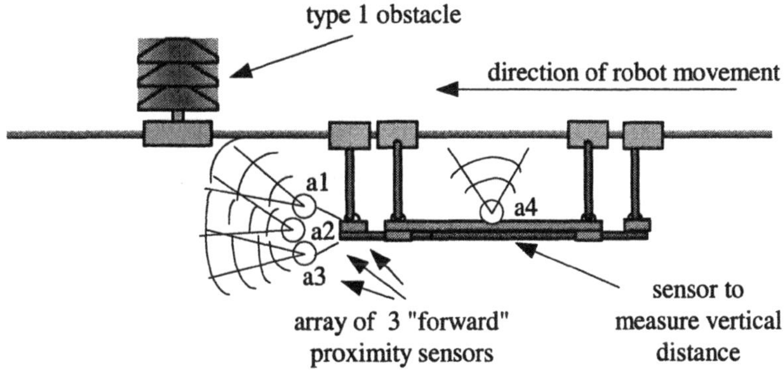

Fig. 12. Sample proximity Sensors of LCR

We want to train a neural network so that the LCR can make a decision about which type of movement it can safely make based on readings from its sensors. Table 1 gives some sample navigation decisions based on aggregate information from fusion of various sensors. Let d denote a decision class. Sensor measurements are separated in collections (a form of sensor fusion) used to construct convex sets. Each convex set contains sensor measurements (often with some noise) either close to a preset threshold or significantly greater than {less than} the threshold. In effect, this form of convex set represents what is known as an upper approximation in rough set theory. In addition, each convex set is associated with a LCR navigation decision that initiates a set of many movements of the robot parts to carry a LCR maneuver. Because of noisy signals (a high-voltage power transmission line is a hostile environment with a severe electromagnetic field surround the conductor), it is not possible to use simple if-then statements to make navigation decisions.

Table 1. Sample Decision Classes for Robot

convex set of measurements	decision	explanation of decision
{a1, a2, a3}	d = 1	avoid small obstacle protruding down from conductor, maintain leg extension, slide legs under obstacle
{a2, a3, a4}	d = 2	avoid large obstacle protruding down from conductor, increase leg extension, slide legs under obstacle
{a1, a2, a3, a4}	d = 3	avoid small obstacle protruding down from conductor, increase leg extension, slide legs under obstacle
{a1, a2, a3, a4}	d = 4	avoid large obstacle protruding down from conductor, maintain leg extension, slide legs under obstacle
{a1, a2, a3, a4}	d = 5	move forward (normal state)

4 Rough Measures and Integrals

Because of the complexity of information and high number of inputs from LCR sensors, a neural computing approach to object classification offers a fairly straightforward solution to the LCR navigation problem. In this section, a brief presentation of rough measures and discrete rough integrals underlying the design of the particular form of rough neural networks used to classifying types of obstacles encountered by a robot is briefly presented. It is assumed that the reader is familiar with the basic rough set methods described in [34].

4.1 Indistinguishability Relation

To begin, let IS = (U, A) be an infinite information system where U is a non-empty subset of the reals $\Re$ and A is a non-empty, finite set of attributes, where a: $U \rightarrow V_a$ for every $a \in A$. Let $a(x) \geq 0$, $\delta > 0$, $x \in \Re$ (set of reals) and let $\lfloor a(x)/\delta \rfloor$ denote the greatest integer less than or equal to a(x)/δ ("floor" of a(x)/δ) for attribute a. If $a(x) < 0$, then $\lfloor a(x)/\delta \rfloor = -\lfloor |a(x)|/\delta \rfloor$. The parameter δ serves as a "neighborhood" size on real-valued intervals. Reals within the same subinterval bounded by kδ and (k+1)δ are considered indistinguishable. For each $B \subseteq A$, there is associated an equivalence relation $\mathrm{Ing}_{A,\delta}(B)$ defined in Eq. (2).

$$Ing_{A,\delta}(B) = \left\{ (x, x') \in \Re^2 \,\middle|\, \forall a \in B.\ \lfloor a(x)/\delta \rfloor = \lfloor a(x')/\delta \rfloor \right\} \tag{2}$$

If $(x, x') \in \mathrm{Ing}_{A,\delta}(B)$, we say that objects x and x' are indistinguishable from each other relative to attributes from B. Further, a partition U/ $\mathrm{Ing}_{A,\delta}(B)$ denotes the family of all equivalence classes of relation $\text{Ing}_{A,\delta}(B)$ on U. For $X \subseteq U$, the set X can be approximated only from information contained in B by constructing a B-lower and a B-upper approximation denoted by $\underline{B}X$ and $\overline{B}X$,respectively, where $\underline{B}X = \left\{ x \mid [x]^{\delta}_{B \cup M} \subseteq X \right\}$ and $\overline{B}X = \left\{ x \mid [x]^{\delta}_{B \cup M} \cap X \neq \varnothing \right\}$.

Proposition 4.1 $\text{Ing}_{A,\delta}(B)$ is an equivalence relation.

4.2 Rough Membership Set Function

To deal with point sets (i.e., reals), it is necessary to consider an alternative to the traditional rough membership function [21]. A set function form of the rough membership function was introduced in [23]. The rough membership function is the fundamental computation component of a rough neuron. It is responsible for "weighting" of aggregated inputs either from sensors or from other neurons.

Definition 4.1 Let $S = (U, A)$ be an information system with non-empty set U and non-empty set of attributes A. Further, let $B \subseteq A$ and let $[y]^{\delta}_{B}$ be an equivalence class of any sensor reading $y \in \Re$. Let ρ be a measure on a set $X \subseteq \wp(U)$, where $\wp(U)$ is a class (set of all subsets of U). The rough membership set function is then defined in Eq. (3).

$$\mu_{y}^{B,\delta} : \wp(U) \rightarrow [0,1], \text{ where } \mu_{y}^{B,\delta}(X) = \frac{\rho\left(X \cap [y]^{\delta}_{B}\right)}{\rho\left([y]^{\delta}_{B}\right)} \tag{3}$$

for any $X \in \wp(U)$ is called a *rough membership set function (rmf)*. In this chapter, the rmf in (1) is interpreted as a ratio of integrals in Eq.(4).

$$\mu_{y}^{B,\delta}(X) = \frac{\rho\left(X \cap [y]^{\delta}_{B}\right)}{\rho\left([y]^{\delta}_{B}\right)} = \frac{\int_{X \cap [y]^{\delta}_{B}} 1 \, dx}{\int_{[y]^{\delta}_{B}} 1 \, dx} \tag{4}$$

4.3 Rough Measures

Let S = (U, A) be an information system where $X \subseteq U$, $B \subseteq A$, and let $\text{Ing}_{A,\delta}(B)$ be the δ-indistinguishability relation on U. Then consider the introduction of a rough measure based on classical measure theory [41].

Definition 4.2 The tuple $(X, \wp(X), U/\text{Ing}_{A,\delta}(B))$, where $U/\text{Ing}_{A,\delta}(B)$ denotes a set of all equivalence classes determined by $\text{Ing}_{A,\delta}(B)$ on U, and is called an δ-*indistinguishability space* over X and B.

Definition 4.3 Let u ∈ U. A non-negative and additive set function ρ_u : $\wp(X) \rightarrow [0, \infty)$ defined by $\rho_y(Y) = \rho'(Y \cap [y]_B^\delta)$ for $Y \in \wp(X)$, where $\rho' : \wp(X) \rightarrow [0, \infty)$ is called a *rough measure* relative to U/Ing$_{A,\delta}$(B) and u on the d-indistinguishability space (*X*, $\wp(X)$, U/Ing$_{A,\delta}$(B)).

Definition 4.4 Let ρ_y for y=a(x) and xäU be a *rough measure* on the d-indistinguishability space (*X*, $\wp(X)$, U/Ing$_{A,\delta}$(B)) relative to U/Ing$_{A,\delta}$(B) and y. The tuple (*X*, $\wp(X)$, U/Ing$_{A,\delta}$(B), ρ_y) is a *rough measure space* over *X* and *B*.

Proposition 4.2 Let *S* = (*U*, *A*) be an information system, $B \subseteq A$, and let $[y]_B^\delta$ be an equivalence class of an object *u* = *a(x)* for xäU of Ing$_{A,\delta}$(*B*). The rough membership function $\mu_y^{B,\delta}$ as defined in Definition 3.1 is non-negative, additive on *U*.

An important consequence of Prop. 4.2 is that we obtain the basis for a rough measure space, and makes it possible to define an integral useful in measuring noisy sensor signals. This provides a foundation for rough integrals used in the design of neurons in rough neural networks. This is explained in more detail in [39].

4.4 Discrete Rough Integral

The discrete rough integral was introduced in [39], and elaborated in [22]. In the earlier study of rough integrals, only rough measures define relative to finite sets were considered. In this paper, the discrete rough integral is defined relative to the measure of uncountable sets and real-valued functions. That is, we integrate f: X → $\Re$ with respect to a rough measure of a set of reals. In the case where the domain of a function is a powerset, then the function is called a set function.

Definition 4.5 **Discrete Rough Integral.** Let ρ_y for y=a(x) and x ä U be a *rough measure* on the d-indistinguishability space (*X*, $\wp(X)$, U/Ing$_{A,\delta}$(B)) relative to U/Ing$_{A,\delta}$(B) and y, and let f: X → $\Re^+$ be a function such that $f(x_{(1)}) \leq \ldots f(x_{(n)})$ (monotonic non-decreasing), $X_{(i)} := \{x_{(i)}, \ldots, x_{(n)}\}$, and $f(X_{(0)}) = 0$, where •(i) is a permuted index. Then the discrete rough integral of f with respect to rough measure ρ_y is defined in (x).

$$\int_X f \; d\rho = \sum_{i=1}^{n} \left[\left(f\left(x_{(i)}\right) - f\left(x_{(i-1)}\right) \right) \rho_u\left(X_{(i)}\right) \right]$$

It has been shown that the discrete rough integral is a form of ordered weighted average [22]. This integral value can be useful in evaluating the relevance of sensors, if we enforce the criterion that the integral of a sensor must have a value greater than some pre-set threshold. This idea is illustrated in the example in the next section.

4.5 Sample Integration of a Sensor

Intuitively, we want to identify those sensors with discrete rough integral values closest to some threshold. That is, consider a sensor a ∈ A in an information system S = (U, A) with non-empty universe U such that sensor a is a functional. Let t_a be a threshold used to judge the relevance of a sensor *a*. By way of illustration of this idea, a sensor *a* is considered relevant to the classification of a class of obstacles, if $\int_X a\, d\mu_B^\delta \geq t_a$. Consider, for example, threshold $t_a = 0.2$ and the discrete rough integral values of three sensors, namely, a_1, a_2, and a_3. Next, we make two simplifying assumptions in the example computations in Table 2 in looking for relevant sensors (i.e., those sensors with integral values above the threshold $t_a = 0.2$). First, the three sensors are evaluated over the same subintervals. Second, for each of the three sensors, $[y]_B^{0.1}\Big|_{y=0.5}$ is the same.

Table 2. Permuted Sensor Values

$X_i \setminus \{a\}$	a_1	a_2	a_3	Steps in Computing Integrals
$X_1 = [1,1.8]$	0.6	0.55	0.5	$X = \mathbf{[1, 1.8]} \cup [2.6, 3.4] \cup [4.8, 5.0]$
$X_2 = [2.6,3.4]$	0.75	0.65	0.7	$[y]_B^{0.1}\Big\|_{y=0.5} = \mathbf{[1.32, 1.8]} \cup [3.42,$
$X_3 = [4.8,5.0]$	0.21	0.4	0.1	
permuted X	permuted sensor values			$3.72]$
$X_{(1)} = [4.8,5.0]$	0.21	0.4	0.1	$X_{(1)} \cap [y]_B^{0.1}\Big\|_{y=0.5} = X_{(3)} \cap$
$X_{(2)} = [1,1.8]$	0.6	0.55	0.5	
$X_{(3)} = [2.6,3.4]$	0.75	0.65	0.7	$[y]_B^{0.1}\Big\|_{y=0.5} = 0$
				$X_{(2)} \cap [y]_B^{0.1}\Big\|_{y=0.5} = [1.32, 1.8]$

Then from Table 2, we obtain $\int_X a_1\, d\mu_B^{0.1} = 0.24$, $\int_X a_2\, d\mu_B^{0.1} = 0.092$, $\int_X a_3\, d\mu_B^{0.1} = 0.615$ and $\mu_{0.5}^{B,0.1}(X_{(1)}) = \mu_{0.5}^{B,0.1}(X_{(3)}) = 0$, $\mu_{0.5}^{B,0.1}(X_{(2)}) = 0.615$.

Then, for example, the value of $\mu_{0.5}^{B,0.1}(X_{(2)})$ is computed as follows:

$$\mu_{0.5}^{B,0.1}(X_{(2)}) = \frac{\int_{X_{(2)} \cap [y]_B^{0.1}\big|_{y=0.5}} 1\, dx}{\int_{[y]_B^{0.1}\big|_{y=0.5}} 1\, dx} = \frac{1.8 - 1.32}{1.8 - 1.32 + 3.72 - 3.42} = \frac{0.48}{0.78} = 0.615$$

The discrete rough integral is a form of ordered weighted average. Next, we consider convex sets useful in a neural classification system for a robot control system.

4.6 Convex Sets

When sensor values for different sensors are close to each other, it becomes difficult to make LCR navigation decisions, especially when one considers the influence of noisy signals on classification decisions. Dangerous situations can result. Consider, for example, the plot in Fig. 13 where measurements from three different sensors are represented by the • (solid disk), ♦ (solid diamond) and ■ (solid square). In addition, symbols are used to represent rough integral values: ⊗ ("circle x" for average • value), ⊕ ("circle +" for average ♦value), and ∅ ("empty set symbol" for average ■ value).

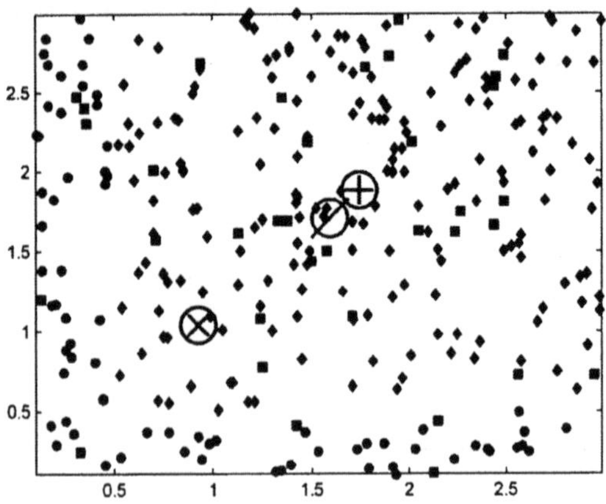

Fig. 13. Sample sensor measurements

The three colours in Fig. 13 denote three decision classes. Recall that the rough integral (RI) value is a mean value. Therefore, for a decision class denoted by a • RI value would be somewhere close to point (1,1). It is easy to see that point (1,1) by no means describes the cluster shape (and location) of • sensor values. There are many more subsets of the given set of points which would better fit to it (like a sphere with the middle in point (1,1)).. Similar observations can be made about the "♦" and "■" sensors. Ideally all decision classes should have Gaussian distribution to facilitate neural classification of these sensor values. Unfortunately, LCR sensor values are usually not distributed normally. One solution to the problem of distinguishing sensor values is to construct convex sets that "enclose" values of a sensor (see, e.g., [40]).

Definition 4.6 A set **A** is said to be *convex* if the straight line segment joining any two points in **A** lies entirely within **A**.

Definition 4.7 The convex hull H of an arbitrary set S is the smallest convex set containing S.

The main problem is the lack of knowledge of where a set starts and ends since all we are given are some points in a space. Our approach is to consider partitioning our set into the least number of convex sets. Two sample partitions of sensor values (one convex hull and one non-convex hull partition) are shown in Table 3.

Table 3. Sample Distribution of Sensor Values

Set S of Sensor Values	Convex Hull	Not a Convex Hull

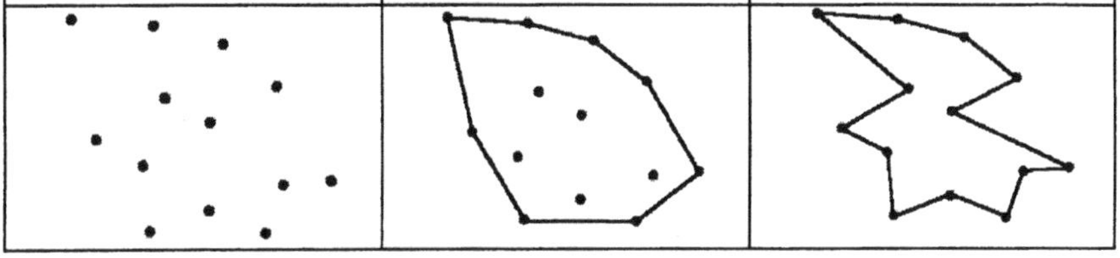

There are points in the convex hull in Table 3 that definitely belong to the "middle" of a set or to the boundary. But we cannot uniquely draw borders of the set S. Two equally probable versions are shown in Table 3.

5 Rough Neural Network

The study of various forms of rough neurons is part of a growing number of papers on neural networks based on rough sets [1-2, 4-15, 18-21, 23-24, 30, 32-33, 42-49]. In what follows, we give a brief description of the basic architecture of the rough neural network used in the design of the classify layer of the LCR control system.

5.1 Architecture of LCR Classification Layer

The basic architecture of a rough neural network for classifying sensor signals using convex sets is shown in Fig. 14. For each cluster, the neural network maintains decision class information that is used to make a classification decision after calibration. The rough neural network (RNN) in Fig. 14 calibrates δ to obtain a training set that matches target information about LCR decision classes. The RNN itself consists of a single layer of neurons. Each neuron constructs a table of rough integral values relative to a single decision class (see Fig. 15).

Definition 5.1 Given a sensor-based information system $S = (U, A)$, define a function $f_u: \wp(U) \rightarrow \hat{A}^k$ for each ***k-dimensional Rough Neuron*** N_k where $[u] \subset \wp(U)$ and $p_i \in \Re^k$ is an element of the vector computed by f_U such that

$$\forall_{1 \le i \le k} \; p_i = \int_U a_i \, d\mu_{[u]}^{a_i} \qquad \text{where } a_i \in A.$$

Input to a rough neuron is a cluster extracted from a convex set $[u]$, and an output is a k-dimensional vector of rough integral values p.

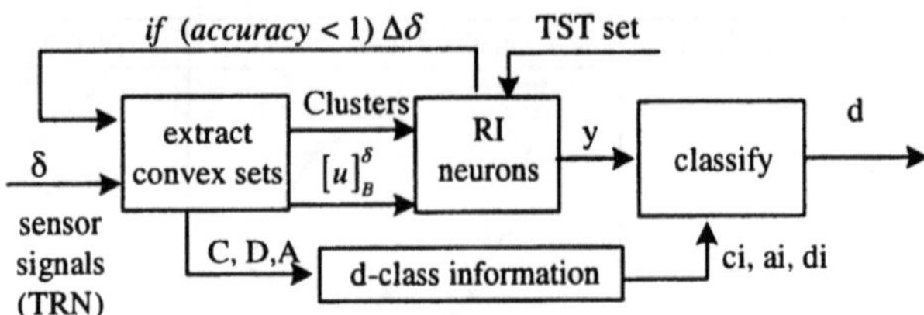

Fig. 14. Basic RNN Architecture

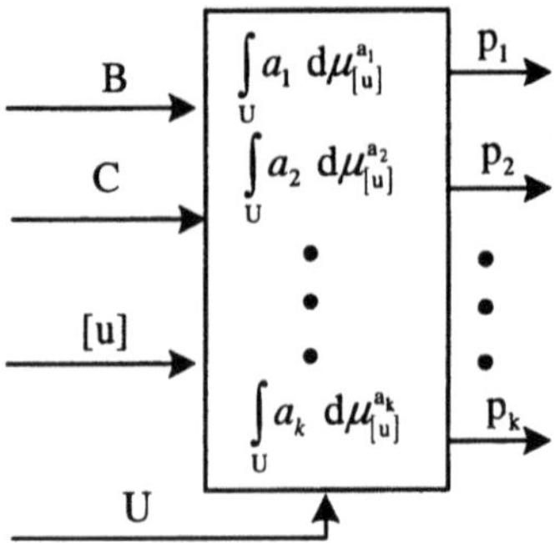

Fig. 15. Architecture of Rough Neuron

The input [u] is defined relative to a δ-indistinguishability relation over a point set containing sensor signal values. The set U denotes the universe (point set defining a region of the Euclidean plane containing possible sensor values), and the set C denotes a cluster extracted from a convex set. Each [u] represents a partition of U. The output of a single rough neuron is a pattern vector $[p_1\ p_2\ \ldots\ p_k]$ that is associated with a row in a navigation decision table. This vector defines a pattern that should be as close as possible to a known pattern (target vector) $[t_1\ t_2\ \ldots\ t_k]$

asociated with a particular decision. Adjustments are made to δ (and [u]) during training to make [p_1 p_2 ... p_k] as close as possible to [t_1 t_2 ... t_k].

6 Experimental Results

The rough neural network used in classifying LCR sensor signals has been implemented in a toolset called MB in Visual C++ (see Fig. 16).

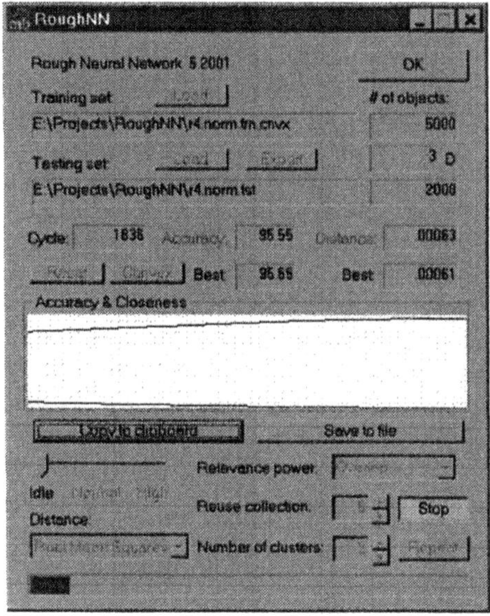

Fig. 16. Sample MB display

MB computes a closeness measure which is the root mean square of all the differences between training and testing tables in the current epoch. MB measures the accuracy of the classification equals the % of correctly classified test data in the current epoch.

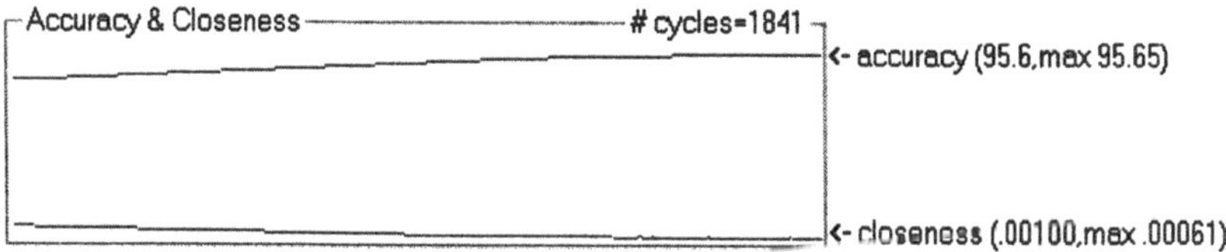

Fig. 17. Sample Classification Accuracy

In each epoch after the first epoch, an adjusted δ to construct convex sets from the sample data until the computation is halted by a user or when 100% of the test table entries are correctly classified. To verify the design of classify layer of the LCR control system, a set of 7000 values output by three proximity sensors relative to 6 decision classes. A classification accuracy of 51.2% is achieved after 501 epochs. After 1841 epochs, the classification accuracy between training and testing tables is 95.6% (see Fig. 17).

7 Concluding Remarks

This chapter has focused on an approach to designing the classify layer of the LCR control system. The classify layer is part of a layered architecture for robot control systems introduced by Brooks. It has been verified that rough neural networks work quite well in classifying noisy sensor signals commonly produced by sensors on a line-crawling robot. Navigation by a robot crawling along a high-voltage transmission line is a complex task because of the many operations that must be performed by the robot in working its way between, round and underneath various obstacles such insulators and girders on towers. Because noisy sensor signals are common in an LCR, it is better to make navigation decisions based on approximation methods commonly used in rough set theory. Inputs to a classifier neural network are in the form of clusters extracted from convex sets relative to robot sensor values in a δ-mesh. Such sets provide input to neurons that compute ordered weighted averages for each robot sensor relative to particular navigation decision class. It has been shown that the proposed approach results in high classification accuracy.

Acknowledgements

The research of James Peters has been supported by the Natural Sciences and Engineering Research Council of Canada (NSERC) research grant 194376 and grants from Manitoba Hydro and the University of Information Technology, Rzeszów, Poland. The research by Sheela Ramanna has been supported by NSERC research grant 185986. The research of Maciej Borkowski and Vitaliy Degtyaryov has been supported by research grants from Manitoba Hydro.

References

[1] M. Banerjee et al., Rough fuzzy MLP: Knowledge encoding and classification, IEEE Trans. on Neural Networks, vol. 9, no. 6, 1998, 1203-1215.

[2] B. Chakraborty, Feature subset selection by neuro-rough hybridization. In: W. Ziarko, Y. Yao (Eds.), Proc. of the Second Int. Conf. on Rough Sets and Current Trends in Computing (RSCTC'00), Banff, Canada, 16-19 Oct. 2000, 481-488.

[3] L. Han, R. Menzies, J.F. Peters, L. Crowe, High voltage power fault-detection and analysis system: Design and implementation. In: Proc. of the Canadian Conference on Electrical & Computer Engineering (CCECE'99), 1999, 1253-1258.

[4] L. Han, J.F. Peters, S. Ramanna, R. Zhai, Classifying faults in high voltage power systems: A rough-fuzzy neural computational approach. In: N. Zhong, A. Skowron, S. Ohsuga (Eds.), New Directions in Rough Sets, Data Mining, and Granular-Soft Computing, Lecture Notes in Artificial Intelligence, vol. 1711. Berlin: Springer Verlag, 1999, 47-54.

[5] P.J. Lingras, Fuzzy-rough and rough-fuzzy serial combinations in neurocomputing, Neurocomputing: An International Journal, vol. 36, Feb. 2001, 29-44.

[6] P.J. Lingras, Rough neural networks. In: Proc. of the 6th Int. Conf. on Information Processing and Management of Uncertainty in Knowledge-based Systems (IPMU'96), Granada, Spain, 1996, p1445-1450.

[7] P.J. Lingras, Comparison of neurofuzzy and rough neural networks, Information Science: An International Journal, vol. 110, 1998, 207-215.

[8] S. Mitra, P. Mitra, S.K. Pal, Evolutionary modular design of rough knowledge-based network with fuzzy attributes, Neurocomputing: An International Journal, vol. 36, Feb. 2001, 45-66.

[9] H.S. Nguyen, M. Szczuka, D. Slezak, Neural networks design: Rough set approach to real-valued data. In: Proc. of PKDD'97, Trondheim, Norway. Lecture Notes in Artificial Intelligence 1263, Berlin, Springer-Verlag, 1997, 359-366.

[11] T. Nguyen, R.W. Swiniarski, A. Skowron, J. Bazan, K. Thagarajan, Applications of rough sets, neural networks and maximum likelihood for texture classification based on singular decomposition. In: Proc. of the Third International Workshop on Rough Sets and Soft Computing, San Jose, CA, U.S.A., 10-12 Nov. 1994, 332-339.

[12] S.K. Pal, P. Mitra, Rough Fuzzy MLP: Modular Evolution, Rule Generation and Evaluation, IEEE Trans. on Knowledge and Data Engineering [to appear].

[13] S.K. Pal, L. Polkowski, A. Skowron (Eds.), Rough-Neuro Computing: Techniques for Computing with Words. Berlin: Springer-Verlag, 2002.

[14] S.K. Pal, J.F. Peters, L. Polkowski, A. Skowron (Eds.), Rough-Neuro Computing: An Introduction. In [18], 16-43.

[15] W. Pedrycz, L. Han, J.F. Peters, S. Ramanna, R. Zhai, Calibration of software quality: Fuzzy neural and rough neural computing approaches, Neurocomputing, vol. 36, Feb. 2001, 149-170.

[16] J.F. Peters, S. Ramanna, A. Skowron, M. Borkowski: Wireless agent guidance of remote mobile robots: Rough integral approach to sensor signal analysis. In: N. Zhong, Y.Y. Yao, J. Liu, S. Ohsuga (Eds.), Web Intelligence, Lecture Notes in Artificial Intelligence 2198. Berlin: Springer-Verlag, 2001, 413-422.

[17] J.F. Peters, S. Ramanna, M. Borkowski, A. Skowron: Approximate sensor fusion in a navigation agent, in: N. Zhong, J. Liu, S. Ohsuga and J. Bradshaw (Eds.), Intelligent agent technology: Research and development. Singapore: World Scientific Publishing, 2001, 500-504.

[18] J.F. Peters, S. Ramanna, Z. Suraj, M. Borkowski, Rough neurons: Petri net models and applications. In [18], 474-493.

[19] J.F. Peters, A. Skowron, Z. Surai, L. Han, S. Ramanna, Design of rough neurons: Rough set foundation and Petri net model, International Symposium on Methodologies for Intelligent Systems (ISMIS'2000). In: Z.W. Ras, S. Ohsuga (Eds.), Foundations of Intelligent Systems, Lecture Notes in Artificial Intelligence, vol. 1932. Berlin: Springer Verlag, 2000, 283-291.

[20] J.F. Peters, A. Skowron, L. Han, S. Ramanna, Towards rough neural computing based on rough membership functions: Theory and Application, Rough Sets and

Current Trends in Computing, Lecture Notes in Artificial Intelligence, vol. 2005, Berlin, Springer-Verlag, 2000.

[21] J.F. Peters, L. Han, S. Ramanna, Rough neural computing in signal analysis, Computational Intelligence, vol. 1, no. 3, 2001, 493-513.

[22] Z. Pawlak, J.F. Peters, A. Skowron, Z. Suraj, S. Ramanna, M. Borkowski, Rough measures and Integrals. In: S. Hirano, M. Inuiguchi, S. Tsumoto (Eds.), *Lecture Notes in Computer Science*, 2002 [to appear].

[23] L. Polkowski, A. Skowron, Rough-neuro computing. In: W. Ziarko, Y. Yao (Eds.), Proc. of the Second Int. Conf. on Rough Sets and Current Trends in Computing (RSCTC'00), Banff, Canada, 16-19 Oct. 2000, 25-32.

[24] A. Skowron, Toward intelligent systems: Calculi of information granules. In: S. Hirano, M. Inuiguchi, S. Tsumoto (Eds.), Bulletin of the International Rough Set Society, vol. 5, no. 1 / 2, 2001, 9-30.

[25] J.F. Peters, A. Skowron, J. Stepaniuk, Rough granules in spatial reasoning. In: Proc. Joint 9th International Fuzzy Systems Association (IFSA) World Congress and 20th North American Fuzzy Information Processing Society (NAFIPS) Int. Conf., Vancouver, British Columbia, Canada, 25-28 June 2001, 1355-1361.

[26] A. Skowron, J. Stepaniuk, Information Granules: Towards foundations of granular computing, International Journal of Intelligent Systems, vol. 16, no. 1, Jan. 2001, 57-104.

[27] A. Skowron, J. Stepaniuk, Information granules and approximation spaces. In: Proc. of the 7th Int. Conf. on Information Processing and Management of Uncertainty in Knowledge-based Systems (IPMU'98), Paris, France, 6-10 July 1998, 1354-1361.

[28] A. Skowron, J. Stepaniuk, J.F. Peters, Hierarchy of information granules. In: L. Czala (Ed.), *Proc. of the Workshop on Concurrency, Specification and Programming*, Oct. 2001, Warsaw, Poland, 254-268.

[29] A. Skowron, J. Stepaniuk, J.F. Peters, Extracting patterns using information granules. In: S. Hirano, M. Inuiguchi, S. Tsumoto (Eds.), Bulletin of the International Rough Set Society, vol. 5, no. 1 / 2, 2001, 135-142.

[30] W. Ziarko, Y.Y. Yao (Eds.), Proc. Int. Conf. on Rough Sets and Current Trends in Computing (RSCTC'2000), Lecture Notes in Artificial Intelligence, vol. 2005, Berlin, Springer Verlag, 2001.

[31] L.A. Zadeh, Fuzzy logic = computing with words, IEEE Trans. on Fuzzy Systems, vol. 4, 1996, 103-111.

[32] P. Wojdyllo, Wavelets, rough sets and artificial neural networks in EEG analysis. In: Proc. of RSCTC'99, Lecture Notes in Artificial Intelligence 1424, Berlin, Springer-Verlag, 1998, 444-449.

[33] R.W. Swiniarski, Rough sets and neural networks application to handwritten character recognition by complex Zernike moments. In: [13], 617-624.

[34] J. Komorowski, Z. Pawlak, L. Polkowski, A. Skowron, Rough sets: A tutorial. In: S.K. Pal, A. Skowron (Eds.), Rough Fuzzy Hybridization: A New Trend in Decision-Making. Berlin: Springer-Verlag, 1999, 3-98.

[35] Z. Pawlak, Rough Sets: Theoretical Aspects of Reasoning About Data. Boston, MA, Kluwer Academic Publishers, 1991.

[36] Z. Pawlak, A. Skowron, Rough membership functions. In: R. Yager, M. Fedrizzi, J. Kacprzyk (Eds.), Advances in the Dempster-Shafer Theory of Evidence, NY, John Wiley & Sons, 1994, 251-271.

[37] R.A. Brooks, A robust layered control system for a mobile robot, vol. RA-2, no. 1, March 1986, 14-23.

[38] T. Gomi, Evolutionary Robotics. Ontario, Canada: AAI Books, 1997.

[39] Z. Pawlak, J.F. Peters, A. Skowron, Z. Suraj, S. Ramanna, M. Borkowski, Rough measures: Theory and Applications. In: S. Hirano, M. Inuiguchi, S. Tsumoto (Eds.), Rough Set Theory and Granular Computing, Bulletin of the International Rough Set Society, vol. 5, no. 1 / 2, 2001, 177-184.

[40] R.C. Gonzalez, R.E. Woods, Digital Image Processing, Prentice Hall, Upper Saddle River, New Jersey 07458, 2002.

[41] P.R. Halmos, Measure Theory. London: D. Van Nostrand Co., Inc., 1950.

[42] R.W. Swiniarski, L. Hargis, Rough sets as a front end of neural networks texture classifiers, Neurocomputing, vol. 36, Feb. 2001, 85-103.

[43] M. Swiniarski, F. Hunt, D. Chalvet, D. Pearson, Prediction system based on neural networks and rough sets in a highly automated production process. In: Proc. of the 12th System Science Conf., Wroclaw, Poland, 1995.

[44] M. Swiniarski, F. Hunt, D. Chalvet, D. Pearson, Intelligent data processing and dynamic process discovery using rough sets, statistical reasoning and neural networks in a highly automated production systems. In: Proc. of the 1st European Conf. on Application of Neural Networks in Industry, Helsinki, Finland, 1995.

[45] M. S. Szczuka, Refining classifiers with neural networks, International Journal of Intelligent Systems, vol. 16, no. 1, Jan. 2001, 39-55.

[46] M. S. Szczuka, Rough sets and artificial neural networks. In: L. Polkowski, A. Skowron (Eds.), Rough Sets in Knowledge Discovery 2: Applications, Cases Studies and Software Systems. Berlin: Physica Verlag, 1998, 449-470.

[47] M.S. Szczuka, Refining classifiers with neural networks, International Journal of Intelligent Systems, vol. 16, no. 1, Jan. 2001, 39-56.

[48] M.S. Szczuka, Rough sets and artificial neural networks. In: L. Polkowski, A. Skowron (Eds.), Rough Sets in Knowledge Discovery 2: Applications, Case Studies and Software Systems. Berlin: Physica-Verlag, 1998, 449-470.

[49] M.S. Szczuka, Rough set methods for constructing artificial neural networks. In: B.D. Czejdo, I.I. Est, B. Shirazi, B. Trousse (Eds.), Proc. of the 3rd Biennial Joint Conf. on Engineering Systems Design and Analysis 7, Montpellier, France, 1-4 July 1996, 9-14.

Hierarchical Planning in a Mobile Robot for Map Learning and Navigation

Cristina Urdiales, Antonio Bandera, Eduardo Pérez, Alberto Poncela, and Francisco Sandoval

Dpto. Tecnología Electrónica, ETSI Telecomunicación,
Universidad de Málaga, 29071 Málaga-Spain
cris@dte.uma.es

Abstract. This chapter focuses on autonomous navigation for mobile robots. We propose a hybrid layered architecture, which is used to navigate in totally or partially explored environments using sonar sensors. Our architecture relies on a hierarchical representation of the environment, which has both a metric and a topological level, which is based on the metric level. High level planning layers work at the topological level deliberatively, while low level navigation layers operate at the metric level reactively. The main advantage of the proposed scheme is that it can operate in both known and unknown environments rapidly and efficiently.

1 Introduction

Although robots were originally conceived as reprogrammable, multifunctional devices, mainly designed to move material in industrial environments, the envisioned potential of robotics systems has risen tremendously over the last few decades. Mobility, particularly, has been greatly improved, adding versatility to these systems. Nowadays, robots can be used for hazardous waste cleanup, spatial exploration or demining, just to mention a few applications. However, mobile robots are no longer bound to a well-known operation frame. Furthermore, they can even be meant to operate in unknown or changing environments. In these cases, robots can no longer be preprogrammed to pursue a fixed course of action like traditional factory robots. Instead, they must plan their actions according to the environment and react to potential unexpected situations [1]. To achieve this adaptive behaviour, robots usually rely on one or several types of sensors to perceive and act according to the outer world. Typically, on-board sensors for a given robot are chosen according to several criteria: field of view, range capability, accuracy and resolution, real-time operation, redundancy, simplicity, size and power comsumption. Popular sensors typically include sonar, tactile, infrared and laser sensors and videocameras. The final selection basically depends on what kind of behaviour is expected of the robot. Navigational behaviours, particularly, often rely on sonar sensors despite their obvious drawbacks. Even though sonar sensors have a wide arc of uncertainty and, in their simplest version, only provide information about the distance to the closest obstacle in the beam direction, they are light, cheap, fast, easy to process and have a long detection range. The advantages of using sonar sensors for navigation are widely discussed in [2].

Basically, the navigation problem consists of answering three simple questions: i) where am I? ii) where am I going? and iii) how do I get there? The answer to the first question is known as the localization problem. It involves determining the agent's position according to what it perceives and where it was previously believed to be. This problem is typically solved by measurement, correlation and triangulation. The second and third questions involve determining a goal and planning a path that leads to the goal. They basically concern path planning and collision avoidance. Hence, the navigation problem can be subdivided into three tasks: i) collision avoidance; ii) robot positioning; and iii) path planning.

The localization problem is not easy to solve. Basically, most systems rely, at least partly, on odometry. However, robot slippage provokes small positioning errors that accumulate unrestrainedly. Consequently, after a while the robot may not know its real position. The problem is even worse if no odometric information is available. In this case, the problem is known as global localization, while localization based on odometric information is known as tracking. Most tracking approaches rely on Kalman filtering to integrate sensor information over time. Global techniques aim instead at locating significant features (landmarks) in the environment in order to determine the robot position with respect to those landmarks. Even though localization is of capital importance to navigation, this chapter does not cover the issue, because, in our case, we rely on well-known localization techniques plus a compass to correct odometric information. Further information on localization can be found in [2] and [3]. Instead, this chapter focuses on techniques to determine a goal and to plan how to get there. Our system is meant to operate in real time even in unknown or changing environments.

The chapter is organised as follows. First, Section 2 presents an overview of classic navigation control techniques and a proposal for a hybrid control architecture meant to combine efficient planning and fast reactive behaviours. Section 3 proposes a new model of the environment, which is required to implement the proposed architecture. The main novelty of the proposed model is that it efficiently combines a topological and a metric representation to allow hierarchical planning. Then, Section 4 presents a navigation technique based on the proposed architecture. This technique allows the robot to move flexiblely and efficiently in order to visit one or more places in the environment. Section 5 presents several results for a real robot running on the proposed scheme. Finally, conclusions and future work are presented in Section 6.

2 A Hybrid Approach to the Navigation Problem

Navigation is clearly a central concern for mobile robotics and, consequently, many navigation techniques have been proposed. Nevertheless, navigation control schemes can be broadly divided into two large groups [4]: reactive and deliberative schemes. Initially, most navigation schemes were based on deliberative planning [5]. Deliberative planning typically relies on a classical top-down methodology known as horizontal decomposition [6][7]. In these cases, the world is represented and processed according to actions and events in a *sense−model−plan−act* cycle. These schemes are typically decomposed hierarchically into several levels, which interchange information by means of well-defined flow paths. While the higher levels are in charge of planning and reasoning, the lower levels support low-level control and direct hardware actions. Briefly, these schemes have the following features:

- A hierarchical structure with well-defined functions for each level.
- Communication between levels is predictable and deterministic.
- The upper levels in the hierarchy decompose each task into subtasks for lower levels.
- The lower levels work locally.
- They are strongly dependent on representations of the environment.

Deliberative schemes are often criticized for their inability to react rapidly. It can be easily observed that the robot must sense, model and plan before acting in these schemes. Although there have been some attempts to fix this latency problem using temporal constraints [8], these schemes have a second problem: they traditionally assume that the environment remains almost static between consecutive observations. In this case, any condition violating this assumption, like mobile or unexpected obstacles, may cause problems for the robot.

To overcome the drawbacks of deliberative systems, reactive schemes rely on directly coupling sensors and actuators [9]. The reactive paradigm is based on animal intelligence models, and it basically produces a global action by combining one or more reactive behaviours. This action is known as emergent behaviour. Unlike classic deliberative systems, reactive schemes can easily deal with several sensors, as well as aiming for several goals. Besides, they are quite robust against sensor errors and noise, and they can be easily modified to deal with changes in hardware or tasks. Briefly, reactive schemes are preferable to deliberative schemes when [4]:

- The environment is dynamic.
- Immediate robot sensing is adequate for the task at hand.
- The robot is not easy to locate with respect to a global coordinate system.
- No valid representation of the environment is available.

The best known reactive scheme is the subsumption architecture [10], which consists of a number of behaviour modules arranged in a hierarchy.

Different layers of the architecture are responsible for different behaviours, which are known as levels of competence. A level of competence may be, for example, to avoid contact with objects, to wander through the environment without hitting things or to build a map. Each level of competence includes all the earlier ones. The final emergent behaviour of the system is the set of reactions that emerge from the modules designed to achieve the different levels of competence. Unfortunately, reactive schemes also have important drawbacks. First, emergent behaviours may be very unpredictable. For this very reason, scheme performance may become inefficient in some cases. They are also typically prone to fall into local traps. Finally, they are quite difficult to debug.

Hybrid systems combine deliberative and reactive schemes in order to achieve better performance. Usually, low level control is performed reactively, whereas high level processing follows a deliberative pattern. Not only are hybrid systems supported by biological evidence [4], but they are also capable of providing efficient navigation in dynamic and totally or partially unknown environments. The main concern when building a hybrid scheme is to find the right boundary for the subdivision of functionality. Lyons [11] proposes three different ways of integrating planning and reaction:

- Hierarchical integration, where planning or reacting depend on the situation at hand.
- Planning to guide reaction, where planning modules are used to configure and set parameters for the reactive control system.
- Coupled planning-reacting, where planning and reacting are concurrent activities, each guiding the other.

Basically, according to the way in which integration is implemented, the most representative hybrid architectures are [4]: i) selection, where deliberative modules configure which reactive modules should be active; ii) advice, where deliberative modules just offer advice that reactive modules may or not may accept; iii) adaptation, where deliberative modules modulate reactive modules according to the world and task requirements; and iv) postponement, where deliberative modules work only when necessary.

Most recent approaches to navigation rely on hybrid architectures because of their advantages with regard to purely reactive or deliberative ones. Thus, in this Section we propose a hybrid architecture to control a sonar-based mobile robot. Even though there are several guidelines for building an efficient architecture for robotic control, most schemes tend to be built *ad hoc*, because a given architecture is heavily influenced by hardware conditions, goals and algorithmic criteria. An excellent review of the most general approaches for build a robotic architecture can be found in [12].

One of the most pressing issues when designing a control architecture is how to manage the growing complexity of interactions both between the system and its environment and among the individual components of the

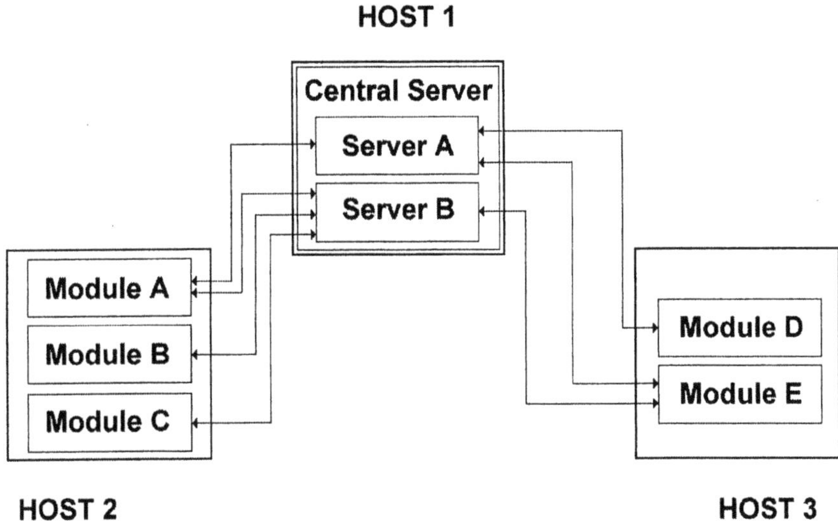

Fig. 1. Proposed architectural style.

system. Modern robotics systems, usually requiring concurrent embedded real-time performance, cannot be operated using conventional programming techniques. The computational concepts underlying a given system are referred to as architectural style [12]. One of the most popular styles for dealing with complexity is to decompose the system into modules which operate concurrently. Each of these modules has a different function, and they exchange information to work as a whole. When a system is decomposed into modules, it is necessary to provide the mechanisms required by the different modules to exchange information in order to cooperate in parallel. There are several approaches to this problem, like end-to-end connections between modules [13], specific routing agents [14] [15], or shared memory-based systems [16]. We have chosen a shared memory-based approach because it suited our scheme better than the others. In our case, the different modules of the architecture are distributed over different machines, which may even have different operating systems. Thus, we used a scheme proposed by Dulimarta [16], which was very suited for transparent information exchange between different machines and operating systems. In this scheme, modules share information by sending it to a specific data server. This server is in charge of controlling the access to all shared data. Thus, when a module needs information, all it has to do is place a request with the server. However, Dulimarta proposed a single server to manage all the information flow. Thus, when many modules send or request information simultaneously, they are queued and response

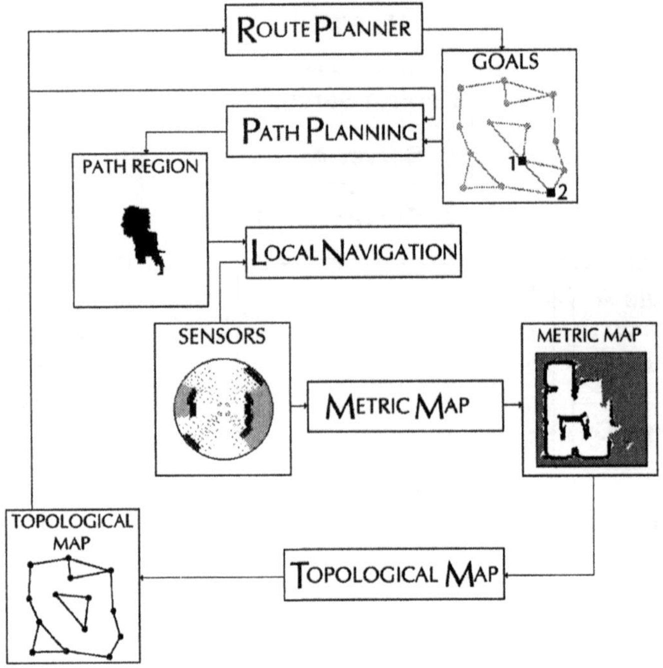

Fig. 2. Proposed architecture

time is increased. Hence, we have modified the original scheme by adding local data servers which are controlled by a central server. When a module requests information from the central server, it redirects the request to one of the local servers. Hence, it is free to receive more requests.

Now we have the architectural style, the next thing is to design the architectural structure. Basically, any given robot needs to accomplish a set of tasks, which must be decomposed into modules. Different applications may need to be decomposed differently and, since systems tend to grow, a structure needs to be flexible enough to accommodate different decomposition strategies. Often, system decomposition is hierarchical, because this approach leads to more modular systems. Indeed, quite a popular structure for navigation is the navigation hierarchy [17], because: i) each level of the hierarchy can be tested experimentally; and ii) it is valid for representing biological and robotic navigation. In this hierarchy, navigation behaviours are classified according to the complexity of the task which they can perform. While local navigation behaviours require only recognition of the goal location, way-finding behaviours require knowledge about several places, as well as their interrelation. We propose a structure based on the navigation hierarchy, where different layers are in charge of progressively less complex behaviors (Fig. 2).

Basically, high level layers are in charge of reasoning about the environment and planning a course of action. Different planning algorithms are used for this purpose, depending on the goal to be achieved. Planning algorithms require models of the environment about which they reason. Thus, on-board sensor readings are used to build these models. The processes required to model the environment are discussed in detail in Section 3. In our case, two different models are required: geometric and topological maps. The advantages of this approach, along with a brief review of other models, are presented in the next Section. We will see that these models are built by means of two different layers (Fig. 2). One layer is in charge of creating and continously updating a geometric map using the sonar readings. A second layer is in charge of creating a topological map. This second layer uses the geometric map as an input. Thus, we keep the topological map grounded.

The highest level layer of the whole structure is the route planner. This layer is in charge of determining which areas of the environment are going to be visited and in which order. If our goal is just to move from one place to another, the output of this layer is a single goal, which is equal to the final location. If, instead, the goal is to visit a set of places, the output is an ordered list of goals, which is created by minimizing factors like the total distance traveled. Then, the path planning layer uses the topological map as an input to calculate an obstacle-free path between any two locations. Finally, the local navigation layer relies on reactive navigation to track the paths calculated by the path planning layer. Section 4 focuses on the practical implementation of all these layers. Note that all these layers share information using the above architectural style. Each time a layer requires information, it sends a request to the central server. This server routes the request to the respective local server, which, in turn, sends the most recently updated data to the querying layer.

The proposed hybrid architecture works efficiently for several reasons: i) deliberative layers propose efficient paths to a goal; ii) these paths are quickly and reactively tracked to handle unexpected situations; iii) the environment is represented at two hierarchical levels so that it can be both updated and processed rapidly; iv) the local navigation layer can work even when the output from higher level layers is outdated. Thus, the robot does not need to stop when a given path needs to be recalculated due to unexpected obstacles in the way.

3 Map Generation

Representations of the environment are usually built according to either the metric or the topological paradigm. Metric approaches generate representations that explicitly reproduce the metric structure of the environment. Metric maps can be either geometric or grid-based. In geometric representations, world features, like walls or corridors, are directly mapped with respect to a global coordinate system. In grid-based representations [18], each grid cell is associated with a specific position in the environment, which is implicitly given by the cell coordinates (x, y). Any cell yields an occupancy value that is equal to the probability of that cell being occupied according to current and past sensor readings. Topological approaches [19] aim at representing the environment as a set of meaningful regions. These regions are usually represented as nodes, which are inserted each time the robot sensors perceive a pattern corresponding to a representative place. Thus, the map becomes a topological graph and nodes become linked if the robot can find an obstacle-free path between them.

Both approaches have complementary strengths and weaknesses [20]. Of all the metric maps, evidence grids have become popular because they are usually fast and easy to build and update and they provide efficient space-time integration of sonar sensors. Besides, the geometry of the grid directly corresponds to that of the real environment, so the robot position within the model can be determined by its position and orientation in the real world. However, most metric approaches rely heavily on dead-reckoning, which is unreliable in large environments. In addition, a grid is only reliable if it is highly decomposed. Hence, these approaches usually involve a huge data load and, consequently, they are computationally expensive to process when medium or large environments are modeled. On the other hand, topological maps are usually more compact, because their resolution is determined by the complexity of the environment [20]. Topological maps are primarily used for robot position estimation and path planning [21], because they allow fast planning and provide more natural interfaces for human instructions. However, topological representations also have several problems. The most important one is known as the dissambiguation problem, which involves distinguishing different places that are characterised by the same sensory pattern. Similarly, it is necessary to determine whether or not two nodes associated with an equal sensory pattern actually correspond to the same place. Odometric information is usually added to topological maps to overcome these problems. However, odometry is not always reliable and maps may become erroneous when updated or after the robot has been moving for a while. Besides, the dissambiguation process tends to be computationally expensive.

Recently, there have been several proposals for combining metric and topological representations so that their advantages can be combined as well. Basically, there are two ways of constructing a topological-metric representation: either a topological representation is annotated with metric information while it is constructed [22] or a topological map is extracted from a metric one [23] [20] [24]. The main problem of approaches relying on annotating topological maps with metric information is that acquired maps usually require further processing. This processing may be intensive and must often be performed off-line. Extracting topological from metric maps appears to be easier. This type of schemes typically rely on splitting metric maps into a set of homogeneous regions, which are the nodes of the graph.

We propose a new mixed approach for integrating the metric and topological paradigms [25]. Our method is related to proposals by Thrun [20], Arleo [23] or Zelinsky [24] in several respects. First, all these methods use a local occupancy grid to model the region surrounding the robot at the metric level. Some of the above-mentioned methods use local grids to model obstacle boundaries by means of straight lines [23]. Others, like ours, just use these local grids to construct a global one [20]. Despite these similarities, all these approaches use quite different methods to construct the topological map. Thrun's method [20] extracts the map off-line using Voronoi diagrams and only after the whole global map is available. Arleo [23] splits the metric map into rectangular regions, and this method cannot deal with irregular shaped regions. Zelinsky [24] performs this partitioning using quadtrees. However, the optimality of the resulting partition is strongly dependent on the obstacle distribution of the environment. Our approach extracts the topological map from the metric representation on-line. Besides, it can deal with irregular regions and it does not have the disadvantages of quadtree decomposition. The following two subsections describe the metric map learning process and the topological map extraction processes, respectively.

3.1 Metric Map Learning

In this chapter, the metric map is a two-dimensional occupancy grid, similar to the one originally proposed by Moravec and Elfes [18]. Each grid cell (x, y) in the map yields the occupancy probability of the respective region of the environment. These cells are modified according to the readings of Polaroid sonar sensors, which have an arc of uncertainty of approximately 25°. A very simple probability distribution is used to model the cells in a sonar scan. The occupancy probability of a cell is modelled as:

$$P(\theta,\rho)\begin{cases}-\epsilon, & \text{if } 0 \le \rho < d-\delta, 0 \le |\theta| \le \beta/2 \\ +\varphi, & \text{if } d-\delta \le \rho < d+\delta, 0 \le |\theta| \le \beta/2 \\ 0, & \text{otherwise}\end{cases} \tag{1}$$

ρ and θ being the distance to the robot and the angle of the main axis of the sonar beam, respectively; d the range measurement returned by the sonar sensor and β the beam aperture. $2 \cdot \delta$ determines the width of the region of uncertainty where the obstacle could be located. Finally, the empty and occupied probability density functions for a cell inside the sonar beam are given by constants ϵ and φ, respectively. The resulting probability distribution for a single scan is shown in Fig. 3.

To build the metric map, sonar readings are integrated into a local occupancy grid, which is the result of the weighted addition of all the sonar sensor models. Fig. 4a shows an example of a sonar scan and its interpretation. The

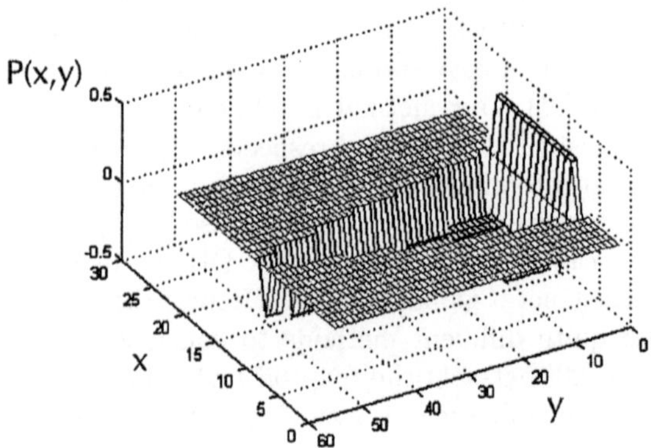

Fig. 3. Current model of the sonar sensor.

darker the value in the circular region around the robot, the larger the internal computed value. Note that the local grid has negative and positive values, which correspond to empty and occupied regions, respectively.

Sonar interpretations must be integrated over time to update the map coherently. For this purpose, the local grid adquired at time t (P^t_{local}) is combined with the available metric map (P^{t-1}_{metric}). In our case, the metric map cells coinciding with the local grid region are updated using the expression:

$$P^t_{metric} = P^{t-1}_{metric} + P^t_{local} \tag{2}$$

The main advantage of the above sensor model and integration algorithm is their low computational complexity. As a field test, the proposed method, the Dempster-Shafer approach [26] and Bayes' method [20] have been used to build different maps of a large real environment, consisting of a corridor and a room of irregular shape. Fig. 4b represents a map created using the proposed approach. The CPU load in the proposed method is 15 % lower than in the other two methods.

Finally, it is important to note that the precision of the resulting metric map depends on the correct alignment of the robot with its map. Hence, slippage and drifting must be detected and corrected [27]. This information is extracted from the localization layer, which uses well-known techniques like correlation of the local map and the respective section of the global map [28].

3.2 Topological Map Building

A new hierarchical structure is proposed in order to extract a topological map from the above metric map. This structure is constructed as follows:

1. Metric map thresholding. Initially, the occupancy value of each cell in the metric map is thresholded. Cells whose occupancy value is below threshold U_1 are considered free space ($P(x,y) = c_F$). Cells whose occupancy

value is above U_1 and below threshold U_2 are considered unexplored ($P(x, y) = c_N$). All other cells are considered occupied ($P(x, y) = c_O$).

2. Hierarchical structure generation. The thresholded metric map becomes the base of a pyramidal structure. Each level l of this pyramid is a reduced map with 1/4 of the cells of the level immediately below. Each pyramid cell (x, y, l) has five associated parameters:
   - Homogeneity, $H(x, y, l)$. $H(x, y, l)$ is set to 1 if the four cells immediately underneath have the same occupancy probability and their homogeneity values are equal to 1. Otherwise, it is set to 0.
   - Occupancy probability, $P(x, y, l)$. If the cell is homogeneous, $P(x, y, l)$ is equal to the occupancy probability value of any of the four cells immediately underneath. If the cell is not homogeneous, the value of $P(x, y, l)$ is set to a fixed value (c_{NH}).
   - Area, $A(x, y, l)$. It is equal to the addition of the areas of the four cells immediately underneath.
   - Parent link, $(X, Y)_{(x,y,l)}$. If $H(x, y, l)$ is equal to 1, the values of parent link of the four cells immediately underneath are set to (x, y). Otherwise, these four parent links are set to a null value.
   - Centroid, $C(x, y, l)$. It is the centre of mass of the base region associated with (x, y, l).

   After generation, the cells at upper levels of the hierarchical structure have an homogeneity value equal to 1. These cells can divide the grid-based map like quadtree approach does [29]. However, the complexity of this decomposition is not directly related to the world complexity. Instead, it depends on the position of the obstacles.
3. Homogeneous cell fusion. In this step, the algorithm tries to link cells whose parent link values are null. Basically, these cells, (x, y, l), are linked to parents of neighbouring cells, $(x_p, y_p, l+1)$, if the following conditions are true:

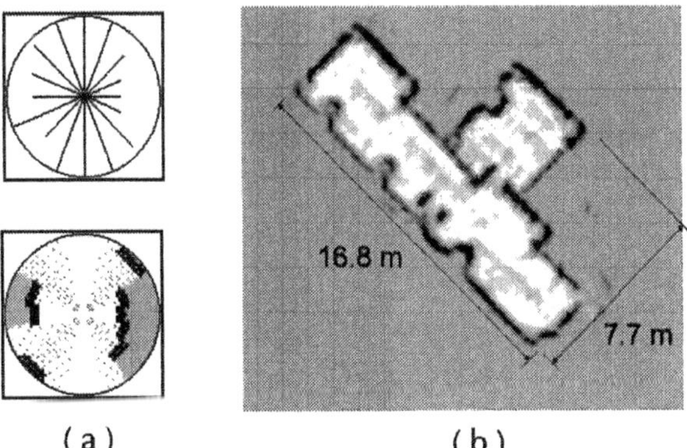

Fig. 4. a) Sensor interpretation: sonar scan (top) and local map (bottom); and b) grid-based map created using the proposed approach.

- $H(x,y,l) = 1 \;\&\; H(x_p, y_p, l+1) = 1$
- $P(x,y,l) = P(x_p, y_p, l+1)$
- $||C(x,y,l) - C(x_p, y_p, l+1)||_2 < DistMax$,
  $DistMax$ being a threshold that establishes the maximum dispersion of the regions at the base.

4. Homogeneous cell classification. Two neighbouring cells, (x_1, y_1, l) and (x_2, y_2, l), are fused if the following conditions are true:
   - $(X,Y)_{(x_1,y_1,l)} = NULL$
   - $(X,Y)_{(x_2,y_2,l)} = NULL$
   - $H(x_1,y_1,l) = 1 \;\&\; H(x_2,y_2,l) = 1$
   - $P(x_1,y_1,l) = P(x_2,y_2,l)$
   - $||C(x_1,y_1,l) - C(x_2,y_2,l)||_2 < DistMax$

The proposed algorithm extracts the topological map from the metric representation on-line. Besides, unlike other approaches, the whole global map does not have to be acquired [20]. Fig. 5 illustrates the process of extracting a topological map from a grid-based map. Fig. 5a shows the thresholded map associated with the map in Fig. 4b. The resulting partitioning is shown in Fig. 5d. Note that cells yielding an area less than 8 are not shown. Finally, Fig. 6 shows the final topological graph. The main advantages of the proposed topological graph extracting algorithm are its low computational load and that it depends only on threshold $DistMax$. The average topological graph extracting times are shown in Table 1.

In these tests, metric maps have 65,536 cells. The low computational time allows the on-line generation of the proposed topological map.

Fig. 7 illustrates the algorithm dependence on threshold $DistMax$. If $DistMax$ is large, the regions tend to be large, but if it is low, the resulting map is excessively partitioned. Fortunately, it is very easy to choose a suitable $DistMax$ and it has been empirically proven that $DistMax \in [50,70]$ works correctly in most cases.

4 Survey Navigation

Survey navigation is the highest level of the navigation hierarchy [17]. In fact, there are many examples of lower level navigation mechanisms, like local navi-

Table 1. Topological map extracting times

Process	Time (typical)
Structure initialisation	0.10 s
Homogeneous cells linking	0.05 s
Node definition	0.05 s
Total time	0.20 s

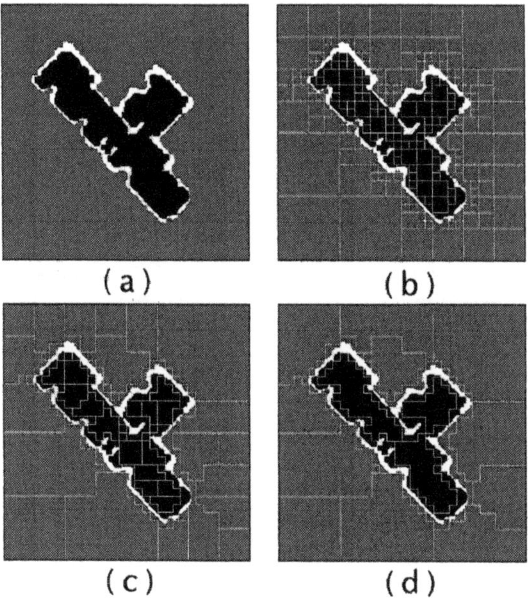

Fig. 5. Extracting the topological map: a) thresholded map; b) base regions generated after the generation step; c) base regions generated after homogeneous cell fusion; and d) base regions generated after homogeneous cell classification (topological regions).

gation behaviours, recognition triggered responses and topological navigation in the animal kingdom, but survey navigation may be limited to vertebrates. Survey navigation specifically requires embedding all known places and their spatial relations into a common framework of reference so that the representation can be manipulated as a whole. When every location is embedded into a common framework of reference, the agent can find novel paths over unknown terrain by inferring its spatial relation to known places. There is

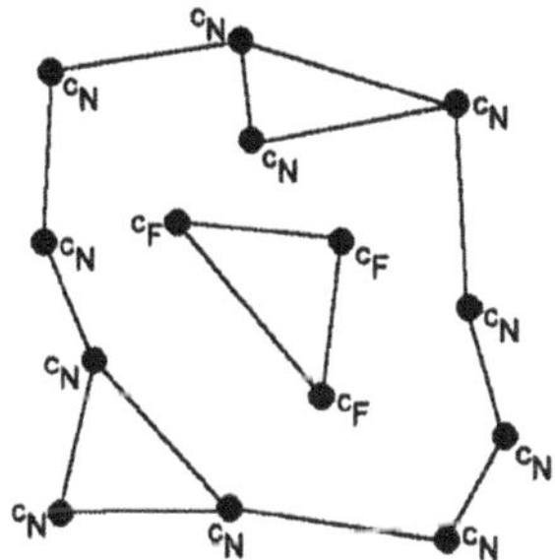

Fig. 6. Topological graph associated with the metric map in Fig. 4.b.

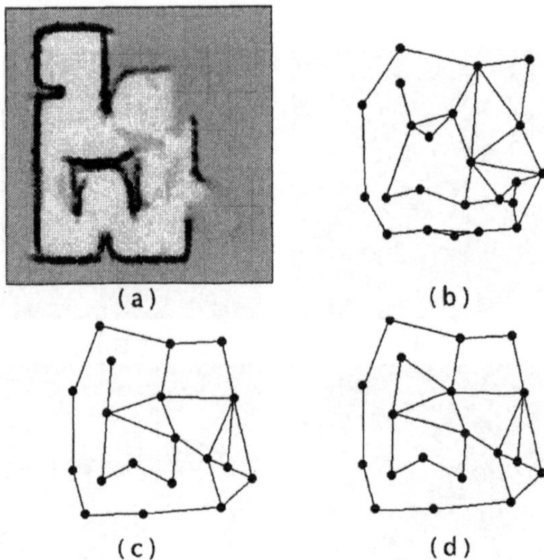

Fig. 7. Threshold *DistMax* evaluation: a) metric map; b) topological graph (*DistMax* = 40), c) topological graph (*DistMax* = 60), and d) topological map (*DistMax* = 80).

not a lot of work on survey navigation. In fact, existing approaches to planning in partially unknown environments rely on metric maps (i.e. [18]), where geometric relations between places are explicitly represented analytically. As mentioned above, metric maps are easy to build and fast to update, but they are also sensitive to all errors affecting metric information, namely robot slippage, and they typically yield a very large data volume. Hence, planning over an average sized environment may be so costly that on-line operation is not possible. Classic topological maps do not include information on unexplored areas, so they are not suitable for this purpose either. However, a combined topological-metric structure like the one proposed can be used for survey navigation. Hence, we propose a navigation scheme for performing survey navigation on a totally or partially unknown environment.

This Section presents the practical implementation of the layers of the architectural structure in Fig. 2 to achieve survey navigation. The metric map layer is in charge of creating and updating the evidence grid. This layer receives the position information from the localization layer and the sonar readings to update the grid each time the sonars are read. The only output of this layer is a global evidence grid (Fig. 8a). It must be noted that the size of the world is not known *a priori*, so this grid may grow in terms of x and y. The grid is the input to the topological map layer, which is in charge of creating a topological map as explained in the previous Section. Even though the proposed topological map can be created rapidly, it cannot be generated as fast as grids are updated. Thus, this layer can be triggered according to two different approaches. If the topological map needs to be updated as often as possible, each time a topological map is finished, the layer starts to create a new one. Obviously, if a planning layer requests a topological map while it is still being created, it will receive the former topological map. However, topological maps are created rapidly, so they are never very outdated. The main problem with this approach is that the resulting computational load may be excessive, specially considering that the topological map does not usually change so often. A second option is to shoot the layer only when the planning module, which uses the topological map as an input, is going to be triggered. The main disadvantage of this approach is that the planning module has to wait for the topological map layer to finish before starting to plan, because, in this case, a given topological map may be quite outdated. These two options may be combined efficiently by predicting how often the environment is going to change in the near future: for example, the metric map is likely to change a lot in unexplored areas. In our current model, the system is only meant to navigate. Thus, the topological map is updated as often as possible. Fig. 8b shows a topological map created by this layer.

As discussed in Section 2, the route planner is in charge of determining which areas of the environment are going to be visited and in which order. Hence, its output is an ordered list of goals which is created by minimizing factors like the total distance traveled. The available topological map is fed to the layer in order to calculate this list. The optimization problem is handled as a classic Traveling Salesperson Problem (TSP). The TSP involves search for the shortest tour of all the nodes, given a finite set of nodes $N = c_1, c_2, ..., c_n$ and a distance $d(c_i, c_j)$ for each pair of nodes. Note that the TSP is one of the most representatives NP-complete problems. Hence, its processing time is drastically increased along with the problem instance. Because of this, the route planner works with the topological map, which yields fewer nodes, rather than with the metric map. Being a well-known problem, many methods have been proposed to solve the TSP. Although the TSP has been traditionally used as a neural network benchmark, its validity for this particular problem has been widely questioned [30]. Thus, we use the method proposed in [31] to solve it. This method is based on a genetic algorithm and its advantages and drawbacks are widely discussed in [31]. Note that this layer only provides the order in which the resulting goal locations should be visited and not a path between them.

The path planning layer is used to calculate an obstacle-free path between two consecutive goal locations of the set given by the route planner. First, a path between the current position of the robot and the goal node is calculated using the topological map as the problem instance. This calculation is performed using the A* algorithm, which is very fast when dealing with a small number of nodes. Then, the resulting node path is propagated to metric level by means of the link structure. At the metric level, the node path is a region of free cells between the robot and the goal location. Fig. 8b shows

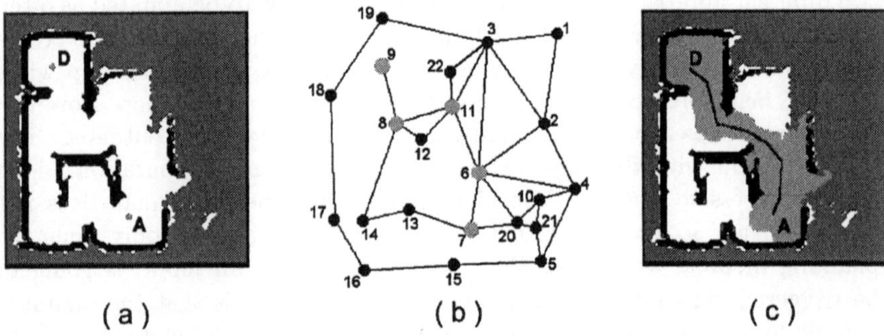

Fig. 8. Path planning: a) metric map (D=departure location; A=arrival location); b) node path at topological level; c) resulting path at metric level

a path calculated at the topological level. Nodes belonging to that path are marked in gray. Fig. 8c shows the path region at metric level, also in gray. The path planning layer is simply providing an efficient path region to the goal. We are not interested in calculating a precise path to the goal, because such a path might be unfeasible in the presence of unexpected obstacles.

We use a potential field approach to implement a low level navigation layer to move along the path provided by the path planning layer. Potential fields were originally proposed by Latombe [32] and they consist of modelling obstacles as repulsors and goals as attractors. The robot moves according to the vector resulting from its position in the field. Note that this approach has several problems, as reported in [33]. Recent proposals to solve these problems [34] [35] rely on more deliberative approaches. Our system relies on modulating low level navigation using the results of the deliberative path planning layer. In this case, the boundaries of the path region, as well as the boundaries of any obstacle inside the path region, become repulsors, while the goal is an attractor. Fig. 8c shows the final path output by the local navigator for the region corresponding to the node path in Fig. 8b. It is important to note that the map of the environment is updated while the robot is navigating. Thus, unexpected objects appearing in the path region may make it impossible to reach the goal. In such cases, the robot keeps moving reactively until a new path region is returned by the path planning layer. It should be noted that, in some cases, the unexpected obstacles could block the path to the goal. In these cases, the route planner can provide a different goal for the path planning layer. Thus, a path region from the current position of the robot to the new goal would be calculated.

The first layer to be triggered when the robot starts to navigate is the route planner. This planner provides a set of goal locations according to the available information about the environment. It should be noted that part of the environment may be unexplored or changed, so this information is likely to change when the robot starts to move. When a set of goals is

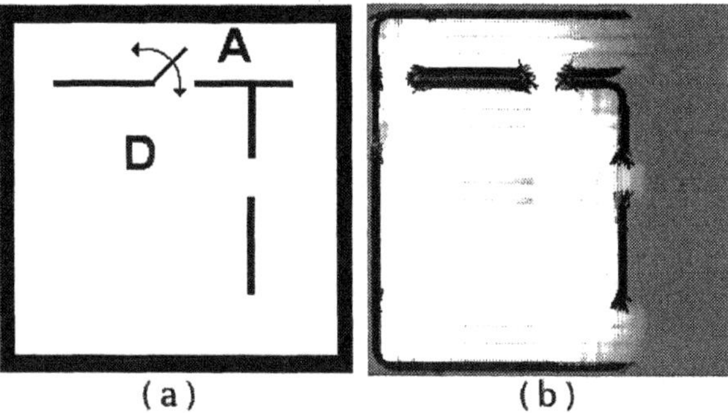

Fig. 9. A survey navigation problem: a) emulated environment with a closing door; b) map available before the closed door is detected.

available, the path planning layer calculates a trajectory between the first two goals. If there is no trajectory to move from one to the other, the route planner must be triggered again. Otherwise, the local navigator tracks the resulting route until it arrives at the goal location. While the robot moves, the environment is being explored and maps are being updated. Thus, if the local navigator cannot reach the goal location, a new path is calculated between the current position of the robot and the goal. It should be noted that the path planning layer is very fast as long as a valid topological map is available, so the new trajectory can be calculated on-line. If no valid trajectory can be found between both locations, the route planner is triggered.

Fig. 9 presents a typical problem for the proposed scheme working on a simulator. The 1056 m^2 environment in Fig. 9a is only partially explored by the robot. Fig. 9b presents an occupancy grid yielding 256x256 cells where obstacles are black, free space is white, and unexplored areas are gray. Initially, the door is open. Thus, the easiest way to go from point D (departure) to point A (arrival) is through the door. Then, we close the door. A recognition triggered response based scheme would fail, because the response to the input sensory pattern at point D would always be to go through the door. A topological approach would find a route across the west side of the room, because no nodes or arcs are available for the unexplored area. Path planning at the metric level would deal with a problem instance of 65536 cells.

Fig. 10 presents the results of applying the proposed survey navigation scheme for the problem in Fig. 9. Three different situations have been tested to prove the efficiency of the scheme. Fig. 10a shows the initial topological metric map when the door is still open. Topological nodes are marked with circles. It should be noted that if there is an arc between two nodes, it means that there is a feasible path from one node to the other. However, the path is not a straight line. It can also be observed that unexplored areas are also represented at the topological level, because their geometric relations to the other nodes can be extracted from the associated metric map. Hence, even

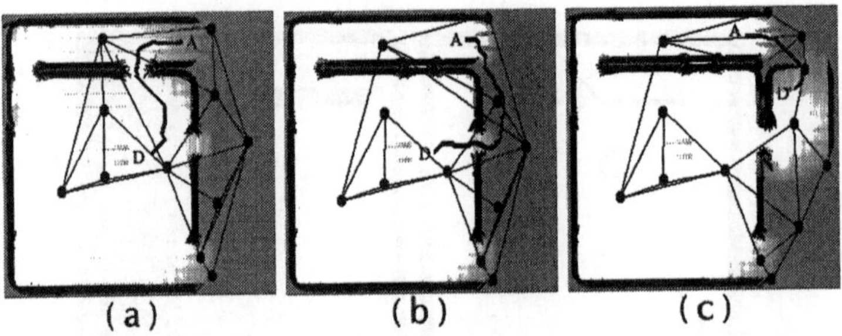

Fig. 10. Maps and resulting paths for: a) original situation with open door; b) situation a) with closed door; c) unexpected obstacle in unexplored regions.

though the robot has not traveled across the east side of the map, it is feasible that there might be a path from D to A crossing the unexplored area. Nevertheless, as long as the door is open, the shortest path from D to A runs through the door as expected. In Fig. 10b, the door is closed. Hence, a new path is required to reach the goal. Note that the metric map is updated to include the closed door. Consequently, the associated topological map is also modified and the path in Fig. 10a is no longer feasible. Note how the upper part of the topological graph is modified to represent this situation. Fig. 10b shows the new route chosen by the robot to reach A. Note that the route crosses the unexplored area, because it would be more costly to border the north wall. Finally, Fig. 10c illustrates a typical problem when planned routes go through unknown regions: there is an unexpected obstacle in the unexplored area that makes the path in Fig. 10b unfeasible. Since the cognitive map can be updated very quickly and path planning based on the topological map is fast as well, the robot can calculate a new route and still reach its goal more efficiently than if it had chosen to border the north wall.

5 Experiments and Results

The proposed survey navigation scheme has been implemented on a Pioneer 2-AT mobile robot in order to test its validity in real conditions. This Pioneer robot is a four-wheeled robot having 6 front sonar sensors plus 2-side sonar sensors. It has an on-board 400 MHz Versak6 PC with PC104+ bus and 32 Mb RAM. This robot is connected to an on-board laptop which is linked to a local area network by means of a 11 Mbps Ethernet link. The architecture central server runs on a 166 MHz Pentium PC with 32 Mb RAM. The other machines in the LAN are also ordinary PCs.

It should be noted that real environments are quite noisy from the sensor point of view. Sonar readings are affected by refractions, reflections and multiple echoes and, consequently, the adquired models of the environment are not as well defined as they were in simulation.

A first test in a small environment is presented in Fig. 12. This environment is the laboratory shown in Fig. 11a. Even though it is not a large place, different kinds of materials, like wood, metal cupboards or glass, as well as tables and chairs, are present. Initially, the robot starts at the northeast corner of the room. After a few movements, the metric map of the place shown in Fig. 12a is available. This map presents 256x256 cells, and its corresponding topological map is presented in Fig. 12b. Fig. 12c shows departure and arrival points (D and A, respectively). D is equal to the current position of the robot and A is equal to the desired goal. Using the topological map, the path planning layer returns a path region to the local planner. Fig. 12c shows the path that the robot would follow according to the local navigation layer. It should be observed that this is a preliminary path, because the final path is influenced by on-going sensor readings.

In order to test the robot's ability to react to unexpected obstacles in its way, a person walked in front of the robot when it was following the path in Fig. 12c. Fig. 12d shows a metric map in which the mobile obstacle is already mapped. Since the former path is unfeasible, the path planning layer calculates a new one using the updated topological map in Fig. 12e. Fig. 12f shows the final path that the local navigation layer plans to follow.

Table 2 shows the processing times of the algorithm on a Pentium 550. It should be noted that no route planning is necessary in these cases, because only one node is going to be visited. Nevertheless, we forced the route planning layer to work to evaluate its processing time. Note that processing times are lower than a second. Even though there is an upper bound of approximately 0.25 s on topological map extraction for 256x256 grids on a PC like the aforementioned one, processing times may vary depending on the layout of the grid, and the topological map is generated much faster if the structure of the environment seems to be easier to analyze.

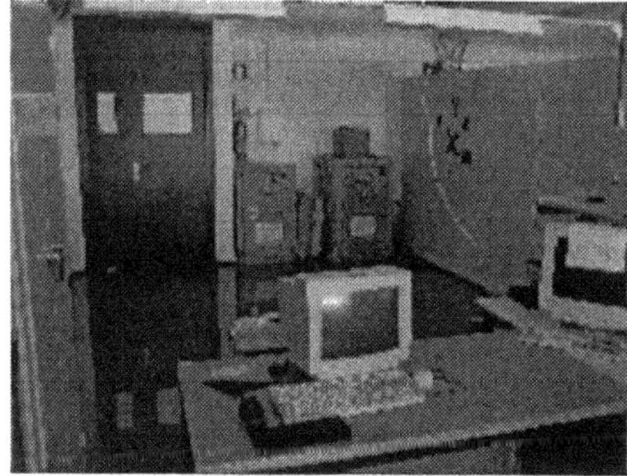

Fig. 11. Real environments.

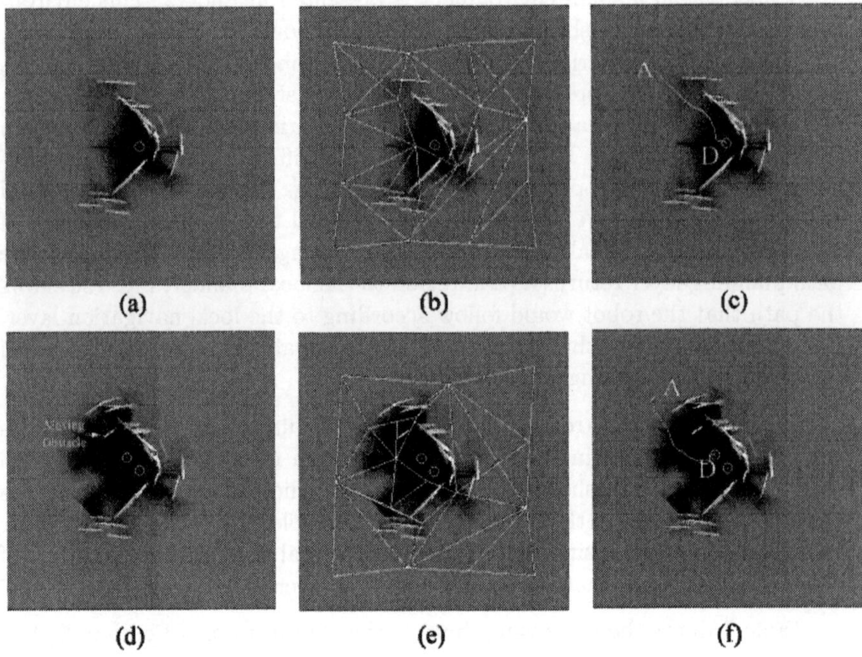

Fig. 12. Point to point navigation: a) original metric map; b) original topological map; c) original route proposed by the local navigator; d) new metric map; e) new topological map; f) final route to goal.

A second test environment is presented in Fig. 13. This environment is the corridor in Fig. 11b, a larger place than the laboratory shown in Fig. 11a. Figs. 13a,d,g,h represent the metric map with 256x256 cells, Figs. 13b,e,h,k represent the topological map, and Figs. 13c,f,i,l represent the original route proposed by the local navigator. As shown in Fig. 13a the initial metric map is completely unknown, the initial topological map of Fig. 13b has four nodes, and the robot begins to move from the actual position (departure point D) to one node of the topological map (arrival point A) in Fig. 13c. Some time

Table 2. Processing times for navigation in a small environment

Process	Time (typical)
Topological map building	0.20 s
Route planning	0.20 s
Path planning	0.10 s
Total time	0.50 s

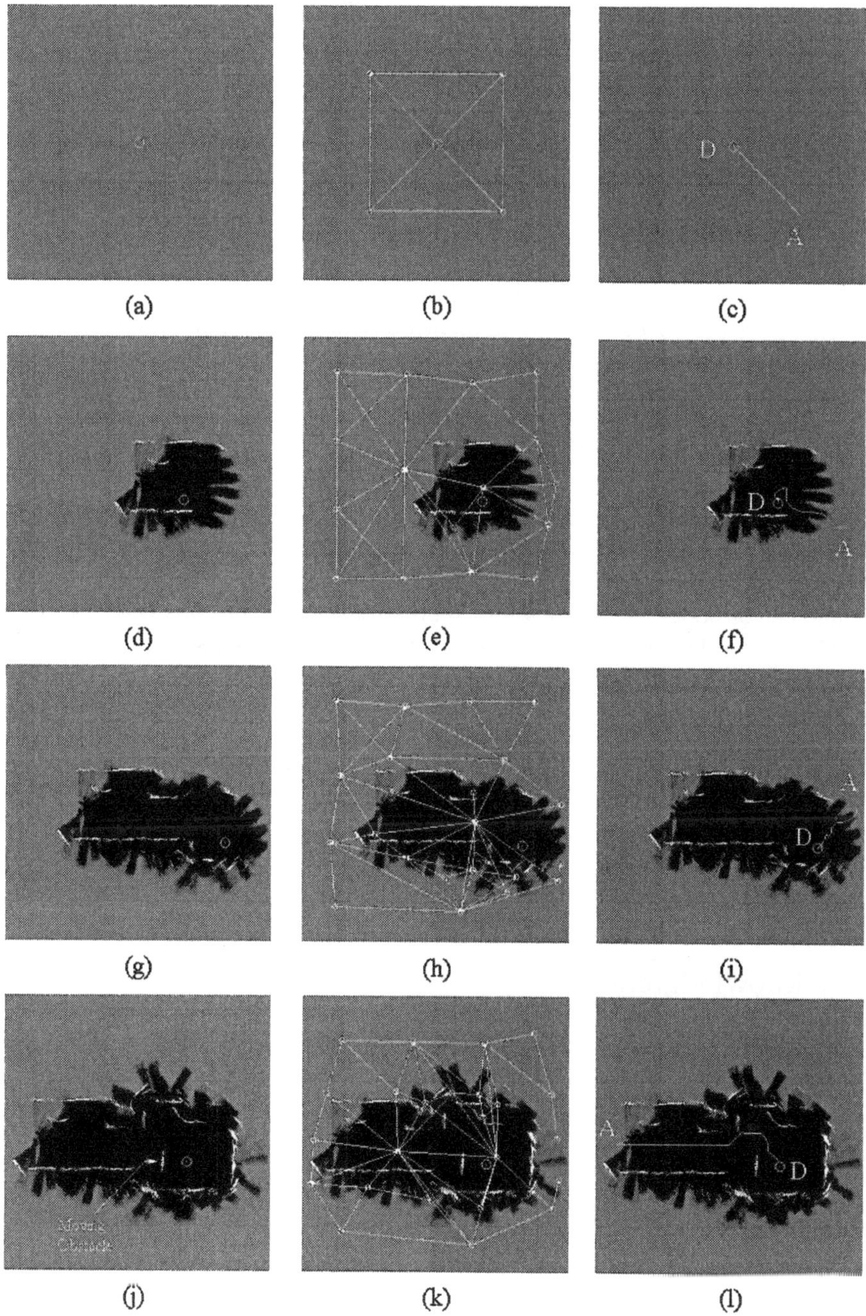

Fig. 13. Point to point navigation: metric maps, topological maps and routes proposed by the local navigator at different times.

later, as shown if Figs. 13d to i, the metric map and the topological map are updated as the robot moves, so the next node to be visited and the route to be followed changes. A mobile obstacle appears in Fig. 13j, and the path planning layer calculates a new path using the updated topological map in Fig. 13k. The final path is shown in Fig. 13l. Processing times are quite similar to the ones presented in Table 2.

6 Conclusions

This chapter has presented a hybrid architecture for sonar-based survey navigation in totally or partially unknown environments. The architecture has a client-server style. Unlike similar approaches, there is a global and several local servers rather than just one server in this style. Thus, data requests are handled faster and more efficiently. The architectural structure consists of several layers. Some of these layers are deliberative, while others are reactive. Basically, they exchange information using a hierarchical representation of the environment. The main advantages of this representation is that sensors are used to quickly and simply create and update a metric map, but high level processing is performed on a grounded topological map with a small number of nodes. Also, the whole representation can be constructed rapidly while the robot is navigating, and it explicitly represents unexplored areas at the topological level. Since deliberative planning tends to be computationally expensive, it is performed at the topological level. Reactive layers, which depend on the local geometry of the environment, work at the metric level instead. The main advantage of the proposed technique is that local navigation is modulated by deliberative planning, so that it is not only fast but efficient as well. Finally, it should be noted that layered modular systems are usually easy to expand. Our future work will focus on expanding the described architecture to add a vision module to the robot.

Acknowledgements

This work has been partially supported by the Spanish Ministry of Science and Technology and FEDER funds, project No. TIC2001-1758.

References

1. Dudek, G., and Jenkins, M. (2000): Computational principles of mobile robotics. Cambridge University Press, Cambridge, USA.
2. Leonard, J.J. and Durrant-Whyte, H.F. (1992): Directed Sonar Sensing for Mobile Robot Navigation. Kluwer Academic Publishers, Massachussetts.
3. Maybeck, P. S. (1990): The Kalman filter: An introduction to concepts. In I. J. Cox, & G. T. Wilfang (Eds.), Autonomous Robot Vehicles, New York: Springer, pp. 194-204.

4. Arkin, R.C. (1998): Behaviour based robotics, MIT Press, Cambridge.
5. Moravec, H.P. (1983): The Stanford Cart and the CMU Rover. Proc. of the IEEE, Vol. 71, No. 7, pp. 872-884.
6. Albus, J. (1991). Outline for a theory of intelligence. IEEE Transactions on Systems, man and Cybernetics, Vol. 3, No. 21, pp. 473-509.
7. Hu, H. and Brady, M. (1996): A parallel processing architecture for sensor based control of intelligent mobile robots. Robotics and autonomous systems, Vol. 17, 235-257.
8. Tsotsos, J.K. (1997): Intelligent control for perceptually attentive agents: The S* proposal. Robotics and autonomous systems, Vol. 21, pp. 2-21.
9. Brooks, R.A. (1991): Intelligence Without Reason. Proc. IJCAI-91', pp. 569-595, Sydney, Australia.
10. Brooks, R.A. (1986): A Robust Layered Control System for a Mobile Robot. IEEE Journal of Robotics and Automation, Vol. 2, No. 1, pp. 1423.
11. Lyons, D. (1992): Planning, Reactive. In Shapiro, S. (Ed.): Encyclopedia of Artificial Intelligence, 2nd Edition. John Wiley & Sons, New York, pp. 1171-1182.
12. Coste-Manière, E. and Simmons, R. (2000): Architecture, the backbone of robotic systems. Proc. of the IEEE International Conference on Robotics and Automation (ICRA), San Francisco, pp. 67-72.
13. Chocon, H. (1992): Object-Oriented Design and Distributed Implementation of a Mobile Robot Control System. Proc. of Workshop on Architecture for Intelligent Control Systems, R. Chatila and S.Y. Harmon, Eds. Nice, France.
14. Wise, J.D. and Ciscon, L. (1992): TelRIP Distributed Application Environment Operating Manual, Version 1.6. Technical Report 9103, Universities Space Automation/Robotics Consortium.
15. Simmons, R.G. (1994): Structured Control for Autonomus Robots. IEEE Transactions on Robotics and Automation, Vol. 10, No. 1, pp.34-43.
16. Dulimarta, H.S. (1996): A Client/Server Control Architecture for Robot Navigation. Pattern Recognition, Vol. 29, No. 8, pp. 1259-1284.
17. Franz, M.O. and Mallot, H.A. (2000): Biomimetic robot navigation. Robotics and autonomous systems, Vol. 30, pp. 133-153.
18. Moravec, H. P. (1988): Sensor fusion in certainty grids for mobile robots. AI Magazine, Vol. 9, No. 2, pp. 61-74.
19. Matarić, M.J. (1994): Interaction and intelligent behavior. Technical Report AI-TR-1495, MIT, AI-Lab, Cambridge-USA.
20. Thrun, S., Bucken, A., Burgard, W., Fox, D., Frohlinghaus, T., Hennig, D., Hofmann, T., Krell, M., and Schimdt, T. (1998): Map learning and high-speed navigation in RHINO. MIT/AAAI Press, Cambridge.
21. Borenstein, J., Everett, H.R. and Feng, L. (1996): Navigating mobile robots: systems and techniques. Wellesley, Massachusetts: A.K. Peters, Ltd.
22. Kuipers, B.J. and Byun, Y.T. (1991): A robot exploration and mapping strategy based on a semantic hierarchy of spatial representation. Journal of Robotics and Autonomous Systems, Vol. 8, pp. 47-63.
23. Arleo, A., Millán, J.R. and Floreano, D. (1999): Efficient learning of variable-resolution cognitive maps for autonomous indoor navigation. IEEE Transactions on Robotics and Automation, Vol. 15, No. 6, pp. 990-1000.
24. Zelinsky, A. (1992): A mobile robot navigation exploration algorithm. IEEE Transactions on Robotics and Automation, Vol. 8, pp. 707-717.

25. Bandera, A., Urdiales, C. and Sandoval, F. (2001): An hierarchical approach to grid-based and topological maps integration for autonomous indoor navigation. Proc. of IEEE/RSJ International Conference on Intelligent Robots and Systems (IROS 2001), pp. 883-888, Maui, Hawaii, USA.
26. Pagac, D., Nebot, E.M. and Durrant-White, H. (1998): An evidential approach to map-building for autonomous vehicles". IEEE Transactions on Robotics and Automation, Vol. 14, No. 4, pp. 623-629.
27. Rencken, W.D. (1993): Concurrent localisation and map building for mobile robots using ultrasonic sensors. Proc. of the IEEE/RSJ International Conference on Intelligent Robots and Systems, Vol. 3, pp. 2192-2197, New York-USA.
28. Schiele, B. and Crowley, J. (1994): A comparision of position estimation techniques using occupancy grids. Robotics and autonomous systems, Vol. 12, pp. 163-171.
29. Chen, D., Szczerba, R. and Uhran, J. (1997): A framed-quadtree approach for determining Euclidean shortest paths in 2-D environment. IEEE Transactions on Robotics and Automation, Vol. 13, No. 5, pp. 668-680.
30. Gee, A.H., and Prager, R.W. (1995): Limitations of Neural Networks for solving the Traveling Salesman Problem. IEEE Transaction on Neural Networks, Vol. 6, No. 1, pp. 1542-1544.
31. Larrañaga, P., Kuijpers, C.M., Murga, R.H., Inza, I. and Dizdarevic, S., (1999): Genetic algorithms for the Travelling Salesman Problem: a review of representations and operators. Artificial Intelligence, Vol. 13, No. 2, pp. 129-170.
32. Latombe, J.C. (1991): Robot Motion Planning. Ed. Kluwer, Academic Publishers, Boston.
33. Koren, Y. and Borenstein, J. (1991): Potential Fields Methods and their Inherent Limitations for Mobile Robot Navigation, IEEE International Conference on Robotics and Automation, pp. 1398-1404, California, USA.
34. Ulrich, I. and Borenstein, J. (1998): VFH+: Reliable Obstacle Avoidance for Fast Mobile Robots. IEEE International Conference on Robotics and automation, pp. 1572-1577, Leuven, Belgium.
35. Minguez, J. and Montano, L. (2000): Nearness Diagram Navigation (ND): A New Real Time Collision Avoidance Approach. Proc. of IEEE/RSJ International Conference on Intelligent Robots and Systems (IROS 2000), pp. 2094-2100, Takamatsu, Japan.

An Analytical Method for Decomposing the External Environment Representation Task for a Robot with Restricted Sensory Information

Félix de la Paz, José R. Álvarez, José Mira

Dpto. de Inteligencia Artificial - UNED, Madrid, Spain.
{delapaz,jras,jmira}@dia.uned.es

Abstract. The problem of modelling and reducing knowledge needed to build an internal representation of the environment is still a milestone in robotics. This representation task is crucial for both understanding perception in humans and programming advanced robots with reasonable navigation skills.

In this chapter, we propose an analytical method for decomposing this representation task in terms of a set of primitive inferences. All these inferences are analytical transformations of the sensory data that expand the input space, use rules to compute the centre of areas and polygons of open space and, finally, build a topological graph with possibilities of being updated by learning and used for navigation.

The methodological approach followed in this chapter is to search for a library of reusable modelling components that could be used to solve other similar problems of topological representation.

1 Introduction

1.1 The Proposals of Craik and Marr

In spite of great efforts by neuroscience, artificial intelligence and robotics, there is still no theory of calculus for the generic task of the external environment representation. We have a set of sensory data and we need to construct an internal representation of this external world to be used in understanding, decision making, prediction and planning. This internal representation is what Craik [1], in "Hypothesis on the Nature of Thought," called model of the environment.

Craik was a clear forerunner of the current work being carried out in the field of robotics. The robot "reasons" on the model of the environment where formal inference is the computational counterpart of causality in the physical world. Craik distinguished three essential processes:

1. *Translation* of external processes into numbers and symbols in a representational space.
2. Arrival at other numbers or symbols by a process of *"reasoning,"* deduction or inference using only the entities and relations of this model of the environment, which parallels external causality.

3. *Retranslation* of these transformed symbols or numbers into external processes or, at least, recognition of the correspondence between these symbols and the external events, "as in realizing that a prediction is fulfilled."

If the model is good, that is to say, if we have been able to capture the external processes in terms of efficient formal mathematical abstractions, then the process of "reasoning" will produce a final result *"similar to that which might have been reached by causing the actual physical process to occur"* [1].

The question for the robot is not what the external environment is, but how it can be modelled (reconstructed in terms of a graph) by creating programs of transformation that properly abstract it to predict the results of actions in the outside world. That is to say, *the only* knowledge available to the robot for whatever task (prediction, learning, planning, etc.) is the model. Consequently, any adaptive task (such as learning) has to be an updating of the model.

Unfortunately, as Marr clearly stated [2], it is not at all easy to construct this internal representation of the external world in one step. We need a set of successive transformations of the raw data. These raw data correspond to a specific set of physically distinct states of the *observable* world, and each observable is a rule associating a number or a label with each physical or abstract state. Finally, we arrive at different models of the environment according to the sequence of intermediate representations and the set of rules used in each one of these intermediate transformations. The global objective is to extract the sort of information that the robot will later need to navigate in this environment.

More recent results concerning the representation of the external environment are now summarised.

1.2 State of the Art

The existing spatial representations of the world can be classified into three principal categories: feature-based representations, grid-based representations and relational representations.

Feature-based representations model the world as a set of features provided by the robot sensors (laser, cameras, sonars etc.). These features (generally, segments or regions) are used to determine the free space in which the robot can navigate, as well as to estimate its position [3].

Grid-based representations involve a tessellation of the space in which the robot is to navigate. Kaiser *et al.* [4] incrementally build a map and cover the grid occupation of each cell, while Borenstein [5] uses this type of representation of the world to store many different types of data: procedural, geometric, sensory, etc.

Finally, *relational representations* are used to try to avoid the accumulation of errors characteristic of the above two representations, storing the relations between signals and markings of the world rather than storing metric information. These models are generally based on graphs; examples can be found in [4].

There are two aspects of this classification that we would like to emphasise. First, most of the spatial representations classified are only concerned with navigation. Second, the three categories are not exclusive, that is, the representations may have characteristics from more than one group. A clear example of this can be found in the work presented in [6], in which a hierarchic, relational and feature-based representation is developed. As far as navigation is concerned there are two fundamental tendencies, both included in McCulloch's original proposal [7]. The first is purely reactive navigation, in which the movement of the robot is the direct result of the sensors reading at any time (reflex arcs). The second is map-based navigation, which typically has large memory requirements, since a detailed metric map of the environment (extended version of Craik's model) has to be stored.

The high computational power required by map-based systems for error correction have led, in the last few years, to an intense development of relational techniques based on qualitative as opposed to quantitative features. Qualitative methods are based mainly on the use of topological maps, where the search for the characteristic features of the robot environment and rules to relate these features take precedence over the metric information. The underlying mathematical model is usually a graph.

The systems using graphs to represent the environment generally apply two different approaches. In the one hand, each node of the graph stores topological features about a region and the arcs represent the connections between regions. On the other hand, some researchers prefer to use the graph nodes to store the system states and the arcs for the transitions between system states. Our proposal in this chapter takes the first approach. The main problem with this approach is selecting an appropriate set of features such that they allow to distinguish different regions correctly.

By way of an example, we can consider Kuipers' work [8, 9] (based on work by Lee [10]), where he uses a graph derived from the sensory data. The graph nodes store what he calls "distinctive places," that is, points in the space with a local maximum value of any feature. This feature is calculated at each point or robot position from the sensor values. The procedure is to move the robot to the point where the feature has a local maximum value using a gradient ascent method (hill climbing).

The relevant feature we have selected for our proposal in this chapter is the *centre of area*, which, as we will discuss later, has some invariance properties that make it advantageous for distinguishing different regions as compared to other proposals. This feature overcomes the typical problem with topological maps of the robot having to be positioned at the characteristic point when it arrives at the region to recognise the node. In our approach, the centre of area position can be calculated from any point in the same local region, so the robot can identify the region without going to that special point (the centre of area).

There are other proposals for solving the recognition problem stated above, such as the work by Thrun [11], using an hybrid method of representation, which builds a topological graph onto a grid-map representation. Our proposal only needs the topological map with partial metric information derived for the node to solve the problem.

In this chapter, we present a simple example of incremental construction of a model of the environment for a generic wheeled robot with specific limitations concerning sensors and repertory of movements.

The remainder of the chapter is organised as follows. In the next paragraph, we introduce the set of constraints concerning the world and the robot sensors. Then, the method used to decompose the representation task is described, and the following sections are given over to each one of the sub-tasks generated by the method: "sensory expansion," "centred extrapolation," "area feature extraction," and "topological map building." We conclude with some remarks on the potential uses of the topological map in decision-making, planning and robot navigation tasks.

1.3 Constraints

The development proposed in this chapter is constrained by the available physical devices.

- We use a holonomic wheeled robot, which moves in 2D surfaces. The robot is cylindrical and has sensor system that can rotate independently of the base.
- The sensor system comprises a sonar ring, an infrared ring and an impact-sensor ring. These range-finder sensors are radially distributed around the periphery of the robot.
- The range-finder sensors detect the property "occupation" at each point of the space along the measured sector, from the position or the sensor to the first obstacle (equivalent to a vector of consecutive obstacle-free points in a direction).
- The typical environment will be indoor premises with empty polygonal rooms surrounded by vertical walls (Figure 1). The rooms are of an appropriate size for sensor ranges so that the robot is able to detect most of the walls of each room from inside it.
- The environment changes slowly with regard to robot movement and the sensors sampling speed.

The robot's sensory arrangement induces the use of polar sensory data, since this favours data processing for representation purposes. Owing to the reduced features of the system, there is one data collection procedure that uses the robot movement together with previous values stored in memory.

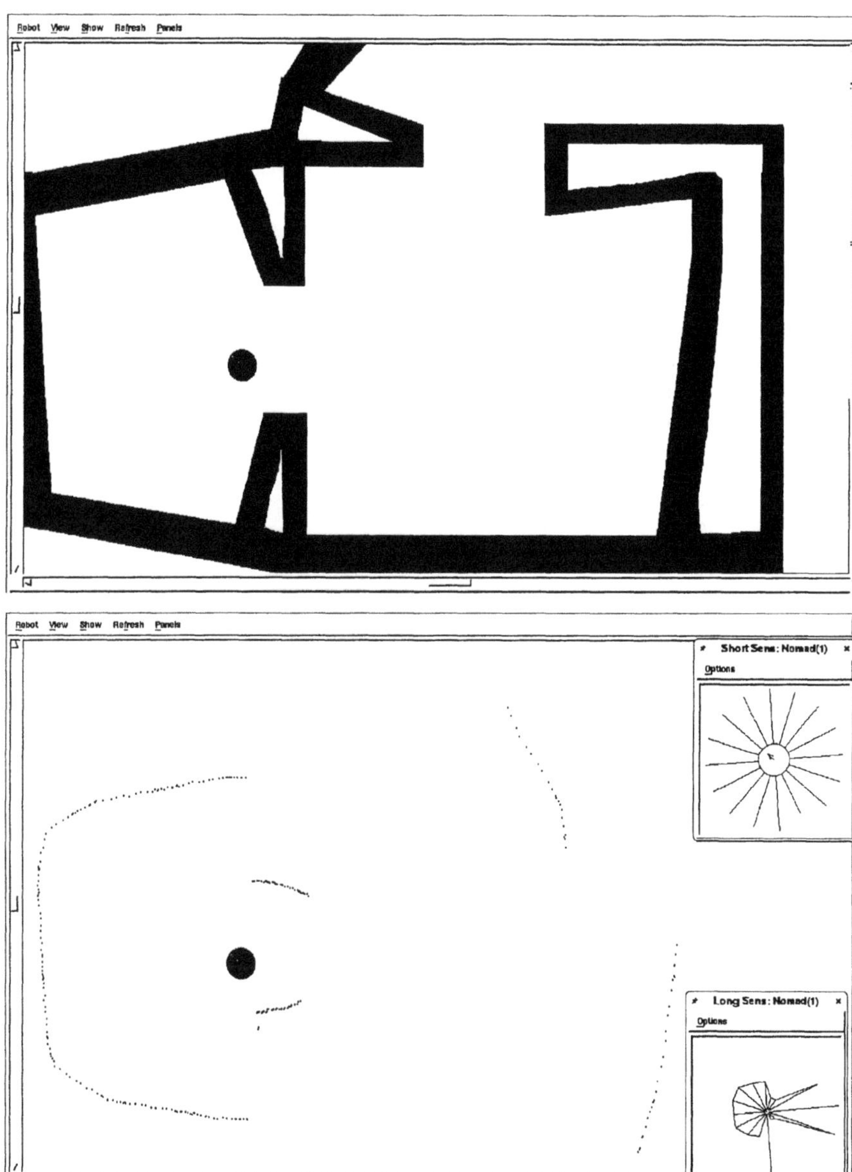

Fig. 1. Example of sensor precision: real map view (upper part) and sensor sampling view (lower part).

1.4 Decomposition of the Representation Task

Since the knowledge level was introduced by Newell [12] and the equivalent "Theory of Calculus" was stated by Marr [2], it is usual practice in knowledge engineering to take advantage of the design stage, starting from a library of generic tasks and problem solving methods used to decompose these tasks in terms of simpler subtasks, until we reach to the level of primitive inferences where we only need data from the specific domain of application.

The representation task consists of a sequence of analytical transformations of sensory data to obtain an endogenous representation in terms of an open space polygons graph in the receptive field of the robot. The obtained open space polygons are invariant to sensor rotations and to local robot movements.

The analytical method used to solve this task is illustrated in Figure 2 in terms of the set of subtasks that sets out the actions to be carried out. The inference structure (in Figure 3) shows the data flow among these inferences. We also need to specify the set of input and output roles played by the knowledge of the application domain in each inference.

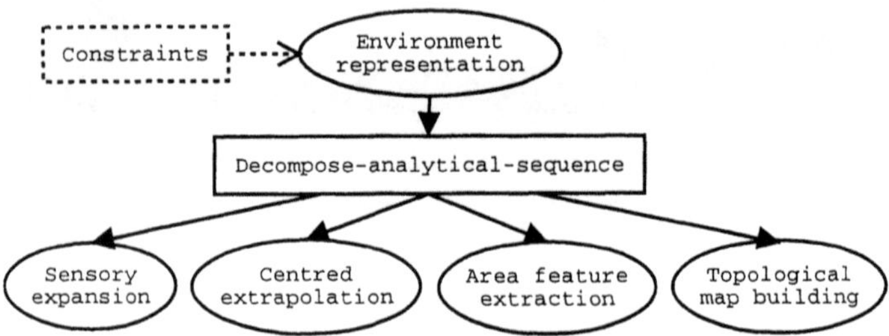

Fig. 2. Diagram of global task decomposition.

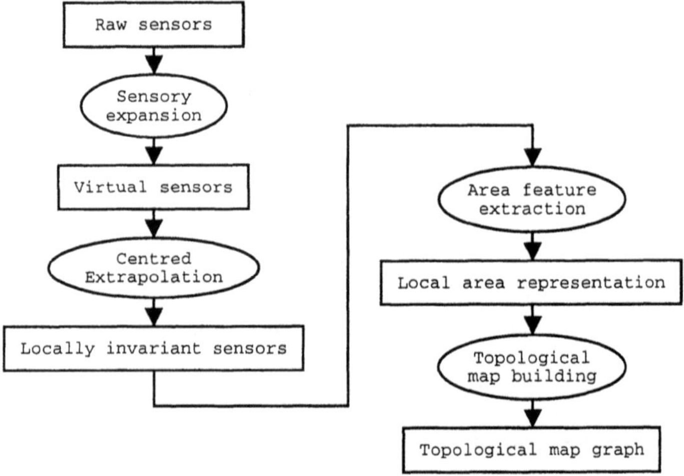

Fig. 3. Inference diagram for the global task.

2 "Sensory Expansion" Subtask

The first subtask from the inference diagram in Figure 3 is "Sensory expansion," which transforms the input data in the "Raw sensors" into "Virtual sensors." The decomposition of this subtask, represented by the inference diagram in Figure 4b, gives the subtasks "Data routing," "Spatio-temporal interaction," and "Movement correction." In the same decomposition, the role "Raw sensors" corresponds to the inputs "Range sensors," "Rotation sensors," and "Movement sensors," and the role "Virtual sensors" corresponds to the output "Dynamic virtual sensors."

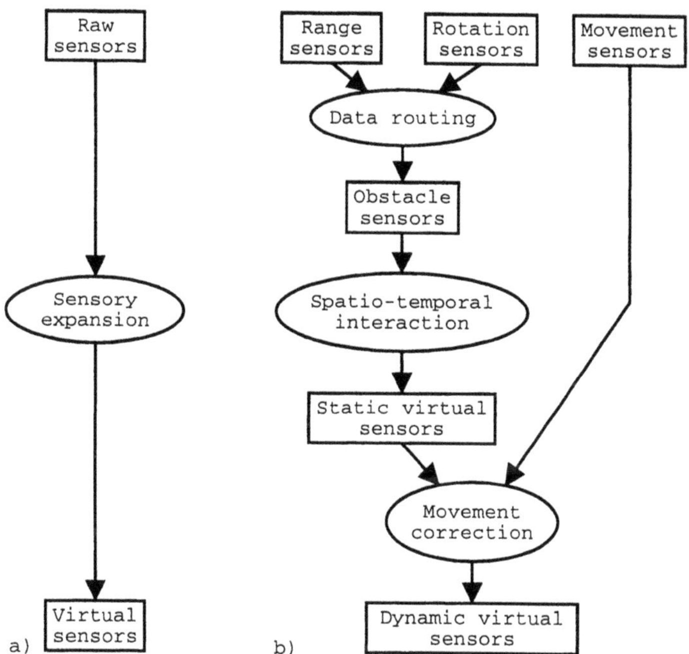

Fig. 4. a) "Sensory expansion" subtask, b) inference diagram for the decomposition of the "Sensory expansion" subtask.

One way to use system movement to improve sensors resolution is to accumulate the instantaneous values of the information at the primary sensors corresponding to many different coordinates in successive sampling intervals. This expansion is achieved in two parts: 1) rotation of the sensors, without movements ("Static virtual sensors") and 2) movement in one direction, without rotation, that gives us the "*corrected"* or "Dynamic virtual sensors."

Given that the information received by the system has to be transformed to represent the environment from the *endogenous point of view*, the first representation, relative to the system position and independent of the direction, defines the properties of the virtual sensors (VSs):

1. The VSs are "placed" in the centre of the system. This means that the distance value they store is relative to this centre.
2. Every VS is assigned to a two-dimensional spatial sector around the system to represent the distance to the closest object in that sector.
3. The VSs receive unified values from the different kinds of real sensors only when one of them is in range (not saturated) and is facing in the same direction as assigned to the VS.
4. The VSs change the stored information concerning distances when the system rotates or when it moves, and thus the representation of the external obstacles is kept more or less invariant.

2.1 Raw Sensory Data

Distance Sensors. The system is composed of a collection of distance sensors, which can move as a whole and are organised with planar radial symmetry. The sensors can be of several types with different properties. From a formal point of view, these sensors are characterised by the following properties (Figure 5):

1. Each sensor is fixed in a point at a distance R^t (where t means the type) from the system centre. This distance is the same for all the sensors of the same type.
2. The sensory field of each sensor faces outward from the system. The position of a sensor i (of type t) is determined by an angle θ_i^t (relative to the system) in the same direction as the axis of its sensory field. As a first approach, we suppose that the sensors of each type are distributed uniformly around the system such that $\theta_i^t = i \cdot \Delta\theta^t$, where $\Delta\theta^t = \frac{2\pi}{N^t}$ and N^t is the total number of sensors of type t in the system.
3. The sensor has a sensitivity sector defined by the angle δ^t centred on its axis.
4. The sensor can detect objects within its sensory field, between a minimum distance (d_{min}^t) and a maximum distance (d_{max}^t) far from its position. The value given by the sensor represents this distance relative to the sensor from the closest object within range. The sensor can inform about saturation (all objects are out of range, far away or too close). The precision of the returned value can be limited. This can be represented by the value belonging to a finite set only. The most common distribution of these values represents a *linear* range.
5. Each type of sensors has an accuracy given by a function depending on the distance and the angular position of the object relative to the sensor. There is also a minimum size of detectable object (i.e. its projection), again depending on the distance and the angle.

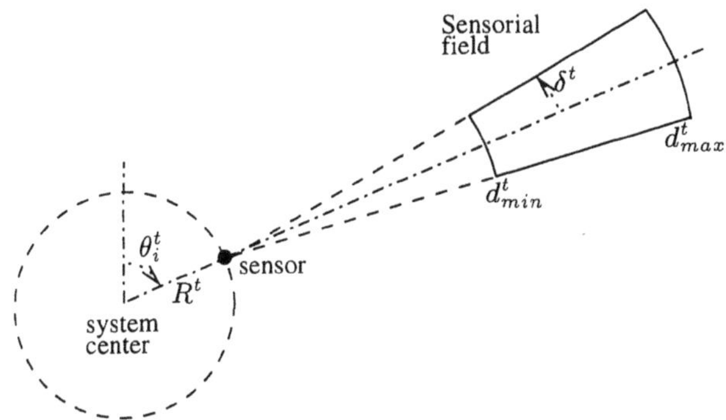

Fig. 5. Sensory field and geometrical characterisation of a distance sensor.

System Movement and Inner Sensors. The system can move in a two-dimensional space in any direction relative to the sensors orientation. The set of sensors can rotate around the centre of the system and independently of it, but provided the relative positions between the sensors remain unchanged.

Apart from the set of main sensors, the system has a way to measure its own movements and rotations. This means that the system also has a linear movement sensor and a register indicating the direction (angle of movement), plus a sensor to detect the angle between the system reference and the main sensors set. These three sensors form the inner sensors set.

The inner sensors are defined by the set of the returned values (precision) and the error due to dead reckoning (i.e. the value can be smaller or bigger than the true distance or angle of movement).

2.2 From Raw Sensory Data to Virtual Sensors

In the first stage of sensory expansion (accumulation in rotations), the data from the real sensors is distributed to the obstacle sensors as shown in Figure 6. This distribution of data guarantees independence of the representation with respect to sensor rotations referred to the external space. Also, it provides a finer representation (more points).

The main function at this stage is data routing depending on the angular direction of the sensors. Data is distribution by intermediate elements, which group obstacle sensors into zones to allow more modularity and fault tolerance. There are as many groups as there are real sensor sectors for the first step in the distribution. The sector covered by the obstacle sensors of a group belongs to the group. Every real sensor is connected to the group with the respective assigned sector facing in the direction of the real sensor at sampling time, when the measurement is taken.

The second step in the first stage of the transformation is demultiplexing. That is to say, the distribution of the value from each real sensor to a group

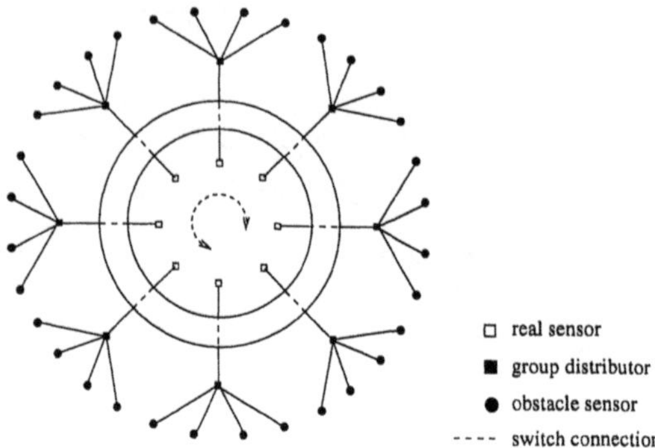

Fig. 6. Distribution of raw sensory data to the obstacle sensors.

and from that group to the obstacle sensor that corresponds to this exact orientation.

2.3 Temporal and Spatial Accumulation

Once the valid data (sensor in range) is in the correct place, each static VS accumulates the new incoming value from the obstacle sensor to the previously stored value with a weight. This temporal accumulation also includes a continuous "forgetting" (increment the distance) if the VS is not activated frequently. The accumulation is also spatial by lateral interaction with the neighbouring virtual sensors. Small contributions from the neighbours are added to the stored value. The contributions are related to the dispersion in the real sensors depending on the overlap between sensory fields.

The "forgetting" consists of a periodic increment inversely proportional to the stored distance. This function allows the correction of isolated erroneous data. The following expression for the distance increment due to forgetting includes all these functional specifications

$$\Delta d = K \cdot d_{min} \cdot \left(\frac{d_{max}}{d} - 1 \right) \tag{1}$$

It is null when $d = d_{max}$ and it is $K \cdot (d_{max} - d_{min})$ when $d = d_{min}$. The constant K must be the part of the complete range corresponding to the number of times that it is activated in a sampling period of the real sensors. So, it must be $K = \frac{N}{\widetilde{N}} \frac{f_r}{f_a}$, where N is the number of real sensors of one kind, $\widetilde{N}$ is the number or virtual sensors, f_r is the sampling frequency of the slowest real sensors, and f_a is the activation frequency of the forgetting increments in the virtual sensor. Normally, $f_a > f_r$ and $\widetilde{N} > N$, so K is a small value.

2.4 Movement Corrections in the Virtual Sensors

The virtual sensors must correct the stored value when the system changes its position, reflecting the changes of the objects relative to the system. This correction has two components depending on the angle of movement relative to the direction the virtual sensor is facing:

- A longitudinal correction, corresponding to an increment or decrement of the distance due to movement away from or toward the represented objects.
- A transversal correction, due to part of the neighbouring sensor field coming into the sensory field and part of the stored value going to the other neighbour.

The two corrections are proportional to the projection of the movement over the facing angle of the sensor. To avoid the global calculation, that depends on the angle of each sensor relative to the movement, we distribute computations and connections between the virtual sensors in a local and modular way. We also assume, to simplify the calculations, that the movement is in the nearest direction corresponding to one virtual sensor.

The process starts at the sensor facing in the same direction as movement. This sensor has a longitudinal correction equal to the movement and a null transversal correction. The sensor transmits this information to and activates the two neighbouring sensors. The sensors receive information from one side, compute their corrections, and transmit new information to the other side. This cascade process ends with the last sensor (pointing in the opposite direction of movement), which receives two activations, balancing the transversal corrections (the number of virtual sensors must be even). This method of computation allows all the sensors to use the same formulae, irrespective of the relative angle to the movement.

We now compute the correction of a sensor at place number k, counting from the first sensor activated (in the same direction as the movement) with index 0. We denote the distance stored before the movement in the k-th sensor as d_k and the corrected distance stored after the movement as $\widehat{d_k}$, and it will be the interpolation (or extrapolation) between d_k and d_{k-1}. The correction depends on the angle of the sensor relative to the direction of movement, θ_k, that can be substituted by $\theta_k = k \cdot \Delta\theta$, where $\Delta\theta = \frac{2\pi}{\widetilde{N}}$, where $\widetilde{N}$ is the number of virtual sensors. We will denote the distance travelled by the system as a.

The diagrams in Figure 7 is an aid for developing the expressions for the new corrected value. There are two possible geometric configurations, depending on the forward movement and the value of the previous sensor (d_{k-1}). The first one is calculated by *interpolation* and the second one by *extrapolation*. The results are the same in both cases, as we will prove. We use a shortened notation denoting $S \equiv \sin(\Delta\theta)$, $C \equiv \cos(\Delta\theta)$, $s_k \equiv a \cdot \sin(\theta_k)$ and $c_k \equiv a \cdot \cos(\theta_k)$.

The *interpolation* correction gives the new value of the distance $\widehat{d_k} = d_{k-1}C - c_k + p$, where the last term p can be solved by similar triangles (see

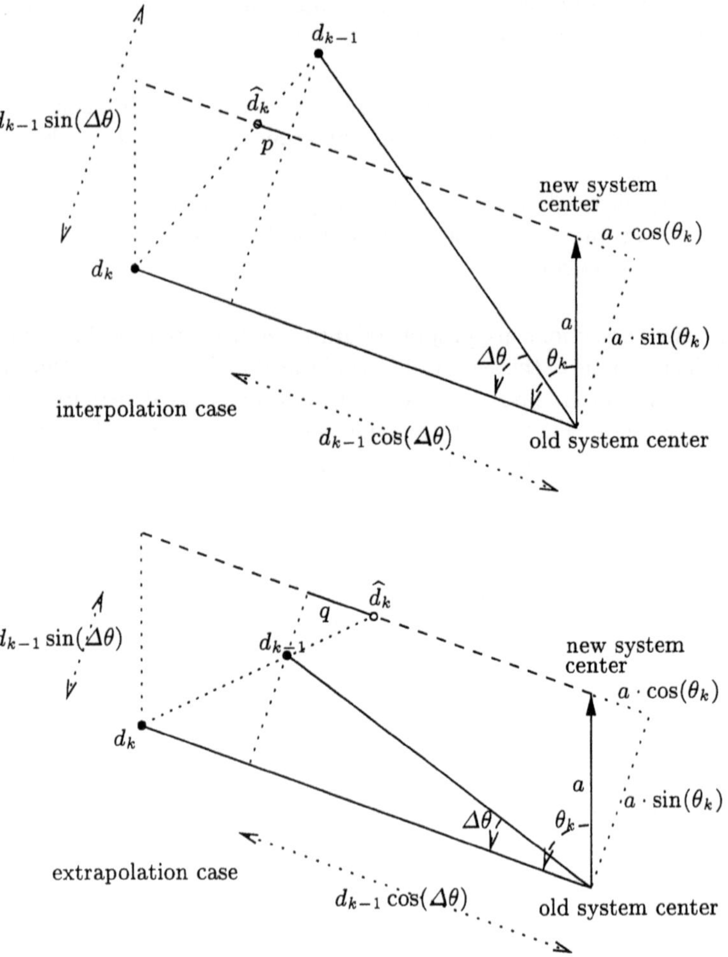

Fig. 7. Diagram for the projections used in the calculations of the longitudinal and transversal corrections of the virtual sensors in a system movement. The two possible configurations are represented (interpolation above and extrapolation below).

Figure 7) as

$$p = \frac{d_k - d_{k-1}C}{d_{k-1}S}\,(d_{k-1}S - s_k) \tag{2}$$

The extrapolation correction is similar, except that now the last term is $-q$ (see lower part of Figure 7), $\widehat{d_k} = d_{k-1}C - c_k - q$, where the term q solved by similar triangles is

$$q = \frac{d_k - d_{k-1}C}{d_{k-1}S}\,(-d_{k-1}S + s_k) \tag{3}$$

which is equal to p with the opposite sign.

We can substitute $p = -q$ in any of the above expressions to obtain the same correction expression for both the interpolation and extrapolation cases

$$\widehat{d_k} = d_{k-1}C - c_k + \frac{d_k - d_{k-1}C}{d_{k-1}S}(d_{k-1}S - s_k) \tag{4}$$

and simplification with term reordering yields

$$\widehat{d_k} = d_k + \frac{s_k}{S}\left(C - \frac{d_k}{d_{k-1}}\right) - c_k \tag{5}$$

where the second term on the right-hand side is the transversal correction and the last term is the longitudinal correction.

Expression (5) as shown and using only the definitions for s_k and c_k is directly dependent on θ_k, which varies with the direction of movement . We are looking for an expression that depends only on sensor values and the neighbour that activates this sensor. We use the trigonometric properties for the angle sum and assume that $\theta_k = \theta_{k-1} + \Delta\theta$, with s_k and c_k, giving the result:

$$\begin{aligned} s_k &= C \cdot s_{k-1} + S \cdot c_{k-1} \\ c_k &= -S \cdot s_{k-1} + C \cdot c_{k-1} \end{aligned} \tag{6}$$

where both expressions only depend on the previous sensor values and on constants (S and C). That is to say, every virtual sensor is corrected using the three values sent by the adjacent sensor (s_{k-1}, c_{k-1}, and d_{k-1}) and using formulae 5 and 6. Every sensor sends the three values s_k, c_k, and d_k to the next one. In the first sensor activated, the initial values are $s_0 = 0$ (null transversal correction) and $c_0 = a$ (full longitudinal correction).

The result of the "Sensory expansion" subtask described in this section is illustrated in Figure 8. A sample of raw sensor representation is shown on the left-hand side. The right-hand side shows the same environment represented using the expanded sensors (dynamic virtual sensors) resulting after the "Sensory expansion" subtask. The resolution is increased and the match between the representation and the real map is better after expansion of the receptive field.

3 "Centred Extrapolation" Subtask

The next subtask from the inference diagram in Figure 3 is "Centred extrapolation," which transforms the results from the previous subtask, "Virtual sensors," into "Locally invariant sensors." The decomposition of this subtask, represented by the inference diagram in Figure 9b, gives the subtasks "Modular accumulation" and "Extrapolation." In the same decomposition, the role "Virtual sensors" corresponds to the inputs "Dynamic virtual sensors" and the role "Locally invariant sensors" corresponds to the output "Local centre extrapolated sensors."

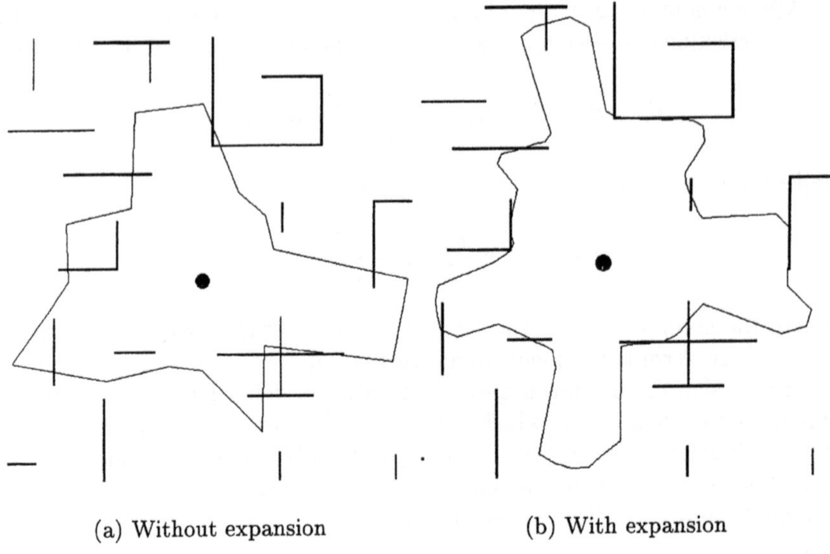

Fig. 8. Sensor resolution enhancement due to the expansion of the receptive field (virtual sensors).

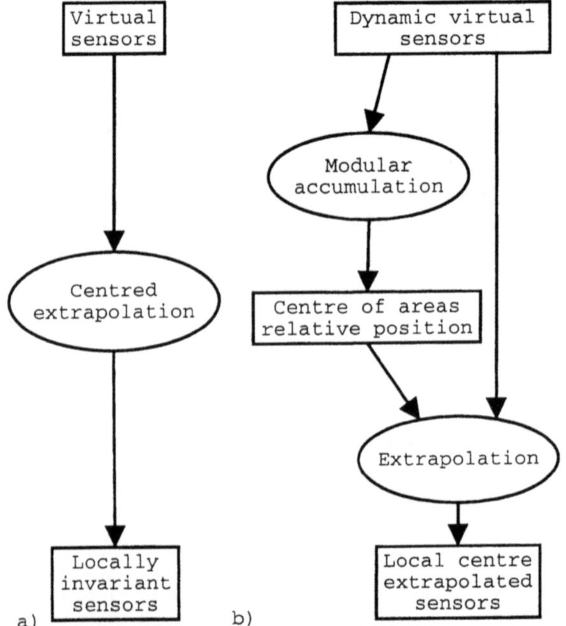

Fig. 9. a) "Centred extrapolation" subtask, b) inference diagram for the decomposition of the "Centred extrapolation" subtask.

From the representation obtained in the first transformation, (independent of sensor set rotations relative to the centre of the system and which adapts to the movements), we set another representation that is also independent of the position of the system within sensors range zone (zone of local homogeneity). For this purpose:

1. We calculate a position that is "centred" (relative to the obstacles detected around and represented in the virtual sensors).
2. We build a second set of virtual sensors (local centre extrapolated sensors) that represent the external environment from this previously computed centred position.

3.1 Local Computation of the Centre of Area

At this stage we have a representation of the obstacles around the system in terms of a set of radial distances. We can assume that the distances stored in the virtual sensors represent the open space polygon vertexes around the system. In Figure 10 (right-hand side), we have represented an imaginary triangle formed by the distances stored in two neighbouring VSs (r_a and r_b) and the union between them, with angles α_a and α_b relative to a common origin. The difference between these angles remain constant, $\alpha_a - \alpha_b = \Delta\theta$, as in the previous section.

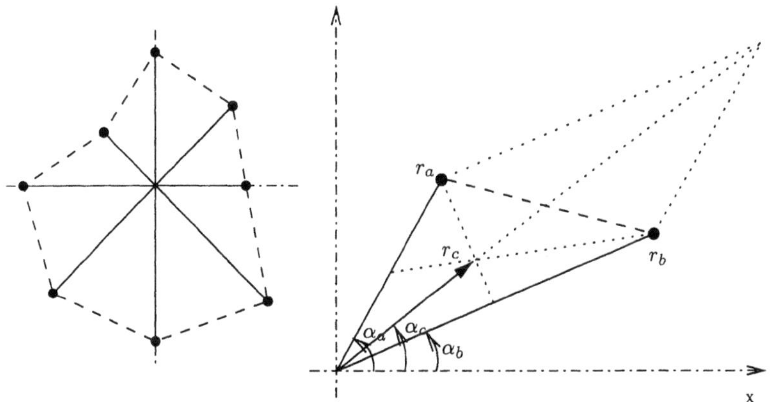

Fig. 10. Diagram used to illustrate the calculus of the centre of areas in triangular sectors. On the left-hand side, we have the panorama of eight virtual sectors of open space around the system. On the right-hand side, we have the illustration of the calculus of r_c.

An intermediate layer of computing elements should be used to obtain the centre of areas of all the triangular sectors, where each element of the layer is connected with two adjacent VSs and receives the information concerning the local values of area and centre of area from both. Well-known geometrical considerations give us the following expressions for the whole area (s_c) and the coordinates (x_c, y_c):

$$\begin{aligned} s_c &= \tfrac{1}{2} r_a r_b \sin \Delta\theta \\ x_c &= \tfrac{1}{3}\left(r_a \cos\alpha_a + r_b \cos\alpha_b\right) \\ y_c &= \tfrac{1}{3}\left(r_a \sin\alpha_a + r_b \sin\alpha_b\right) \end{aligned} \tag{7}$$

where we have taken into account that $\sin \Delta\theta$, α_a and α_b are constant for each pair of neighbouring VSs and that $x_a = r_a \cos\alpha_a$, $y_a = r_a \sin\alpha_a$, $x_b = r_b \cos\alpha_b$, and $y_b = r_b \sin\alpha_b$.

For the local and accumulative computation of the whole centre of area, we need a module that receives the data (s_1, x_1, y_1; s_2, x_2, y_2) concerning the centres of area input from two VSs or other modules and which computes the weighted sums

$$\begin{aligned} s_c &= s_1 + s_2 \\ x_c &= \tfrac{s_1 x_1 + s_2 x_2}{s_1 + s_2} \\ y_c &= \tfrac{s_1 y_1 + s_2 y_2}{s_1 + s_2} \end{aligned} \tag{8}$$

Taking into account the symmetry of these equations, we can easily get a repetitive and local formulation of the calculus. Each module receives and transmits each coordinate multiplied by the corresponding area ($\widehat{x}_1 \equiv s_1 x_1$, $\widehat{y}_1 \equiv s_1 y_1$, $\widehat{x}_2 \equiv s_2 x_2$, $\widehat{y}_2 \equiv s_2 y_2$, $\widehat{x}_c \equiv s_c x_c$, $\widehat{y}_c \equiv s_c y_c$) and only at the end of the process (binary tree) is the normalisation performed. So, Equation 8 is just additions,

$$\begin{aligned} s_c &= s_1 + s_2 \\ \widehat{x}_c &= \widehat{x}_1 + \widehat{x}_2 \\ \widehat{y}_c &= \widehat{y}_1 + \widehat{y}_2 \end{aligned} \tag{9}$$

3.2 Local Centre Extrapolated Sensors

The last transformation (from $\widehat{d}_k$ to $\widehat{\widehat{d}}_k$) completes the change of representation, as summarised in Figure 11. Starting from the virtual sensors ($\widehat{d}_k$) computed in Section 2.4 and changing the point of reference to the centre of areas computed above, we get a representation independent of local movements. This change is accomplished by a correction (interpolation or extrapolation) of the VSs corresponding to an imaginary movement of the centre of areas. This correction is formally identical to the one computed in Section 2.4 for the real system movements and the local procedures of calculus also match. We assume that the system "moves" in the same direction as the angle of the closer sensor and by an amount equal to the distance from the centre of areas. All the adjacent sensors perform the same computation as in Equations 5 and 6. The result of this transformation is the local centre extrapolated sensors that represent the open space area from the point of view of the centre of areas.

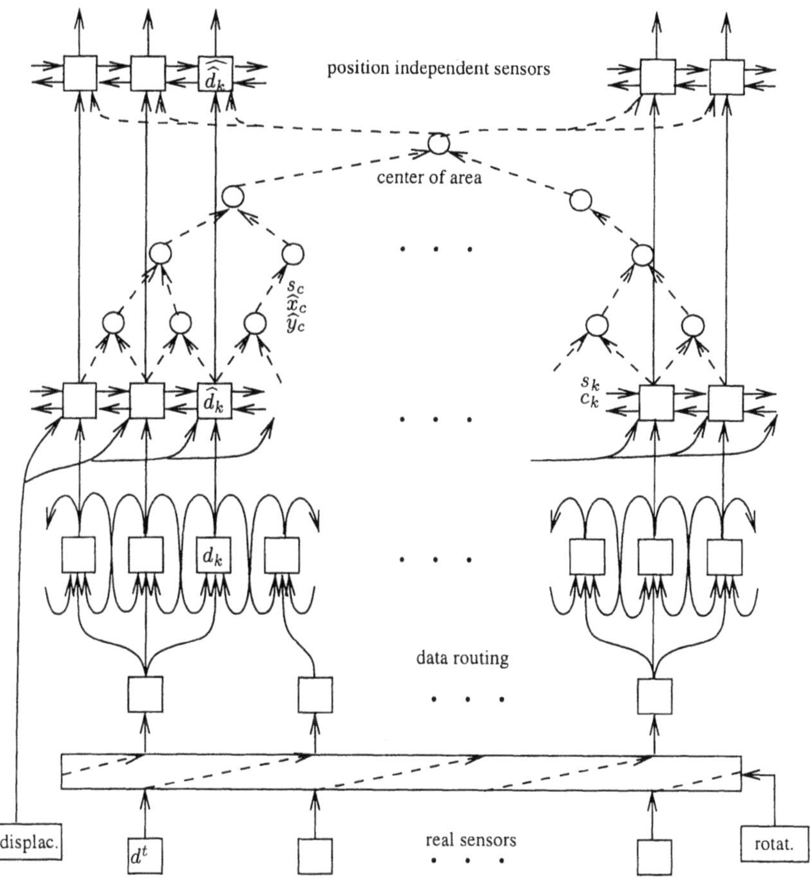

Fig. 11. Connectionist view of the virtual sensor transformations. $d^t$are the real sensor inputs, d_k are the virtual sensors, $\widehat{q}_k$ are the corrected virtual sensors (Equation 5), $\widehat{\widehat{q}}_k$ are the position independent virtual sensors (Equation 5 again), and $\widehat{s}_c$, $\widehat{x}_c$, $\widehat{y}_c$ are the area and coordinates of the centre of area from Equation 9.

3.3 Centre of Areas and Its Stability

Starting from the polygon that represents the virtual sensors, we have been able to calculate the position of the CA modularly [13] as discussed above. Each region can be represented by its own CA given that the position of this centre is invariant to translations of the point from which it is calculated. This theoretical construction allows us to simplify the spatial information used for navigation purposes, since these invariant points are usually obstacle-free regions. Navigation will take place preferably there.

Figure 12 shows the representation of space points distributed over a square area with a lateral aperture (Figure 12a) and the respective CAs calculated from the sensors at those points (Figure 12b). These calculations have been made by placing a real robot in the positions marked in Figure 12a and, hence, the results in Figure 12b are affected by some errors inherent to the measurement process. Figure 12b also shows how the lateral aperture influences the CA position. In Figure 12, some points near the aperture and their respective CA have been surrounded with dotted ellipses to stress the influence of the lateral aperture. If we were to represent the above mentioned correspondence mathematically, not including the errors due to the measurement process, we would have obtained all the points symmetrically aligned only in the strictly orthonormal direction toward the lateral aperture. This is because of increments of values of open space in front of the aperture.

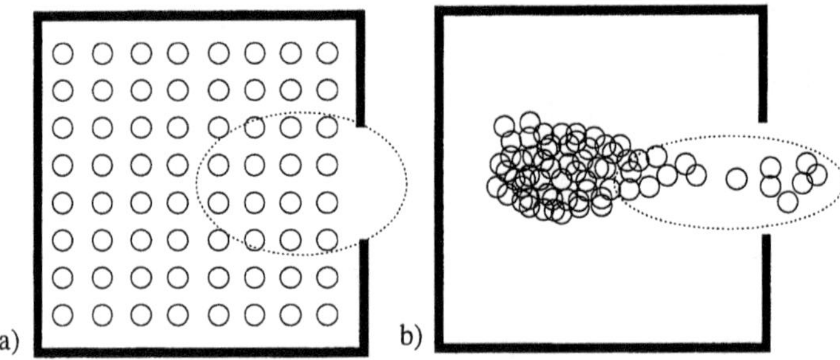

Fig. 12. Correspondence of space points with the centre of areas using data from a real robot. a) Points from which the centre of areas has been calculated. b) Position of the respective centres calculated.

The centre of area is (with some degree of accuracy) an invariant point for a given region of the open space around the robot. If the environment does not change, small movements of the robot will result in nearly the same centre of areas. However, if the region is not homogeneous, the CA changes for small robot movements. This movement of the CA, now termed *instability*, tells us two things: first, we are leaving a region and a new one with different properties will emerge; and second, the direction in which the new region appears.

The relation between the movement of the CA and the robot is a measure of how close or far from a new stable centre the robot is and, hence, how far away a new stable region is. If the robot is in a stable centre of areas, this relation will be very little. Per contra, if the robot moves away from that point, the value of the relation will increase up to a maximum, which is reached midway between two stable regions.

4 "Area Feature Extraction" Subtask

The third subtask from the inference diagram in Figure 3 is the "Area feature extraction" that transforms the results from the previous subtask, "Locally invariant sensors" into "Local area representation." The decomposition of this subtask, represented by the inference diagram in Figure 13b gives the subtasks "Spatial lateral interaction," "Continuous zone labelling," and "Zone properties computation." In the same decomposition, the role "Locally invariant sensors" corresponds to the inputs "Local centre extrapolated sensors" and the role "Local area representation" corresponds to the output "Set of zone characteristics."

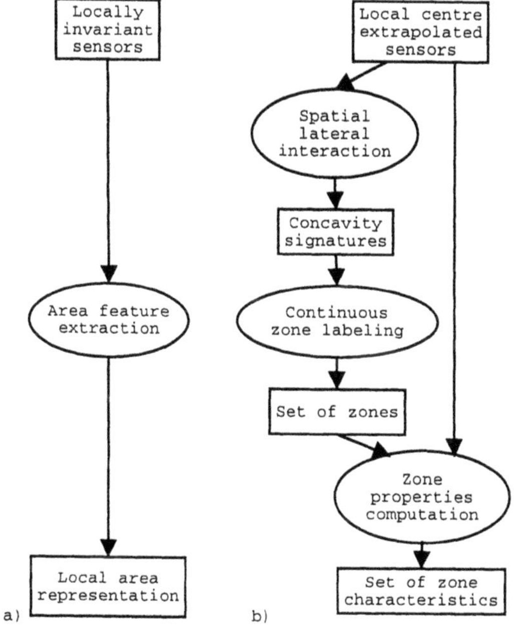

Fig. 13. a) "Area feature extraction" subtask. b) Inference diagram for the decomposition of the "Area feature extraction" subtask.

4.1 Transformation of Centred Equivalent Polygons

The representation of open space polygons obtained in the sensory space (by means of the local centre extrapolated sensors) is independent of sensory set rotations in relation to the centre of the system and adapts to robot movements. Having built this representation, we obtained another one also independent of the robot position within a region defined by the sensor range (region of similitude or homogeneity)

After obtaining the CA, we can translate any open space polygon to the equivalent polygon that would be seen if the robot were placed in the CA (Figure 14). The polygons can be clustered in groups (equivalence classes) which are centred in the representative CA of these classes (canonical polygons).

So, storing the polygons translated to the CA (centred), any other equivalent polygon can be compared (by means of specific features extracted in the next stage) and identified with the centred polygon used as a canonical representative of the region.

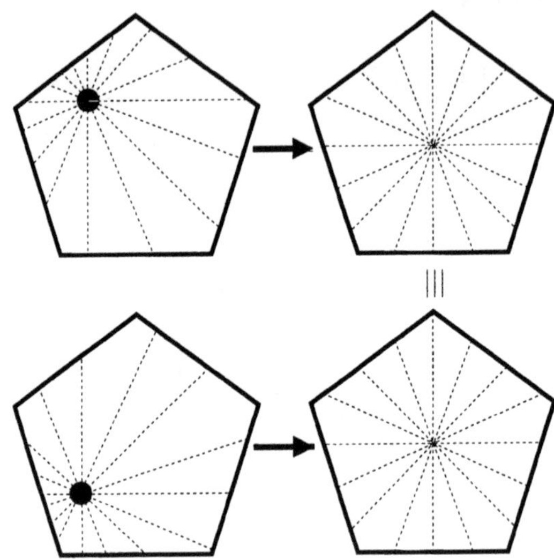

Fig. 14. Polygons from two different positions which are equivalent in the representation of centre of area.

4.2 Modular Detection of Concavity Zones

We transformed the sensory data into centred polygons above. We now need to store some properties of these polygons, which are invariant in relation to the CA and robust enough in the face of small sensor changes, the measured position and environment. The main property, which is always maintained, is the general shape of the polygon. That is to say, the succession of concavities and convexities of the perimeter that correspond to obstacles or apertures in the real world. These homogeneous and distinctive parts of the polygon are called *zones.*

Zones can be detected by means of differential operators that compute local changes of the polygon for each vertex. The second-order spatial derivative gives us information concerning the concavity of the zone (potential access to another region) or convexity (potential wall). Finally, the accentuation of differences in concavity, which delimit different zones or parts of the polygon, is noteworthy.

As in the previous calculus of the virtual sensors and CA (see paragraphs 2.4 and 3.1), the detection of concavity is also computed in a connectionist or modular manner [13]. We use a network with two layers, as shown in Figure 15, where the inputs to the first layer are the data defining the centred polygons. These data are in fact measures of the distances from the centre to the vertex, which represent detected obstacles. A discrete non-recurrent lateral inhibition layout with receptive fields of a central element and two neighbours in the periphery generates the respective zone marks.

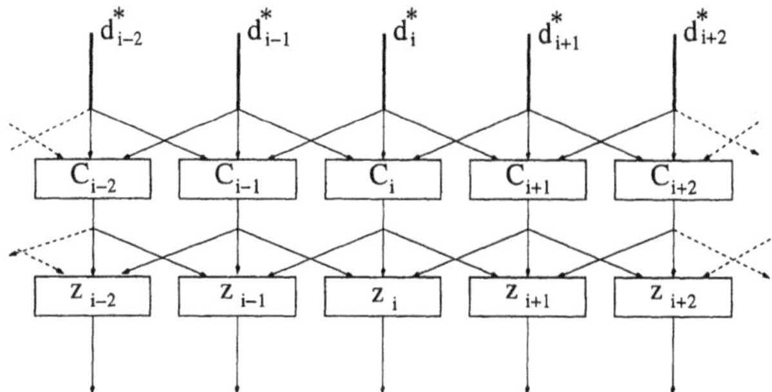

Fig. 15. Detection of concavity zones by means of a lateral interaction network.

Input data d_i^* comes from the transformation of polygons centred in the centre of areas. Each module C_i in the first layer calculates the second-order spatial derivative, taking as a basis the distances d_i^*, d_{i-1}^* and d_{i+1}^* and producing an output of type $\{+1, 0, -1\}$, which indicates the sort of local convexity of its receptive field. That is to say, $C_i = \text{sign}\left(\Delta^2 d_i^*\right) = \text{sign}\left(\Delta d_i^* - \Delta d_{i+1}^*\right)$, with $\Delta d_i^* = d_i^* - d_{i+1}^*$. These C_i values are then taken as inputs to the second layer, where each module z_i performs a comparison between local concavity values in order to detect changes of zone. Thus, each point is classified as belonging to: the inner part of a concave zone (IZ), an edge (SZ or EZ) or a transition zone (DC). This classification is attained by comparing C_{i-1}, C_i and C_{i+1} inputs using the look-up table shown in Table 1. These outputs are later used to identify the border of the zones and, consequently, to detect specific characteristics of each zone. Figure 16 represents an example of the results for the layers applied to a polygon.

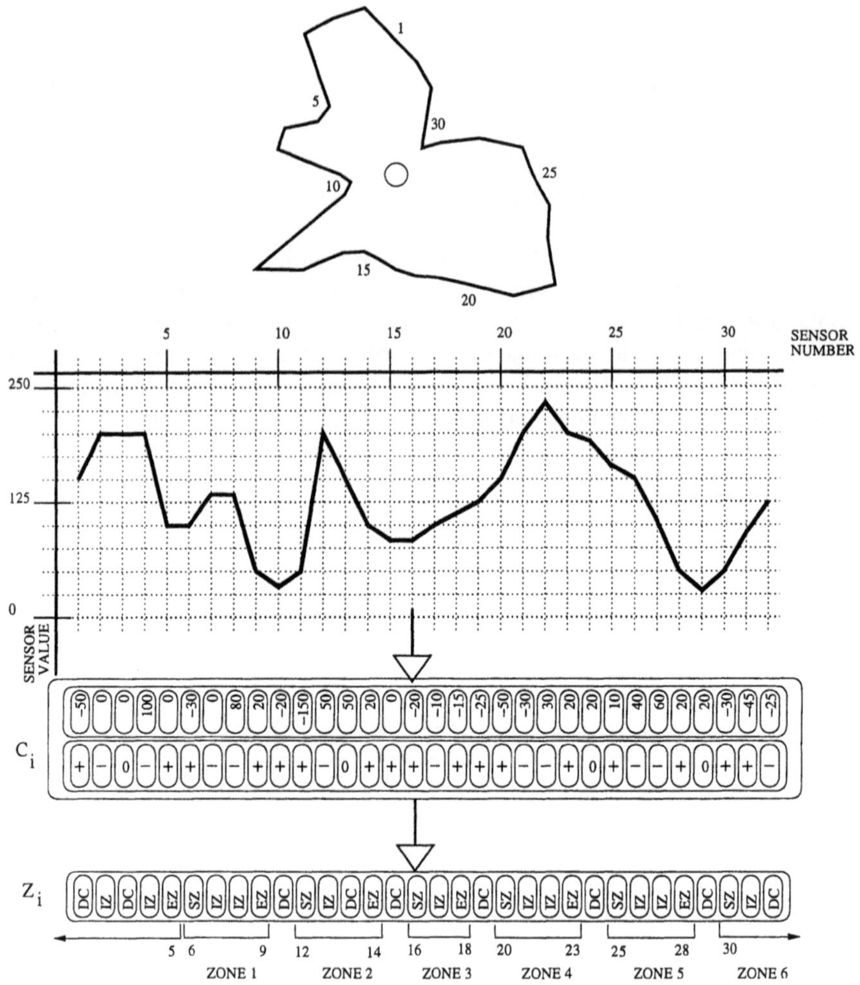

Fig. 16. Example of calculations for zone detection in a polygon.

Table 1. Look-up table for zone labelling.

C_{i-1}	C_i	C_{i+1}	zone label
+ or 0	+	−	SZ (start zone)
− or 0	+	+ or 0	EZ (end zone)
+ or 0 or −	−	+ or 0 or −	IZ (inside zone)
other combinations			DC (don't care)

5 "Topological Map Building" Subtask

The last subtask from the inference diagram in Figure 3 is "Topological Map Building" that transforms the results from the previous subtask, "Local area representation" into "Topological map graph." The decomposition of this subtask, represented by the inference diagram in Figure 17b, gives the subtasks "Equivalent node search," "Update node connection," and "Insert node." In the same decomposition the role "Local area representation" corresponds to the inputs "Set of zone characteristics" and the role "Topological map graph" corresponds to the output "Graph of area representations."

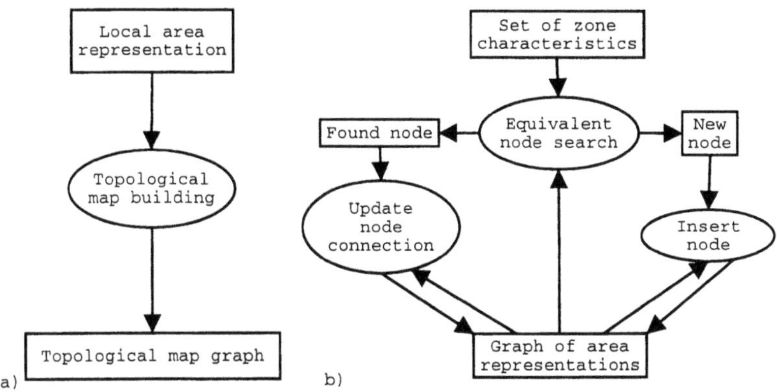

Fig. 17. a) "Topological map building" subtask. b) Inference diagram for the decomposition of the "Topological map building" subtask.

Once we have detected the zones within a centred polygon, we extract and store the following parameters: distance to the centre d_k, distance between both borders a_k, depth p_k and orientation, ψ_k, relative to a fixed direction in the space, as depicted in Figure 18. To complete the topological information to be encoded in the graph, we add other fields of information concerning the state of the connectivity of each zone with other regions (e_k, c_k), where e_k reports whether or not a region has been visited, and c_k identifies the node with whose zone it is connected. Finally, we add the total surface of the centred polygon s_i, and coordinates in an absolute system of reference, (x_c, y_c). Thus, we obtain all potential nodes characterised by the numerical frame

$$N_i = (\{(d_k, a_k, p_k, \psi_k, e_k, c_k)\}, s_i, (x_c, y_c))$$

After this information has been compiled, we proceed to compare it with every node included in the string that constitutes the graph

$$\widetilde{G} = \{N_1, \ldots, N_i, \ldots, N_k\}$$

This information is either added, if there is no match (see Figure 19), or identified, if it matches with any pre-existing nodes. In the event of no match, a new equivalence class appears, which is a new label and, consequently, a new node with the characteristics derived from the new sort of zone. In this case,

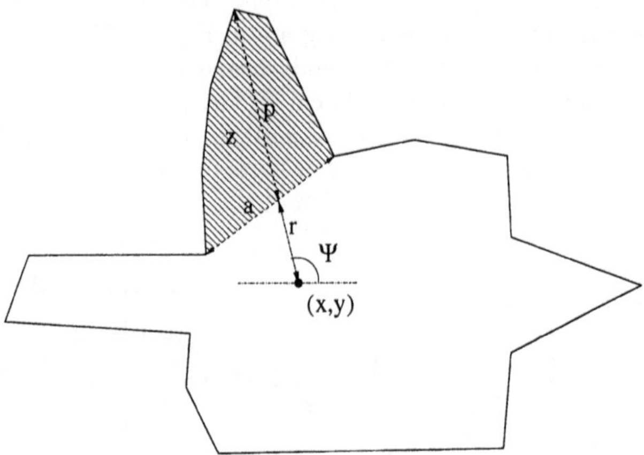

Fig. 18. Illustration of the characterisation frame N_i for a centred polygon (with attributes of one zone).

the matching connection (arc pending of unexplored paths) is also associated with the arc between the new node and the one that the robot came from in its path. In the event of a match, it can be said that region in which the robot is located was already known, and we only need to update connectivity parameters between regions.

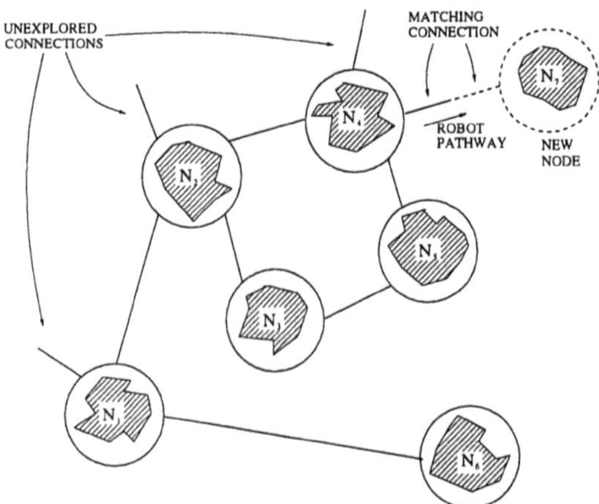

Fig. 19. Insertion of a new node (connecting a previous node) during the construction of the graph for topological space representation.

The formalisation and building of graphs for topological maps (graph-like worlds) have been discussed by other authors, like Dudek et al. [14].

6 Conclusions

In this chapter, we have presented an analytical method for decomposing the environment representation task for a mobile robot. The ultimate goal is for the robot to build an endogenous model of the external environment and use it for navigation and other tasks, including learning (understood here as "updating the model"). For this purpose, we have proposed a systematic set of transformations of the data received from the robot sensors into a topological graph representation. Each stage of this process extracts useful information from the data obtained in the previous one to refine the representation. The result is a graph where nodes store features of an area ("distinctive place") that are invariant under rotations and small translations around a representative point that can be calculated from any point in the same area. Any arc in the graph connects two nodes and represents the path between them.

The description of levels or stages for the representation task is given in terms of the usual methodological orientation in Knowledge Engineering, starting from a conceptual model at the knowledge level and in the external observer domain. The transformations are described using natural language and basic knowledge of Mechanics and Geometry. Throughout the decomposition of the task, the given inference diagrams (data flow) describe the relations between the data from raw sensors and the graph, with the inferences of the transformations. This form of modelling, along the lines proposed by Clancey [15], Chandrasekaran [16] and in the CommonKADS methodology [17], among others, aims to develop a set of reusable modelling components. The emphasis on the environment model comes from Craik and McCulloch, who stated back in the 40s that this was the entire world from the robot point of view. All that exists for the robot is what is in its model.

To achieve the invariance of area representation stored in the nodes of the graph, we have selected the centre of area (CA) as the representative point. The CA has the required properties of invariance and it can be computed from any point in the surroundings. Once we have the position of the CA, the open space detected from the expanded robot sensors is translated into a representation from the CA point of view, such that the features extracted from it are invariant.

As summarised in Table 2, apart from the possible disadvantages, such as it being difficult to build and use in real time, the advantages of the analytical method proposed in this chapter are associated with the generic character of the inferential scheme and the simplicity in the map-building process. Also, there are other positive aspects such as the invariance of the representation under local translations, which makes the model more robust.

Finally, we would like to believe that this idea of virtual expansion of raw sensory data, and the subsequent incremental building of the map graph, may be considered to be of some value to neurophysiologists in the understanding of the cortical neural processes underlying external space representation (from transducers to thalamic nuclei and then to layers and columns in the cortex).

Table 2. Summary of the advantages and disadvantages (in the lower part) of the method proposed in this chapter as compared with other usual methods of representation in robotics.

	Grid-based (metric) methods	Other topological graph methods	*Our proposal in this chapter*
Robustness	low	medium	*high*
Fault tolerance	low	high	*high*
Planning use	worse	better	*better*
Localization ambiguity	no	yes	*no*
Optimal path finding	yes	no	*yes*
Precision	high	low	*medium*
Map building computational cost	low	high	*high*
Construction complexity	simple	complex	*medium*
Real-time use	fast	slow	*slow*

Acknowledgements This work has been partially supported by Spanish Interministerial Commission of Science and Technology (CICYT) project TIC-97-0604.

References

1. Craik K (1943): The Nature of Explanation. Cambridge University Press
2. Marr D (1982): Vision. W. H. Freeman
3. González J, Ollero A, Reina A (1994): Map Building for a mobile robot equipped with a Laser Range Scanner. In: IEEE International Conference on Robotics and Automation, IEEE, San Diego, CA
4. Kaiser M, Klingspor V, Millán JdR, Accame M, Wallner F, Dillmann R (1995): Using Machine Learning Techniques in Real-World Mobile Robots. *IEEE Expert Intelligent Systems and their Applications* , 37–45
5. Borenstein J (1989): Real Time Obstacle Avoidance for Fast Mobile Robot. *IEEE Transactions on System, Man and Cybernetics* 19, 5, 1179–1187
6. Fennema C, Hanson A, Riseman E, Beveridge JR, Kumar R (1990): Model-Directed Mobile Robot Navigation. *IEEE Transactions on Systems, Man and Cybernetics* 20, 6
7. Kilmer WL, McCulloch WS (1969): The Reticular Formation: Command and Control System. In: Leibovic KN (Ed.) Information Processing in the Nervous System, Springer-Verlag, 297–307

8. Kuipers B, Byun YT (1991): A Robot Exploration and Mapping Strategy Based on a Semantic Hierarchy of Spatial Representations. *Journal of Robotics and Autonomous Systems* 8, 47–63, reprinted in Walter Van de Velde (Ed.) Towards Learning Robots, Bradford/MIT Press, 1993, 47–63
9. Kuipers B (2000): The Spatial Semantic Hierarchy. *Artificial Intelligence* 119, 191–233
10. Lee WY (1996): Spatial Semantic Hierarchy for a Physical Mobile Robot. Phd thesis, The University of Texas at Austin
11. Thrun S (1997): Learning Maps for Indoor Mobile Robot Navigation. *Artificial Intelligence*
12. Newell A (1982): The Knowledge Level. *Artificial Intelligence* 18, 1, 87–127
13. Álvarez Sánchez JR, de la Paz López F, Mira Mira J (1999): On Virtual Sensory Coding: An Analytical Model of Endogenous Representation. In: Mira Mira J, Sánchez-Andrés JV (Eds.) Engineering Applications of Bio-Inspired Artificial Neural Networks, no. 1607 in Lecture Notes in Computer Science, International Work-Conference on Artificial and Natural Neural Networks, Springer-Verlag
14. Dudek G, Jenkin M, Milios E, Wilkes D (1997): Map Validation and Robot Self-Location in a Graph-Like World. *Robotics and Autonomous Systems* 22, 2, 159–178
15. Clancey WJ (1985): Heuristic classification. *Artificial Intelligence* 27, 289–350
16. Chandrasekaran B (1986): Generic tasks in knowledge-based reasoning: High level building blocks for expert system design. *IEEE Expert* 1, 23–29
17. Schreiber G, Wielinga B, de Hoog R, Akkermans H, van de Velde W (1994): CommonKADS: A comprehensive methodology for KBS development. *IEEE Expert* 9, 6, 28–36

Evolutionary Artificial Potential Field - Applications to Mobile Robot Path Planning *

Prahlad Vadakkepat, Tong-Heng Lee, and Liu Xin

Department of Electrical and Computer Engineering 4 Engineering Drive 3, National University of Singapore, Singapore 117576
{prahlad, eleleeth}@nus.edu.sg

Abstract. This chapter discusses the application of the evolutionary artificial potential field (EAPF) in mobile robot path planning. The parameters of the evolutionary artificial potential field are optimized with the multi-objective evolutionary algorithm. The EAPF is utilized in a robot soccer system.

1 Introduction

Present industrial systems are much complex and multiple mobile robots are increasingly preferred. Multiple mobile robots are not spatially constrained and performance benefits are manifold. The robots/agents equipped with knowledge, motivations, reasoning and planning capabilities may communicate each other and, share data and information. Cooperation protocol by distributed control; effective communication and fault tolerance while having efficiency of cooperation, adaptation and robustness are some of the research directions associated with multi-robot/agent systems [1]. Position correction and communication congestion are some of the difficulties to tackle with. Autonomous behaviors, interaction with humans or other robot agents, goal directed actions and adaptation to environmental changes, bring in new features to applications.

Autonomous agents vary from software agents (softbots) [2] to autonomous mobile robots or vehicles. Autonomous agents are categorized based on the degree of interaction with real world. Agents that interact least with the real world are neither embodied nor situated (computer simulation of actual agents). Traditional industrial robots are embodied (have physical bodies) and not situated (robot behaviors do not depend on the current state of the environment). Autonomous mobile robots are embodied and situated.

An agent can learn other agents' intentions and beliefs as well as the characterization of the task environment. Modeling of other agent's goals and beliefs enhances the ability to reason about and to coordinate the activities. In multi-agent systems (MAS), the autonomous interacting intelligent

* This work is supported by the University Faculty Research Fund R-263-000-087-112. Part of this work was presented at the Joint 9th IFSA World Congress and 20th NAFIS International Conference, Vancouver, Canada, July 25-28, 2001.

agents coordinate actions to achieve the goal(s) jointly or competitively. The agents may have homogeneous or heterogeneous structures. In a homogeneous system the agents have identical structures (goals, domain knowledge, and set of actions). The agents may differ by way of the sensor input and effector output. Agents differing in goals, domain knowledge and actions constitute a heterogeneous system. Heterogeneity imparts more power to the MAS at the cost of complexity. The agents may be friendly (benevolent) or inhibitory (competitive) in nature.

The hierarchical and behavior structures are the two general types of multi-agent system designs. The two structures differ on the type of information processed and interconnections. In a hierarchical structure the control is divided along functional lines into progressive levels of abstraction of data. In a behavior structure the control problem is broken into behaviors without any central intelligent agent. Through interaction among competing constituents emergent behaviors result. Such systems require action arbitration mechanisms as different actions arise from different behaviors.

The basic issue in a multi-agent/robot system is to determine the action should an agent/robot take when situated in a specific state. Complex and difficult scenarios arise when the environment is dynamic and the robots/agents compete. Action selection procedures are more complex and challenging in a dynamical environment with competing robots/agents [3]. Adaptive, robust, tactical and versatile behaviors are the requirements for agents to be autonomous [4], though it is not possible to impart all with the present state of the technology. With the adaptive behavior an agent is able to accommodate the variations in environment. Robustness imparts fault tolerance to environmental properties. Tactical behavior is required to maintain multiple goals and to pursue a particular goal based on the prevailing situation. Versatility allows carrying out a variety of tasks (multi-purpose behavior).

A robot agent in a multi-robot setup is required to cooperate, with capabilities to adapt and learn. The evolution of a robot agent is necessary to perform certain tasks even under unexpected environmental changes. Cooperation of heterogeneous mobile robots and a collection of particle-like-small-cell-concept micro-robots, called cellular robots, are reported in [5]. The multiple robotic systems proposed in [6,7] have limited number of agents resulting in narrow fields of application domains.

The micro-robot soccer platform serves as a test bed to study and research on the issues pertaining to cooperative mobile robots. A robot team needs to coordinate its actions while competing with another. The robot soccer system also has an explicit performance measure, the match score. Cooperation, decision making, planning, modeling, learning, robot architecture, vision tracking algorithm, sensing, communication, and so forth are some of the directions of study. Different schemes exist for soccer robot control [8]. In a robot-based scheme, each autonomous robot takes a decision based on the information collected with its sensors. If needed the robots communicate with others. The control structure, behaviors and actions of the robots influence the performance of the robot-soccer teams.

The chapter is organized as follows. Section 2 deals with Evolutionary Robotics. The multi-objective evolutionary algorithm is outlined in Section 3. Section 4 introduces the robot soccer system. The evolutionary artificial potential field based robot path planning within a robot soccer setup is discussed in Section 5 with the implementation results. Conclusions are drawn in Section 6.

2 Evolutionary Robotics

Autonomous path planning plays an important role in mobile robot systems. The path-planning problem can be stated as seeking a collision free path between two locations with certain optimization criteria. The two major directions in collision free navigation are the Artificial Intelligence (AI) and the potential field approaches. The AI approach focuses on global path planning with optimization algorithms. The AI approach involves complex computing, and can be resorted to when the information on the environment is ambiguous.

The potential field approach provides freedom in selecting the potential field functions and is simple in realization. Currently the robot vision systems are capable of efficient image processing on the workspace in real time. As a result, the positions of the objects in the workspace are easy to identify and the subsequent potential field generation is straightforward.

Potential field functions have evolved from physical potential field functions in the beginning to the Artificial Potential Field (APF) functions. Research on APF problems ranges from building artificial potential functions to overcoming the restrictions on the workspace [9,10]. In [11] an artificial potential field method for the path planning of non-spherical single-body robots is proposed which simulates the steady-state heat transfer with variable thermal conductivity. Harmonic potential field method can be utilized to navigate well in complex environments like maze; however, the involved computation is comparably complicated [12]. A tradeoff between the navigation performance and involved computation is required in real time applications. Simple and effective functions are preferred in dynamic environments where absolute accuracy is of less importance.

Recently the topic of Evolutionary Robotics (ER) has generated much attention as a tool for the creation and programming of robot control systems [13]. ER is an attempt to develop robots and their sensorimotor control system through an automatic design process involving artificial evolution [14]. The core technique behind ER is Evolutionary Algorithm (EA), which is aimed at a coherent population-oriented methodology for structural and parametric optimization of a diversity of systems. The stochastic algorithms model the natural phenomena like, genetic inheritance and Darwinian strife for survival. EAs offer an important ability to cope with realistic goals and design objectives reflected in the form of relevant fitness functions. From biological sciences it is learned that the animal brains and bodies have developed in parallel. By evolving the robot structure and control program in parallel (co-evolution), it is hoped that ER can begin to solve more complex problems.

Research on APF with genetic algorithms is presented in [15]. Blended with EAs, the Evolutionary Artificial Potential Field (EAPF) is capable of real-time robot path planning [16]. The artificial potential field method is combined with genetic algorithms to derive optimal potential field functions. The multi-objective evolutionary algorithm (MOEA) is resorted to, to deal with the multiple objectives associated with mobile robot navigation. MOEA is a stochastic search technique inspired by the principles of natural selection and genetics. MOEA has attracted significant attention from researchers in various fields due to its ability to search for a set of pareto optimal solutions. The solution is not guaranteed to be the best, but it brings out fine control results in most of the time.

The EAPF functions are utilized for real time path planning in a robot soccer system setup, which is one of the test beds in the study on multi-agent systems [1,17]. In a robot soccer system, the active environment is placid and continuous, with fixed bounds and goals. For such a known environment, APF approaches are convenient for path navigation and collision avoidance.

3 Multi-Objective Evolutionary Algorithm

Evolutionary algorithms (EAs) mimic the mechanics of natural selection and evolution. EAs are effective in searching for a set of globally optimized trade-off solutions simultaneously. In contrast with the traditional gradient-guided search techniques, no derivative information of the search points is required in EAs.

As a result no stringent conditions like well behaved or differentiable are attached to the objective functions in EAs. Many real world problems involve multiple measures of performance or objectives for simultaneous optimization [2]. In certain cases, the objective functions may be optimized separately, however a suitable set of solutions to the overall problem can seldom be obtained. Optimal performance according to one of the objectives may result in unacceptable low performance in other objective dimensions. Evolutionary techniques for multi-objective optimization have attracted wider attention from researchers in various fields recently. Evolutionary algorithm (EA) is a well-suited tool in multi-objective optimization. Multiple individuals can search for multiple solutions in parallel, eventually taking advantage of any similarities available in the family of possible solutions to the problem.

Multi-objective evolutionary algorithm (MOEA) is a global search technique based on the mechanics of natural selection. MOEA is effective to solve complex multi-objective optimization problems where conventional optimization tools fail. The existing methods linearly combine multiple attributes to form a composite scalar objective function. MOEA incorporates the concept of Pareto's optimality or modified selection schemes to evolve a family of solutions at multiple points along the trade-off surface. Real-world applications of MOEA include cancer treatment, control engineering and physiological processes of biological plants.

The different approaches in MOEA are the plain aggregating approaches, the population based non-Pareto approaches and the Pareto-based approaches. In the plain aggregating approach a combination of the objectives is optimized to produce a single compromise solution. If the optimal solution is unacceptable new runs of the optimizer are required till a suitable solution is found. In the population-based non-Pareto approach appropriate fractions of the old generation are selected according to the separate objectives to form the subsequent generation. After shunting all of the sub-populations together the cross-over and mutation operators are applied as usual. Each objective is effectively weighted proportionally to the size of each sub-population and, more importantly, proportionally to the inverse of the average fitness (in terms of that objective) of the whole population at each generation. In the Pareto-based approach all the non-dominated individuals in the population are assigned with equal reproduction probability.

4 The Robot Soccer System

The micro-robot soccer system platform (Fig. 1) is utilized to test the EAPF navigation approach. The system consists of three bi-wheel type robots with RF receiver, a host computer and RF-transmitters.

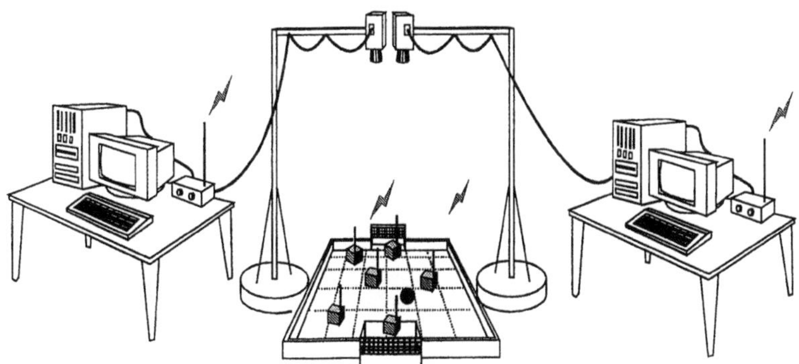

Fig. 1. An artist's view of the robot soccer system setup

The robots used in the setup are 7.5 (cm) cubic in size (Fig. 2). The robots have driving mechanism, communication parts and computational parts. The computational part controls the robot's velocity according to the command data received from the host computer. The robots receive two parameters (right and left wheel velocities) via the RF-receivers. All the calculations for vision data processing and position control of robots are done on the host computer. A CCD camera (vision system) grabs the positions of the ball (goal) and other robots (obstacles). The color patches on top of the robots are utilized to identify the coordinates of the robots (Fig. 3).

To control the robots accurately, the vision sampling time must be very small. As all algorithms of the system are centralized in the host computer, the communication protocol for multi agent cooperative system is simple. However the burden on the host computer increases with the increase in the number of agents.

Fig. 2. The robot soccer system

Fig. 3. The soccer robot and ball

5 Evolutionary Artificial Potential Field

In the traditional artificial potential filed methods an obstacle is considered as a point of highest potential and a goal as a point of lowest potential. The attractive force ($\overline{F}_a$) is inversely proportional to the distance from the robot to the target/goal.

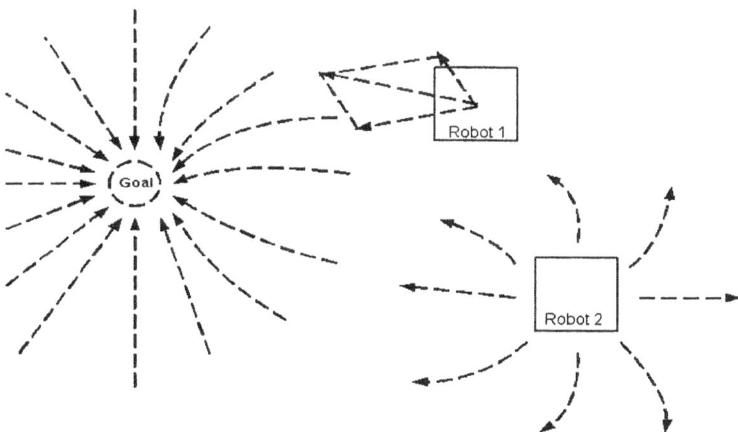

Fig. 4. Artificial potential field

The mobile robot field is mapped onto a 2-D coordinate system (Fig. 5). The home robot positions are presented by the vector $\overline{P}_R(x_r, y_r)$, the opponent robot positions by $\overline{P}_{Oi}(x_{Oi}, y_{Oi}), i = 1, 2, 3$, and of the ball by $\overline{P}_G(x_g, y_g)$. Distances of the robot from the goal point and obstacles are calculated with Eqs. 1 and 2.

$$D_{rg} = ||\overline{P}_G - \overline{P}_R|| \tag{1}$$

$$D_{roi} = ||\overline{P}_R - \overline{P}_{Oi}|| \tag{2}$$

The attractive and repulsive forces are defined in Eqs. 3 and 4 respectively. The potential field angle θ (Eq. 5) is the control signal to the robot.

$$||\overline{F}_a|| = \frac{1}{||\overline{P}_G - \overline{P}_R||} \tag{3}$$

$$||\overline{F}_r|| = \frac{1}{(p||\overline{P}_G - \overline{P}_R||)^n} \tag{4}$$

$$\theta = \angle\overline{F}_a - \angle\overline{F}_r \tag{5}$$

Where D_{rg} is distance between the robot and the goal point, D_{ro} is the distance between the robot and an obstacle; p and n are positive parameters that are to be optimized.

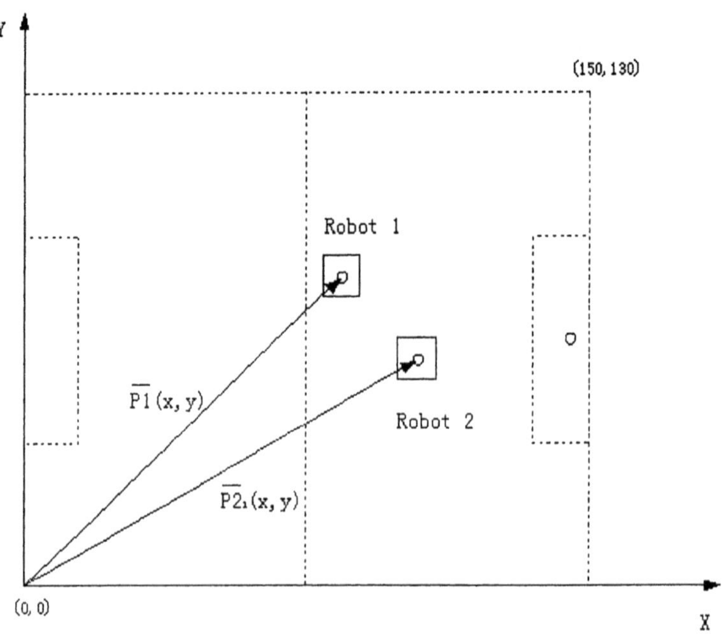

Fig. 5. Position representation of robots

When the attractive and repulsive forces balance out, the robot is trapped. To avoid such a scenario an escape force $\overline{F}_e$ (Eq. 6) is utilized [18]. When a robot is in escape status, the escape force, $\overline{F}_e$, is executed three times to ensure that the robot escapes that area. The reason for such a design is that if the escape force becomes zero at the exact time interval when the robot is out of the escape area, then the robot cannot move effectively out of the null-force area. It may possibly fall into the escape status again and may take longer time to move towards the target.

$$||\overline{F}_e^{(i)}|| = \frac{|cos(\angle\overline{F}_a^{(i)} - \angle\overline{F}_r^{(i)}) - cos(c)|}{D_{min}} \quad (6)$$

Where, $D_{min} = min(||\overline{P}_R - \overline{P}_{Oi}||), i = 1, 2, 3; \angle\overline{F}_e = \pi/4$

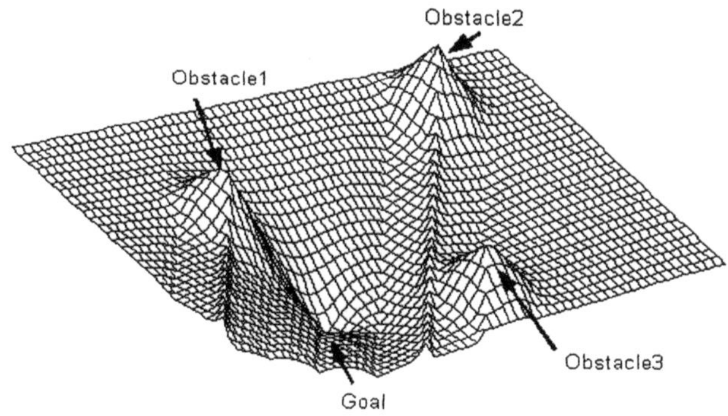

Fig. 6. Potential field distribution

The multi-objective evolutionary algorithm (MOEA) is used to optimize the parameters (p, n and c) associated with the potential field functions (Eqs. 4, 6). The parameters involved in potential field configuration affect the field layout. With an in-crease in p, the repulsive forces from obstacles decrease exponentially resulting in a smooth field surface (Fig. 7). With a higher value of n, the repulsive forces decrease (Fig. 8). A steeper gradient refers to a larger force.

5.1 Optimization of the EAPF Parameters with MOEA

Fitness selection is the preliminary problem in optimization. A population of parameters through the Evolutionary Process creates the next generation.

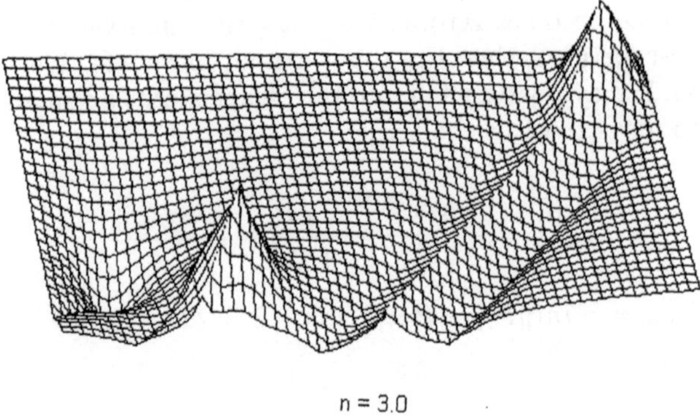

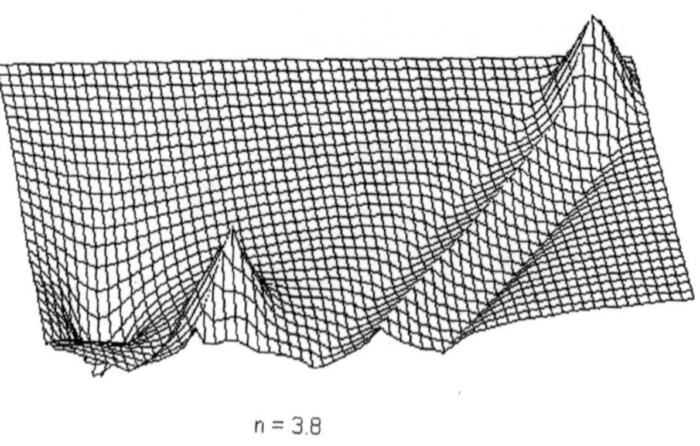

Fig. 7. Potential fields with different values of n

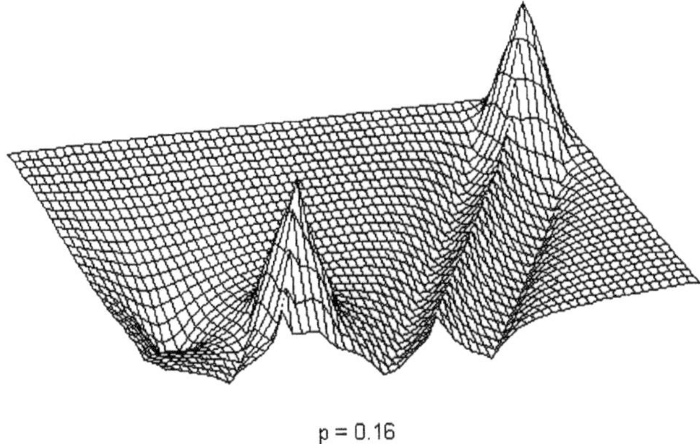

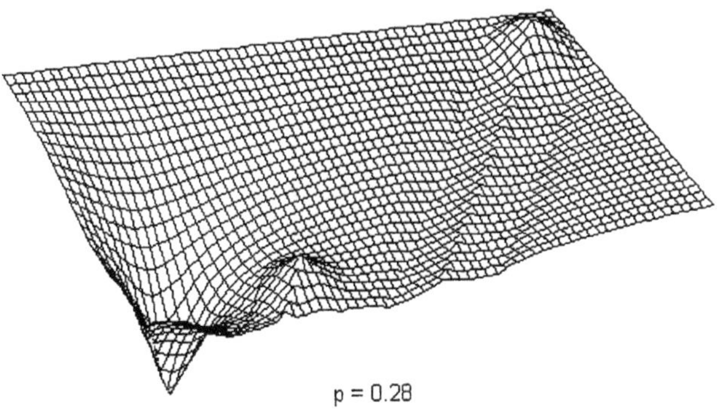

Fig. 8. Potential fields with different values of p

Pairs of strings are selected from this temporary population for the crossover process and subsequent mutation to form the new population.

The MOEA program [19] utilizes the Elitists strategy, where the best strings from the previous generation are added to a new population. Before adding these strings, the MOEA program checks whether these strings already exist in the new population. If a reserved string is not present in the new population, it is added to that population. This may result in a population that is larger than the user-specified population size. The new population is then fed to the model to generate new cost functions and new ranks are generated. If the population size is larger than the user-specified population size, the population is evaluated and the inferior strings are removed. In the Elitist strategy, good strings are set aside to include in the subsequent population. With this strategy, good strings are preserved, and the population always evolves to the better. For multiple objectives problems, the strings with the rank of one, which are not inferior to any other strings, are preserved. If too many strings are ranked one, the cost after niching is considered. Selection of appropriate fitness functions is crucial in the optimization process. The influence of cost terms in control policy and evolutionary program techniques is presented in [13,20].

The fitness function is formulated in two ways: as a linear combination of penalty functions and through prioritization. The robot is desired to arrive at the goal point (kicking the ball - C_1) through a collision free path (C_2). The fitness function C_1 represents the penalty value associated with the distance between the robot and the goal point. When the robot reaches the goal point, C_1 is zero (Eq. 8). C_2 denotes the penalty value considered on collision. C_2 is set to zero when the distance between the robot and an obstacle (D_{ro}) is more than the Limit. Limit is defined as the accepted safe value of D_{ro}. In this work the Limit is set to the sum of the radii of 2 robots. C_3 represents the path length. C_1 and C_2 have higher priority than the path length (C_3). The robot is programmed to follow the potential force direction to approach the goal with a set of parameters. At the end of the session, three fitness values are calculated to evaluate the effectiveness of the set of parameters.

$$\begin{aligned} C_1 &= min(D_{rg}) \\ C_2 &= \begin{cases} 0 & \text{if } min(D_{ro}) > Limit \\ Limit - min(D_{ro}) & \text{otherwise} \end{cases} \\ C_3 &= \int ds \end{aligned} \tag{7}$$

5.2 Experimental Results

The Micro-Robot Soccer System platform (Fig. 1) is used to test the evolutionary artificial potential field based navigation approach. The potential field angle (Eq. 3) is the control input signal and the appropriate velocities proportional to the turn-angle are transmitted through RF to the robots.

The actual motor velocities of the left (v_{left}) and right (v_{right}) wheels are determined by the on-board micro-controller through a classical PID controller with velocity feedback [18]. The wheel velocity range of the robot is (-127, 127). The PID controller gains are K_p, K_i and K_d. Larger values of K_p lead to quicker turning. The ranges of the different parameters are provided in Table 1.

Table 1. Robot controller parameters

K_p	$0.12 \sim 0.14$	K_i	$0.14 \sim 0.20$	K_d	$0.50 \sim 0.80$

Fig. 9 shows the scenario where robot 1 is tracking the ball while facing competition from robot 2. The robot 1 is able to avoid the moving robot effectively. Robot 3 is stationary.

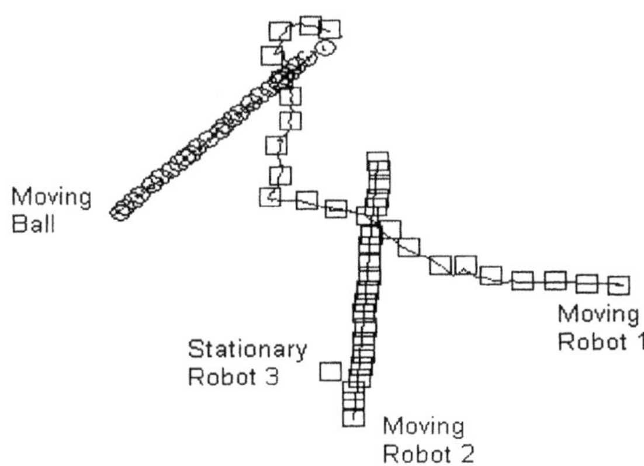

Fig. 9. Ball tracking while in competition with a non-stationary robot

In Fig. 10 there are 3 moving obstacles (opponent robots) and the robot 1 is able to kick the ball to the left. Robot 3 took a path to intercept the ball and robot 2 tried to block the path of robot 1. The path of robot 1 is modified as robot 2 moved towards it. Robot 1 is successful to avoid the moving obstacle (robot 1) and to reach the goal (ball) before robot 3 could (competition).

One of the main problems in EAPF is that the robot cannot pass between two obstacles; even the space in between the two is enough for the robot to pass through. This happens as the direction of the sum of the two repulsive forces point away from the opening between the two close obstacles [10]. This problem also affects the smoothness of the robot motion.

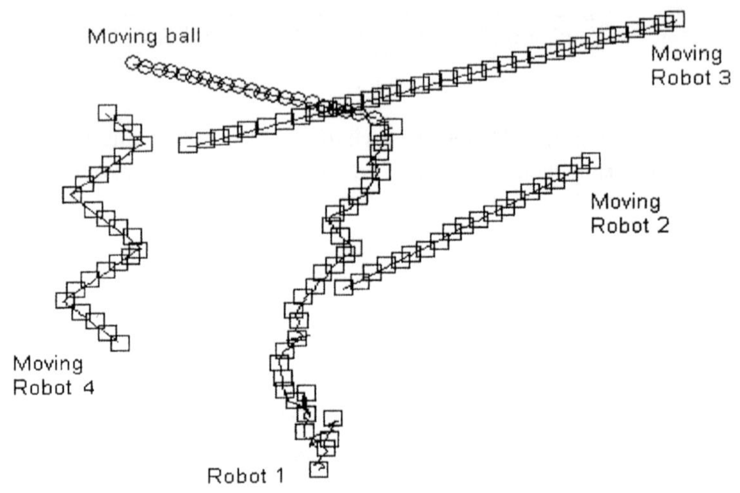

Fig. 10. Ball kicking while competing with 3 opponent robots

6 Conclusion

In this work the application of the evolutionary artificial potential field (EAPF) method in a micro-robot soccer environment is presented. The EAPF functions proposed are tested in different scenarios in ball tracking and kicking, while facing competition from other robots. Both stationary and non-stationary obstacles are considered in the test setup. The goal point (ball) is a passive non-stationary target that is being tracked.

The discussed path planning algorithm can be modified for multiple mobile robots operating in dynamic environments. Vision processing and sampling is a crucial issue in a robot soccer setup. Fast vision processing and higher sampling rates provide better performance.

For more accurate solutions, it is required to optimize the parameters associated with the EAPF functions in real-time. Further research is needed to impart cooperative behaviors and learning capabilities to the mobile robots.

References

1. Kim J.-H. and Vadakkepat P. (2000) Multi-agent systems: A survey from the robot soccer perspective. Int. J. of Intelligent Automation and Soft Computing 6(1), pp 3-17
2. Jennings NR and Wooldridge M (1996) Software Agents. IEE Review 42(1), pp 17-21
3. Tyrrell T (1993) The use of hierarchies for action selection. From Animals to Animats 2, The MIT Press, pp 138-147
4. Brooks RA (1991) Intelligence without representation. Artificial Intelligence, 47(1), pp 139-159
5. Kawauchi Y, Inaba M and Fukuda T (1993) A Principle of Distributed Decision Making of Cellular Robotic System (CEBOT). IEEE Proc. Int. Conf. Robotics and Automation, vol 3, USA, pp 833-838
6. Kube CR and Zhang H (1996) The Use of Perceptual Cues in Multi-Robot Box-Pushing. IEEE Proc. Int. Conf. Robotics and Automation, pp 2085-2090
7. Brumitt BL and Stentz A (1996) Dynamic Mission of Planning for Multiple Mobile Robots. IEEE. Proc. of Conf. Robotics and Automation, pp 2396-2401
8. Kim JH, Shim HS, Kim HS, Jung MJ, Choi IH, and Kim JO (1997) A cooperative multi-agent system and its real time application to robot soccer. Proc. IEEE Int. Conf. on Robotics and Automation, Albuquerque, New Mexico, vol 1, pp 638-643
9. Masoud AA (1998) Integrating Directional Constraints in Motion Planning Using Nonlinear, Anisotropic, Harmonic Potential Fields. Proc of the 1998 IEEE ISIC/CIRA/ISAS Joint Conference, Gaithersburg, MD, pp 14-18
10. Tsuji T, Morasso PG and Kaneko M (1996) Trajectory Generation for Manipulators Based on Artificial Potential Field Approach with Adjustable Temporal Behavior. Intelligent Robots and Systems'96, IROS 96, Proc. of the 1996 IEEE/RSJ International Conference, vol 2, pp 438-443
11. Wang Y and Chirikjian GS (2000) A New Potential Field Method for Robot Path Planning. Proc. of IEEE Int. Conf. on Robotics and Automation, vol 2, pp 977-982
12. Jin-Oh Kim, Pradeep K. Khosla (1992) Real-Time Obstacle Avoidance Using Harmonic Potential Functions. IEEE Trans. Of Robotics and Automation, vol 8, no. 3, pp 338-349
13. Timothy ER and McCartney R (2000) A Cost Term In An Evolutionary Robotics Fitness Function. Proc. of Congress on Evolutionary Computation, vol 1.1, pp 125-132
14. Nolfi S (1998) Evolutionary Robotics: Exploiting the full power of self-organization. Self-Learning Robots II. Bio-robotics, Digest No. 1998/248, IEE pp 3/1 -3/7

15. Dozier G, Homaifar A, Bryson S and Moore L (1998) Artificial potential field based robot navigation, dynamic constrained optimization and simple genetic hill-climbing. Proc. Evolutionary Computation, IEEE World Congress on Computational Intelligence, pp 189-194
16. Vadakkepat P, Tan KC and Wang ML (2000) Evolutionary artificial potential fields and their application in real time robot path planning. Proc. Congress on Evolutionary Computation, vol 1, pp 256-263
17. Shim HS, Jung MJ, Kim HS, Kim JH and Vadakkepat P (2000) A Hybrid Control Structure for Vision Based Soccer Robot System. Intelligent Automation and Soft Computing, vol 6, no 1, pp 89-101
18. Vadakkepat P, Lee TH and Xin L (2001) Application of Evolutionary Artificial Potential Field in Robot Soccer System. Joint 9th IFSA World Congress and 20th NAFIS International Conference, Vancouver, Canada, pp 2781-2785
19. Tan KC, Wang QG, Lee TH, Khoo TT and Khor EF (1999) A Multi-Objective Evolutionary Algorithm toolbox for Matlab (http://vlab.ee.nus.edu.sg/~kctan/moea.htm)
20. Rana AS and Zalzala AMS (1995) An evolutionary algorithm for collision free motion planning of multi-arm robots. First International Conference on Genetic Algorithms in Engineering Systems: Innovations and Applications, pp 123-130

Part 3

LEARNING, ADAPTATION AND CONTROL

Using Hierarchical Fuzzy Behaviors in the RoboCup Domain

Alessandro Saffiotti and Zbigniew Wasik

Center for Applied Autonomous Sensor Systems
Dept. of Technology, Örebro University
S-70182 Örebro, Sweden
{asaffio,zbych}@aass.oru.se
http://www.aass.oru.se

Abstract. An important reason for the popularity of the behavior-based paradigm in autonomous robotics is the possibility to design complex robot behaviors in an incremental way. We propose a fuzzy hierarchical behavior-based architecture, in which rules and meta-rules are used in a uniform way at all levels of the control hierarchy. This architecture has been successfully used in a number of robots performing autonomous navigation tasks. In this paper, we show the use of hierarchical fuzzy behaviors to implement a set of navigation and ball control behaviors for a Sony four-legged robot operating in the RoboCup domain. We also show that the logical structure of the rules and the hierarchical decomposition simplify the design of very complex behaviors, like the "GoalKeeper" behavior.

1 Introduction

The commercial interest in autonomous robots is rapidly growing. Robots which are able to operate in natural, unmodified environments have an increasing number of potential applications, including: services in home, offices, and industrial sites; help to disabled or elderly people; monitoring and manipulation of hazardous sites and materials; emergency rescue operations; and entertainment.

Autonomous robot operation in natural environments, however, poses a number of challenges which are only partially solved by existing technologies. In particular, the control program of an autonomous robot must be able to cope with high degrees of uncertainty and unpredictability in the environment, with the presence of multiple and time-dependent goals, and with limited perceptual and computational resources. Such a control program necessarily has a high degree of complexity.

A popular way to address this complexity is to design the robot controller in a modular way. According to the behavior-based paradigm [1,2], complex controllers are built by combining in an appropriate way a number of simple behavior-producing units, or *behaviors*. The key to design simplicity is that each behavior is only meant to achieve a simple, elementary goal under a limited set of conditions. Complexity emerges from the combination of different behaviors, resulting in an overall controller that can cope with a larger number of goals under a larger set of conditions. However, the fundamental problem of *how* different behaviors should be combined together is not well understood yet. This seriously limits the scalability of behavior-based systems. The two main problems here are: (i) how to effectively ***design*** effective behavior combination strategies in complex domains, and (ii) how to ***prove*** that these strategies achieve the intended goals. The second point has been addressed, e.g., in [3]. This chapter contributes to the first point.

We propose a hierarchical approach to the incremental design of complex robot behaviors based on fuzzy logic. (See [4] for an overview of the uses of fuzzy logic in autonomous robotics.) The main points of our approach are:

- Behaviors are defined in terms of fuzzy "if-then" rules; this makes it easier to write complex control strategies based on heuristic knowledge.
- Complex behaviors are obtained by combining simpler ones using fuzzy meta-rules and a mechanism for behavior blending based on fuzzy logic.
- Behaviors inform perception so that the perceptual resources can be used effectively in a task-dependent way.

This approach has been used to implement complex navigation, perception, and ball manipulation behaviors on a team of Sony AIBO robots, and it has been tested in the RoboCup domain in the context of Team Sweden.[1]

The rest of this chapter is organized as follows. In the next section, we introduce the application domain and the overall architecture used to control our robots. Section 3 describes how basic behaviors are defined and implemented within this architecture, while Section 4 shows how complex behaviors are built by combining simpler ones. Section 5 deals with the use of behaviors to control perceptual resources. Section 6 gives a concrete example of how a very complex behavior (a full goal keeper) can be implemented using our approach. Section 7 briefly discusses the main benefits and problems of our approach, and concludes.

2 The RoboCup Domain

RoboCup is an international robot soccer competition held every year since 1996. The RoboCup competition is divided into several leagues, which differ in the type of robots and environment involved. The approach described in this paper was used in the legged robot league. This league is peculiar in that all competing teams use exactly the same physical platform, a special version of the Sony four-legged robot AIBO [6]. Two teams of four robots each compete on a field of approximately 3×5 meters, where all the relevant objects (the ball, the nets, the robots, and six landmarks around the field) can

[1] "Team Sweden" [5] is the Swedish national team that entered the Sony legged robot league at the RoboCup 1999, 2000, and 2001 competitions. Team Sweden currently consists of four universities: Örebro University, Lund University, Umeå University, and the Blekinge Institute of Technology.

Fig. 1. A snapshot from the 2001 RoboCup competition.

be identified by their color. The robots operate in fully autonomous mode, but can use a radio link to exchange data between them. Fig. 1 shows a snapshot from a game.[2]

Fig. 2 shows one of the robots used in our work. The robot has three degrees of freedom in each leg, and additional degrees of freedom in the neck, head and tail. The principal sensor in this domain is a color camera that allows the robot to detect the objects in the environment and to estimate their position. The estimated position is affected by uncertainty and imprecision due to errors in image segmentation and to partial occlusions. The on-board perceptual and computational resources are limited, so it is important to rely on a cognitive architecture that can best use these resources while effectively copying with the uncertainty and umpredictability that characterize the RoboCup domain.

In our work, we use the layered architecture sketched in Fig. 3. This is a variant of the Thinking Cap, the autonomous robot architecture based on fuzzy logic in use at Örebro University [7]. We outline below the main elements of this architecture.

The lower layer (commander module, or CMD) provides an abstract interface to the sensori-motor functionalities of the robot. The CMD accepts abstract commands from the upper layer, and implements them in terms of actual motion of the robot effectors. In particular, CMD receives set-points for the desired displacement velocity $\langle v_x, v_y, v_\theta \rangle$, where v_x, v_y are the forward and lateral velocities and v_θ is the angular velocity, and translates them to an appropriate walking style by controlling the individual leg joints. This

[2] The snapshot was taken at the RoboCup 2001 games in Seattle. According to the RoboCup regulations at that time, the size of the field was about 2 × 3 meters, teams only had three robots, and no radio communication was allowed.

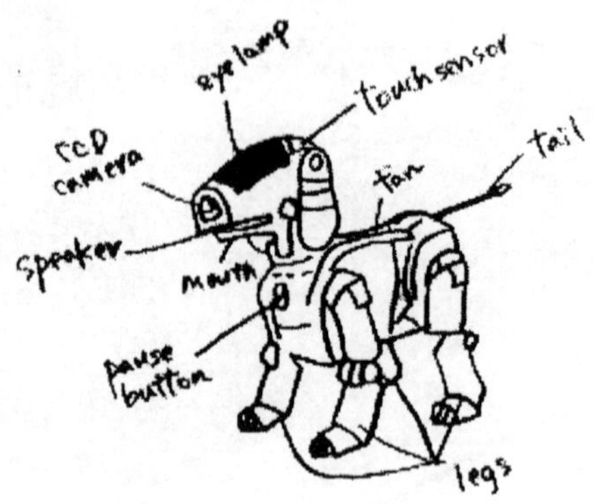

Fig. 2. The Sony AIBO legged robot. ©1999 Sony Corporation.

abstraction strategy allows us to write motion behaviors that can be easily ported across different physical platforms, including both legged and wheeled robots.

The middle layer maintains a consistent representation of the space around the robot (Perceptual Anchoring Module, or PAM), and implements a set of robust tactical behaviors (Hierarchical Behavior Module, or HBM). The PAM acts as a short term memory of the location of the objects around the robot: at every moment, the PAM contains an estimate of the position of these objects based on a combination of current and past observations with self-motion information. For reference, objects are named Ball, Net1 (own net), Net2 (opponent net), and LM1–LM6 (the six landmarks). The PAM is also in charge of camera control, by selecting the fixation point according to the current perceptual needs [8]. The HBM realizes a set of navigation and ball control behaviors, and it is described in greater detail in the following sections.

The higher layer maintains a global map of the field (GM) and makes real-time strategic decisions (RP). Self-localization in the GM is based on fuzzy logic, as reported in [9]. The RP implements a behavior selection scheme based on the artificial electric field approach [10]. We attach sets of positive and negative electric *charges* to the nets and to each robot, and we estimate the heuristic value of a given field situation by measuring the resulting electric potential at some *probe* position — for instance, the position of the ball. This heuristic value is used to select the behavior that would result in the best situation.

In the rest of this chapter we focus on the definition and implementation of the HBM. Additional information about the Team Sweden architecture and its other components can be found on the team web site [5], which also points to the relevant on-line publications.

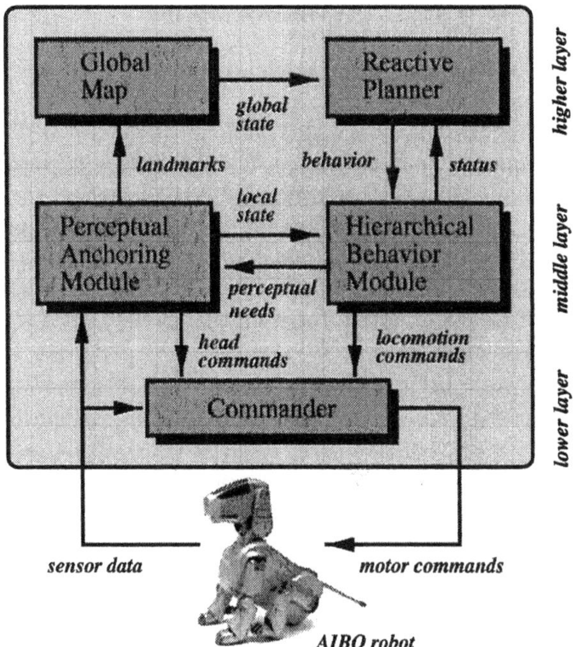

Fig. 3. The variant of the Thinking Cap architecture used by Team Sweden.

3 Basic Behaviors

In this section, we show how we define and implement *basic* behaviors, that is, behaviors that perform elementary types of actions. In our domain, these include turning toward the ball, going to the ball, or moving to a given position. These behaviors constitute the basic building blocks from which more complex types of actions are obtained by hierarchical composition.

3.1 Desirability Functions

In most behavior-based approaches found in the robotic literature, each behavior B acts as a regulator that takes its input from the robot's internal state variables (or directly from the robot sensors) and produces as output set-points for the robot actuators:

$$B : \text{State} \to \text{Control}\,. \tag{1}$$

Each behavior, then, can be thought of as a decision making agent that generates, in every situation, one "most preferred" control value. Different behaviors *compete* for the control of the actuators, and arbitration is often performed by some sort of winner-take-all mechanism.

Our formal approach to behavior definition is different [3,13]. We describe each behavior B in terms of a *desirability function*

$$Des_B : \text{State} \times \text{Control} \to [0,1], \tag{2}$$

that measures, for each state vector x and control vector u, the desirability $Des_B(x,u)$ of applying the control u when the state is x. Intuitively, B expresses desirable behavioral traits as quantitative *preferences*, defined over the set possible control actions, from the perspective of the goal associated with that behavior. For example, a behavior for avoiding obstacles could map configurations of sonar readings that correspond to the presence of an obstacle on the left of the robot into a function that prefers actions that steer the robot to the right.

What is important in Equation (2) compared to (1) is that in Equation (2) several control actions can be desirable, to a different extent, for a behavior. This approach recognizes that many alternative controls can generate, to a greater or lesser extent, the same *type* of behavior. Each behavior, then, can be thought of as an agent that generates, in every situation, a set of preferences as to which command to apply. As we shall shortly see, this allows different behaviors to *cooperate* in the control of the actuators by combining their individual preferences into a tradeoff control value. This view is grounded in the formal semantic characterization given by Ruspini [11] after the seminal work by Rescher [12].

In our system, the input space (State) used by all behaviors is the *local state* provided by the PAM (see Fig. 3), which contains the current estimates of the position of all the objects in the field. The output space (Control) consists of the velocity set-points $\langle v_x, v_y, v_\theta \rangle$ which are transmitted to the CMD module (Fig. 3). An additional control variable k is used to indicate that a kick of a given type should be performed.

3.2 Implementing Desirability Functions

In practice, we implement a desirability function Des_B for a given behavior B by a set of *fuzzy control rules* of the form

$$\text{IF } A_i \text{ THEN } U_i, \quad i = 1, \ldots, n. \tag{3}$$

where U_i is a fuzzy set on the universe of control values, and A_i is a propositional formula in fuzzy logic whose truth value depends on the current values of the local state variables, e.g., the estimated distance to the ball. Fuzzy control rules allow us to easily express our heuristic knowledge about which actions should be performed in each situation in order to promote the goal of that specific behavior, e.g., to get close to the ball.

For instance, the following fuzzy control rules implement the *GoToBall* basic behavior.

IF (BallOnLeft ∧ ¬ BallHere)	TURN(Left)
IF (BallOnRight ∧ ¬ BallHere)	TURN(Right)
IF (BallAhead ∨ BallHere)	TURN(Ahead)
IF ¬ BallHere	GO(Fast)
IF BallHere	GO(Stay)
ALWAYS	SIDE(None)

The left hand side of these rules contain fuzzy formulas obtained from a set of fuzzy predicates, like BallOnLeft, by the standard fuzzy connectives ∧ (min), ∨ (max) and ¬ (complement to 1). ALWAYS denotes a precondition which is always true. Each fuzzy predicate is defined by a function that computes its truth value, in the $[0, 1]$ real interval, from the value of the internal state variables. Figure 4 (left) shows the truth values of the fuzzy predicates BallOnLeft, BallOnRight and BallAhead as a function of the estimated angle θ between the robot and the ball. Note that the use of general fuzzy formulas to express rule preconditions makes our approach more expressive than standard fuzzy control, where only conjunctions of positive literals are allowed in the antecedent of the control rules.

The right hand side of each rule indicates which control variable should be affected by that rule: the GO and SIDE keywords respectively indicate the v_x variable, corresponding to forward-backward motion, and the v_y variable, corresponding to side-to-side motion. The TURN keyword indicates the v_θ variable, corresponding to rotational motion. The parameters Left, Right and so on are linguistic labels that denote fuzzy sets of control values for each control variable. Fig. 4 (right) shows the five fuzzy sets for the v_θ (TURN) variable.

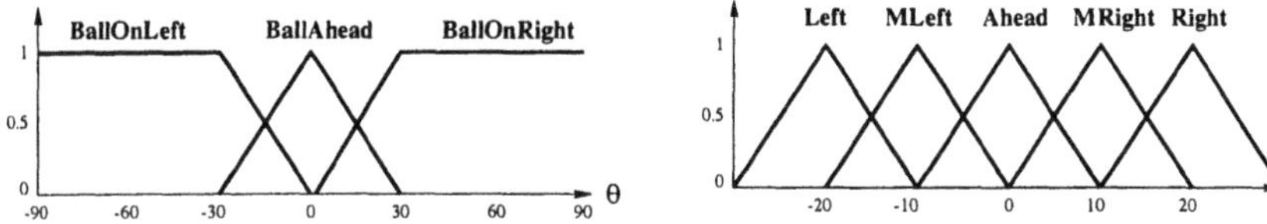

Fig. 4. Left: the three fuzzy predicates related to the ball angle. Right: the five fuzzy sets for the TURN control values.

Given a set R of n fuzzy rules of the form (3), our fuzzy controller computes a corresponding desirability function Des_R by:

$$Des_R(x, u) = [A_1(x) \wedge U_1(u)] \vee \cdots \vee [A_n(x) \wedge U_n(u)] . \qquad (4)$$

Intuitively, Equation (4) characterizes a control action u as being desirable in state x if there is some rule in R that supports u and whose antecedent is true in x. This interpretation of a fuzzy rule-set is that of a classical (Mamdani type) fuzzy controller, generalized so as to allow each antecedent A_i to be an arbitrary fuzzy-logic formula. The Des_R desirability function can be directly used to select a most desired control action $\hat{u}$ by applying a defuzzification technique. In our system, we use Center of Gravity (CoG) defuzzification:

$$\hat{u} = \frac{\int u \, Des_R(x,u) \, du}{\int Des_R(x,u) \, du} . \tag{5}$$

In general, however, this desirability function is first combined with the ones produced by other concurrent behaviors, as discussed in Section 4 below.

3.3 A Suite of Basic Behaviors for RoboCup

Fig. 5 lists the basic behaviors that we have implemented in our robots for the RoboCup domain. Behaviors are classified into three categories: *navigation* behaviors (marked by N) modify the position of the robot in the field; *manipulation* behaviors (M) modify the position of other objects, typically the ball; and *perceptual* behaviors (P) acquire information. The table also shows the number of fuzzy rules used in each behavior. Usually, only a small number of rules is needed to define each behavior, which makes behaviors simpler to write, tune, and maintain. The two key factors for this simplicity are: (i) an accurate design choice about which behaviors should be implemented as basic, and (ii) the fact that we allow arbitrary fuzzy formulas in the rule preconditions.

Type	*Behavior name*	*# rules*	*Typical instance*
N	*GoTo(X)*	7	GoTo(Ball)
N	*GoToStar(X)*	6	GoToStar(Target)
N	*Face(X)*	7	Face(Ball)
N	*Align(X, Y)*	13	Align(Ball,Net2)
M	*Kick*	5	Kick
M	*StealLeft*	3	StealLeft
M	*StealLeft*	3	StealLeft
M	*Block*	2	Block
P	*Search(X)*	4	Search(Ball)
P	*SelfLocalize*	5	SelfLocalize

Fig. 5. The basic behaviors implemented for the RoboCup domain.

Most behaviors take as arguments one or more objects in the environment. For instance, the generic *GoTo* behavior is typically used to go close to the ball by giving it the ball as argument. In a similar way, we typically use *Face(Ball)* and *Align(Ball,Net2)* to perform local position adjustments in order to achieve an adequate posture for kicking. (Net2 denotes the opponent net, and Net1 denotes our own net.) Obviously, the control program cannot be given a pointer to an external physical object. Instead, it is given a pointer to an internal data structure, called *anchor*, whose properties are kept in synch with the properties of the physical object using perception. The PAM module (Fig. 3) is in charge of implementing this anchoring process — see [8] for more on this point.

The difference between the *GoTo* and the *GoToStar* behaviors is that the latter performs obstacle avoidance while trying to go to the target position. This is done by setting up a local fuzzy occupancy grid around the robot, and performing A* path planning on this grid. The grid contains 15 × 17 cells, each representing a square of 10 × 10 cm, and it is filled with occupancy information about the objects to be avoided. Each object is blurred to account for the uncertainty in its estimated position — see Fig. 6. Since the grid is small, filling and planning are inexpensive and they are re-computed at every control cycle (200 msec in our implementation). The *GoToStar* behavior is typically used as a sub-behavior (see below). The grid is filled by the caller behavior, since that behavior knows which objects should be avoided and which ones should not. For instance, navigation behaviors usually need to avoid the ball when the robot is moving in the direction of its own net, but not when it is moving toward the opponent net.

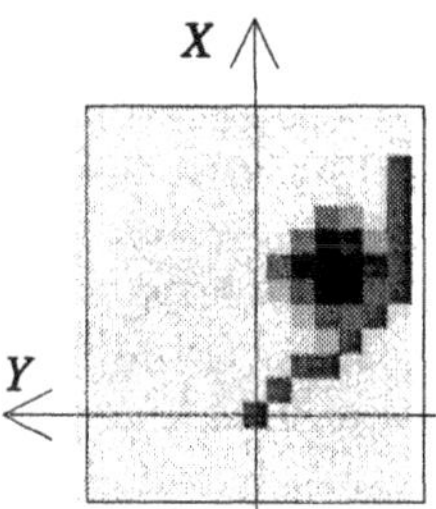

Fig. 6. The local occupancy grid used by the *GoToStar* behavior for obstacle avoidance, and a path leading to a target point behind the ball. The robot is at the origin, in top-view and directed as the X axis.

The *Kick*, *StealLeft* and *StealRight* behaviors respectively kick the ball in the forward direction and to the left/right of the robot. A combination of leg and head motions is used in the CMD to implement each kick. The *Block* behavior causes the robot to stand in a crab-like posture, which is useful to block an incoming ball.

Finally, the *Search* and *SelfLocalize* behaviors visually scan the field until the object is seen, or until enough landmarks are seen to allow establish the robot's location in the field, respectively. These behaviors use a combination of head motion and body rotation.

4 Complex Behaviors

In our approach, we build complex behaviors by combining simpler ones using fuzzy meta-rules. This procedure can be iterated to build a full hierarchy of increasingly complex behaviors. In this section, we discuss the main mechanism used to realize behavior composition, called *Context-Dependent Blending*, and show how we have implemented it in our robots.

4.1 Context-Dependent Blending

In behavior-based approaches, the overall behavior of the system is the result of the coordinated activity of several independent behavior-producing units. While this divide and conquer strategy is the key to the power of these approaches, it is also its main source of difficulty. In fact, the problem of how to coordinate behaviors so that they result in the performance of the intended task still constitutes the Holy Grail of behavior-based robotics.

Following [14], we split the behavior coordination problem into two conceptually different sub-problems: (i) how to decide which behaviors should be activated at each moment; and (ii) how to combine the results from different behaviors into one command to be sent to the robot's effectors. We call these the *behavior arbitration* and the *command fusion* problems, respectively.

The arbitration policy determines which behavior(s) should influence the operation of the robot at each moment, and thus ultimately determines the task actually performed by the robot. Many proposals in the literature use a winner-take-all crisp switching schema: in each situation, one behavior is selected and is given complete control of the effectors (e.g., [1,15]). This simple scheme may be inadequate in situations where several criteria should be taken into account. To see why, consider a robot that encounters an unexpected obstacle while following a path, and suppose that it has the option to go around the obstacle from the left or from the right. This choice may be indifferent to the obstacle avoidance behavior. However, from the point of view of the path-following behavior, one choice might be dramatically better than the other. In most implementations, the obstacle avoidance behavior alone could not know about this, and would take an arbitrary decision.

More flexible arbitration policies can be obtained using fuzzy meta-rules of the form

$$\text{IF } \textit{context} \text{ THEN } \textit{behavior}, \tag{6}$$

meaning that *behavior* should be activated with a strength given by the truth value of *context*, a formula in fuzzy logic. The use of fuzzy meta-rules to express behavior arbitration policies has two main advantages: (i) the ability to express partial and concurrent activations of behaviors; and (ii) smooth transitions between behaviors.

If several behaviors can be simultaneously activated, we need to solve the problem of how to combine the output of different behaviors that refer to the same control variable. That is, we have to solve the command fusion problem. Our definition of behaviors as preference-producing modules suggests that command fusion can be thought of as the problem of aggregating individual preferences: if Des_{B1} and Des_{B2} denote the desirability functions produced by two behaviors B1 and B2, respectively, then the combined preferences of B1 and B2 are represented by the function Des_B given, for all state x and control value u, as

$$Des_B(x,u) = Des_{B1}(x,u) \wedge Des_{B2}(x,u)\,, \tag{7}$$

where $\wedge$ denotes as usual the minimum operator.

In practice, we represent the output of each behavior $B_i, i = 1,\ldots,n$, at a given state x by a fuzzy set U_i over the space U of control values. We then use a fuzzy conjunction operator to combine the preferences of different behaviors, represented by fuzzy sets on U, into a collective preference U_{fuse} given by

$$U_{\text{fuse}}(u) = \bigwedge_{i=1,\ldots,n} U_i(u)\,. \tag{8}$$

Finally, we chose a command from this collective preference according to a defuzzification strategy, e.g., the CoG Equation (5). Fig. 7 (left) illustrates this process on two behaviors $B1$ and $B2$ both concerned with the turning angle.

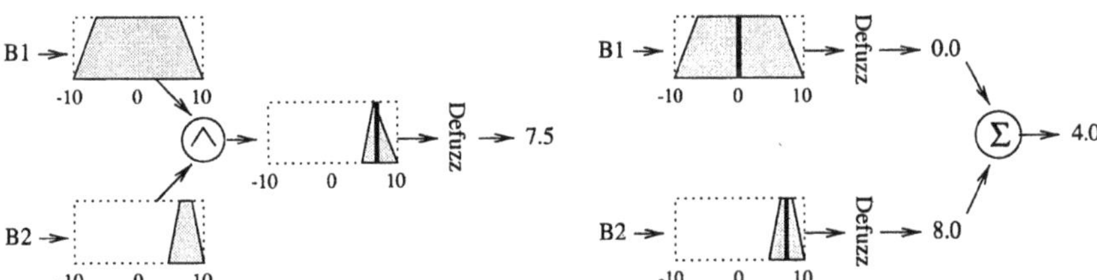

Fig. 7. Two ways to fuse commands: combining individual preferences (left); combining individual decisions (right). The final result may be different.

The fact that defuzzification is performed after combination is crucial, since the decision taken from the collective preference can be different from the result of combining the decisions taken from the individual preferences. Fig. 7 graphically illustrates this point. Intuitively, the individual decision issued by a behavior tells us which is *the* preferred command according to that behavior, but does not tell anything about the desirability of alternatives. Preferences contain more information, as they give a measure of desirability for each possible command. This observation explains why fuzzy command fusion is fundamentally different from potential field methods [2,16], in which force vectors represent individual decisions.[3] A similar distinction between combining preferences and combining decisions is commonly made in the field of data fusion [17].

It should be noted that blind application of averaging defuzzification strategies, like CoG, to the combined fuzzy set may produce problematic results. In particular, when this set is not unimodal, defuzzification may result in the selection of an undesirable control value, i.e., a value which lies in the gap between two peaks of the combined set. In the case of robot control, this may mean that the robot, having the option to avoid an obstacle from the right or from the left, decides to go straight. It is the responsibility of the behavior designer to make sure that the rule-set is free from inconsistencies and ambiguities: e.g., rules that propose drastically different controls should have mutually exclusive pre-conditions. Other authors address this problem by defining *ad-hoc* defuzzification schemes (e.g., [18]).

In our system, we use fuzzy meta-rules (6) to represent the arbitration policy, and fuzzy fusion (7) to perform command fusion. The overall desirability function *Des** is then given by

$$Des^*(x,u) = \bigwedge_{j=1,\ldots,m} \left(C_j(x) \wedge Des_{B_j}(x,u)\right) , \tag{9}$$

where C_j and B_j are the *context* and the *behavior* of the jth meta-rule. This general form of behavior combination, originally proposed in [13], is called *context-dependent blending*, or CDB for short.[4] CDB has been used, sometimes under different names, by several authors in several robots. An overview of the uses of CDB in the literature can be found in [14].

4.2 Implementing Context-Dependent Blending

We have implemented CDB in a very simple way. Any behavior can include fuzzy meta-rules of the form

$$\text{IF } C_j \text{ USE } B_j, \quad j = 1, \ldots, m . \tag{10}$$

where C_j is a propositional formula in fuzzy logic, called the *context*, and B_j is a sub-behavior to be called. For example, the following rule set implements a simple PenaltyKick behavior, which brings the robot in front of the ball, aligns it with the opponent net (Net2), and kicks.

[3] Vector approaches can be simulated as a special case of fuzzy command fusion, e.g., by using fuzzy numbers for preferences, prod-sum combination, and COG defuzzification. In this case, the order in which defuzzification and combination are performed is irrelevant.

[4] Context-depending blending can be given more general definitions in terms of arbitrary T-norms [3]. Our implementation, however, relies on the distributive property, which only holds when using the min/max pair of norms.

IF (BallFar) USE GoTo(Ball)
IF (BallNear $\wedge$ $\neg$ Aligned) USE Align(Ball,Net2)
IF (BallNear $\wedge$ Aligned) USE Kick(Ball)

This behavior will achieve the intended goal under the assumption that the robot starts somewhere in the part of the field which is behind the ball with respect to Net2. Note that in some situations several sub-behaviors can be (partially) activated and combined by CDB. For instance, as the robot approaches the ball, the Align behaviors becomes more and more active while the GoTo behavior is progressively deactivated. As a result, the robots starts the alignment maneuver while smoothly slowing down as it gets closer to the ball.

Calls to sub-behaviors can be iterated, so that we can build a full hierarchy of behaviors. For instance, the following rules use the above PenaltyKick behavior to implement a simple Score behavior where the robot can start anywhere in the field.

IF (BehindBall) USE PenaltyKick()
IF ($\neg$ BehindBall) USE GoBehind(Ball,Net2)

The GoBehind behavior uses the GoToStar behavior to reach a location behind the ball while avoiding collisions with ball. Fig. 8 shows the corresponding behavior hierarchy.

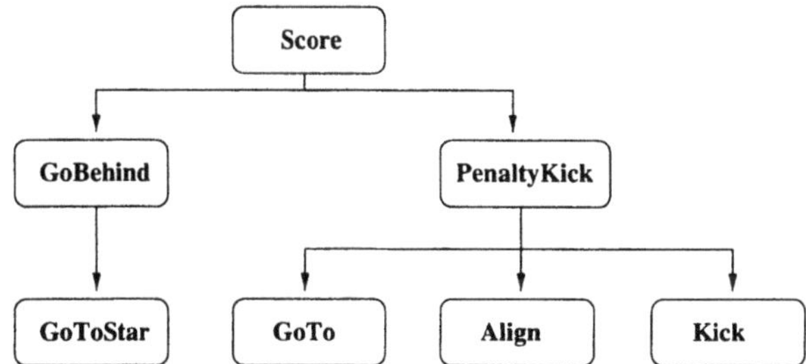

Fig. 8. Behavior hierarchy for a simple Score behavior.

In principle, all of the fuzzy rules in a called sub-behavior B_j are treated as if they were rules inside the caller behavior, except that their consequent is weighted by the value of the context C_j. Suppose that a behavior B contains the meta-rule

$$\text{IF } C' \text{ USE } B' \,,$$

where B' is composed of the rules

$$\text{IF } A_i \text{ THEN } U_i, \quad i = 1, \ldots, n \,.$$

Then, the above meta-rule in B is (virtually) replaced by the set of rules

$$\text{IF } (C' \wedge A_i) \text{ THEN } U_i, \quad i = 1, \ldots, n\,.$$

If B' itself contains some meta-rules, the expansion process is repeated recursively. The overall desirability functions (one for each controlled variable) are computed from the resulting set of rules using Equation (4). The final control values are obtained from these functions by using COG defuzzification, as per Equation (5). This solution is equivalent to performing CDB according to Equation (9), provided that the min and max pair of T-norm and T-conorm are used to evaluate $\wedge$ and $\vee$ [3].

The above implementation strategy has the advantage that object-level control rules and meta-rules can be intermixed in the same behavior. For instance, the following set of fuzzy rules is used to implement the *PushBall* behavior:

```
IF (BallInFront)     USE GoTo(Ball)
IF (BallOnLeft)      SIDE(Left)
IF (BallOnRight)     SIDE(Right)
IF (¬ BallInFront)   GO(Stay)
IF (¬ BallInFront)   TURN(Ahead)
```

The first rule makes the robot move toward the ball when the ball is in front of it, thus pushing the ball. The second and third rules add some side-to-side motion intended to keep the ball centered in front of the robot. The last two rules inhibit forward and rotational motion when the ball is not in front of the robot, and until the lateral (SIDE) motion brings it to a central position again. (Stay and Ahead are linguistic labels that denote zero forward and rotational velocity, respectively.)

4.3 CDB Versus FSM

CDB provides a purely reactive way to combine behaviors: at every moment, the behaviors which are activated depend only on the current perceptual input, as reported by the PAM, and they do not depend on the previous history of behavior activations. Another popular way to define complex behaviors is to use finite state machines (FSM), where states are associated with behaviors and transitions with perceptual events. In a FSM, the behavior(s) to be activated depend on the perceptual input *and* on the previous history. Typically, in these approaches exactly one behavior is activated in every state, and it is given complete control of the effectors.

Fig. 9 shows a FSM implementation of a Score behavior based on the same building blocks as the previous example (Fig. 8). If the robot starts somewhere in the area between the ball and Net2, it will activate in sequence GoToStar, GoTo, Align and Kick. The same would happen using the

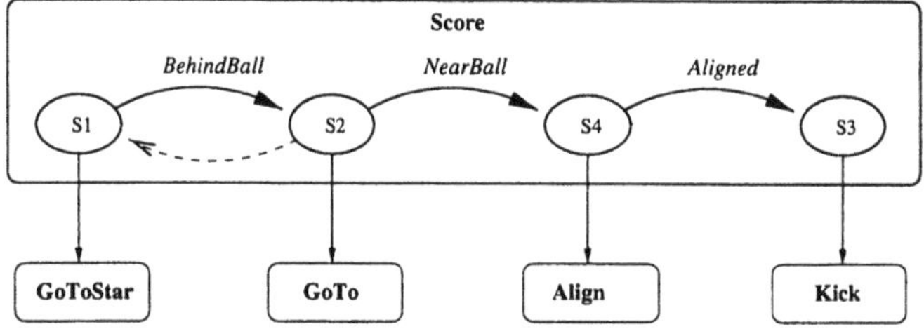

Fig. 9. A FSM for a simple Score behavior.

hierarchical behavior shown in Fig. 8 and implemented by CDB. CDB, however, is intrinsically more robust with respect to sensor noise and unexpected events than the FSM approach. For instance, suppose that after the robot has reached the area behind the ball and has started the GoTo behavior, the ball is moved so that the robot is not behind it any more. The CDB solution naturally copes with this situation, since the BehindBall condition becomes false thus causing the GoBehind behavior to be re-activated. By contrast, in the FSM a new transition should be designed in order to account for this exception (dashed arrow in Fig. 9). A similar problem would arise if a transition were fired by a spurious perceptual event. In general, the FSM needs to incorporate explicit transitions to account for any possible sequence of events.

No matter what the advantages of CDB are, in some cases the introduction of internal state is unavoidable. For instance, we cannot write a purely reactive *Patrol* behavior that goes from one net to the other and back: if the robot is in the middle of the field and perpendicular to the axis between the nets (say, while avoiding an obstacle) it cannot know which way to turn unless it has an internal state to indicate to which net it was going.

Internal states are easily incorporated into our fuzzy behaviors by maintaining a state variable inside the behavior, and deciding which behavior to call according to the value of this variable. Here is a simple example implementing a Patrol behavior.

```
IF (State(0))               SetState(1)
IF (State(1) ∧ AtNet2)      SetState(2)
IF (State(2) ∧ AtNet1)      SetState(1)
IF (State(1))               USE GoTo(Net2)
IF (State(2))               USE GoTo(Net1)
```

The first three rules encode the state transitions of the FSM, the remaining rules associate a behavior with each state. This type of complex behavior can be seen as an instance of a hybrid automaton: a FSM in which each state is associated with a specific control law [19].

5 Behaviors and Perception

Basic and complex behaviors in the Hybrid Behavior Module (HBM) take their input from the local state provided by the PAM (see Fig. 3). This acts as a short term memory of the location of the objects around the robot, stored in robot-centered coordinates. The local state is updated and sent to the HBM at a regular rate (5 Hz in our current implementation). Updating is done by three mechanisms: by *perceptual anchoring*, whenever the object is detected by the robot's camera and its position is measured from the image data; by *global information*, for static objects (e.g., the nets) whenever the robot knows its location in the global map; and by *odometry*, modifying the previous position of the object whenever the robot moves.

In general, estimates based on fresh perceptual data are more reliable than estimates based on odometric update. This is especially true in our domain where: (i) objects can move in unpredictable ways and (ii) odometry is very unreliable when using legged locomotion. Unfortunately, because of limited perceptual and computational resources, the robot cannot keep all the objects in the field under observation at all times. In our AIBO robots, for instance, the head-mounted camera has a limited field of view and limited pan-tilt speed. Keeping track of an object, then, necessarily implies losing track of many others. The design of effective allocation strategies for the perceptual resources is therefore of paramount importance in this domain.

The key observation here is that the robot typically needs to access information about different objects at different stages of execution. For example, in order to kick the ball into the opponent's net, the robot only needs to know the position of the ball and of the net; when it later returns to its home position, it needs the position of the landmarks in order to correctly self-localize, but it does not need to know the position of the ball or of the net any more. In short, perception should be *task-dependent* [20].

To achieve task-dependent perception, we introduce a link in our architecture that feeds the perceptual needs of the controller back to the perceptual module (see Fig. 3) . At every control cycle, the currently active behaviors in the HBM inform the PAM of which objects are "needed" for execution and which ones are not. The perceptual processes in the PAM use this information to decide where to point the camera and which perceptual routines to activate.

Behaviors specify their perceptual needs using fuzzy rules of the form

$$\text{IF } A_i \text{ NEED}(O_i), \tag{11}$$

where A_i is a fuzzy formula, and O_i is the name of one of the objects in the environment. The effect of this rule is to assert the perceptual need for object O_i at a degree that depends on the truth value of A_i.

For example, we can extend the *PenaltyKick* behavior described in the previous section with the following perceptual rules:

IF (BallNear) NEED(Net2)
ALWAYS NEED(Ball)

These rules say that the controller needs to have fresh perceptual data for the ball at all times, but only needs to have an accurate estimate of the position of the opponent's net (Net2) when the robot is close to the ball in order to correctly perform the *Align* and *Kick* maneuvers. In a situation where the ball is at 400 mm from the robot, the truth value of BallNear is 0.7, and these rules assert a degree of need of 1.0 for the ball object and of 0.7 for the opponent's net.

We can think of (11) as a rule controlling an additional control variable $needed_i$ that expresses the degree of perceptual need for object O_i. The NEED rules of different behaviors are combined through CDB in the same was as the rules for the other control variables. Since the consequents of these rules are numbers in $[0,1]$, the result of the combination is simply given, for each variable $needed_i$, by the maximum value computed by any rule that has NEED(O_i) as its consequent.

The values of the $needed_i$ variables so computed are sent to the PAM, which uses them to decide which object should constitute its perceptual focus at every given moment. Intuitively, this is an object which is needed by the currently active behaviors and which has not been updated recently. Full details on our active perception strategy can be found in [8].

During the robot activity, behaviors are dynamically activated and deactivated according to the fuzzy meta-rules. Correspondingly, the perceptual needs change continuously. The combined action of the perceptual rules in the HBM and of the focus selection mechanism in the PAM directs the perceptual resources towards the needs with are most important with respect to the current task at each moment.

Fig. 10 shows how perceptual processes are directed by perceptual needs during an execution of the *PenaltyKick* behavior using the two perceptual rules above. The upper part of the figure plots the temporal evolution of the estimates of the distance and angle (θ, ρ) to the ball, and of the angle θ to Net2. The circles mark the fixation points sent to the camera pan joint by the PAM; the dotted vertical lines mark visual scans done in order to find an object. The lower part plots the degree of perceptual support, on a $[0,1]$ scale, for the ball and the net, respectively.

Until time 118 the robot was approaching the ball (as testified by the decreasing ρ) while tracking it. Since the "NearBall" predicate was false, the net was never selected as the focus of attention; as a consequence, the camera was never pointed at the net, and θ maintained its default zero value. At time 118, the truth value of "NearBall" was high enough for the net to become the focus of attention. The robot then started a visual scan to search for the net, momentarily losing sight of the ball. The net was soon seen, and the camera was then pointed to the expected position of the ball to re-acquire it. When the robot got even closer to the ball (around time 130) it executed

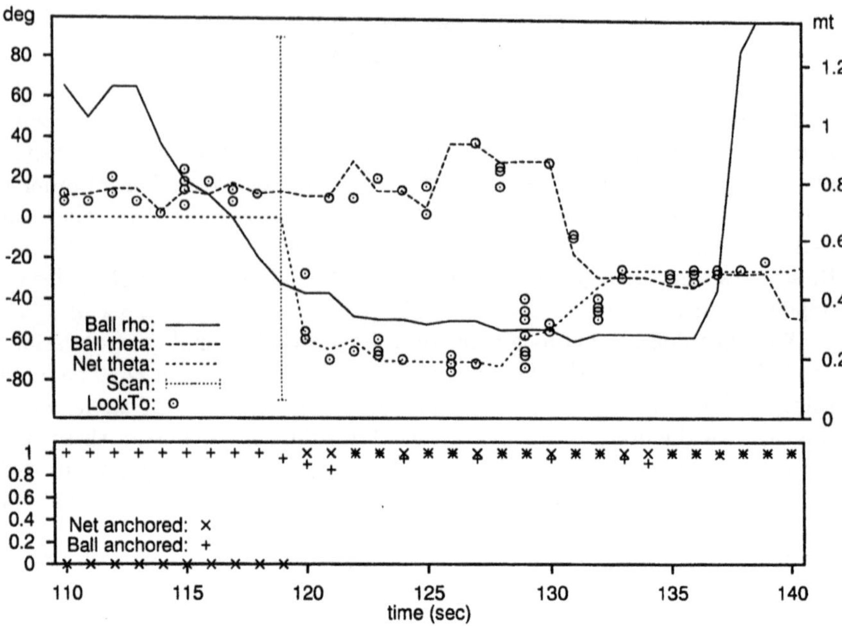

Fig. 10. Task-dependent perception. The robot only tracks the ball until it gets close to it, and then tracks both the ball and the net (see text).

the alignment maneuver while using the camera to alternatively anchor the ball and the net. The alignment maneuver brought the θ of both the ball and the net close to zero. Finally, at time 137, the robot kicked the ball, which headed away from the robot and toward the net.

6 A Case Study: the Goal Keeper

We now give an example of a particularly complex behavior: the *GoalKeeper* behavior. This behavior is run by a robot who is designated as the team goal keeper, and it is intended to perform the full goal keeping task. This behavior is entirely implemented in the HBM using a complex hierarchy of sub-behaviors: the Reactive Planner is not used by the goal keeping robot (see Fig. 3).

6.1 Overall Strategy

The task of the goal keeper is to defend its net from possible scores from any robot. We have adopted a rather defensive strategy: the robot always stays close to its net, tries to be on the way between the ball and the net, and only

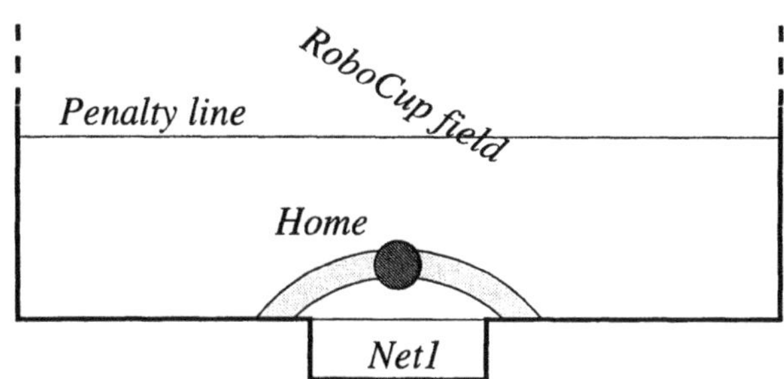

Fig. 11. The workspace of the goal keeper.

acts to send the ball away when there is a good opportunity to do so. Fig. 11 shows the workspace of the goal keeper.

The goal keeper operates in two distinct modes. Under normal conditions, the robot stays in the zone between the net and the penalty line, and it operates in "Play" mode. What is done in this mode depends on the position of the ball. If the ball is far away, or not seen at all, the robot stays at its home position (the circle in Fig. 11) pointing to the opponent net. From this position the robot can see most of the field and it has better chances to localize the ball. Occasionally, the robot turns to the left and to the right to check the blind spots behind it. If the ball is in view, the robot moves left and right at a fixed distance from the center of the net (the circular arc in Fig. 11) in order to be on the line between the ball and the net. If the ball comes too close, it may decide to push or kick it away.

If the robot finds itself well beyond the penalty line, it switches to the "Home" mode of operation. This may happen, for instance, after the robot has performed a long foray to push the ball away, or if it has been displaced by the referees in application of a game rule. Under this mode, the robot tries to reach its home position as quickly as possibly while avoiding collisions with the ball or the other robots on its way.

In addition to implementing the above strategy, we have made a design choice to develop a set of specialized walking and kicking routines for the goal keeper robot, implemented in the CMD module (Fig. 3). We have used walking styles that keep a very low posture so as to improve stability and grip, and we have increased the speed of lateral motion since the goal keeper needs to quickly move left and right. We also keep the legs far apart whenever possible, so as to cover a larger portion of the net. For instance, in the "Block" position, the robot stays still in a crab-like posture that covers about 50% of the size of the net.

6.2 The Behavior Hierarchy

The above strategy has been implemented as a set of specialized behaviors, organized according to the hierarchy shown in Fig. 12.

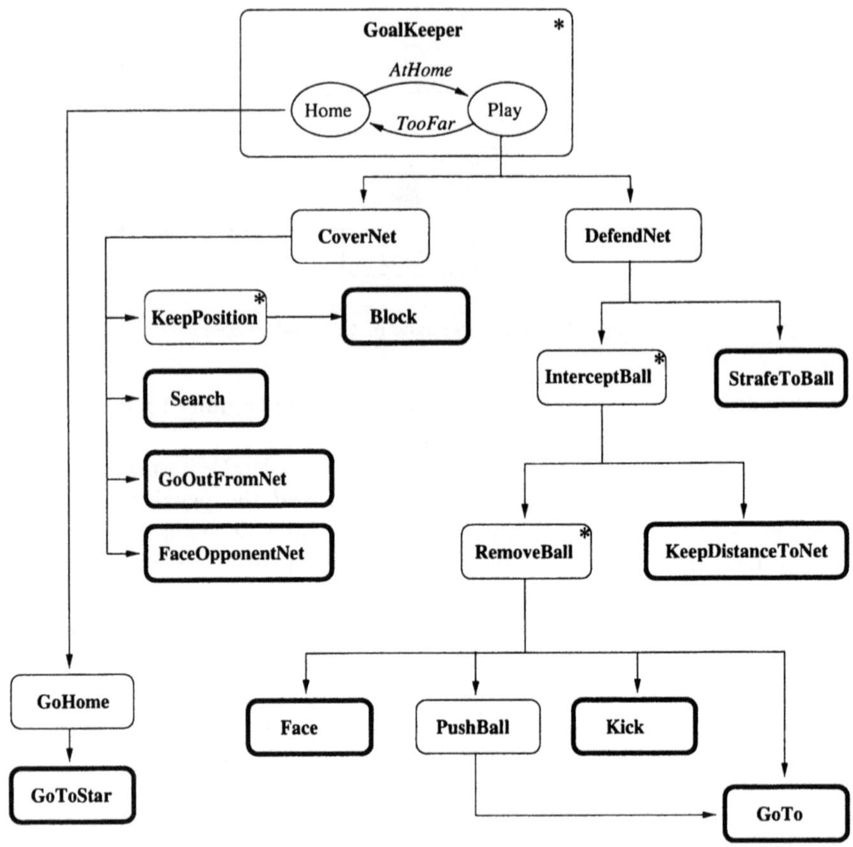

Fig. 12. Behavior hierarchy for the GoalKeeper behavior. The boxes with thick borders indicate basic behaviors. Behaviors with a '*' are detailed in the text.

The boxes with thick borders indicate basic behaviors, that is, behaviors that directly control motion and do not call any sub-behaviors. These are the same basic behaviors listed in Fig. 5 above, plus the following goal-keeper specific behaviors:

FaceOpponentNet Orients the robot towards the opponent's net.
GoOutFromNet Gets the robot out from inside its own net.
KeepDistanceToNet Moves the robot forward or backward in order to keep a fixed distance from the net.
CrabToBall Gets the robot in front of the ball using side-to-side motion.

Notice that the task of *FaceOpponentNet* could be performed by a call to *Face(Net2)*. However, the *Face* behavior is tuned to perform the precise but slow maneuvers typically needed in order to achive a correct position for an effective kick. We have therefore preferred to implement a specialized behavior that provides faster, albeit less accurate, motion.

The other behaviors in the hierarchy are complex behaviors intended to perform the following tasks.

GoalKeeper This is the top-level behavior, described in the next section.
CoverNet Maintains a protective position while observing the field and occasionally checking the left and right corners.
DefendNet Intercepts the ball when it comes close, and possibly gains control of it.
GoHome Reaches the home position from any place on the field while avoiding collisions.
KeepPosition Maintains the home position.
InterceptBall Places the robot on the line between the ball and the net.
RemoveBall Gains possession of the ball and gets it out of the penalty area.

In addition, the *PushBall* behavior described in Section 4 above is used by *RemoveBall.*

Some of the above behaviors express specific perceptual needs by way of perceptual rules of the form (11) above. For instance, most behaviors express a need for the ball position. The *KeepPosition* and *GoHome* behaviors both need to have accurate information about the robot's own location in the field, and hence they express a need for the most probably visible landmarks, including the opponent's net. In addition, the *GoHome* behavior also needs to know the position of all the robots in order to perform obstacle avoidance while navigating to the home position. In general, the overall perceptual needs of the *GoalKeeper* behavior depend on which sub-behaviors are activated at every moment, and are used to direct perception as discussed in Section 5.

6.3 The Top-Level Behavior

The top-level *GoalKeeper* behavior is implemented by the rules shown in Fig. 13. Rules 1–3 encode a Finite State Machine with two internal states, corresponding to the two modes of operation described above: Home (state 1) and Play (state 2). The use of a FSM here is necessary since in general the action to perform does not only depend on the current perception, but also on the current mode. For instance, consider the situation in which the robot is close to the penalty line and oriented toward its own net and no ball is seen: in the Home mode, the robot should keep going forward toward its home position (using *GoHome*); in the Play mode, however, the robot should turn toward the opponent's net and go backward to its home position (using *CoverNet*).

Each mode is implemented by a call to one or more sub-behavior(s). In the Home mode (rule 4) the *GoHome* behavior is called: this fills up the local occupancy grid with information about the position of all the robots and the ball, and then calls the *GoToStar* behavior to navigate to its home position.

1.	IF (State(0))	SetState(1)
2.	IF (State(1) ∧ AtHome)	SetState(2)
3.	IF (State(2) ∧ FarFromHome)	SetState(1)
4.	IF (State(1))	USE GoHome()
5.	IF (State(2) ∧ (¬ BallInView ∨ BallOnOtherSide))	USE CoverNet()
6.	IF (State(2) ∧ BallInView ∧ BallOnThisSide)	USE DefendNet()

Fig. 13. The *GoalKeeper* behavior.

In the Play mode, the robot reactively chooses a play strategy depending on the position of the ball. If the ball has not been seen for a while or it is in the opposite half of the field, then the *CoverNet* behavior is used (rule 5). This behavior uses the *KeepPosition* and *FaceOpponentNet* to maintain a good observation position. In addition, the *Search(Ball)* behavior is used to turn and inspect blind spots at regular intervals.[5] The *GoOutFromNet* behavior may also be needed if the robot ends up inside its own net — an event which is not uncommon during the RoboCup games!

If the ball is detected on our side of the field, the *DefendNet* behavior is used (rule 6). This in turn uses the *CrabToBall*, *InterceptBall* and *Keep-DistanceToNet* sub-behaviors in order to quickly position the robot on the line between the ball and the net at a fixed distance from the net. If the ball comes close to the robot, the *RemoveBall* behavior is used to try and send it out of the penalty area.

6.4 Other Behaviors

In order to give a more concrete impression of how the goal keeper behaviors are implemented, we discuss here three representative behaviors in the hierarchy shown in Fig. 12:

- *KeepPosition*,
- *InterceptBall*, and
- *RemoveBall*.

The *KeepPosition* behavior uses six fuzzy rules to control forward, lateral and rotational motion in order to keep the robot located at its home position and oriented toward the opponent net. These rules are invoked if the robot has moved from its home position, e.g, after a defensive action. In addition, it uses a meta-rule to call the *Block* sub-behavior when this position is achieved. The resulting behavior will cause the robot to move to its home position and wait for the ball in a posture that covers about 50% of the net. The corresponding rules are given as follows:[6]

[5] An internal timer is used for this purpose. In a sense, then, the *CoverNet* behavior also includes a FSM.

[6] For sake of clarity, all the rules shown in this section are given in a slightly simplified form with respect to the actual rules implemented in the robot.

1.	IF (ForwardFromHome)	GO(Back)
2.	IF (BackwardFromHome)	GO(Slow)
3.	IF (RightOfHome)	SIDE(Left)
4.	IF (LeftOfHome)	SIDE(Right)
5.	IF (OpponentNetOnLeft)	TURN(Left)
6.	IF (OpponentNetOnRight)	TURN(Right)
7.	IF (AtHome ∧ FacingOpponentNet)	USE Block()
8.	IF (¬ WellLocalized)	NEED(LM2)
9.	IF (¬ WellLocalized)	NEED(LM4)
10.	ALWAYS	NEED(Net2)
11.	ALWAYS	NEED(Ball)

Fig. 14. The *KeepPosition* behavior.

The intuitive meaning of the fuzzy predicates in the conditions of rules 1–7 should be clear from the drawing in Fig. 11 above. (The actual fuzzy sets used to implement these predicates are not important here.) The meaning of the linguistic labels used in the TURN rules is as shown in Fig. 4 above, the meaning of the other linguistic labels is similar. Note that if no rule for a given control variable is applicable, that variable is given a default "no-motion" value.

In addition to motion control rules, this behavior incorporates perceptual rules 8–11. In order to accurately measure the home position, the robot needs to observe the two opposite landmarks (LM2 and LM4). Rules 8 and 9 say that these should be re-acquired if the self-localization is deteriorating. Rules 10 and 11 say that information about the opponent's net and the ball is always needed.

In contrast to the *KeepPosition* behavior, the *RemoveBall* behavior only uses sub-behaviors to perform its task. This behavior is called when the ball is close to the robot and the robot is positioned on the line between the ball and the net (an effect produced by the *InterceptBall* behavior). It consists of the meta-rules shown in Fig. 15.

1.	IF (¬ BallInFront)	USE Face(Ball)
2.	IF (¬ BallNear)	USE GoTo(Ball)
3.	IF (BallNear ∧ BallInFront ∧ ¬ CloseToNet)	USE Kick(Ball)
4.	IF (BallNear ∧ BallInFront ∧ CloseToNet)	USE PushBall()

Fig. 15. The *RemoveBall* behavior.

The first two rules are used to get possession of the ball. Rule 3 and 4 are used to send the ball away from the robot, hopefully out of the penalty area. The choice between these two rules depends on the position of the robot with respect to its own net. If the robot is close to its net, it chooses the *Push* behavior, which will not send the ball very far away but does not incur the risk of a self-score if, for instance, the ball bounces off the legs of another robot. If the robot is far from its net, it can safely kick the ball away. This strategy was chosen after observing the risk of self-scores in our experiments. Note that no perceptual rules are used in this behavior, since all the relevant perceptual needs are asserted by the sub-behaviors.

Our last example is the *InterceptBall* behavior. This behavior combines the use of control rules to directly control the robot's motion and the use of meta-rules to call sub-behaviors. The purpose of this behavior is to keep the robot on the line between the ball and the net at a fixed distance from the latter. The rules for this behavior are shown in Fig. 16.

1.	IF (BallOnRight)	TURN(Right)
2.	IF (BallOnLeft)	TURN(Left)
3.	IF (BallOnRight $\wedge$ AtRightDistance)	SIDE(Right)
4.	IF (BallOnLeft $\wedge$ AtRightDistance)	SIDE(Left)
5.	IF (Aligned)	USE RemoveBall()
6.	IF ($\neg$ Aligned)	USE KeepDistanceToNet()
7.	IF ($\neg$ WellLocalized)	NEED(LM2)
8.	IF ($\neg$ WellLocalized)	NEED(LM4)
9.	ALWAYS	NEED(Ball)

Fig. 16. The *InterceptBall* behavior.

The AtRightDistance fuzzy predicate is true is the distance between the robot and the net is the correct one (see Fig. 11). Aligned is true if the robot is on the line between the ball and its own net. The alignment motion is obtained by a combination of *KeepDistanceToNet* and of the basic control rules 1–4. As in the case of the *FaceOpponentNet* behavior, we prefer to implement alignment directly instead of using the *CrabToBall* and *FaceBall* sub-behaviors since the latter are tuned to provide accurate but slow motion. Finally, when the robot is aligned with the ball the *RemoveBall* behavior is activated to send the ball away.

The *InterceptBall* behavior needs to know the position of the ball and the location of the robot itself. The latter is used to establish the position of the net, which is behind the robot and therefore is not directly observable. The perceptual rules 7–9 express these needs.

6.5 Critical Assessment

The *GoalKeeper* behavior is probably one of the most complex reactive robot behaviors reported in the literature to this date. It involves the use of navigation, manipulation, and perceptual actions in a highly dynamic and unpredictable environment. The rule-based approach and hierarchical organization allowed us to design, implement and test this behavior with a limited amount of effort. The full *GoalKeeper* behavior has been decomposed into 17 behaviors, which invove a total of 78 fuzzy rules (including those in the basic behaviors) plus 15 perceptual rules. The development of these rules required about 4 weeks of work by one person. This behavior was used in the RoboCup 2000 and 2001 competitions, resulting in a satisfactory performance.

In our experience with the *GoalKeeper* behavior we have noticed a number of positive and negative points of our approach to building complex behaviors. The main positive ones are as follows.

- The use of fuzzy "if-then" rules made it easy to design and tune several motion strategies involving complex combinations of forward, lateral and rotational motion.
- The hierarchical structure greatly simplified the top-down design process, by allowing us to first concentrate on higher-level behaviors that implement a given strategy, and then on lower-level behaviors that realize the necessary movements.
- The hierarchical structure also simplified the bottom-up tuning and debugging, by allowing us to first focus on the tuning of simple individual behaviors, and then debug the way in which these are combined.
- Behaviors are modular, which simplifies the design of new complex behaviors once a sufficient set of simpler ones is available; for instace, we could build a *GoalKeeper* based on a different strategy by combining the behaviors described in the last section in a different way.
- The use of reactive rules to perform behavior arbitration allowed us to concentrate on *what* should be done in each situation, while ignoring *in what sequence* the events will happen.
- Perceptual rules of the form (11) provided an easy and effective mechanism to orient perception according to the needs of the controller.

As for the observed limitations, the main ones are:

- The tuning of the fuzzy rules used in the behaviors has to be made by trial & error; this is time consuming, and there is no guarantee that the best tuning has been achieved.
- There is currently no way to connect fuzzy predicates which are defined in different behaviors but which should logically related; for example, the *GoToBall* and the *KickBall* behaviors both define internally a predicate called CloseToBall; if these two predicates are tuned differently, then *GoToBall* might stop in front of the ball at a distance which is too large for the *KickBall* behavior to kick it.
- Dealing with potential conflicts between behaviors can be tricky; for instance, two sub-behaviors might suggest drastically different control actions in some situation; in this case, we must make sure that the caller behavior never calls both sub-behaviors simultaneously in this situation; however, there is no systematic way to identify such critical situations.

- The mechanism to express perceptual needs is somewhat limited; in particular, we can specify how much we need information about a given object, but not how often this information should be updated; this limitation should be easy to fix.

In addition to these general considerations, it is interesting to note a few specific limitations in the current *GoalKeeper* behavior. First, the performance of the goal keeper strongly depends on the quality of the perceptual data. Although the fuzzy rules tolerate some noise in the input data thanks to the smooth switching between different rules, bad decision can still be taken in case of transient large errors (e.g., a bad estimate of the ball position) or persistent smaller errors (e.g., inaccurate self-localization). Probably the only solution to this problem is to improve the quality of the perceptual data.

A second limitation is related to our overall strategy. The actions performed in the "Play" mode rely on the assumption that the goal keeper robot is somewhere between the net and the ball. If the ball ends up between the net and the robot, then our strategy (try to reach a location on the line between the ball and the net) might result in hitting the ball and causing a self-score. The solution to this problem would be to extend our strategy.

As a third limitation, the *GoalKeeper* behavior only looks at the current situation, and does not try to make any prediction. In order to improve the performance, especially in a fast game, the goal keeper should be able to predict likely future state, reason about them, and take actions that prevent possible future scores. This should probably be done using the Reactive Planner (Fig. 3), which include mechanisms for prediction and short-term planning.

Finally, the *GoalKeeper* behavior currently does not take into account the position of the other robots in the field. The introduction of this element would considerably increase the number of possible types of situation that have to be considered, calling for an even more complex behavior.

We speculate that the *GoalKeeper* behavior presented here is close to the frontier of complexity that can be managed by using reactive rules and a hierarchical organization: more complex behaviors would probably need the use of state-based systems to reason about the past states (e.g., a FSM) or the future states (e.g., the Reactive Planner) or both. However, these systems could still use complex fuzzy behaviors as atomic actions. In fact, this is how the player robots have been implemented in Team Sweden: the Reactive Planner decides between high-level actions like *GoBehindBall* or *GoToBall*, and these actions are implemented by complex reactive behaviors in the HBM. Making high-level actions available to the planner has two advantages: to reduce the complexity of the planner since this only has to plan at a coarse level of granularity; and to alleviate the need for careful monitoring and replanning since behaviors already provide a good deal of reactivity to unexpected events.

7 Conclusions

The hierarchical approach presented in this chapter greatly simplifies the design of complex reactive behaviors. In this approach, design, implementation and debugging are done in a modular and incremental way based on heuristic knowledge. We have shown how we have used this approach to build a full set of navigation, ball manipulation and perceptual behaviors for a team of soccer playing four-legged robots. These behaviors include particularly complex ones, like the *GoalKeeper* behavior analyzed in the last section. We believe that the use of fuzzy rules and hierarchical organization are the keys factors that allowed the behavior designer to manage this complexity. In particular, the hierarchical organization answers the criticism of non-scalability often moved to fuzzy control techniques [21–23].

Our approach has been used in several other platforms and domains besides the ones described in this work, including wheeled robots for office navigation [3,13] and a robot manipulator performing pick-and-place tasks [24]. In all cases, the use of fuzzy rules and of CDB to define behavior combination strategies provided ease of design and high reactivity to unexpected events.

The critical assessment in the last section revealed several points in our approach that still need to be improved. Of these, the most critical one is perhaps the development of principled techniques to simplify the tuning and testing of the fuzzy rules. Since unpredictable domains like RoboCup are hard or impossible to model, we cannot hope to use classical design and validation tools for this. Instead, we are currently considering the use of statistical simulation techniques to estimate the properties of a set of behaviors.

Acknowledgements

The work reported in this chapter was partly supported by the Swedish KK Foundation. Tom Duckett provided useful comments on a draft of this chapter. Our warmest thanks to all the past and present colleagues of Team Sweden, listed at http://www.aass.oru.se/Agora/RoboCup/.

References

1. Brooks, R.A. (1986) A robust layered control system for a mobile robot. IEEE Journal of Robotics and Automation, **2**, 1, 14–23
2. Arkin, R.C. (1998) Behavior-Based Robotics. MIT Press, Cambridge, MA
3. Saffiotti, A., Konolige, K., and Ruspini, E.H. (1995) A multivalued-logic approach to integrating planning and control. Artificial Intelligence, **76**, 1-2, 481–526. Online at http://www.aass.oru.se/~asaffio/
4. Driankov, D. and Saffiotti, A. (2001) Fuzzy Logic Techniques for Autonomous Vehicle Navigation. Physica-Springer Verlag, Berlin, DE. Online at http://www.aass.oru.se/Agora/FLAR/

5. Saffiotti, A. Team Sweden web site. http://www.aass.oru.se/Agora/RoboCup/
6. Sony Corporation. Entertainement robot AIBO. http://www.aibo.com/
7. Saffiotti, A. The Thinking Cap. http://www.aass.oru.se/~asaffio/Software/TC/
8. Saffiotti, A. and LeBlanc, K. (2000) Active perceptual anchoring of robot behavior in a dynamic environment. In: Proc. of the IEEE Int. Conf. on Robotics and Automation (ICRA), San Francisco, CA, 3796–3802. Online at http://www.aass.oru.se/~asaffio/
9. Buschka, P., Saffiotti, A., and Wasik, Z. (2000) Fuzzy landmark-based localization for a legged robot. In: Proc. of the IEEE/RSJ Int. Conf. on Intelligent Robots and Systems (IROS), Takamatsu, Japan, 1205–1210. Online at http://www.aass.oru.se/~asaffio/
10. Johansson, S. and Saffiotti, A. (2001) Using the Electric Field Approach in the RoboCup domain. In: Proc. of the Int. RoboCup Symposium, Seattle, WA. Online at http://www.aass.oru.se/~asaffio/
11. Ruspini, E.H. (1991) Truth as utility: A conceptual synthesis. In: Proc. of the 7th Conf. on Uncertainty in AI, Los Angeles, CA, 316–322
12. Rescher, N. (1967) Semantic foundations for the logic of preference. In: The Logic of Decision and Action (ed. Rescher, N.). Univ. of Pittsburgh Press, Pittsburgh, PA
13. Saffiotti, A., Ruspini, E.H., and Konolige, K. (1993) Blending reactivity and goal-directedness in a fuzzy controller. In: Proc. of the 2nd IEEE Int. Conf. on Fuzzy Systems, San Francisco, CA, 134–139. Online at http://www.aass.oru.se/~asaffio/
14. Saffiotti, A. (1997) The uses of fuzzy logic in autonomous robot navigation. Soft Computing, **1**, 4, 180–197. Online at http://www.aass.oru.se/~asaffio/
15. Payton, D.W. (1986) An architecture for reflexive autonomous vehicle control. In: Procs. of the IEEE Int. Conf. on Robotics and Automation, San Francisco, CA, 1838–1845
16. Khatib, O. (1986) Real-time obstacle avoidance for manipulators and mobile robots. International Journal of Robotics Research, **5**, 1, 90–98
17. Bloch, I. and Hunter, A., editors (2001) Special issue on data and knowledge fusion. International Journal of Intelligent Systems, **16**, 10
18. Yen J. and Pfluger, N. (1995) A fuzzy logic based extension to Payton and Rosenblatt's command fusion method for mobile robot navigation. IEEE Trans. on Systems, Man, and Cybernetics, **25**, 6, 971–978
19. Henzinger, T.A. (1996) The theory of hybrid automata. In: Proc. of the 11th Symp. on Logic in Computer Science (LICS), New Brunswick, New Jersey, 278–292
20. Ballard, D.H. (1991) Animate vision. Artificial Intelligence, **48**, 57–87
21. Berenji, H., Chen, Y-Y., Lee, C-C., Jang, J-S., and Murugesan, S. (1990) A hierarchical approach to designing approximate reasoning-based controllers for dynamic physical systems. In: Proc. of the 6th Conf. on Uncertainty in AI, Cambridge, MA, 362–369
22. Raju, G.V.S., Zhou, J., and Kisner, R.A. (1991) Hierarchical fuzzy control. Int. J. Control, **54**, 5, 1201–1216
23. Berenji, H.R. (1994) The unique strength of fuzzy logic control. IEEE Expert, August, page 9. Response to Elkan's "The paradoxical success of fuzzy logic," same issue
24. Wasik, Z., and Saffiotti, A. (2002) A fuzzy behavior-based control system for manipulation. Submitted, manuscript available from the authors

A Bio-Inspired Robotic Mechanism for Autonomous Locomotion in Unconventional Environments

Darío Maravall and Javier de Lope

Department of Artificial Intelligence
Faculty of Computer Science, Universidad Politécnica de Madrid
Campus de Montegancedo, 28660 Madrid, Spain
{dmaravall,jdlope}@fi.upm.es

Abstract. This chapter presents a robotic mechanism aimed at navigating in unconventional environments like rigid aerial lines – power, telephone, railroad – and reticulated structures – ladders, grills, bars, etc –. A novel method of obstacle avoidance for this mechanism is also introduced. The computation of collision-free trajectories generally requires the analytical description of the physical structure of the environment and the solution of the kinematic equations. For dynamic, uncertain environments with unknown obstacles, however, it is very hard to get real-time collision avoidance by means of analytical techniques. The main strength of the proposed method resides, precisely, in that it departs from the analytical approach, as it does not use formal descriptions of the location and shape of the obstacles, nor does it solve the kinematic equations of the mechanism. Instead, the method follows the perception-reason-action paradigm and is based on a reinforcement learning process guided by perceptual feedback, which can be considered as biologically inspired at the functional level. From this perspective, obstacle avoidance is modeled as a multi-objective optimization problem. The method, as shown in the chapter, can be straightforwardly applied to real-time collision avoidance for articulated mechanisms, including conventional manipulator arms.

1 Introduction

Before embarking upon the design of a robot able to navigate in unconventional environments like rigid aerial lines and reticulated structures, it is interesting to observe how this problem has been solved by nature. In particular, how human beings or primates move within these environments. Fig. 1 shows a sequence of movements performed by a primate to move along an aerial rope.

The main characteristic of the sequence of movements displayed in Fig. 1 is that it takes place in the 3D space, as the primate rotates around the vertical axis OZ in order to progress along the horizontal axis OX. Given the kinematic configuration of a primate, roughly the same as humans, the movement sequence is very efficient, both mechanically and energetically. There is a second way to perform the same operation, which consists of successive

Fig. 1. Sequence of movements performed by a primate along a liana

contractions and expansions of the anterior and posterior extremities. As for reticulated environments, Fig. 2 shows some examples. Due to their kinematics, primates and humans may have some trouble in negotiating these environments.

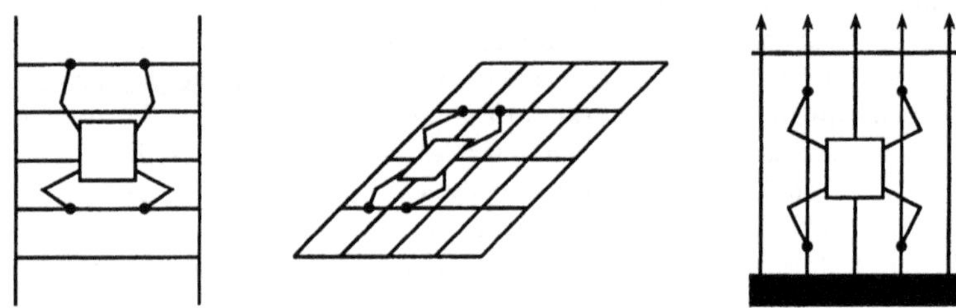

Fig. 2. Instances of reticulated structures and an artist's view of mechanisms moving along

Nature's wisdom notwithstanding, it is feasible to design a much less complex robotic mechanism for navigating in these unconventional environments. Obviously, having much more complex kinematics, primates and humans are endowed with a sophistication of movements that is totally prohibitive for our proposed mechanism. For instance, they can switch from one plane to another, a capability that would call for our robot to have additional degrees of freedom, not to mention the locomotion through many other different environments.

The chapter is organized as follows. In the first place, the robotic mechanism is introduced and its two basic movements are analyzed: (1) grasp and (2) release. The hard computing approach to the movement control problem is described in some detail. Afterwards, a novel method for dynamic collision avoidance is introduced. The proposed method does not need to know the position and shape of the existing obstacles, neither does it use direct or inverse kinematics for the generation and execution of collision-free trajectories. The method is, somewhat, bio-inspired, in the sense that it imitates the way in which primates or humans perform collision-free movements without using mathematical descriptions of the obstacles and without employing analytical expressions of their kinematics. Rather, they achieve such complex sensorimotor tasks, seemingly, by means of a reinforcement learning process, in which their movements are controled by perceptual feedback, which provides the basic information they need about the effects of their previous actions on the state of the environment. Of course, we claim only a functional, not physiological, inspiration for the proposed method.

2 Robotic Mechanism for Locomotion Along Rigid Aerial Lines and Reticulated Structures

The simplest mechanical structure for performing forward and backward movements along rigid aerial lines consists of two rigid links, L_1 and L_2, united by an intermediate revolute joint, R_0, and with two revolute joints R_1 and R_2, at its two extremities, both of which are equipped with grippers G_1 and G_2 to grasp the line. The mechanism is shown in Fig. 3. Note that all the joints rotate around the OY axis, so that the mechanism is constrained to stay in the XOZ plane.

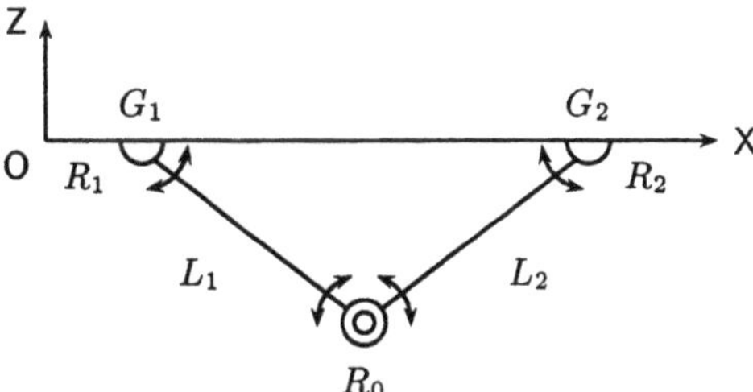

Fig. 3. Robotic mechanism aimed at moving along OX axis

A similar mechanism, intended to move up the rungs of a ladder, was proposed in [1]. More recently, this same mechanism has been extensively analyzed and tested, including simulated locomotion along a rope [2]. In this case, only the central joint was motorized, so that the underactuated mechanism had to swing up to move up a ladder, which was succesfully performed by a physical prototype.

Our robotic mechanism somewhat resembles an articulated bipendulum, although it is equipped with two rather sophisticated extremities. Let us now discuss the two basic movements of this particular double pendulum.

2.1 Basic Movements of the Robot

As mentioned above, the mechanism can move forward and backward along the horizontal axis OX by means of two basic steps, release and grasp, as illustrated in Fig. 4.

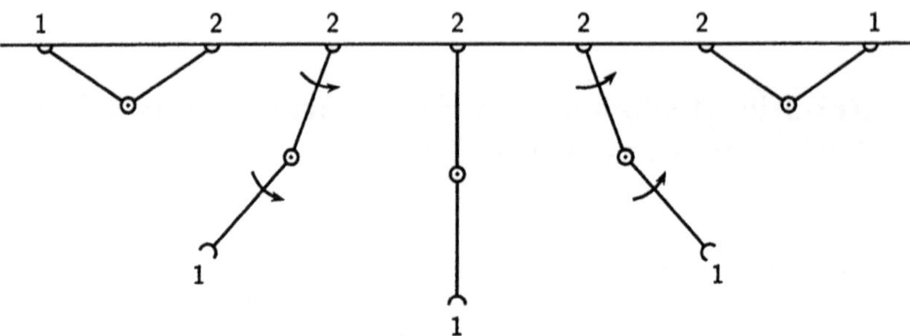

Fig. 4. Sequence of the basic release and grasp steps. Note that the robot is always fixed to the line by one of the two grippers

The three revolute joints are motorized to effect the basic movements. Note the simplicity of this mechanism, which performs its movements with perfect symmetry and some elegance. Although basically similar, more complex mechanisms could have been designed to perform the same tasks. For instance, Fig. 5a shows an articulated triple pendulum, which we have actually built and tested for the implementation of the second above-mentioned type of locomotion, which is contraction and expansion of both extremities. A quadruple pendulum is actually preferable for this kind of navigation. Fig. 5b represents a double pendulum with two additional prismatic joints, P_1 and P_2. The prismatic joints are intended to change the length of the robot steps, which could be of interest for some reticulated structures of variable size and shape. Fig. 5c shows a multiple articulated pendulum with two additional prismatic joints.

Continuing with this speculative line of thought, note that all these mechanisms are planar and confined to movements within the ZOX plane. To be able to switch to different planes and, thus, become a 3D robotic mechanism, an additional degree of freedom would have to be introduced at each hand or extremity, which would convert our particular double pendulum into a spherical double pendulum. In this chapter, however, we confine our study to the simple 2D robotic mechanism. We start with the conventional, hard computing control of its basic movements.

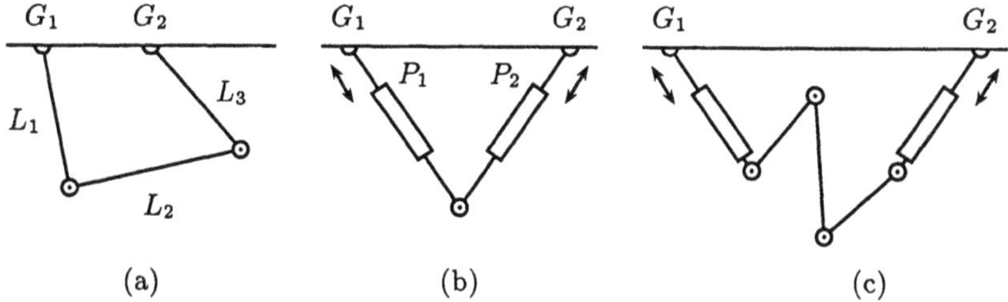

Fig. 5. Conceptual generalizations of the robotic mechanism, with additional revolute and prismatic joints

3 Hard Computing Control of the Basic Movements

The two basic movements of the robotic mechanism can be effected by generating suitable trajectories of its two generalized coordinates, θ_1 and θ_2, as shown in Fig. 6. Note the change in the notation of the vertical axis, as we have taken the standard convention of representing the plane in which the mechanism is constrained to move as XOY.

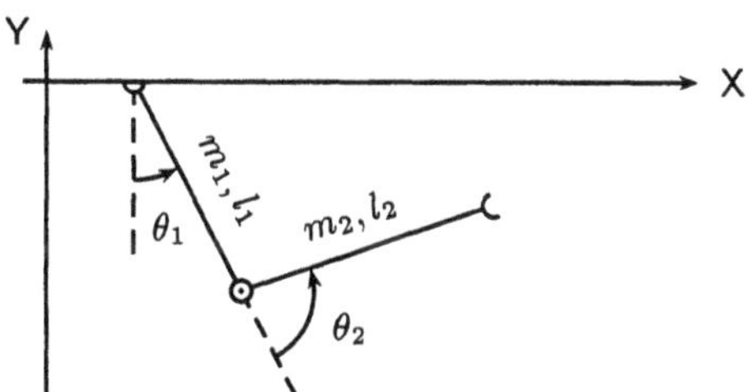

Fig. 6. Robotic mechanism and the chosen angles convention

For the execution of the joint-link angle trajectories, we can use the conventional feedback loop given by Fig. 7.

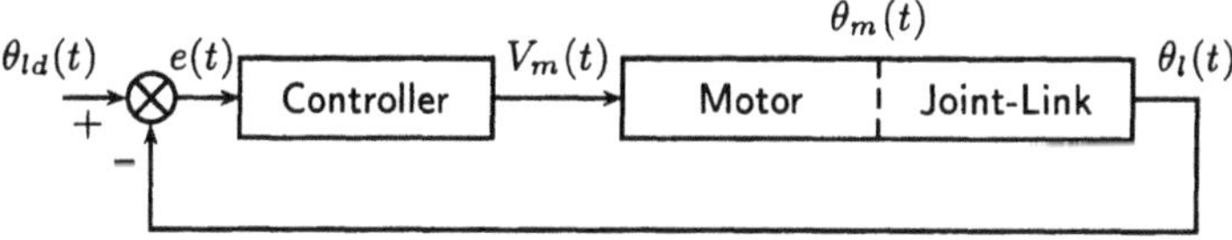

Fig. 7. Feedback control loop of a motorized joint

In Fig. 7, $V_m(t)$ stands for the voltage applied to the motor as a result of the control algorithm, which in turn is based on the error between the desired trajectory, $\theta_{ld}(t)$, and the current angle, $\theta_l(t)$, of the joint-link pair. $\theta_m(t)$ is the angular motor rotation, such that $\theta_l(t) = \eta\,\theta_m(t)$, where η is the gear ratio.

Assuming that the total torque of the motor-link pair is the sum of the motor torque, without mechanical load, plus the link torque, we have:

$$\tau_m = J_m\ddot{\theta}_m + b_m\dot{\theta}_m + \frac{1}{\eta}\tau_l \tag{1}$$

where τ_m and τ_l are, respectively, the motor and the link torques. J_m is the inertia of the motor rotor and b_m is the viscous friction coefficient.

It is straightforward to obtain the relationship between the voltage V_m applied to the motor and its angular displacement:

$$V_m(t) = \frac{R_a}{J_m}K_m\ddot{\theta}_m + \left[\frac{b_m J_m}{K_m} + K_b\right]\dot{\theta}_m + \frac{1}{\eta}\frac{R_a}{K_m}\tau_l \tag{2}$$

where we have assumed the usual approximations concerning the electrical components of the control motor. In (2), R_a is the resistance of the armature windings – we have ignored the inductance –; K_m is the motor torque constant provided by the manufacturer and which relates the motor torque with the armature current as $\tau_m = K_m I_a$. Finally, K_b stands for the back electromotive force constant.

If we ignore the link torque – to be modeled and introduced later on – expression (2) is a second-order linear differential equation with, hopefully, constant parameters, which is easily controled by conventional hard computing techniques. Although attenuated by the factor η, the non-linear, variable torque produced by the link, τ_l, is an important disturbance element that must be considered in detail. Furthermore, as our robotic mechanism is based on two articulated links that sort of *fly* through space, the coupled links may produce considerable torque effects that, if not taken into account, could eventually deteriorate the execution of the desired joint-link trajectories.

Thus, let us obtain the two links torques by applying the Lagrange equation to our mechanism, with the angles convention given in Fig. 6:

$$\frac{d}{dt}\left(\frac{\partial L}{\partial \dot{q}_i}\right) - \frac{\partial L}{\partial q_i} = \tau_i \quad ; \quad i = 1,2 \tag{3}$$

where L is the Lagrangian and $q_i = \theta_i(t)$ are the generalized coordinates. Note that we have dropped the subindex l denoting a link to simplify the notation. As the Lagrangian is $T-V$, i.e. the kinetic energy minus the potential energy, we have:

$$\begin{aligned} T &= E_{c_1} + E_{c_2} = \frac{1}{2}\,m_1\,(\dot{x}_1^2 + \dot{y}_1^2) + \frac{1}{2}\,m_2\,(\dot{x}_2^2 + \dot{y}_2^2) \\ V &= E_{p_1} + E_{p_2} = m_1\,g\,y_1 + m_2\,g\,y_2 \end{aligned} \tag{4}$$

where m_1 and m_2 are the masses of the two robot links.

The Cartesian coordinates of the two links are:

$$\begin{array}{ll} x_1 = l_1 \sin\theta_1 \quad ; & x_2 = l_1 \sin\theta_1 + l_2 \sin(\theta_1 + \theta_2) \\ y_1 = l_1 \cos\theta_1 \quad ; & y_2 = l_1 \cos\theta_1 + l_2 \cos(\theta_1 + \theta_2) \end{array} \tag{5}$$

where l_1 and l_2 are the length of both links.

By operating, we finally obtain the Lagrangian:

$$\begin{array}{c} L = \frac{1}{2}(m_1 + m_2)\, l_1^2\, \dot\theta_1^2 + \frac{1}{2}\, m_2\, l_2^2 (\dot\theta_1^2 + \dot\theta_2^2) + \\ + m_2\, l_1\, l_2 \cos\theta_1 (\dot\theta_1^2 + \dot\theta_1 \dot\theta_2) - \\ -(m_1 + m_2)\, g\, l_1 \cos\theta_1 - m_2\, g\, l_2 \cos(\theta_1 + \theta_2) \end{array} \tag{6}$$

which, substituted in the Lagrange equations (3), gives the final dynamic equations of the robotic mechanism:

$$\begin{array}{c} \ddot\theta_1 \left[(m_1 + m_2) l_1^2 + m_2\, l_2^2 + 2 m_2\, l_1\, l_2 \cos\theta_2\right] + \\ + \ddot\theta_2 \left[m_2\, l_2^2 + m_2\, l_1\, l_2 \cos\theta_2\right] - \\ - m_2\, l_1\, l_2 \sin\theta_2 (2\, \dot\theta_1 \dot\theta_2 + \dot\theta_2^2) - \\ -(m_1 + m_2)\, g\, l_1 \sin\theta_1 + m_2\, g\, l_2 \sin(\theta_1 + \theta_2) = \zeta_1 \end{array} \tag{7}$$

$$\begin{array}{c} \ddot\theta_1 \left[m_2\, l_2^2 + m_2\, l_1\, l_2 \cos\theta_2\right] + \ddot\theta_2\, m_2\, l_2^2 + m_2\, l_1\, l_2 \sin\theta_2 \dot\theta_1^2 - \\ - m_2\, g\, l_2 \sin(\theta_1 + \theta_2) = \zeta_2 \end{array}$$

These dynamic equations can be expressed in the standard, compact form:

$$\begin{array}{c} M(\boldsymbol{q})\ddot q + \boldsymbol{V}(\boldsymbol{q}, \dot{\boldsymbol{q}}) + \boldsymbol{G}(\boldsymbol{q}) = \boldsymbol{\zeta} \\ \boldsymbol{q} = \binom{q_1}{q_2} \quad ; \quad \boldsymbol{\zeta} = \binom{\zeta_1}{\zeta_2} \end{array} \tag{8}$$

where $M(\boldsymbol{q})$ is the symmetric inertia matrix; $\boldsymbol{V}(\boldsymbol{q}, \dot{\boldsymbol{q}})$ is the coriolis/centripetal vector and $\boldsymbol{G}(q)$ is the gravity vector. However, the expanded form (7) provides a more intuitive understanding of the behavior of the mechanism, which is our main concern at this descriptive stage.

After substituting the two link torques given by (7) in their respective motor-link equations, expression (2), the control of the mechanism movements is straightforward and, basically, a matter of tracking the desired trajectories of the two generalized coordinates, as the execution of such trajectories is relatively trivial, provided that the link torques given by (7) can be counter-balanced. Appendix B presents some results concerning the trajectories control.

In summary, the control of the robotic mechanism can be efficiently solved by means of hard computing techniques, and it is, essentially, a matter of generating the appropriate trajectories of the link angles, which eventually leads to the obstacle avoidance problem.

4 The Perception-Reason-Action Cycle

Let us now consider the problem of real-time collision avoidance for the articulated mechanism. Our approach is inspired in the way that humans and some primates control and guide their articulated extremities to avoid obstacles while performing a specific task, which is based on adaptation and learning through sensory feedback. Obviously, human beings and animals do not move their articulated members using the kinematics and dynamics models associated with their respective extremities. Rather, they accomplish this complex task by an action/reaction process based on perceptual feedback, which is mainly visual when there is no physical contact between their articulated members and the objects. To generalize our discussion, we shall use the term of physical agent or even, sometimes, intelligent agent, when referring to a generic living being on which our model for collision avoidance is based.

An intelligent agent possesses a set of input variables or actuators responsible for their articulated member movements. Following the usual notation in control systems, let us denote these input variables as u_1, u_2, ..., u_N. When these variables coincide with the generalized variables q_1, q_2, ..., q_M that define the degrees of freedom of the respective articulated member, then we have a direct control system. The set of generalized or internal variables may or may not coincide with the state variables appearing in the mathematical description of the system in the state-space representation. Furthermore, there is a set of external and observable variables, y_1, y_2, ..., y_P, which conform the basic sensory information needed by the agent to modify its behavior in order to control the performance of the task at hand. With the perceptual feedback, the control loop is closed and, then, the control or input variables u_1, u_2, ..., u_N are updated as a result of the evaluation of the sensory information and the *reasoning* and decision-making processes. This conceptual model is shown in Fig. 8.

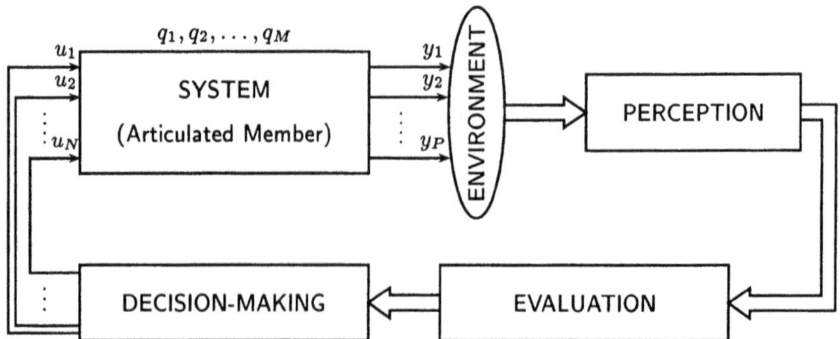

Fig. 8. Block-diagram of an intelligent agent with perceptual feedback

There are three fundamental elements interposed between the agent and its environment: perception, evaluation and decision-making. The perceptual element is the quantitative measurement of the effects of agent actions on the state of the environment which is embodied in the observed variables $y_1, y_2, \ldots, y_P$. This sensory information is vital for activating the reactive chain inside the physical agent aimed at changing the actions $u_1, u_2, \ldots, u_N$.

The reasoning or thinking capability of the agent is, actually, situated in-between the agent perception and action components. This reasoning activity has been divided into two stages: (1) evaluation and (2) decision-making. The idea is that, in the evaluation phase, the agent analyzes what is happening in the environment as regards its final goals or purposes. At this point, it is essential to recall that the concept of purpose or intentionality is of tremendous importance within the agent's existence and, without it, the agent's actions would be erratic, unpredictable or even absurd [3]. Therefore, the agent is endowed with a plan – i.e. a set of goals –, and it can assess, through its perception of the environment and its own states, whether its actions are conducive to fulfilling its plan. It is here that reasoning comes into the play as a mean of deciding which actions to apply, once the effects of its previous actions on the environment have been evaluated. As a consequence of its reasoning activity, the agent is able, again, to move its articulated members by means of the input variables. Thus, we can see now that reasoning or thinking is placed between perception and action. Note, also, that we have not identified reasoning with intelligence, as the latter clearly includes perception and action as well. *De facto*, there is such a deep interaction and overlapping between perception and reasoning that it is often impossible to separate the two: we reason using live images, physical perception, and we also reason with abstract, imaginary perception. Furthermore, when action is involved, an intelligent agent is immersed in what is known as real-time situations, in which perception and thought are subject to time restrictions. Thus, this is a fundamental argument for not confining intelligence to the reasoning realm: the frontier between perception and thought is not at all clearly defined. Whatever intelligence could be, the essential role played by perception in any intelligent agent is unquestionable.[1]

Although there is also an abstract component in perception, whose scope is beyond the engineering approach to intelligence we are taking here, an important conclusion is that sensory or physical perception is a basic pillar in the development and the very existence of any physical agent. Actually, it is impossible to speak about intelligence without taking into account perception.

[1] It is no wonder that the formerly dominant *symbolic* approach to artificial or computational intelligence, with its worship of disembodied intelligence, as illustrated by the Turing test, has been strongly criticized and challenged in practice, by the connectionist community, and chiefly in the autonomous robotics field, where researchers are engaged in designing and building *physical* intelligence.

Furnished with the general conceptual model illustrated in the block-diagram of Fig. 8, we can now try to particularize it to the collision avoidance problem of our robotic mechanism. More specifically, we are going to develop in detail the different elements of the perception-reason-action chain that will enable the autonomous locomotion of our robot without colliding with any existing object.

5 Collision Avoidance Based on Human and Animal Behavior

The general objective of the articulated mechanism is, as we know, to advance through its specific environment – probably while doing an inspection work – without putting itself into danger, that is, to move forward without colliding. This general objective can be divided into two sub-goals: advance and collision avoidance.

5.1 Planning *versus* Improvisation

The sub-goal of collision avoidance raises an interesting basic question for any physical agent – not only the particular one considered in this chapter – and which is related to the general question about the relationship between theory and experience. We are referring to the fundamental question to be addressed by any designer when designing any particular system; i.e. what is going to be incorporated, by design, into the final implementation of the system and what is going, if anything, to be learned by the system itself during its operating life. This is the dichotomy between *a priori* knowledge – theory – and *a posteriori* knowledge – experience –.

Thus, it is a crucial question in the design of our robot mechanism to establish a balance between the formalized, *a priori* available, knowledge about the environment in which the robot is going to navigate, and the flexibility that it must possess in order to adapt – even, sometimes, to learn and to acquire new knowledge of its environment by means of experience – to unpredictable, or simply unknown, peculiarities of the environment. Obviously, the balance between previous and acquired knowledge strongly depends on the characteristics of each particular environment. For instance, in a simple, structured environment, in which the *a priori* modeling is usually exhaustive, the robot will rely almost exclusively on pre-designed knowledge and, therefore, adaptive and learning capabilities play a secondary role. On the contrary, for complex, unstructured and dynamic environments, in which it is virtually impossible to model all the possible situations to be tackled by the robot in advance, learning and adaptation play a fundamental role. Therefore, the characteristics of each environment determine the balance between *planning* and *improvisation*.

Roughly speaking, the collision avoidance problem belongs to the second category of situations, as it is virtually impossible to know what type of obstacles and how and when they will appear in advance. More specifically, for the application at hand – i.e. the navigation of our robotic mechanism – there are a diversity of particular environment states that make the aprioristic formalization of these states unfeasible, preventing the robot from acting in a blind, pre-designed way. Thus, if robot survival is a primary concern, it must be endowed with as much flexibility as possible. Furthermore, it must also be endowed with an advanced perceptual capability, absolutely vital for its adaptation to unknown and unpredictable environments. At the same time, its reasoning ability is not so stringent, as it can be relatively primitive and based on a set of practical rules of the *if-then-else* or reactive kind. In summary, intelligence in our robot is primarily perceptual.

5.2 Bio-inspired Obstacle Avoidance

After these general remarks, let us proceed with the discussion of the obstacle avoidance problem of our mechanism, which can be modeled as the generic problem of collision avoidance for a planar robot with two revolute joints or a R-R planar manipulator. For this reason, we are going to describe our bio-inspired approach to obstacle avoidance by considering the generic case of a planar R-R arm and, afterwards, we shall particularize the model to our mechanism. Thus, Fig. 9 represents a generic situation of obstacle avoidance for a R-R planar robot.

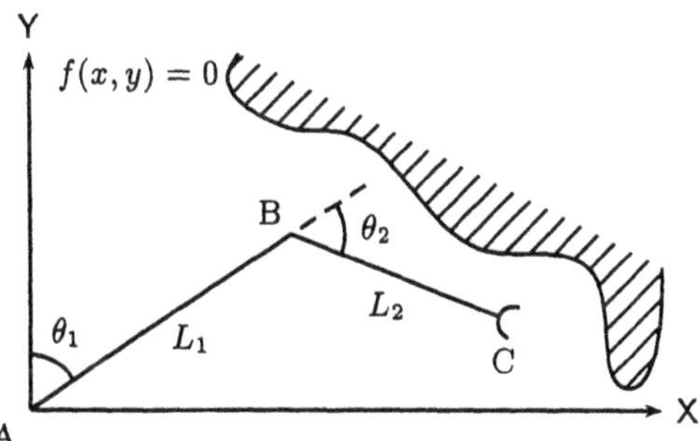

Fig. 9. R-R planar manipulator in front of a solid obstacle

If the obstacle contour function, $f(x, y) = 0$, in front of the robot was known, then the computation of a pre-designed collision-free trajectory, if any, for a particular robot task would, in principle, be possible. Let us suppose for a moment that we have full knowledge of the physical structure of the environment, so that the collision-free trajectory problem can be modeled as the generation of trajectories in the plane of articulated rigid segments – in our case, the two segments $\overline{AB}$ and $\overline{BC}$ –, such that they never touch any physical object given by its contour equation $f(x, y) = 0$. This is a rather difficult computational geometry problem, which can be tackled either in the work space, or, more abstractly, in the configuration space. Apart from the theoretical difficulties that can appear in many practical situations, even for a simple R-R planar robot, the main and definitive reason for abandoning this analytical approach is that it is based on the generation of collision-free trajectories obtained as solutions to particular geometrical problems, in which full knowledge of the physical structure of the environment given by the contour equations $f_i(x, y) = 0$, for $i = 1, 2, \ldots, N$ obstacles is required. Instead, we shall introduce a completely different approach, in which the knowledge of these expressions and the direct or inverse kinematic equations of the robot are not needed to generate collision-free trajectories.

To our knowledge, the cerebellar model articulation controller (CMAC) proposed in the seventies [4,5] is the first example of a robot controled without explicitly employing its kinematic expressions. It is based on learning the mapping functions, given as look-up tables, between the working space – i.e. Cartesian coordinates – and the configuration space – i.e. control variables –, by means of visual feedback. This line of research was pursued mainly in the field of cognitive science – see [6] for a general reference – and produced interesting simulated robots like DARWIN I–III [7], and physical prototypes like MURPHY [8], formed by a Rhino arm and a video-camera. Its reliance on look-up tables, mapping the sensory information to the control actions, makes this paradigm very prone to combinatorial explosion, unless a local search in the mapping functions is applied.

The method proposed in this chapter, however, differs significantly, as it is based on temporal and spatial utility functions, so that the robot performs the mapping between sensory information and actions by optimizing the utility functions, which dramatically reduces the search space.

The question is, how can any automatic robot planner generate collision-free trajectories without knowing the analytical expression of the shape of the obstacles and, especially, without applying kinematic equations? Well, this is a question that is efficiently solved by humans and primates without using mathematics at all while skilfully avoiding obstacles. Then, how do they solve this problem that is so hard to deal analytically?

Analyzing how humans and primates avoid obstacles when performing a specific task with their arms, we find that, basically, they apply small movements following a reinforcement process based on sensory, mainly visual, feedback. Let us consider the apparently simple case of trying to catch an object with one hand without touching neighboring objects. This is a constant trial-and-error process in which the individual applies small torques of the involved joints, in such a way that successful movements are reinforced and failures are abandoned. Throughout the whole process, perceptual feedback makes the evaluation of the failures and successes possible.

After this general discussion, we are now ready to formalize a collision avoidance method, basically inspired in the above-mentioned observations of human and animal behavior.

6 Collision-free Trajectories Using Reinforcement Learning

Let us consider again the generic problem of generating collision-free trajectories for the planar R-R manipulator represented in Fig. 9. Later on, we shall discuss a sensor-based method which does not need the complete reconstruction of the functions that determine the shape of the environment. Let us suppose, for a moment, however, that we have been able to compute the environment function $f(x, y) = 0$ for the neighborhood of the current robot position, so that we can compute some performance indices that evaluate the current state of the robot-environment pair, as illustrated in Fig. 10, which is the adaptation of the general block-diagram of Fig. 8 to the R-R planar robot.

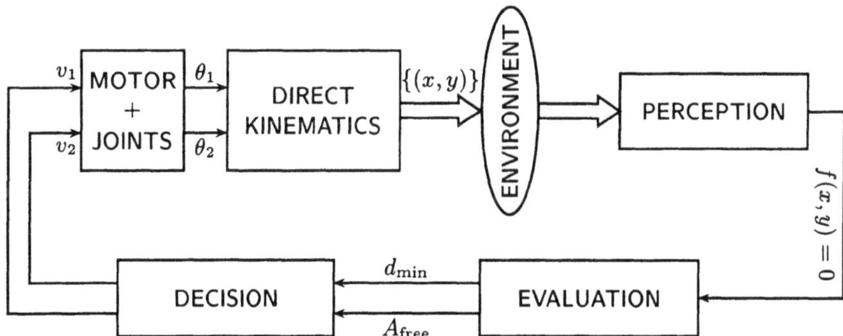

Fig. 10. Block-diagram of the interaction between the robot and the environment, with specific reference to the evaluation of the global system by means of the parameters $d_{\min}$ and A_{free}, which are explained in the text

We can observe from this block-diagram, as compared with the general diagram appearing in Fig. 8, that the perception of the environment has now been transformed into the reconstruction of the environment profile given by the function $f(x, y) = 0$. The analytical expression of the environment shape is not actually necessary for the evaluation of the state of the robot-environment pair, which can be estimated, as explained later on, by the two parameters $d_{\min}$ and A_{free}. These performance indices constitute the basic information needed by the control system to actuate the robotic mechanism in order to execute collision-free trajectories. In conclusion, we have two fundamental components: (1) the perception and evaluation of the global robot-environment state and (2) the control of the robotic mechanism. Let us discuss both of them in order.

6.1 Evaluation of the Robot-Environment State

The two essential objectives of our mechanism are to advance and, simultaneously, avoid obstacles. We have already commented the two basic movements to accomplish the first goal: release and grasp. As regards the obstacle avoidance goal, we shall describe our proposed method by considering the general case of a multiple revolute joint manipulator, of which our R-R robot is a particular case. We have also remarked that for any planar articulated robotic mechanism, obstacle avoidance can be modeled as a geometric problem of moving a set of connected segments without touching a given set of planar curves given by the functions $\{f_i(x,y) = 0\}$. Fig. 11 shows an example of a generic N-R planar articulated mechanism and two obstacles.

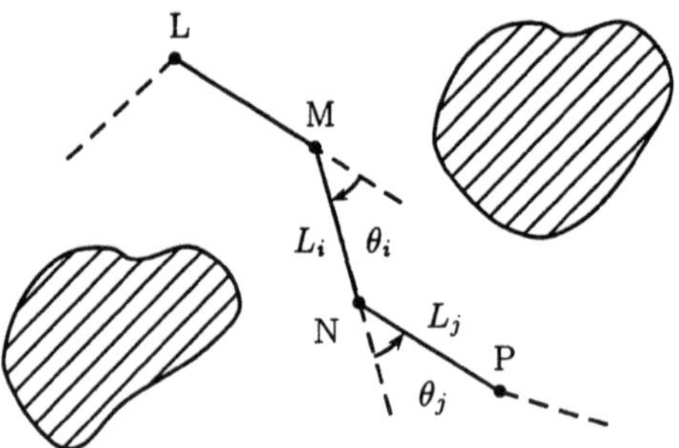

Fig. 11. A N-R planar robot and two obstacles

The segment $\overline{\text{MN}}$ of Fig. 11 corresponds to one generic link and segments $\overline{\text{LM}}$ and $\overline{\text{NP}}$ are two adjacent links. Focusing on the collision avoidance problem for link $\overline{\text{MN}}$, we see that there is a single control variable or generalized coordinate, the joint angle θ_i, for controling the link movements. However, the location in the plane of this generic link depends not only on its control variable, θ_i, but also on the preceding control variables $\theta_1, \theta_2, \ldots, \theta_{i-1}$. Furthermore, it is very hard to solve the problem of generating collision-free trajectories of the N generalized coordinates analytically as it requires full knowledge of the environment functions $\{f_i(x,y) = 0\}$. To avoid the analytical approach, as discussed in the previous section, we propose a bio-inspired method based on the reinforcement learning paradigm. One key question in our method is the evaluation of the robot-environment state, for which we propose two performance functions: d_{min} and A_{free} for each obstacle, which considerably simplifies the problem of obtaining the set of functions

$\{f_i(x,y) = 0\}$. Fig. 12 shows several examples of these performance indices for a R-R planar manipulator.

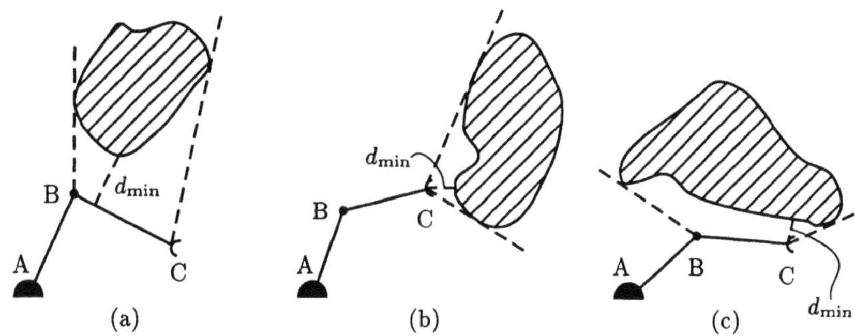

Fig. 12. Several instances of the proposed performance functions for a R-R robot

The performance index d_{min} measures the minimum distance from the second link, i.e. the segment $\overline{BC}$, to the nearest obstacle. A_{free} is the area covered by the two extreme radials traced from each of the two link ends to the nearest obstacle, which gives the free area in front of the link. Both performance functions can be easily computed using the range sensor described in Appendix A. This sensor provides the radial distance measurements shown in Fig. 13. Of course, each robot link has its own pair of performance indices.

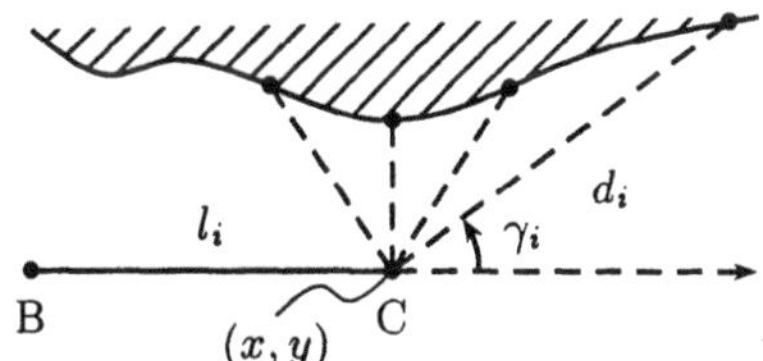

Fig. 13. The radial range sensor

This sensor is motorized and fixed to the robot, so that its position (x, y) is always known. The angle γ_i associated with each distance reading d_i is also known. Thus, it is straightforward to compute the performance index A_{free} for each link, as illustrated in Fig. 14, where F_1 and F_2 are the two visible extreme radials traced from the sensor at C to the obstacle.

Using the polygonal contour approximation, it is possible to compute the partial triangles T_1, T_2, ..., T_N formed by each couple of consecutive readings. All the parameters needed to compute the area of each triangle, γ_i, $\Delta\gamma_i$, d_i and d_{i+1}, are known. Furthermore, the side of each triangle corresponding to the polygonal approximation of the contour can be estimated as $\Delta O_i \cong d_{i+1} \sin \Delta\gamma_i$. Thus, by applying the Heron formula, we finally obtain

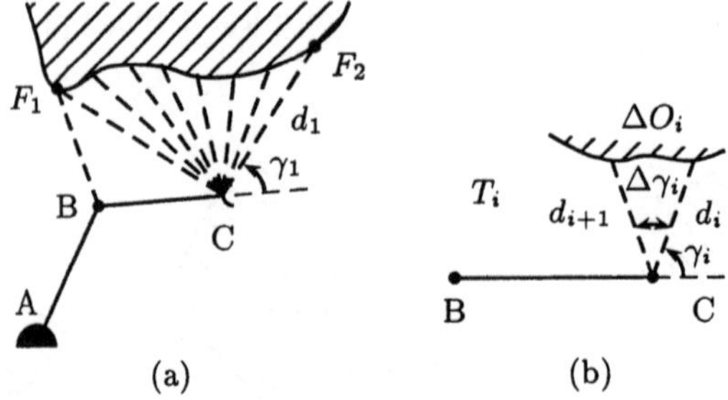

Fig. 14. Computation of A_{free} for the second link by a polygonal approximation of the obstacle contour. Note the generic triangle T_i on the right side

the performance function:

$$
\begin{aligned}
A_{\text{free}} &= \sum_{i=1}^{N} T_i + A' \\
\sum_{i=1}^{N} T_i &= \sum_{i=1}^{N} \sqrt{s_i(s_i - d_i)(s_i - d_{i+1})(s_i - \Delta O_i)} \\
s_i &= \frac{1}{2}\left(d_i + d_{i+1} + \Delta O_i\right)
\end{aligned}
\tag{9}
$$

and the remaining area A' of the triangle BF_1C is easily computed from the readings of the sensors placed at the two link ends.

The other performance function, $d_{\min}$, can also be computed in a forward manner using the same range sensor, as illustrated in Fig. 15.

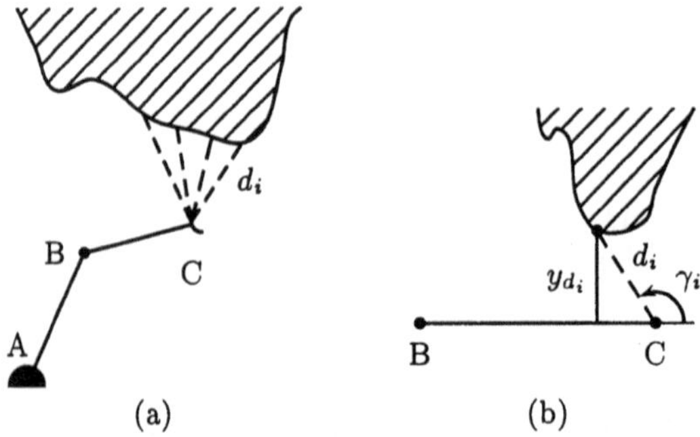

Fig. 15. Calculation of performance index $d_{\min}$

For each reading d_i of the range sensor, the distance between the link and the corresponding obstacle point is:

$$y_{d_i} = d_i \sin\left(\frac{\pi}{2} - \gamma_i\right) = d_i \cos\gamma_i \tag{10}$$

The set of projections y_{d_i} is a good approximation of the distance function from the obstacle to the segment defined by the robot link, so that the performance index is:

$$d_{\min} = \min\{y_{d_i}\} \tag{11}$$

Having computed the two performance indices providing an evaluation of the state formed by the robot-environment pair, we can now proceed with the design of the control system.

6.2 Reinforcement Control for Collision Avoidance

Looking at the block-diagram in Fig. 10, the general objective of the control or decision-making component is to generate the trajectories of the control variables, $\theta_1(t)$ and $\theta_2(t)$, that optimize the performance indices. Although we have introduced two particular performance functions in the previous section, $d_{\min}$ and A_{free}, let us take a general approach for the moment and denote by $J_i(\theta_1, \theta_2)$, for $i = 1, 2, \ldots N$, a set of generic performance functions.[2] If they were known by the designer, it would be possible, in principle, either by predesign or by real-time adaptive computation, to obtain analytical trajectories of the independent variables optimizing the performance functions. One of the most powerful analytical techniques available for tackling this problem is dynamic programming [10,11]. Unfortunately, this is not our case due to the high uncertainty about, or rather ignorance of, the analytical expressions of the performance functions, which, as we know, depend on the shape and position of the objects and obstacles existing in the robot environment.

For some favorable cases of unknown analytical relations between the performance indices and the control variables, it is possible, however, to define and measure some error functions, obtained as the deviation of certain system outputs or states and their respective desired values. These error functions can be minimized using a battery of well-known supervised techniques [12,13].

Our collision avoidance problem belongs to the unsupervised category, in which the designer does not know the analytical expression of the performance functions and cannot even introduce analytical error functions to be minimized. The reinforcement learning paradigm is an attractive and powerful method for tackling such difficult situations [14–16]. In this respect, it is no wonder that the reinforcement paradigm is closely related to the functional and physiological studies of animal behavior, where there is strong empirical evidence that many complex sensorimotor tasks can be explained by this paradigm. Fig. 16 shows the conceptual block-diagram of the reinforcement

[2] The letter J stands for "judgment" and originates from early work during the sixties on optimal control, as discussed in [9].

learning principle. Apart from the system or plant – in our case, the robotic mechanism – and the environment, there is the evaluation block, which is actually further divided into two components: evaluation itself and learning. The former is responsible for assessing the performance currently achieved by the system actions and the latter produces the next action, according to the evaluation results, by means of the reinforcement/punishment principle. Although we have maintained our previous notation, the usual convention is to use the term *critic* instead of evaluation.

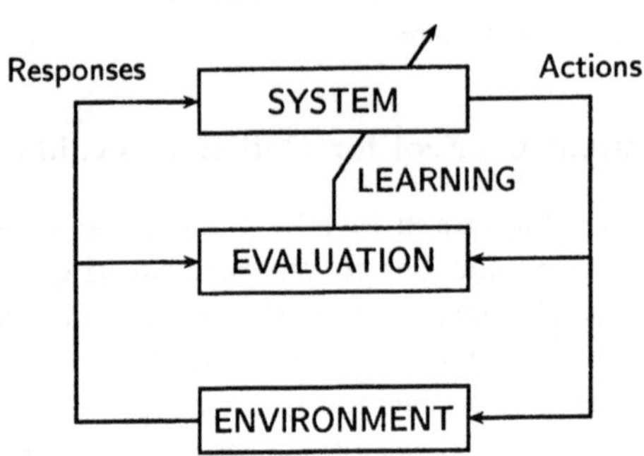

Fig. 16. Conceptual structure of the reinforcement learning paradigm

Focusing on the evaluation or critic block, which in the specialized literature is known as the critic assignment problem, several cases can appear [17], which, basically, depend on the available information about the global robot-environment pair state, as well as on the type of reasoning or decision-making for the generation of the control actions. This latter question is related to the dichotomy between global and local search, or exploration *versus* exploitation, in the control actions space, to be discussed further on. Without claiming to be exhaustive, we can distinguish two main categories concerning credit assignment and the relationship between evaluation and actions: (1) proportional to the magnitude of the current values of the performance functions and (2) proportional to the derivatives of these functions.

There are some interesting questions concerning the reinforcement learning paradigm that are worthwhile noting. First, there is the issue of local *versus* global search in the control or generalized variables space – the configuration space – which is of enormous interest. As is well known, the local/global dichotomy in optimization problems is a strategic issue that has generated immense interest and even inspired novel computing and algorithmic approaches, like heuristics programming and evolutionary computing. Our collision avoidance topic is a local search problem, as in the configuration space defined by the two generalized coordinates, θ_1 and θ_2, we can only apply local reinforcement actions to optimize the analytically unknown performance functions.[3] The second aspect of interest is related to the nature of the reinforcement strategies, which can be, basically, either direct or indirect. In direct reinforcement learning, the learning process takes place in the actions space, whereas in the second case, learning occurs in an abstract space defined by certain measures – usually probabilities or fuzzy memberships – of the action preferences.

The first category mentioned above, i.e. the global critic category, concerns situations in which there is extreme uncertainty about or ignorance of the search space for performance optimization, and, therefore, the actions are updated as a proportion of the current magnitudes of the performance indices in order to expand the search as much as necessary. Temporal difference, TD, reinforcement learning [14] intends to exploit the temporal evolution of a performance function to guide the choice of the control actions.

In the second category, there is more information available about the effects of the actions on the plant-environment pair states, so that the optimization search can be much more specialized and local, with a strong bias toward exploitation of action efficiency. In such situations, well-known supervised methods can be attempted, as the derivatives of the performance functions act like supervised information about the current performance of the control actions. However, overspecialized supervised methods may produce a loss of the generalization necessary to cope with highly dynamic performance functions. For this reason, we have used a simple gradient descent algorithm in the obstacle avoidance problem of our robotic mechanism. This algorithm is based on a single-layer perceptron, whose generalization capability is stronger than other more complex artificial neural networks (ANN) structures, like the multi-layer perceptron trained with the very specialized and highly local backpropagation algorithm. Anyway, the choice of the particular supervised optimization method depends on the application at hand.

Summarizing, the collision avoidance problem for our mechanism belongs to the general category of credit assignment based on the performance index gradients. As regards the control actions, they are obtained, in turn, as a function of these gradients.

Before proceeding with the particularization of the proposed method to our robotic mechanism, we should recall that there are several performance functions, which means that we are faced with a multi-objective optimization situation that is considerably more complex than the usual uni-objective optimization cases [18].

---

[3] Local search in obstacle avoidance means that our physical agent cannot move its articulated members abruptly, unless it does not care about taking too much risk.

6.3 Multi-Objective Reinforcement Learning for Obstacle Avoidance

Let us now particularize these ideas to our robotic mechanism, endowed as we know with two degrees of freedom, θ_1 and θ_2, and with at least the two performance indices $d_{\min}$ and A_{free} defined previously. Later on we shall introduce a third performance index. Thus, there is more than just one performance function, which is the usual situation in most cases [19]. Therefore, our collision avoidance issue is a multi-objective optimization problem [20,21], which raises additional questions about the coordination of the different performance indices.

As regards the reinforcement control system, the only information available about the performance functions is just an improvement/deterioration signal. Due to the strong local character of the optimization problem our robot is faced with, we shall consider two possible actions for each current control variable: increment or decrement. By way of an illustration, let us suppose a time-discrete version of the system, so that at a generic instant kT, each generalized coordinate is updated as follows:

$$\theta_l(k+1) = \begin{cases} \theta_l(k) \pm \Delta\theta_l & ;\text{reinforcement} \\ \theta_l(k) & ;\text{inhibition} \\ \theta_l(k) \mp \Delta\theta_l & ;\text{punishment} \end{cases} \qquad (12)$$

The interpretation of these changes in the control variables is straightforward: whenever performance improves, the current control actions are reinforced; inversely, if performance deteriorates the control actions are punished. Inhibition rarely occurs, as it strictly concerns situations where the performance functions remain unchanged. However, it serves to re-write the learning scheme given by (12) in a compact way. Indeed, let us formalize our control problem for obstacle avoidance as an optimization problem in which the performance functions depend on the two generalized coordinates of the robotic mechanism: $J_i(\theta_1, \theta_2)$, for $i = 1, 2, \ldots, N$. As mentioned *ad nauseam*, the analytical expressions of the performance indices to be optimized are unknown, although the partial derivatives of these functions can be estimated at each epoch of time, so that we can apply a gradient-based search for the optimum of each function:

$$\theta_l(k+1) = \theta_l(k) \pm \omega_{il} \left.\frac{\partial J_i}{\partial \theta_l}\right|_{\theta_l(k)} \quad ; \quad l = 1,2 \ , \ i = 1,2 \qquad (13)$$

where the double sign is due to the objective of maximization or the minimization, respectively. Of course, other techniques for function optimization can be attempted. For instance, ANN offer a series of such techniques. However, we propose a much simpler technique for two reasons: generalization capacity, which is absolutely vital for highly dynamic environments, and simplicity of our collision avoidance method layout, in which we wish to keep the formal development as simple as possible. In (13), ω_{il} plays the strategic role of determining the magnitude of the control action changes, which also depends on performance-relative changes. In fact, as shown further on, this parameter is related to the coordination of multiple existing objectives or performance indices.

The partial derivatives appearing in (13) can be estimated as follows:

$$\left.\frac{\partial J_i}{\partial \theta_l}\right|_{\theta_l(k)} \approx J_i\left[\theta_l(k), \theta_m(k-1)\right] - J_i\left[\theta_l(k-1), \theta_m(k-1)\right] = \left.\Delta J_i(k)\right|_{\theta_l(k)}$$

$$l = 1,2 \quad ; \quad m \neq l \quad ; \quad i = 1,2,\ldots,N \tag{14}$$

Thus, the partial derivatives are obtained as the difference of the consecutive values of each performance function, as a result of changing one and only one of the control variables. Physically speaking, this means that the robot must move each of its joint angles in turn in order to evaluate the performance functions. Therefore, considering the reinforcement control rules given by (12), we can re-write (13) directly as updating control actions:

$$\begin{array}{c} \Delta\theta_l(k) = \omega_{il} \left.\Delta J_i(k)\right|_{\theta_l(k)} = \omega_{il} \Delta J_{il}(k) \\ l = 1,2 \quad ; \quad i = 1,2,\ldots,N \end{array} \tag{15}$$

in which the performance functions increments are given by (14).

Now, let us discuss the important question of multiple objectives or performance functions for obstacle avoidance. Up to now, we have proposed two indices, $d_{\min}$ and A_{free}, and very soon we shall introduce a new one to guarantee that, apart from not colliding, our robot moves forward through the environment. In conclusion, we have got a multi-objective optimization problem. This multi-criteria problem is further complicated by the fact that for each link there are, *de facto*, two performance indices associated with the sub-goal of collision avoidance. For the general case of a robot with n joints or degrees of freedom, there are n generalized coordinates $q_1, q_2, \ldots, q_n$ and, consequently, n links $l_1, l_2, \ldots, l_n$. Usually, all the links share the same set of conceptual performance indices, let us say $J_1, J_2, \ldots, J_N$ and in turn each link has its own set of particularized indices. For instance, in our mechanism we defined two conceptual indices $d_{\min}$ and A_{free}, so that each of the two links has its own set of indices: $d^i_{\min}$, A^i_{free}, with $i = 1, 2$.

Another question of interest concerns the dependence of each performance function on the control variables. As a robotic mechanism is always an articulated chain, each generalized coordinate only influences its own links and other posterior links. Fig. 17 illustrates this kind of serial dependence for a 3R planar manipulator, in which, for instance, the first link performance functions are independent of the last two joint angles θ_2 and θ_3, while all the performance indices depend on the two control variables for the third link.

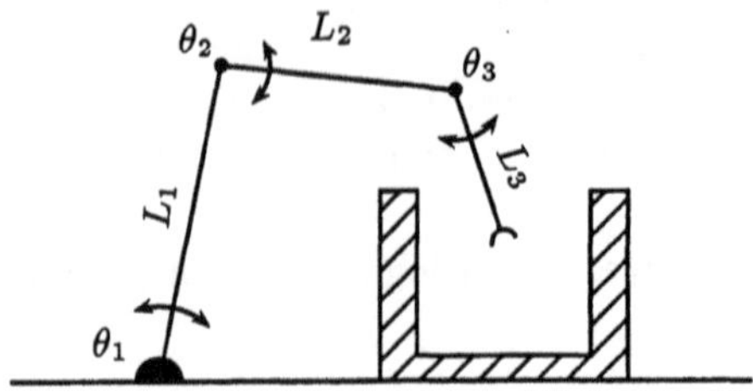

Fig. 17. Illustration of the serial dependence of control variables and performance functions

Particularizing to our robotic mechanism, we have the following expressions for the two control actions:

$$\Delta\theta_1(k) = f_1\left[\omega_{11}\left.\frac{\partial J_1^1}{\partial\theta_1}\right|_{\theta_1(k)}, \omega_{12}\left.\frac{\partial J_2^1}{\partial\theta_2}\right|_{\theta_1(k)}, \omega_{13}\left.\frac{\partial J_1^2}{\partial\theta_1}\right|_{\theta_1(k)}, \omega_{14}\left.\frac{\partial J_2^2}{\partial\theta_2}\right|_{\theta_1(k)}\right]$$
$$\Delta\theta_2(k) = f_2\left[\omega_{21}\left.\frac{\partial J_1^2}{\partial\theta_1}\right|_{\theta_1(k)}, \omega_{22}\left.\frac{\partial J_2^2}{\partial\theta_2}\right|_{\theta_2(k)}\right] \tag{16}$$

in which the superindices refer to the respective link.

Note that we have expressed the control actions, at each joint angle, as a general function of the local gradients of the performance functions, of which the generalized coordinates are independent variables. Further on we shall approach, very roughly, the question of weighting the different gradients appearing in (16), which is, in essence, the problem of coordinating the existing performance functions.

At this point, we can conclude that collision avoidance is a complex problem of multi-objective, some of them opposite, optimization with absolutely no analytical knowledge about the states of a dynamic environment, of which the robot only has occasional estimates of the objectives or performance functions.

As regards the performance indices, so far, we have only considered the collision avoidance sub-goal. We should recall that our robot has, at least, one more sub-goal: to advance through the environment. Thus, we should define the respective performance functions of this sub-goal. Observing Fig. 6, it is straightforward to set out the following performance index for the advance sub-goal:

$$J_3(\theta_1, \theta_2) \triangleq x_k - x_{k-1} = \Delta x_k \tag{17}$$

which quantifies the progress made by the robot with regard to its advance along the aerial line, as x is the horizontal coordinate of the flying end of the robot.

Now, let us focus on the combination of the different performance functions, which corresponds to the coordination problem in multi-objective optimization. A common, straightforward way of combining the objectives is by means of a set of weights:

$$J_T = \sum_{i=1}^{N} \omega_i J_i \tag{18}$$

in which case the control actions generated by each individual objective are weighted by the respective weights. This coordination policy can be made more complex by introducing dynamic weighting parameters as a result of the evolution of the global system formed by the robot and the environment. Concerning our mechanism, we have three performance functions, two for the collision avoidance sub-goal and one for the advance sub-goal, so that we can compose the two existing control actions in the following way:

$$\begin{aligned} \Delta\theta_1(k) &= \omega_{11}\Delta J_{11}^1(k) + \omega_{12}\Delta J_{12}^1(k) + \\ &\quad + \omega_{13}\Delta J_{11}^2(k) + \omega_{14}\Delta J_{12}^2(k) + \omega_{15}\Delta J_{13}(k) \\ \Delta\theta_2(k) &= \omega_{21}\Delta J_{21}^2(k) + \omega_{22}\Delta J_{22}^2(k) + \omega_{23}\Delta J_{23}(k) \end{aligned} \tag{19}$$

Usually, for each control action, the weights are normalized to one. The most simple coordination scheme is to apply the same weight for each performance component. However, there are situations in which some of the sub-goals may be critical. For instance, the collision avoidance sub-goal is of maximum priority when the robot is too close to an obstacle, so that its respective weight should be much greater than those of the other sub-goals: advance, inspect the line, etc. For critical situations, it could be advisable even to inhibit the non critical sub-goals.

The dynamic coordination of the performance functions can be modeled as a finite state automaton, in which the states correspond to sub-goals and the transitions are fired by external events related to the robot-environment pair situations. Nevertheless, the implications and possibilities of the coordination of the performance functions are beyond the scope of this introductory presentation of our method for collision avoidance, and readers are referred to [22] for a more detailed discussion.

Fig. 18 shows a sequence of movements performed by the simulated mechanism while avoiding an obstacle. In particular, note that, when it detects the proximity of the obstacle, the robot changes its usual forward movement to a series of advance/return movements of its two joints aimed at simultaneouly avoiding the obstacle and progressing along the line. The robot is always able to accomplish the avoidance maneuver provided that there is a physical trajectory compatible with both objectives, without either computing an analytical collision-free trajectory or solving its kinematic equations: it just navigates by means of reinforcement control and perceptual feedback.

7 Conclusions

A robotic mechanism designed for locomotion along rigid aerial lines and reticulated structures has been introduced. Conventional hard computing

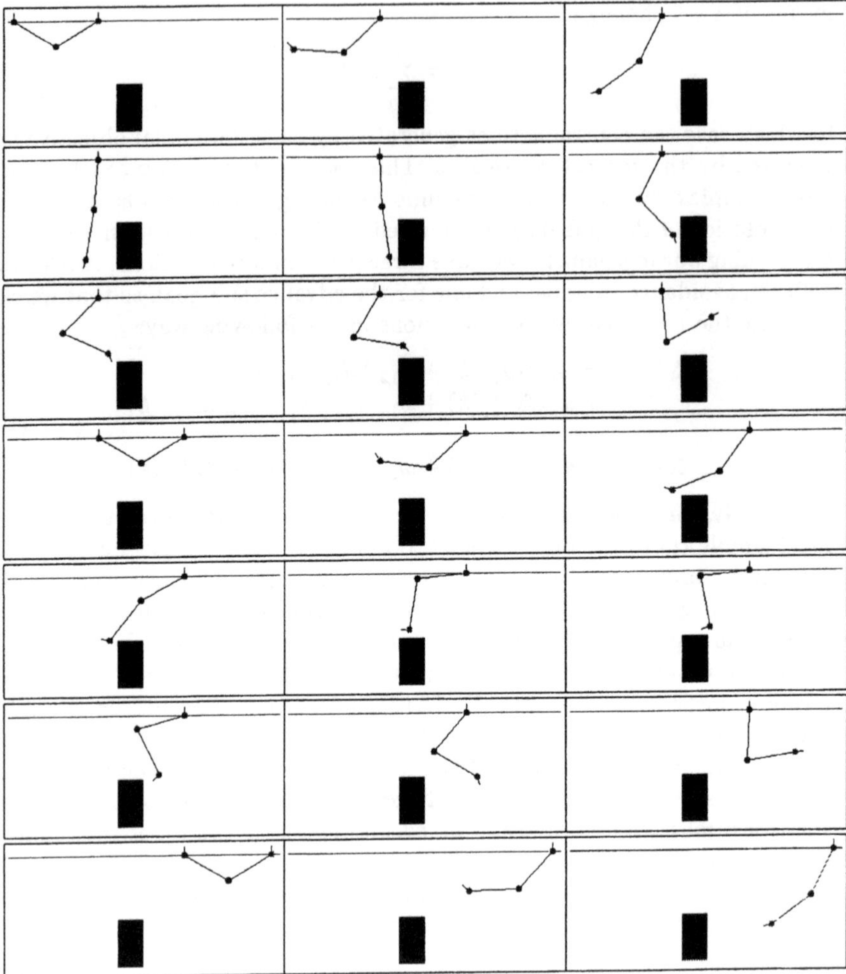

Fig. 18. Sequential movements of the robot when simultaneously advancing along the line and avoiding an obstacle

techniques have been applied to control the basic movement of the mechanism. Afterwards, a bio-inspired method for dynamic obstacle avoidance has been introduced using the perception-reason-action cycle. The symbiosis between perceptual feedback and reinforcement learning control is at the bottom of the method. The justification of employing perceptual feedback is straightforward, as the mechanism needs to evaluate the effects of its actions on the environment state. Two simple performance functions for collision avoidance have been proposed and the respective range sensor has been built. As there is absolutely no information about the analytical structure of the environment and the existing obstacles, the robotic mechanism performs its locomotion task by means of a reinforcement learning process guided by the perception of the robot-environment state. The method can be directly extended to any type of articulated mechanisms, including manipulator arms. Simplicity is the main advantage of the method, as the collision-free trajectories are generated without knowledge of the analytical expression of the physical structure of the environment and, also, without solving the robot kinematic equations.

Appendix A: Omnidirectional Range Sensor

The Sharp GP2D12 infrared sensor has been integrated in a Microbótica CT6811 card as shown in Fig. 19. This card is based on the Motorola 68HC11 microcontroller.

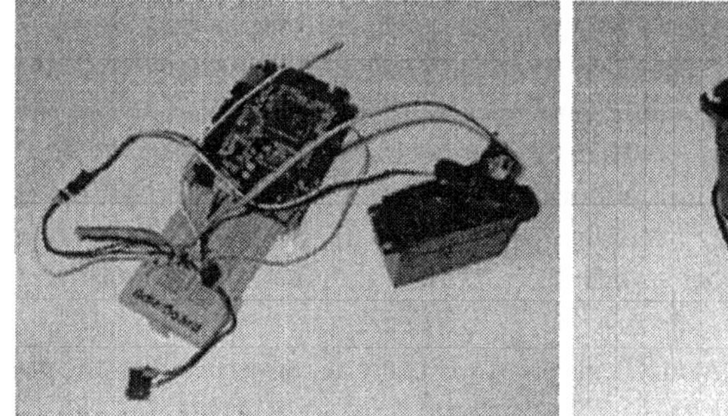

Fig. 19. Infrared sensor, microcontroller and servomotor

The range sensor has been experimentally calibrated in order to get the actual transformation from the sensor readings in Volts to distances in millimeters. Fig. 20 shows the theoretical curve provided by the manufacturer and the calibrated curve, which are quite similar for the range of interest – i.e. from 100 mm to 800 mm –. However, it should be noted that the noisy character and the errors associated with range sensors based on infrared and ultrasound justify the use of filtering techniques or, still better, the application of fuzzy techniques to cope with these problems.

As explained in section 6.1, the method proposed for collision avoidance employs two performance indices, $d_{\min}$ and A_{free}, which are computed using the distances from a sensor placed at each robot joint to the existing obstacles. The radial configuration of the proposed sensor is displayed in Fig. 13 to Fig. 15.

This radial configuration has been implemented by adapting a Futaba S3801 servomotor to the range sensor, which provides as many as 2960 dis-

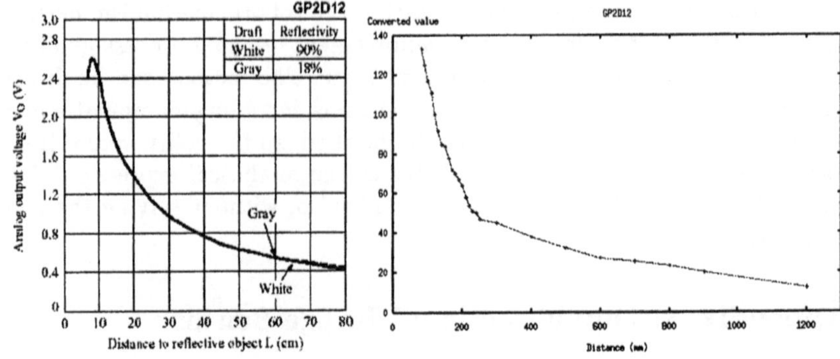

Fig. 20. Conversion curves: (a) given by the manufacturer and (b) obtained after calibration

tance readings for a complete rotation of 180°. The corresponding resolution of 0.06^o is far beyond any practical necessity, even if we wished to accurately reconstruct the obstacle contour functions $f_i(x, y) = 0$. Furthermore, the GP2D12 range sensor has a 40 ms measurement/reading cycle, so that a compromise between speed and accuracy is not problematic. For instance, a typical resolution of around 5° – accurate enough for most of the practical situations faced by the robotic mechanism in avoiding obstacles – is achieved with a cycle of around 3 s.

Fig. 21 shows a polygonal environment and three reconstructions obtained with different angular resolutions: 20°, 14° and 8.5^o, respectively.

Appendix B: Simulation of the Mechanism Trajectories Control

In this appendix, we discuss some results concerning the control design of the robotic mechanism, obtained through computer simulations, in which we have considered the following, realistic parameter values for the two link motors:

$$\begin{aligned} R_a &= 1\,\Omega & K_m &= 0.1\,\mathrm{N \cdot m/A} \\ J_m &= 0.0002\,\mathrm{N \cdot m \cdot s^2/rad} & K_b &= 0.05\,\mathrm{V \cdot s/rad} \\ b_m &= 0.002\,\mathrm{N \cdot m \cdot s/rad} & \zeta &= 100 \end{aligned}$$

And for the links:

$$\begin{aligned} m_1 &= 2\,\mathrm{Kg} \\ m_2 &= 3\,\mathrm{Kg} \\ l_1 = l_2 &= 0.6\,\mathrm{m} \end{aligned}$$

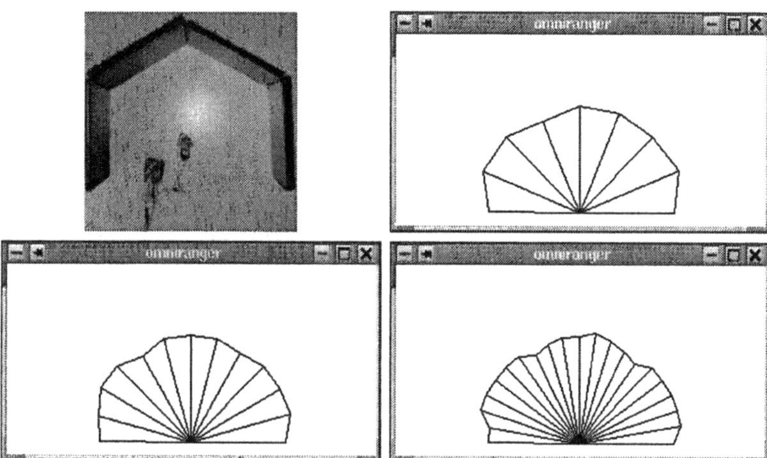

Fig. 21. A polygonal environment and three reconstructions at different angular resolutions: 20°, 14° and 8.5°, respectively

B.1 Linear Trajectories

We have assumed, in the simulations, that the sensors are perfect and that there is no delay time for the motor rotation readings and the application of the control voltages is virtually instantaneous. The sampling period was 10 ms and the basic simulation time unit, 1 ms. After extensive experimentation, the best results were achieved with the PD algorithm:

$$\begin{aligned} V_m(t) &= K_p\, e + K_d\, \dot{e} \\ K_p &= 10 \\ K_d &= 6 \end{aligned}$$

The trajectories of the joint-link angles are given in Fig. 22, where the respective errors and the motor voltage profiles are also shown. The desired trajectories are expressed in radians and the release operation is performed by the mechanism in 1 second. As the results for the two basic steps, grasp and release, are very similar, only the latter have been shown.

The trajectories of the joint-link angles are very satisfactory, as they perfectly track the desired trajectories – note that the real and the desired trajectories are almost indistinguishable –. However, the motor voltages obtained are not acceptable, as they rise to prohibitive values at the initial and final instants, so that a voltage limiter should be applied to both link motors.

B.2 Refined Trajectories

This problem can be solved by introducing a refinement in the desired or target trajectories. Indeed, the linear trajectories generate, at the initials instants, considerable velocities and accelerations in the dynamics of the robotic

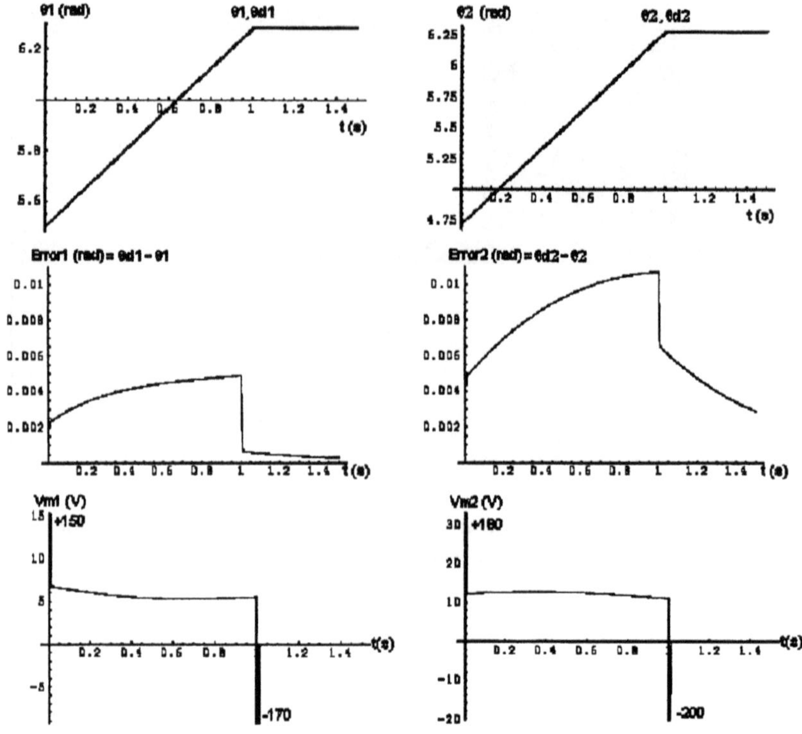

Fig. 22. Results of the simulated release movement for a desired linear trajectory

mechanism, which will eventually produce mechanical wear. This problem is even more serious for the grasp operation, as the robot hand could crash into the aerial line. Therefore, let us consider a refined target trajectory with the following additional restrictions on the initial and final speeds:

$$\dot{\theta}_1(0) = \dot{\theta}_2(0) = 0 \quad ; \quad \dot{\theta}_1(t_f) = \dot{\theta}_2(t_f) = 0 \tag{20}$$

which can be guaranteed by a third-order trajectory:

$$\theta_{d_i}(t) = a_{i_0} + a_{i_1}\, t + a_{i_2}\, t^2 + a_{i_3}\, t^3 \quad ; \quad i = 1, 2 \tag{21}$$

By applying the initial and final conditions to both position and velocity, we obtain the following smooth desired trajectory:

$$\begin{aligned} a_{i_0} &= \theta_i(0) \\ a_{i_1} &= \dot{\theta}_i(0) = 0 \\ a_{i_2} &= \frac{3}{t_f^2}\,[\theta_i(t_f) - \theta_i(0)] \\ a_{i_3} &= -\frac{2}{t_f^3}\,[\theta_i(t_f) - \theta_i(0)] \end{aligned} \tag{22}$$

Fig. 23 shows the excellent results achieved in the release operation for both joints using a simple PD algorithm, thanks to the refined trajectory. In particular, note the limited motor voltages obtained.

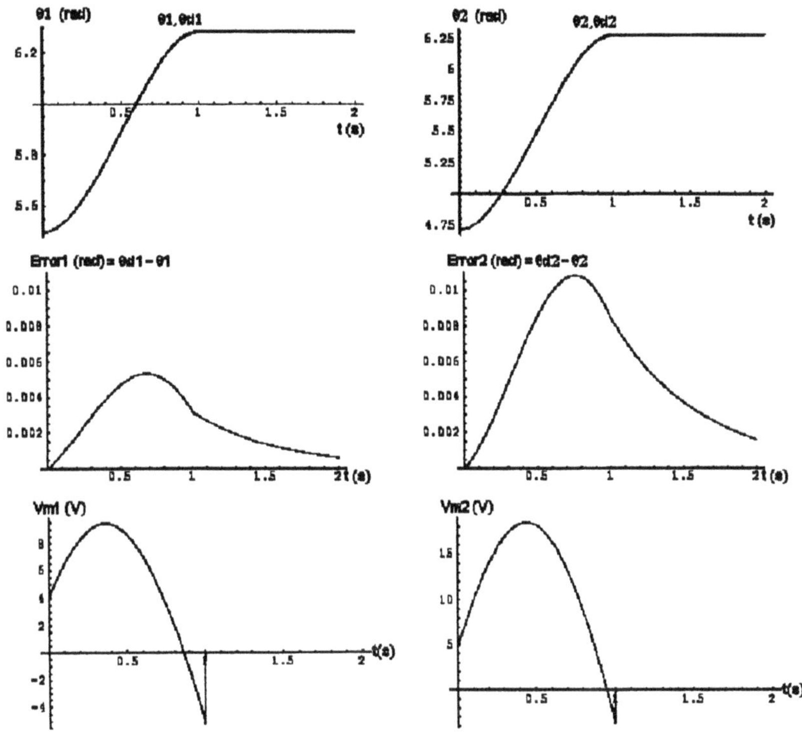

Fig. 23. Results obtained with the refined target trajectory

References

1. Saito, F., Fukuda, T., Arai, F. (1994) Swing and locomotion control for a two-link brachiation robot. IEEE Control Syst. Mag, **14**, 5–12
2. Nakanishi, J., Fukuda, T., Koditschek, D.E. (2000) A brachiating robot controller. IEEE Trans. on Robotics and Automation, **16**(2), 109–123
3. Maravall, D., Baumela, L. (1996) Robotic systems with perceptual feedback and anticipatory behavior. In R. Moreno-Díaz, J. Mira-Mira (eds.), Brain Processes, Theories and Models. MIT Press, Cambridge, Massachusetts, 532–540
4. Albus, J.S. (1975) A new approach to manipulator control: The cerebellar model articulation controller (CMAC). ASME J. of Dynamics Systems, Meas., & Control, **97**, 220–227

5. Meystel, A.M., Albus, J.S. (2002) Intelligent Systems: Architecture, Design and Control. John Wiley & Sons, New York
6. Kawato, M. Cerebellum and motor control. (1995) In M.A. Arbib (ed.), The Handbook of Brain Theory and Neural Netwoks. MIT Press, Cambridge, Massachusetts, 172–178
7. Franklin, S. (1995) Artificial Minds. MIT Press, Cambridge, Massachusetts
8. Mel, B.W. (1990) Connectionist Robot Motion Planning. Academic Press, Boston
9. Werbos, P.J. (1990) A menu of designs for reinforcement learning over time. In W.T. Miller III, R.S. Sutton, P.J. Werbos (eds.), Neural Networks for Control. MIT Press, Cambridge, Massachusetts, 67–95
10. Bryson, A.E., Ho, Y.C. (1969) Applied Optimal Control: Optimization, Estimation and Control. Hemisphere, Massachusetts
11. Westphal, L.C. (1995) Sourcebook of Control Systems Engineering. Chapman & Hall, London
12. Jang, J.-R.R., Sun, C.-T., Mizutani, E. (1997) Neuro-Fuzzy and Softcomputing. Prentice-Hall, Upper Saddle River, New Jersey
13. Lu, Y.-Z. (1997) Industrial Intelligent Control. John Wiley & Sons, New York
14. Sutton, R.S., Barto, A.G. (1998) Reinforcement Learning, MIT Press, Cambridge, Massachusetts
15. Zhou, C. (2000) Neuro-fuzzy gait synthesis with reinforcement learning for a biped walking robot. Soft Computing, **4**, 238–250
16. Zhou, C., Yang, Y., Jia, X. (2001) Incorporating perception based information in reinforcement learning using computing with words. In J. Mira, A. Prieto (eds.), Bio-Inspired Applications of Connectionism, LNCS 2085, Springer Verlag, Berlin, 476–483
17. Barto, A.G. (1995) Reinforcement learning in motor control. In M.A. Arbib (ed.), The Handbook of Brain Theory and Neural Networks, MIT Press, Cambridge, Massachusetts, 809–813
18. Jacob, C. (1999) Stochastic search methods. In M. Berthold, D.J. Hand (eds.), Intelligent Data Analysis, Springer-Verlag, Berlin, 299–350
19. White, D.A., Sofge, D.A. (1992) Handbook of Intelligent Control, Van Nostrand Reinhold, New York
20. Chankong, V., Haimes, Y.Y. (1987) Multiple objective optimization: Pareto Optimality. In M.G. Singh (ed.), Systems & Control Encyclopedia, Vol. 5, Pergamon Press, Oxford, 3156–3165
21. Kang, D.-O. *et al.* (2001) Multiobjective navigation of a guide mobile robot for the visually impaired based on intention inference of obstacles. Autonomous Robots, **10**, 213–230
22. Maravall, D., De Lope, J. (2002) A reinforcement learning method for dynamic obstacle avoidance in robotic mechanisms. 5th International Conference on Computational Intelligence Systems for Applied Research, Gent, Belgium, September 16–18, 2002 (to appear)

Online Learning and Adaptation for Intelligent Embedded Agents Operating in Domestic Environments

Hani Hagras, Victor Callaghan, Martin Colley, Graham Clarke, Hakan Duman

Department of Computer Science, University of Essex, Wivenhoe Park, Colchester CO4 3SQ, England.
hani@essex.ac.uk

Abstract. In this chapter we show how intelligent embedded agents situated in an intelligent domestic environment can perform learning and adaptation. A typical domestic environment provides an environment where there is wide scope for utilising computer-based products to enhance living conditions. Intelligent embedded agents can be part of the building infrastructure and static in nature (e.g. lighting, HVAC etc.), some will be carried on the person as wearables, others will be highly mobile, as with robots. Both non-intrusive and interactive learning modes (including a mix of both) are used, depending on situation of the agent. For instance mobile robotic agents use an interactive learning whilst most building based agents use non-intrusive background learning modes. In this chapter we will introduce the learning and adaptation mechanisms needed by the Building and Robotic embedded agents to fulfil their missions in intelligent domestic environments. We also present a high-level multi embedded-agent model, explaining how it facilitates inter-agent communication and cooperation between heterogeneous sets of embedded agents within a domestic environment.

1 Introduction

The variety of computer-based goods, and their capabilities, is growing at an unprecedented rate fuelled by advances in microelectronics and Internet technology. Cheap and compact microelectronics means most everyday artifacts (e.g. shoes, cups) are now potential targets of embedded-computers, while ever-pervasive networks will allow such artifacts to be associated together in both familiar and novel arrangements to make highly personalized systems.

A typical domestic environment provides an environment where there is wide scope for utilizing computer-based products to enhance living conditions. For instance it is possible to automate building services (e.g. lighting, heating etc), make use of computer based entertainment's systems (e.g. DVDs, TV etc), install work tools (e.g. robot vacuum cleaners, washing machines, cookers etc), or enhance peoples safety (e.g. security and emergency measures). Some of these artifacts will be part of the building infrastructure and static in nature (e.g. lighting, HVAC etc.), others will be carried on the person as wearables or mobiles, or temporarily installed by people as they decorate their personal space (e.g. mobile phones, TVs etc) [1].

In order to realise the intelligent domestic environments, technologies must be developed that will support ad-hoc and highly dynamic (re) structuring of such artifacts whilst shielding non-technical users from the need to understand or work directly with the technology "hidden" inside such artifacts or systems of artifacts. For this vision to be realized in domestic environments, people must be able to use computer-based artifacts and systems without being cognitively aware of the existence of the computer within the machine. Clearly in many computer-based products the computer remains very evident as, for example, with a video recorder, whose user is forced to refer to complicated manuals and to use his own *reasoning* and *learning* processes to use the machine successfully. This situation is likely to get much worse as the number, varieties and uses of computer based artifacts increase. We argue that if some part of the reasoning, planning and learning normally provided by a gadget user, were embedded into the artifact itself, then, by that degree, the cognitive loading on the user would reduce and, in the extreme, disappear (i.e. a substantial part of the computer's presence would disappear). However, this is far from easy as such "intelligent artifacts" operate in a computationally complex and challenging physical unstructured environment which is significantly different to that encountered in more traditional PC programming or AI. A major challenge is the large amount of uncertainty that characterizes real world environments. On the one hand, it is not possible to have exact and complete prior knowledge of these environments: many details are usually unknown. On the other hand, knowledge acquired through sensing is affected by uncertainty and imprecision [2].

In this chapter, we describe an innovative multi heterogeneous agent environment consisting of a domestic environment inhabited by a variety of agents. Intelligent embedded agents can be part of the building infrastructure and static in nature (e.g. lighting, HVAC etc.), some will be carried on the person as wearables or mobiles (termed PA), others will be mobile robotic agents (termed RA). The class of agents we term Building Agents (BA) are situated in the building services and try to learn the occupant's habitual behaviour and preemptively adjust the environment to satisfy him via a non-intrusive learning mode. Intelligent Robotic Agents differ in that they learn behaviours through interaction with the environment. An essential feature that characterizes all our work is that intelligent habitat technology needs to be centered on the individual, tailoring themselves to an individual wherever possible, rather than generalizing across a group of individuals.

In the next section we introduce our heterogeneous multi-agent architecture for intelligent domestic environments that features a hierarchical fuzzy genetic system for online learning and adaptation. We then describe interactive learning in the mobile Robotic Agents and non-intrusive learning in the Building Agents. Finally we offer experimental results, our findings to date and plans for future work.

2 Heterogeneous Multi-Agent Domestic Intelligent Environments Application

We have chosen the Essex Intelligent Dormitory (iDorm) shown in Fig. 1a to form the experimental framework for the domestic environments. Being an intelligent dormitory it is a multi-use space (i.e. contains areas with differing activities such as sleeping, working, entertaining etc) and can be compared in function to a room for elderly or disabled people or an intelligent hotel room. Because this room is of an experimental nature we are fitting it with a liberal placement of sensors (e.g. temp. sensors, presence detectors, system monitors etc) and effectors (e.g. door actuators, equipment switches etc), which the occupant can configure and use. The room looks like any other but above the ceiling and behind the walls hides a multitude of networks and networked devices.

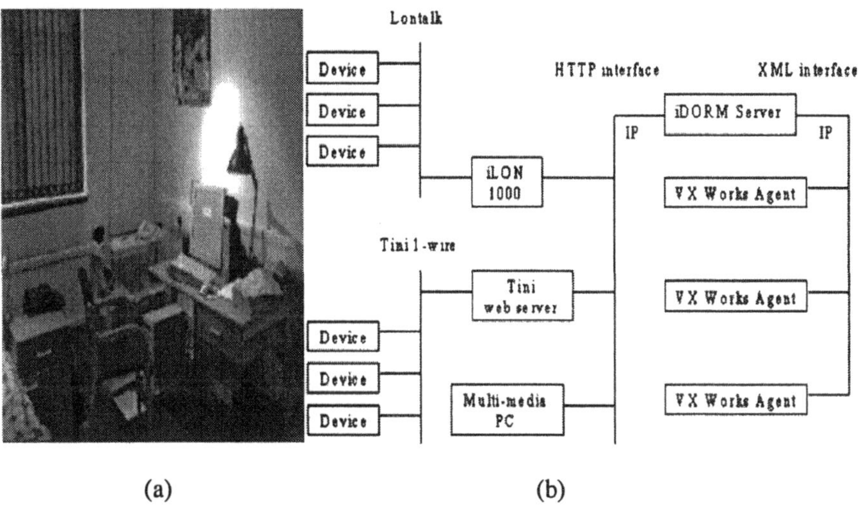

(a) (b)

Fig. 1. a) Photograph of the iDorm. b) iDorm logical infrastructure

The iDorm is based around three networks, Lontalk, Tini 1-wire and IP. This provides a diverse infrastructure and allows the development of network independent solutions. It also gives us an opportunity to evaluate the merits of each network.

To create a standard interface to the iDorm we have an iDorm gateway server. This exchanges XML formatted queries with the entire principal computing components, which overcomes many of the practical problems of mixing networks. The communications architecture is being extended to allow devices to be 'Plug N Play' (enabling automatic discovery and configuration). The iDorm logical infrastructure is shown in Fig. 1b.

The embedded agent used for the intelligent buildings shown in is based on a 68000 Motorola processor with 4 Mbytes of RAM and an Ethernet network connection. It runs the VxWorks Real Time Operating System (RTOS).

Our mobile robots are based around a distributed field bus control system. In particular, we use the CANbus (Controller Area Network) developed for automotive industry, Motorola processors and the VxWorks Real Time Operating System (RTOS). The current design is influenced largely by the requirements for both parallel and distributed processing in a real-time environment. We will use different sizes of robots for our experiments to verify that our learning algorithms are robot independent. We will also perform the robot experiments in difficult outdoor unstructured environments to test online learning and adaptation.

The RA can also take the form of a Manus robot arm, which can even be located at a remote place. We are currently involved in a collaborative project with the Korea Advanced Institute of Science and Technology (KAIST) supported by UK-Korea S&T collaboration fund. In this project our BA in the University of Essex, UK will be cooperating and communicating with RA located in Essex and also remotely to RA in Korea.

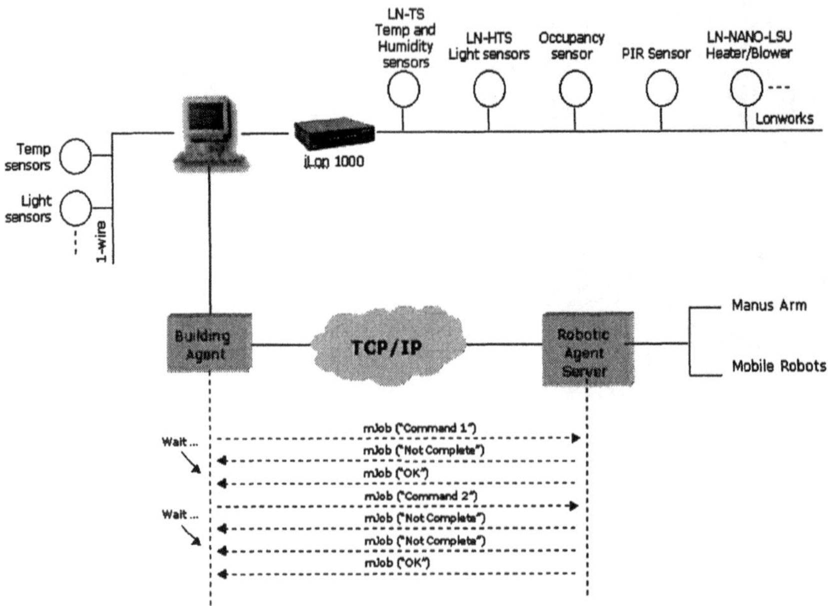

Fig. 2. Framework of the heterogeneous Multi Embedded Agents communication.

Currently the communication between the BA and the RA is established by initiating a request command from the BA to the RA server. The server creates the link between the computers and waits for incoming commands from the BA. Depending on the current situation in the building environment the BA sends out commands to the RA such as moving the Manus Arm to a certain position or commanding the mobile robot to pick up the mail. Once the command has been sent the server passes the request to the responsible RA to fulfill the task and informs the 'commander' (Building agent) about the current status of the robots. If the task is not complete then the server sends a message indicating that the job is "not complete". Every time when the BA wants to send out a new command, it waits until the previously requested job has been successfully finished.

Internet-based control systems rely on the available communication protocols to exchange real-time data between two computers. Most network protocols nowadays provide a reliable and transparent support for data exchange among computers by using protocols such as the Transmission Control Protocol (TCP). Real-time control is used in systems that must react to external stimuli with minimal delay in order to maintain stability. The issue of time delay is not the main subject in this chapter but it has been addressed by applying a feedback system. The designed system allows the RA (Manus arm and mobile robot) to continuously execute new coming commands while transmitting.

The communication between the BA and the RA are implemented by applying a TCP/IP stream socket. The BA has several I/O interfaces. One of them is used to connect to an IP network. To establish the communication the agent uses *Stream Sockets* to communicate with a TCP port within the node. In other words *Stream sockets* use TCP to bind to a particular port number. Another process, on any host in the network, can then create another stream socket and request that it will be connected to the first socket by specifying its host Internet address and port number. After the two TCP sockets are connected, there is a *virtual circuit* set up between them, allowing reliable socket-to-socket communications. Fig. 2 shows the Framework of the heterogeneous Multi Embedded Agents communication.

3 Interactive and Non-Intrusive Learning

In the field of RA, it is preferred that the learning is performed online interactively with the environment. The robot through trial and error evaluates its performance and assigns fitness values to different solutions and it can improve through our patented evolutionary process [3]. The robot by discovering its environment can learn by itself the controller needed to achieve the high level objectives and goals specified by the humans and it can update its controller to any environmental and robot kinematics changes it might encounter with no need to repeat the learning cycle. Such online interactive autonomous learning is desired for RAs operating in unstructured, dynamic and changing environments, which is the case of intelligent domestic environments. Such interactive learning allows the robots to program themselves which results in cutting down the costs of reprogramming and making the robots totally autonomous as they need only a high level mission from the humans.

For the BA the situation is different, as the agent needs to autonomously particularise its service to an individual. Building based learning is focused around the actions of people. Buildings are, largely, occupied by people who for a variety of reasons (e.g. time, interest, skills, etc) would not wish, or be able to cope with much interaction with the building systems. Thus in general, learning should as far as possible, be non-intrusive and transparent to the occupants (i.e. requiring minimal involvement from the occupants). The BAs are sensor rich and it is difficult to be prescriptive about which sensor parameter set would lead to the most effective learning of any particular action. Thus, to maximise the opportunity for the agent to find an optimum input vector set, whilst containing the processing overloads, the ideal agent would be able to learn to focus on a sub-set of the most relevant inputs.

There are two kinds of online learning and adaptation in intelligent domestic environments one is interactive for RA which is called the Associative Experience Engine presented in Section 4 and the other is non intrusive for the BA which is called Incremental Synchronous Learning (ISL) presented in Section 5.

As the human user is the center of our model the BA agent will use the ISL in a non-intrusive mode to capture the user behaviours. Some of the user behaviours will include the agent identifying any change in the person's behaviour that might signal a need for specific forms of help available via RA. The BA will receive high level inputs from the RA such as the robot is near the charger and it will produce high level outputs such as go and fetch a drink based on other input states which can be composite of the building and robot states. The high level output from the BA will be used as a high level objective function for the RA. The RA will learn and coordinate their basic behaviours such as obstacle avoidance, edge following and goal seeking and they will use AEE to learn and adapt their controllers to achieve the high level objective.

Fig. 3 shows a domestic environment for elderly and patient care which involves cooperation between the BA and the RA which is implemented in the project entitled Care Agents supported by the UK-Korea S&T collaboration fund. In this environment we consider a bed-bound person in a MANUS equipped bed who is served by a Mobile Agent (MA) bringing two commodities to the person's bedside when required. The role of the MANUS robot arm is to serve the patient with the delivered consumables that the MA has brought to a prescribed bedside location. The MA would carry items, such as post, newspapers, tissues, food, drinks, personal items (e.g. hairbrush), tissues and medicines etc. for the use of the patient and deliver them to a fixed point at the bedside so that the goods were readily available to the MANUS arm. The inputs provided to the BA include the current temperature and the lighting and entertainment levels in the room. Also the physical state of the person will be supplied to the BA e.g. prone or sitting up. Another aspect might be certain body signs such as the health monitoring aspects e.g. blood pressure, heart rate etc. Depending on the preferences of the patient and current conditions of the temperature and light, these can be adjusted by the BA. For these preferences to be learnt there would have to be some way of indicating the preferred levels e.g. switches/voice control. The MA provides the BA with information about its power level and about the relative (fuzzy) position of the robot to the bed and to the chargers and the commodities (A and B).

The BA has a rich set of data so that the conditions under which the person elected to ask for the commodity can be captured and recognised. The patient actions will be monitored and his behaviours will be learnt in a non-intrusive manner according to the whole input vector collected from the rest of the sensors. The commands are going to be passed as high level objective functions to the RA, which uses AEE to learn the required controllers to do the job as well as possible.

An example of such high level objective is fetch the mail in which the RA will learn and coordinate its basic behaviours such as obstacle avoidance, goal seeking and edge following to go and fetch the mail quickly while avoiding any obstacles.

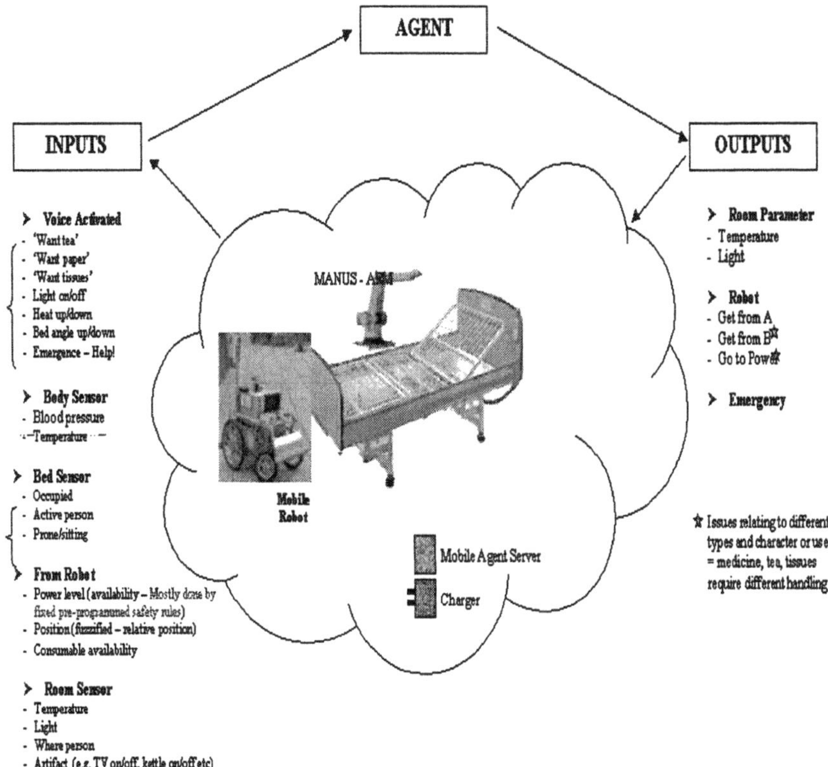

Fig. 3. A scenario showing heterogeneous multi embedded agent cooperation in which the robots implement AEE for learning and the building agents implement ISL

In general we divide the behaviours available to the BA into fixed or dynamic sets, where the dynamic behaviours are learnt from the person, and the fixed behaviours are pre-programmed and include safety and emergency behaviours. These latter behaviours need to be predefined because they cannot easily be learnt. For instance if medicine was going to be administered on a regular basis this would be described explicitly as one of the fixed rules. There will be other safety and emergency behaviours based upon other significant considerations like the temperature at which pipes freeze or what to do in the case of fire and so on. For dynamic behaviours we are going to use a monitoring system, which we call an ISL to record the patient actions and learn to generate rules from this information. These will then be fine-tuned in an incremental and life long mode. This set of behaviours will be able to control, light, temperature, tilt of the bed, sound volume etc and will interface to the RA through high level web based messages.

As most commercial Fuzzy Logic Control (FLC) implementations feature a single layer of inferencing between two or three inputs and one or two outputs. For embedded agents, however the number of inputs and outputs are usually large and the desired control behaviours are more complex. However, by using a *hierarchical* assembly of fuzzy controllers (HFLC), the number of rules required can be significantly reduced [3]. We use a variant of the method suggested by Saffiotti [4] and Tunstel [5]. In this we apply fuzzy logic to both implement the individual behaviour elements and the related arbitration (allowing both fixed and dynamic arbitration policies to be implemented) [3]. To achieve this we implement each behaviour as an independent FLC aimed at a simple task, with a resultant small set of inputs and outputs to manage. Fuzzy co-ordination facilitates expression of partial and concurrent activation of behaviours, thereby allowing behaviours to be active concurrently to differing degrees which gives a smoother control characteristics than switched counterparts [4].

In our design each behaviour uses a FLC using singleton fuzzifier, triangular membership functions, product inference, max-product composition and height defuzzification. The selected techniques were chosen due to their computational simplicity and real-time considerations.

$$Y_t = \frac{\sum_{p=1}^{M} y_p \prod_{i=1}^{G} \alpha_{Aip}}{\sum_{p=1}^{M} \prod_{i=1}^{G} \alpha_{Aip}} \tag{1}$$

Where M is the total number of rules, y_p is the crisp output for each rule, $\prod\alpha_{Aip}$ is the product of the membership functions for each rule's inputs and G is the number of inputs.

In case of using fuzzy numbers for preferences, product-sum combination and height defuzzification, the final output equation, provided by Saffiotti [3], is given below:

$$Y_{ht} = \frac{\sum_i (mm_y * y_t)}{\sum_i mm_y} \tag{2}$$

Where i represents the behaviours activated by context rules, which can be right/left edge-following behaviour, obstacle-avoidance or goal seeking in case of RA and comfort, safety, economy in case of BA. Y_t is the behaviour command output (robot speed and steering in case of RA and room lighting and heating in case of BA). These vectors are fused in order to produce a single vector Y_{ht} to be applied to the embedded agent. mm_y is the behaviour weight.

4 Associative Experience Engine

In our hierarchical learning procedure we start using a set of working (but not necessarily optimum) fixed membership functions. We then commence learning general rules in each individual behaviour by relating the input sensors to the actuator outputs. In this phase the membership values are not important as the agent learns general rules such as, *if the obstacle is close then turn left.* However in order to achieve a sub-optimal solution for the individual behaviour (a subset of the large search space) we need to find next the most suitable membership functions for the learnt rules. After finding a sub-optimal solution for each behaviour we combine these behaviours and learn the best co-ordination parameters that will give a "good enough" solution for the large search space to satisfy a given mission or plan. Readers are referred to [6] and [7] for more information about learning MF and the co-ordination parameters online. After learning the system parameters, the controller then operates in its environment where the online adaptation technique is triggered to adjust the learnt controller to any environmental or kinematics changes or to any new situations it might encounter. If the controller fails to maintain the desired states, the adaptation technique modifies the poor rules in the relevant behaviours to adjust to the embedded agent differing environmental and kinematics conditions without the need to restart the learning cycle. This is called life long learning where the agent can adapt itself to any new situation and it can update its knowledge about its environment. This hierarchical procedure results in a fast learning time for finding a solution for learning and adaptation in changing unstructured environments.

Our learning and adaptation techniques are inspired from Nature as in biology, most scientists agree that the remarkable adaptation of some complex organisms comes as a result of the interaction of two processes, working at different time scales: evolution and life long learning. Evolution takes place at the population level and determines the basic structures of the organism. It is a slow process that works by stochastically selecting the better individuals to survive and to reproduce. The life long learning is responsible for some degree of adaptation at the individual level. It works by tuning up the structures, built in accordance with the genetic information, by a process of gradual improvement of the adaptation to the surrounding environment [8]. Also condensed learning scenarios over short periods of time differ drastically from continuous learning or life long learning ones as life long learning presents the agent with very different perceptual stimuli than learning in a condensed period of time [9]. We emulate the natural process by using evolution and online learning to develop a good enough controller of the agent and we use our patented Fuzzy-Genetic system (the *Associative Experience* Engine) to speed the slow evolution process. Then we use our online adaptation technique to implement the life long learning where the agent is always updating its knowledge and gaining experience and is able to adapt to its changing environment.

Fig. 4 provides an architectural overview of our techniques, which we term as *Associative Experience Engine (AEE).* This forms the learning engine within the control architecture and is the subject of British patent application 99-10539.7. The behaviours are represented as parallel Fuzzy Logic Controllers (FLC) and form the hierarchical fuzzy control architecture presented in the previous section. Each FLC has two modifiable parameters, the *Rule Base* (RB) for each behaviour and the *Membership Functions* (MF). The behaviours receive their inputs from sensors. The output of each FLC is then fed to the actuators via the *Co-ordinator,* which weights their effect. When a behaviour, or collection of behaviours, fail to respond correctly to a situation, a learning cycle is initiated. The learning cycle performed is dependent upon the *Learning Focus.*

When learning or modifying the rule bases, the learning cycle is sub-divided into local situations. This reduces the size of the model to be learnt. The accent on local models implies the possibility to learn by focusing at each step on small parts of the search space only. The interaction among local models, due to the intersection of neighbouring fuzzy sets causes the local learning to reflect on global performance [10]. To further reduce the search space, the system determines if it had encountered similar situations before by checking the stored experiences in the *Experience Bank.* The embedded agent tests different solutions from the experience bank by transferring the "remembered" experiences, which are stored in a queue to the corresponding behaviours. If any of these experiences show success, then they are stored in the FLC and we avoid generating a new solution for our system. An *Experience Assessor* assigns each experience solution a fitness value to indicate the importance of this solution.

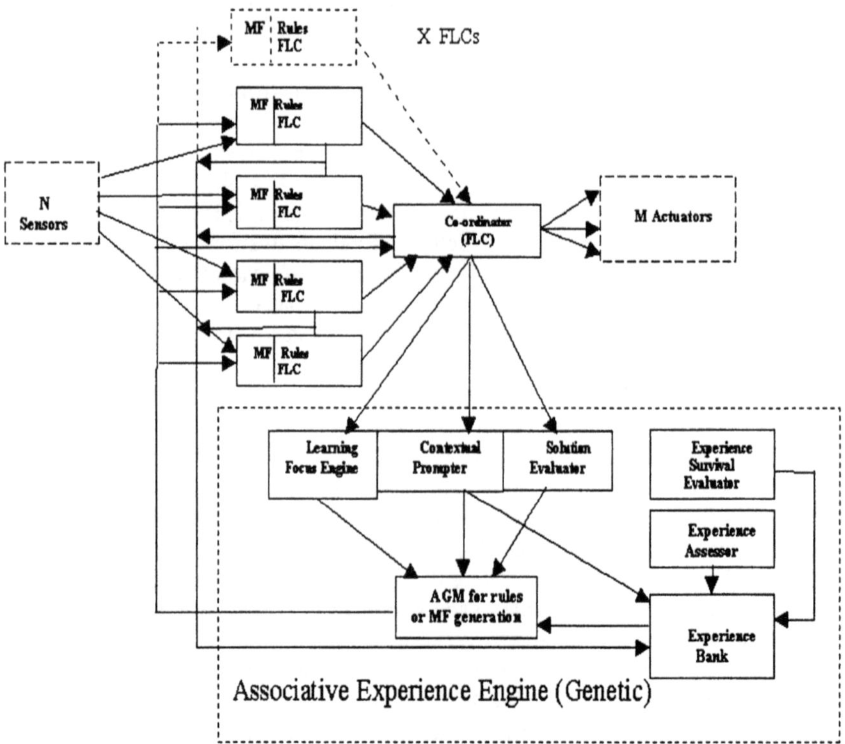

Fig. 4. Architectural Overview of Associative Experience Learning Engine (British patent No 99-10539.7)

When the Experience bank becomes full it is the role of the *Experience Survival valuer* to determine which parameters are retained and which are discarded, according to the parameter importance. If the use of past experiences did not solve the situation, we use the highest fitness experience as a starting point for the new learning cycle. We then fire an *Adaptive Genetic Algorithm (AGA)* mechanism using adaptive crossover and mutation parameters, which helps to speed the search for new solutions. The AGA is constrained to produce new solutions in a certain range defined by the *Contextual Constraints* supplied by sensors and defined by the co-ordinator according to the *learning focus.* This avoids the AGA searching places where solutions are not likely to be found. By doing this we narrow the AGA search space to where we are likely to find solutions. The AGA search converges faster as it started from a good point in the search space supplied by the experience recall mechanism and it used adaptive learning parameters and avoided searching regions where solutions are not likely to be found. After generating new solutions wither rules or MF or co-ordination parameters the system tests the new solution and gives it fitness through the *Solution Evaluator.* The AGA generates new solution until reaching a satisfactory solution.

The online learning mechanism, in addition to the fuzzy behaviours, is also organised as a hierarchy thus leading to one description of this architecture as being a "double-hierarchy". The online learning mechanisms can be regarded as a hierarchy because there is a tiered set of actions. At the highest level a population of solutions is stored in the *Experience Bank* and tested in a queue. If one of these stored experiences leads to a solution then the search ends, if none of these stored experiences leads to a solution then each of these experiences acquires fitness by the *Experience Assessor* depending how well each solution performed in the situation. The highest fitness experience is used as a starting position to the lower level Genetic Algorithm (GA) that is used to generate new solutions to the current situation. This hierarchy preserves the system experience, and speeds up the genetic search by starting the genetic algorithm from the best point found in the space.

4.1 Learning General Behaviours Rules

The rule base of the behaviour to be learnt is initialised randomly. The designer supplies a preliminary input membership function for each behaviour. As was explained earlier the values of the membership functions are not important as we are seeking general rules. In the following sections we will introduce the various steps of the algorithm to learn the rule base of behaviours that receive immediate reinforcement such as edge following and goal seeking in the robot navigation domain.

After the rule-base initialization, the embedded RA starts operating. If the rule-base contains poor rules then it will begin deviating from its objective. In this case our algorithm is fired to generate new set of rules to correct this deviation. The GA population consists of the most two effective rules in the failure situation. Each chromosome will represent the consequents of one of the effective rules during the bad action. We will use a binary representation. As the case with classifier systems,

in order to preserve the system performance, the GA is allowed to replace a subset of the classifiers (the rules in our case). The worst m classifiers
are replaced by the m new classifiers created by the application of the GA on the population [11]. The new rules are tested by the combined action of the performance and apportionment of credit mechanisms. In our case, only two rule actions will be replaced (those already identified with being predominantly responsible for the deviation).

The system fitness is determined by the *Solution Evaluator* and is evaluated by how much the agent reduces the normalised absolute deviation (d) from the normal value as well as maintaining minimum system oscillation. This is given by:

$$d = \frac{|\,normal.value - deviated.value\,|}{\max.deviation} \tag{3}$$

Where the normal value will correspond to the value desired by the human designer, which corresponds to the value that gives the maximum normal input membership function. The deviated value is any value deviating from the normal value. The maximum deviation corresponds to the maximum deviation that can occur as was specified by the human designer membership functions.

The fitness of the solution is given by d1 - d2, where d2 is the normalised absolute deviation before introducing a new solution, and d1 is the normalised absolute deviation following the new solution. The deviation is measured using the robotic agent's sensors, which gives the agent the ability to adapt to the imprecision and noise found in the real sensors rather than relying on estimates from previous simulations.

The fitness of each rule at a given situation is calculated as follows. We can write the crisp output Y_t as in (1). If the robotic agent has N output variables, then we have $Y_{t1} \ldots, Y_{tn}$ The normalised contribution of each rule p output ($Y_{p1}, Y_{p2} \ldots. Y_{pn}$) to the total output $Y_{t1}, Y_{t2} \ldots, Y_{tn}$ can be denoted by $S_{r1}, . S_{rn}$ where $S_{r1} .. S_{rn}$ are given by:

$$S_{r1} = \frac{\dfrac{Y_{p1} \prod_{i=1}^{G} \alpha_{Aip}}{\sum_{p=1}^{M} \prod_{i=1}^{G} \alpha_{Aip}}}{Y_{t1}}, \qquad S_{rn} = \frac{\dfrac{Y_{pn} \prod_{i=1}^{G} \alpha_{Aip}}{\sum_{p=1}^{M} \prod_{i=1}^{G} \alpha_{Aip}}}{Y_{tn}} \tag{4}$$

We then calculate each rule's contribution to the final action $S_c = \dfrac{S_{r1} + S_{r2} + \cdots S_{rn}}{N}$.

The most two effective rules are those that have the two greatest values of S_c, we use only mutation to generate new solutions because of the small population formed by the fired rules.

4.1.1 Memory Application

After determining the rule actions to be replaced according to Eq. 4, the robot then matches these rules to sets of rules stored in an *Experience Bank* containing each rule and its best fitness value up to date. The fitness of each rule fired in a given solution is supplied by the *Solution Evaluator* and is given by:

$$S_{rt} = \text{Constant} + (d_1 - d_2) . S_c .(1-F). \quad (5)$$

$d_1 - d_2$ is the deviation improvement or degradation caused by the adjusted rule-base produced by the algorithm. If there is improvement in the deviation, then the rules that have contributed most will be given more fitness to boost their actions. If there is degradation then the rules that contributed more must be punished by reducing their fitness w.r.t to the other rules. F is the normalised oscillation in the system actions that can be the system steering and velocity in case of the robots. This makes S_{rt} maximised by maximising the deviation improvement compared to the previous action with minimum adjustments to system output. The variable F was introduced to penalise instant differences between the system outputs to give a steady output and also to preserve the actuators by not making large adjustments (ON and OFF) (that might lead to the reduction of the life time of the actuators). The *Experience Survival Valuer* keeps the best fitness solution for each rule in the *Experience Bank.*

After determining the rule actions to be replaced, the RA then matches the effective rules to sets of rules stored in a memory containing each rule and its best fitness value up to date. For each rule action to be replaced, we will replace the current actions in the behaviour rule-base by the best fitness actions stored in the *Experience Bank.* If the deviation decreases, then the RA will keep the best rules in the behaviour rule-base. If the deviation remains the same or increases, the agent uses the GA to produce a new set of solutions. These are obtained by mutating the best fitness chromosomes from the *Experience Bank* to create new actions to replace the most effective rules actions until the deviation begins decreasing or the rule is proved ineffective. This is then considered a solution for the current situation and the rule fitness according to Eq. 5 is calculated and is compared with the best fitness rule stored in the memory. If its fitness is greater than the best kept one in the *Experience Bank* then it replaces the best one, otherwise the best one still is kept in the *Experience Bank.* The memory action is supposed to speed up the GA search as it starts the GA from the "best-found" point in the solution space instead of starting from a random point.

4.1.2 Using GA to Produce New Solutions

The GA begins its search for a new rule action to replace those identified with poor performance by mutating the two best fitness chromosomes from the *Experience Bank* to replace the most two effective rules actions to generate new solutions. The chromosome consists of the effective rule consequents (which can be speed and steering in case of RA). The population consists of two chromosomes corresponding to the two most effective rules. We use a small population, which will give faster convergence to a good enough solution (which is important in online learning) due to the smaller number of individuals to be evaluated and reproduced each generation. However, small populations run the risk of stagnating genetically, i.e. the population doesn't contain sufficient genetic diversity to properly explore the search space. This is why it is important to introduce high mutation rates to introduce new genetic material in the population without running into the risk of ending by a random search [12]. A mutation rate of 0.5 was chosen after gathering empirical evidence gathered from experimenting with different mutation rates from 0 to 1.0. This was achieved by monitoring the time the embedded robotic agent needed to achieve its goal. The agent also uses the *Contextual Constraints* supplied by the sensory information to narrow the search space of the GA and thus reduces the learning time. For example, if the robot is implementing left wall following and it is moving towards the wall, then any action that suggests going to the left will not lead to a solution. Hence, if we use a left side sensor and the robot senses that it is going towards the wall then the GA solutions will be constrained not to go left. Hence the solutions generated for the temperature are constrained to be less than the current temperature. This is similar to using a constrained GA, where the sensor information constrains the GA search space and forces it to look at areas where solutions are likely to be found and avoids doing blind searches in the whole search space, also avoiding abnormal outputs which results from mutation. We have also conducted learning experiments without the *Contextual Constraints.* It was noticed that this increased the learning time compared to those using *Contextual Constraints* by up to four times (average). This is because the GA now has to search in the entire search space, instead of just searching the limited regions where solution are likely to be found.

Fig. 5 gives an example of a robot deviating from its objective and after trying the memory actions with no success it triggers the GA. In this example the robot has only two output actions (left and right wheel velocities) each one is represented by only 4 fuzzy sets which are *Very Low, Low, Normal, High.* We need only two bits to represent the consequent of each rule thus we have a four bit chromosome. The actions are decoded as follows, *Very Low* is 00, *Low* is 01, *Medium* is 10, *High* is 11. In this example rule 1 and rule 2 were the most effective rules according to Eq. 4 and their actions are to be replaced by mutating the actions that obtained the maximum fitness according to Eq. 5 which are stored in the *Experience Bank.*

The generation of the new solution is constrained according to the contextual constrains and the mutation continues till the deviation decreases or the effective rules are not effective anymore. After this the fitness of this solution is evaluated using Eq. 5 and is compared with the one stored in the *Experience Bank* if it is greater than it, then it replaces the actions stored in the *Experience Bank.* The robot learns this situation and other similar local situations involving different rules. The accent on local models implies the possibility to learn by focusing at each step on small parts of the search space only. The interaction among local models, due to the intersection of neighboring fuzzy sets causes the local learning to reflect on global performance [10]. This reduces the size of the model to be learnt. The RA finishes the learning of the whole behaviour when it reaches a stopping criterion.

4.1.3 Stopping Criteria

The RA assumes it has learnt the rule-base for the behaviour if it succeeds in satisfying the goals specified to its behaviour. In case of the robot navigation, the robot assumes it has learnt the rule-base for the behaviour if it succeeds in maintaining the nominal value for the behaviour for an optimum learning distance.

The optimal learning distance has been related to units of length of the robot, so that the algorithm can be applied in an invariant manner to different size robots. In order to determine the optimal learning distance we have conducted numerous experiments using different learning distances corresponding to the robot's length (e.g. 1x robot's length, 2x robot's length, etc.). We then follow the same track that was used during the learning phase to determine the absolute deviation at each control cycle from the optimum value which is maintaining a constant distance from a wall in case of edge following and going towards a goal in case of goal seeking. The average and standard deviation of this error are calculated over 10 experiments over different tracks. It was found that the average and standard error for the wall following stabilizes at three times the robot's length at average value of 4 cm and standard deviation of 3 and become almost the same for a larger number of experiments. Thus we use three times the robot length as our learning length criteria.

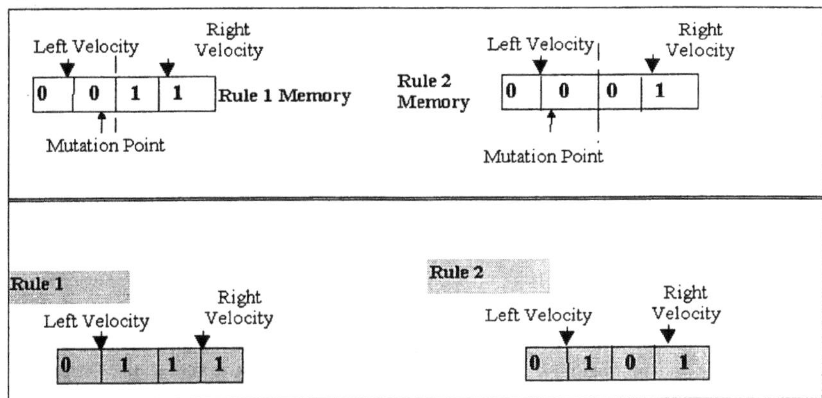

Fig. 5. An example of GA processes in which rule 1 and rule 2 consequents stored in Experience Bank and having the best fitness are generating new consequent for rules 1,2 which are the most effective rules responsible for behaviour failure.

4.2 Online Adaptation of the Learnt Controller and Life Long Learning

In the previous sections we explained how to learn the embedded RA controller online and through interaction with the environment using our evolutionary techniques, imitating the natural evolution and using our patented techniques to speed it up so that we can learn online. In this section we discuss how can we adapt the RA to any environmental changes and implement a life long learning scheme where the agent gains and updates its knowledge by encountering new situations. This means that the RA can add or delete or modify rules according to the situations it encounters.

Also this approach is useful in prototyping as we can learn the embedded RA controller on a prototype agent in a controlled environment and then we transfer it to the agent running in the real environment where we only run a short adaptation cycle with no need to repeat the learning cycle [3]. Perhaps more importantly, this adaptation and life long learning session is important in the case that environmental or the RA characteristics changes occur and the agent needs to rapidly adapt to these changed circumstances.

Although fuzzy logic allows a degree of imprecision to exist within the environment, significant environmental or agent differences will prevent the rule sets from operating correctly. One solution to this problem is to provide a new rule set for each of the different environments, although prior knowledge of the different environments would be needed (which is difficult if not impossible for dynamic unstructured environments). An alternative solution is to allow the existing co-ordinated rule set to be adapted to compensate for the environmental differences. This latter approach only requires a basic rule set to be present from the prototype robot controller and does not require prior knowledge of all possible environments. We start the adaptation from the best-learnt HFLC by the RA, instead of starting from a random point. If the system was started with random rule bases, the system could still modify its actions but would take approximately six times the time needed when started from a good rule base containing some inappropriate rules for the new environment. This figure was found by practical experimentation.

The system discovers the rules (in different behaviours) that, if modified, can lead the embedded agent to adapt to its environment. Whilst it might seem a good idea to change the co-ordination parameters, or the membership functions of the individual behaviours, to adjust the agent behaviour when it fails, this is not the best approach. This can readily be illustrated by considering a case where the rules in the individual rule bases become inappropriate, then clearly changing the co-ordination parameters will never correct the RA behaviour (as was proved by experimentation). Also changing the MF is a difficult task, as it needs to be done to each individual behaviour necessitating the robot to be taken away from its environment for MF calibration. However, our method allows the poor rules to be found and corrected for each behaviour, online and in-situ without the need to repeat the whole learning cycle, modifying only the actions of a small number of rules that performed poorly.

Imagine a situation where the robot had failed to do its mission. In this case we will modify all the rule bases whose behaviours were active. The RA then uses its short-term memory to return back to its pre-failure position, identifying the two most dominant rules (which can belong to different behaviours). The RA then replaces the actions of these rules to solve this situation. The way the agent determines the dominant rules is done by calculating the contribution of each rule to the final output in a given situation as follows: Apply Eq. 2 and substitute Y_t from Eq. 1. Then we can write the crisp output Y_{ht} as:

$$Y_{ht} = \frac{\sum_{y=1}^{4} mm_y \frac{\sum_{p=1}^{M} y_p \prod_{i=1}^{G} \alpha_{Aip}}{\sum_{p=1}^{M} \prod_{i=1}^{G} \alpha_{Aip}}}{\sum_{y=1}^{4} mm_y} \tag{6}$$

Where M is the total number of rules, Y_p is the crisp output for each rule, $\Pi\alpha_{Aip}$ is the product of the membership functions of each rule inputs. G is the number of the input variables, mm_y is firing strength of each of the four behaviours.

Because we have N output variables, then we have $Y_{ht1}, \ldots Y_{htn}$. The contribution of each rule p in the behaviour y to the total output $Y_{ht1}, ..Y_{htn}$ is denoted by $S_{ra11}, .\ S_{rann}$ where S_{ra11}, S_{rann} are given by:

$$S_{ra11} = \frac{\frac{mm_y}{\sum_{y=1}^{4} mm_y} \cdot \frac{Y_{p1} \prod_{i=1}^{G} \alpha_{Aip}}{\sum_{p=1}^{M} \prod_{i=1}^{G} \alpha_{Aip}}}{Y_{ht1}}, \quad S_{rann} = \frac{\frac{mm_y}{\sum_{y=1}^{4} mm_y} \cdot \frac{Y_{pn} \prod_{i=1}^{G} \alpha_{Aip}}{\sum_{p=1}^{M} \prod_{i=1}^{G} \alpha_{Aip}}}{Y_{htn}} \tag{7}$$

We then calculate each rule's contribution to the final action Sa_{c1} by:

$$Sa_{c1} = \frac{S_{ra11} + S_{ra22} \ldots + S_{rann}}{N}.$$

The most effective rules are those rules with the greatest values of Sa_{c1}.

In order to implement the Life-Long learning strategy the agent has a long-term memory composed of all the experiences it encountered in the past and how the agent had succeeded to solve the problems it encountered. These experiences are in the shape of rule clusters having the rule numbers and the consequents the agent had learnt to associate with inputs of these rules in a situation where the agent failed to satisfy the desired behaviour. The rule clusters are arranged in a queue starting from the most recent experiences. These clusters represents how the agent have adapted to different changing environments and situations, i.e. in two different rule clusters we can find the same rule having different consequents which means that each environment should have its different actions. So in case of the agent failing to achieve the desired behaviour in the changing unstructured environment then the actions of the most effective rules can be modified to solve this situation. The RA then matches the effective rules to sets of rules stored in the *Experience Bank.* If for example; rules 1,2 from the obstacle-avoidance rule base, rule 3 of the left edge-following behaviour and rule 4 of the right edge following need to be replaced, AND the first rule cluster in the Bank contains the consequences of rules 1 of obstacle avoidance; rule 3 of left edge following and rule 6,7 of right edge following, then the consequences of rules 1 for obstacle avoidance and 3 for left edge following will be changed and rule 2 for obstacle avoidance and 4 for right edge following will remain the same. Then the RA starts operating with this modified rule-base taken from the *Experience Bank.* If it survives and gets out of this situation with no behaviour failures, then these rules are kept in the rule-base of the controller. In this way we have saved the process of learning a solution to this problem from the beginning by using the memorized experience that had worked in different environmental conditions. This is the same as nature where the agent tries to recall its experiences that solved the same situation before. If none of these experiences had solved this situation, the agent assigns to each experience a fitness value indicating how well it performed in the current situation. Then the agent chooses the consequents of the rule clusters that have given the highest fitness and keep them in the rule-bases of the controller in order to serve as a starting position of the Adaptive Genetic Algorithms (AGA) search instead of starting from a random point, these consequents will be the parents of the new rule consequents. This action helps to speed up the genetic search as it starts the search from the best-found point in the search space instead of starting from a random point.

The robot then calculates the fitness value of the solution starting from the modified rule base supplied by the *Experience Bank*. If there is an improvement in the performance produced by the modified rule base, then the rules that contributed most must be given increased fitness to boost their actions. If there is no improvement then the rules that contributed most must be punished by reducing their fitness w.r.t to other rules and the solutions that had small contributing actions are examined. The fitness of each rule is supplied by the *Solution Evaluator* and is given by:

$$\mathbf{S}_{\mathbf{rat}\,1} = \text{Constant} + (d_{new} - d_{old}) \,.\, S_{ac1} \,.\, (1-F) \qquad (8)$$

Where d_{new} is the new performance of the robot (using the modified rule base, d_{old} is the old performance of the robot before modifying the rule base, d_{new}-d_{old} is the performance improvement or degradation caused by the adjusted rule-base produced by the algorithm. F is the normalised oscillation in the system actions. Thus S_{rat1} is maximised by improving the robot performance before behaviour failure with minimum oscillations in the system output. In the first population of the GA, because there is no performance assessed yet, we blame the rules that have contributed most to the behaviour failure. The fitness of each rule is given by:

$$\mathbf{S}_{\mathbf{rat}} = \text{Constant} - S_{ac1} \qquad (9)$$

The performance of the robot is assessed according to the failure of the robot to reach or maintain a certain target. For example in case the robot approaches an obstacle to a dangerous distance then the performance of the robot will be assessed by how it increases the distance moved before collision.

However, a problem occurs as the system begins accumulating experience that exceeds the physical memory limits. This implies that we must delete some of the stored information. To deal with this, for every rule cluster we attach a *difficulty counter* to count the number of iterations taken by the agent to find a solution to a given situation, we also attach a *frequency counter* to count how frequently this rule have been retrieved. The *degree of importance* of each rule cluster is calculated by the *Experience Survival Valuer* based on the product of the *frequency counter* and the *difficulty counter*. This attempts to keep the rules that required a lot of effort to learn (due to the difficulty of the situation) and also the rules that are used frequently. When there is no more room in the *Experience Bank*, the rule cluster that had the least *degree of importance* is selected for replacement. If two rule clusters share the same importance degree, tie-breaking is resolved by a least-recently-used strategy. The rule that has not been used for the longest period of time is replaced. Thus an *age parameter* is also needed for each rule cluster. The value of the age parameter increases over time, but is initialised whenever the associated cluster is accessed. The limit for the memory clusters is set to 2000 rule clusters, so if we exceed this limit the *Experience Survival Valuer* begins using the *degree of importance* and the age operator to optimise the memory.

The AGA is using the Srinivas method [12]. If, for example in the robot domain, rule number 5 of the obstacle avoidance and rule 7 of the left wall following are chosen for reproduction by roulette wheel selection due their high fitness. This implies they have either contributed more with their actions to the final action that caused improvement, or contributed less with their actions to final action that caused degradation. Then we apply adaptive crossover and mutation operators to both chromosomes. The resultant offspring will be used to replace the consequent of rules 1 and 2 for obstacle avoidance that were largely blamed for the behaviour failure. The AGA is constrained to produce offspring according to the *Contextual Constraints* supplied by the input sensors. The robot ends the adaptation when the agent achieves the desired response.

5 Incremental Synchronous Learning for Non-intrusive Learning

The Incremental Synchronous Learning (ISL) architecture is shown in Fig. 6. The BA need to address somewhat different learning needs, short "initialisation" and long "life-long" learning. In general a learning mechanism within a building should be non-intrusive.

The ISL forms the learning engine for the BA. The agent is an augmented behaviour based architecture, which uses a set of parallel Fuzzy Logic Controllers (FLC), each forming a behaviour. The behaviours can be fixed, in the case of an Intelligent Building (IB) these could be - the Safety, Emergency and Economy behaviours or dynamic and adaptable such as, in an IB, Comfort behaviours (i.e. behaviours adapted according to the occupant's actual behaviour).

Each dynamic FLC has one parameter that can be modified which is the *Rule Base* (RB) for each behaviour. Also, at the high level the co-ordination parameters can be learnt [7]. The ISL system aims to provide life-long learning and adapts by adding modifying or deleting rules. It is also memory based in that it has a memory enabling the system to use its previous experiences (held as rules) to narrow down the search space and speed up learning.

The ISL works as follows: - when a new user enters the room he is identified by a key button and the ISL enters a Monitoring initialisation mode where it learns the user's preferences during a non-intrusive cycle. In the Experimental set-up we used a period of 30 minutes but in reality this is linked to how quickly and how complete we want the initial rule base. For example in a care house we want this rule base to be as complete as possible with some fine tuning, in a hotel we want this initialisation period to be small to allow fast learning. The rules and preferences learnt during the Monitoring mode form the basis of the user rules which are retrieved whenever the user renters the room. During this time the system monitors the inputs and users action and tries to infer rules from the user monitored actions. The user will usually act when given an input vector the output vector is unsatisfactory to him. Learning is based on negative reinforcement, as the user will usually request a change to the environment when he is dissatisfied with it.

After the Monitoring initialisation period the ISL enters a Control mode in which it uses the rules learnt during the Monitoring mode to guide its control of the rooms effectors. Whenever a user behaviour changes, so he needs to modify, add or delete any of the rules in the rule base the ISL goes back to the non-intrusive cycle and tries to infer the rule base change to determine the users preferences in relations to the specific components of the rule that has failed. This is a very short cycle that the user is essentially unaware of and is distributed through the lifetime of the use of environment, thus forming a life-long learning phase.

As in the case of classifier systems the worst m classifiers are replaced by the m new classifiers [11]. In our case we will change all the consequences of the rules whose consequences were unsatisfactory to the user. We find these rules by finding all the rules firing at this situation where $\Pi \alpha Ai > 0$. We replace these rules consequents by the fuzzy set that has the highest membership of the output membership function. We have done this replacement to achieve the non-intrusive learning and to avoid direct interaction with the user. The learnt consequent fuzzy rule set is guided by the *Contextual prompter*, which uses the sensory input to guide the learning. The normalised contribution of each rule p output (Y_{pN}) to the total output Y_{ht} can be denoted by S_{ra11} and is given by Eq. 7.

During the non-intrusive Monitoring and life-long learning phases the agent is introduced to different situations, such as having different temperature and lighting levels inside and outside the room with the agent guided by the occupant's desires as it attempts to discover the rules needed in each situation. The learning system consists of learning different episodes; in each situation only small number of rules will be fired. The model to be learnt is small and so is the search space. The accent on local models implies the possibility of learning by focusing at each step on a small part of the search space only, thus reducing interaction among partial solutions. The interaction among local models, due to the intersection of neighbouring fuzzy sets means local learning reflects on global performance [10]. So we can have global results coming from the combination of local models, and smooth transition between close models. It is necessary to point to a significant difference in our method of classifying or managing rules, which is, rather than seeking to extract generalised rules we are trying to define particularised rules.

After the initial Monitoring initialisation phase the system then tries to match the user derived rules to similar rules stored in the *Experience Bank* that were learnt from other occupiers. The system will choose the rule base that is most similar to the user-monitored actions. The system by doing this is trying to predict the rules that were not fired in the initialisation session thus minimising the learning time as the search is starting from the closest rule base rather than starting from random. Also this action will be satisfactory for the user as the system starts from a similar rule-base then fine-tuning the rules.

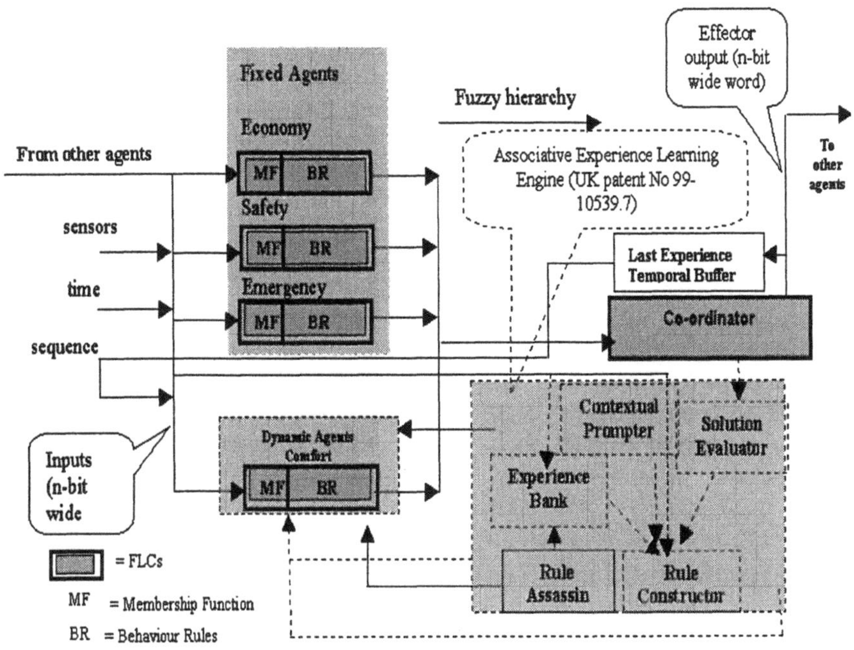

Fig. 6. The ISL Embedded-Agent Architecture

After this the BA will be operating with the rules learnt during the monitoring session plus rules that are dealing with uncovered situations during the monitoring process which are ported from the rule base of the most similar user, all these rules are constructed by the *Rule Constructor.* The system then operates with this rule-base until the occupant's behaviour indicates that his needs have altered which is flagged by the *Solution Evaluator* (i.e. the agent is event-driven). The system can then add, modify or delete rules to satisfy the occupant by re-entering the briefly to Monitoring mode. In this case again the system finds the firing rules and changes their consequence to the desired actions by the users. We also employ a mechanism - *learning inertia* - that only admits rules to the rule base when their use has exceeded some minimal frequency (we have used 3). One of our axioms is that "the user is king" by which we mean that an agent always executes the users instruction. In the case where commands are inconsistent with learned experience learning inertia acts as a filter that only allows the rule-based to be altered when the new command is demonstrated by its frequent use to be a consistent intention. It is in this way that the system implements a life long learning strategy. It is worth noting that the system can be Monitoring for long times to learn as much needed rules which is highly needed in care houses. Also the system can start with short Monitoring time and then jump to the life long learning mode to learn the user behaviour quickly, which is needed in hotel rooms. The weighting is set by adjusting the Monitoring non-intrusive period to the required application [14].

The emphasis on particularisation over generalisation can create problems when the particularised rules storage needs of different users exceed the physical memory limits. This implies that we must remove some stored information. To manage this we employ various mechanisms. One such mechanism is for every individual rule base cluster we have, we attach a *difficulty counter* to count the time taken by the agent to learn a particular rule base. We also attach a *frequency counter* to count how often this rule base has been retrieved. The *degree of importance* of each rule base cluster is calculated by the *Rule assassin* and is given by the product of the *frequency counter* and the *difficulty counter*. This approach tries to keep the rules that have required a lot of effort to learn (due to the difficulty of the situation) and also the rules that are used frequently. When there is no more room in the *Experience Bank*, the rule base cluster that had the least *degree of importance* is selected for removal by the *Rule assassin*. If two rule base clusters share the same importance degree, tie-breaking is resolved by a least-recently-used strategy; derived from a "life invocation" flag that is updated each time a rule is activated.

6 Experiments and Results

This section describes the experiments and results of applying the AEE for online learning and adaptation in mobile RA and the application of ISL to BA and how the Building and Robotic agents communicate and cooperate to do the specified jobs.

All the robots will have the same sensors, there will be three front sonar sensors for obstacle avoidance and two sonar sensors on each side of the robot for edge following, each sensor is represented by three membership (which are Near, Medium, Far) as this was found to be the least number of membership function to give a satisfactory result. Also the robots have sensors to measure the bearing from the goal. Each bearing sensor is represented by seven memberships as this was found to be the least number of membership functions to give a satisfactory result. The output membership functions will be represented by four membership functions (which are Very Low, Low, Medium and High).

The following experiments shows the RA using the AEE to learn their basic behaviours which are goal seeking, edge following and obstacle avoidance. We had used different sizes of robots to show that our techniques are robot independent and also we performed some experiments in outdoor environments to test that the AEE can deal with highly difficult and unstructured environments.

Fig. 7a shows an indoor robot learning the rule base of the goal seeking behaviour, using an imprecise infrared beacon system to emulate a goal. The robot learnt the rule-base in an average of 96 seconds (this time is not the actual time to find a solution but it is bounded by the robot microprocessor speed and the robot speed) over 5 trials using a 68020 microprocessor robot with maximum speed of 13cm/sec starting from different positions with different initialised rule bases.

Fig. 7b shows the indoor robot learning the rule-base for the right edge following behaviour whilst dealing with an irregular edge and a highly imprecise ultra sound sensor. The robot succeeds in learning the desired behaviour in average of 96 seconds.

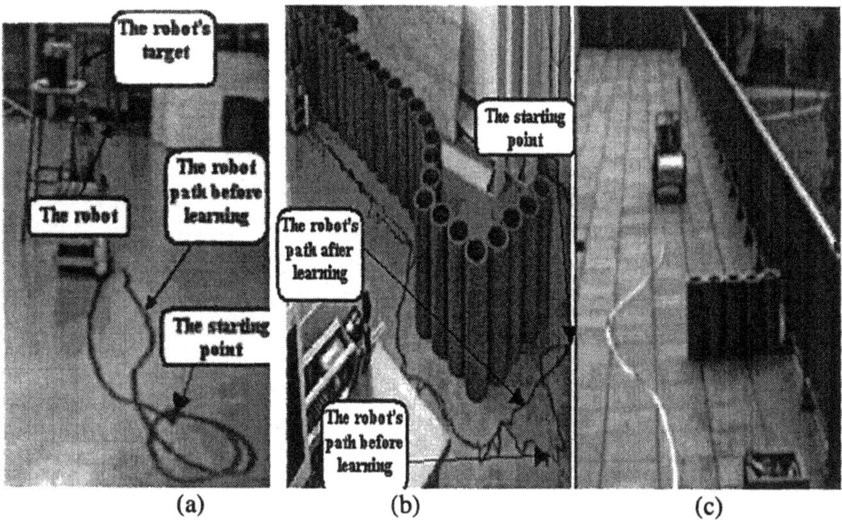

Fig. 7. a) Indoor robot learning goal seeking behaviour. b) The outdoor robot learning the edge following behaviour. c) The outdoor robot learning the obstacle avoidance behaviour.

Table 1 shows the learnt rule base that adapts the robot to the edge irregularity while taking into account the robots kinematics and dynamics. The rule-base was initialised randomly with the first 4 rules actions assigned to move forward. The actions of the fifth and the sixth and the seventh rules were assigned to turn left. And the actions of the remaining rules were assigned to turn right. Some of the rules learnt in the rule base shown in Table 1 look contradictory to human reasoning but they fit the robot. For example in the rule stating *if* Right Front Sonar sensor *(RFS) is Near and the Right Back Sonar sensor (RBS) is Near,* a human consequent would be to turn left to avoid collision with the fence but the learning system had chosen to turn right towards the fence. The reason for this is that the robots dynamics are biased to the left and when the robot is activating rules 2 and 3, the robot turns sharply to the left and the robot takes time in order to stabilise again, thus causing a zigzag path. The first rule consequent had resulted in damping the combined actions composed of these rules and thus resulting in a smooth path with minimum deviation rather than a zigzag path. Also the seventh rule states if *RFS is Far and RBS is Near* which is a corner situation, a human decision would be to turn right to follow the fence and turn around this corner. The learnt rule suggests going to the left away from the fence to give the robot more space until the robot goes to the next rule which is *RSF is Far and RSB Med* then the robot turns around the corner smoothly as it has enough space to perform the turning. This demonstrates the importance of learning online using real agents rather than using simulations as there is a lot of things that cannot be realised till we try the real agent in its environment.

Table 1. The best learnt rule base of the learnt right edge following behaviour using indoor robots.

Rule No	RFS	RBS	Left Speed	Right Speed
1	NEAR	NEAR	MED	LOW
2	NEAR	MED	VERY LOW	HIGH
3	NEAR	FAR	LOW	MED
4	MED	NEAR	HIGH	LOW
5	MED	MED	MED	MED
6	MED	FAR	LOW	HIGH
7	FAR	NEAR	LOW	MED
8	FAR	MED	HIGH	LOW
9	FAR	FAR	HIGH	VERY LOW

Fig. 7c shows the outdoor robot learning the obstacle avoidance behaviour while following an irregular metallic fence full of bumps in an outdoor environment at a desired distance of 120 cm whilst avoiding obstacles at a distance of 1m. The robot converged to a solution after an average of 6 iterations taking 3 minutes of robot time. After learning the rule bases the robot then learns the best membership functions that suits the learnt rules to form a sub optimal behaviour. The robot then learns the best coordination parameters to satisfy the high level mission to form the HFLC. More information about learning the MF and the coordination parameters online can be found in [6,7]

What were learnt are the HFLC, which were sub-optimal under certain environmental and robot kinematics conditions. If the environmental or the robot kinematics changed dramatically these parameters would not be sub-optimal and the learning cycle would have to be repeated from the beginning. Of course this is not practical in highly and unstructured dynamic environments such as the domestic environments where the environment is continuously changing. In our case we apply our life long learning and online adaptation strategy. The online adaptation technique will discover the rules in different behaviours that, if modified online and in-situ, without the need to repeat the whole learning cycle can lead the robot to adapt to its environment [3].

In Fig. 8a the indoor robot was given a high level mission which is to align to center of the corridor while avoiding any obstacles and going to its goal after exiting the corridor, this high level mission can be supplied by the BA. In this figure we had destroyed the rules in all behaviours by setting all the rule consequences to “go right”, thus causing the robot to collide with walls in the corridor, we also tried loosening the left wheel to simulate changing robot conditions. Also the rules in the goal seeking behaviour were destroyed to make the robot always shift to the right. These changes were done to simulate changes in the environmental and robot kinematics conditions and to test if the life long learning strategy within the AEE can adapt the robot online with no need to repeat the learning phase.

Table 2. The modified rules for the experiment in Fig. 8-a.

	LFS	**MFS**	**RFS**	**Left Speed**	**Right Speed**
Obstacle Avoidance	Near	Far	Near	Med	Med
	Med	Near	Near	Very Low	High
	Med	Med	Near	Low	High
	Med	High	Near	Low	Med
	Far	Far	Near	Very Low	High
	High	Med	Near	Low	High
	High	High	Near	Very Low	Med
	LSF	**LSB**	**Left Speed**	**Right Speed**	
Left Edge Following	Med	Near	Low	Med	
	Far	Near	Very Low	High	
	High	Med	Med	High	
	High	High	Med	High	
	RSF	**RSB**	**Left Speed**	**Right Speed**	
Right Edge Following	Near	Near	Very Low	High	
	Low	Med	Low	High	
	Low	High	Very Low	High	
	Bearing		**Left Speed**	**Right Speed**	
Goal Seeking	Very Very Negative		Very Low	High	
	Very Negative		Low	High	
	Med Negative		Med	High	

It took the robot 8 minutes (including reversing time) to get out of the corridor and to modify the rule bases of the co-ordinated behaviours. It modified 7 rules in the obstacle avoidance behaviour, 4 rules in the left wall following behaviour, 3 rules in the right wall following behaviour and 3 rules in the goal seeking behaviour (i.e. 17 rules). It learnt these rules in an average of 20 iterations (episodes) over 4 experiments. The modified rules are shown in Table 2.

In order to test the life long learning strategy in severe conditions we had used the outdoor robots, which navigates in outdoor environments under changing conditions. Fig. 8b shows the outdoor robot after learning the sub optimal HFLC co-ordinating the obstacle avoidance behaviour and the goal seeking behaviour. The robot was required to go up hill to go to target 1 and then go down hill to target 2 and then it had to repeat the path back up hill to target 1, the operation was repeated continuously for 2 hours. This operation is similar to the missions described before in care environment in which the BA gives a high level mission of going to target 1 and target 2 while avoiding obstacles. While going uphill the online adaptation module had to change the goal seeking behaviour speed consequents to High to give the robot more power to go up this steep hill, as the robot was trained on a nearly flat ground. When going down hill the goal seeking consequents had to be changed again to Low. Although the robots are equipped with wheel encoders and a speed controller however there is a need for a high level set point from the goal seeking behaviour to give more power to the robot up hill and to protect it from slipping when going downhill. When the robots goes uphill again it doesn't need to relearn the rule base, thanks to the Experience Bank which allows the robots to recall its past experience rather than relearning. The robot path after two hours is nearly the same with an average deviation from the original path of 2.5 cm and standard deviation of 1.2.

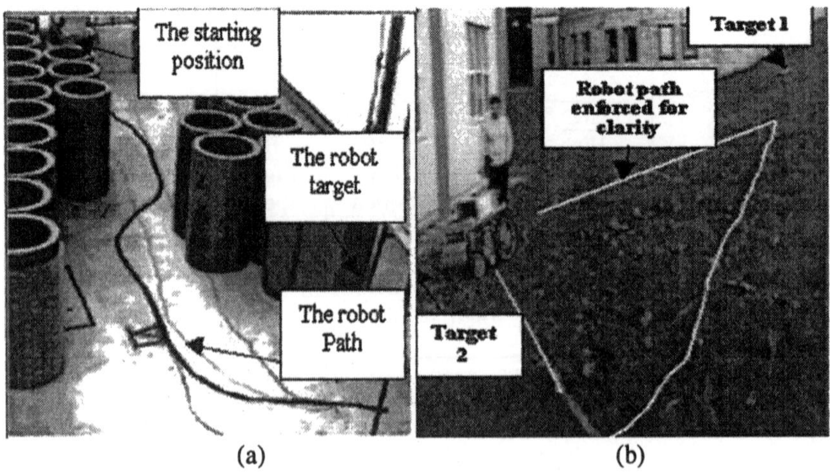

Fig. 8. The indoor robot using life long learning to adapt itself. b) The outdoor robot implementing continuous learning.

We have conducted a number of experiments using the ISL to learn and adapt the behaviours of different users with different tastes and desires. For each experiment the users have spent up to 5 hours in the iDorm.

For the experiments the room had four environmental parameters to control which are the lighting level (I1) represented by 3 triangular fuzzy sets (Low

Medium High), the temperature (I2) represented by 3 triangular fuzzy sets (Low Medium High), the outside level of lighting (I3) represented by three triangular fuzzy sets (Bright Dim Dark) and the user location represented by two fuzzy sets (I4) wither he is on the Desk or lying on the Bed. There were seven outputs to control which are two dimmable spot lights above the desk (O1) represented by five triangular fuzzy sets (Vlow, Low, Medium, High, Vhigh), two dimmable lights above the Bed (O2) again represented by five triangular fuzzy sets, a Bedside Lamp (O3) represented by ON-OFF Fuzzy states, a Desk Lamp (O4) represented by ON-OFF Fuzzy states, automatic blinds whose opening can be controlled (O5) represented by 5 triangular fuzzy sets, where the fuzzy sets Vlow, Low, dealing with blind opening to the left and Vhigh, High deal with blind opening to the right and Medium (Med) is 50 % opening, a Fan Heater (O6) represented by ON-OFF Fuzzy states and a Fan Cooler (O7) represented by ON-OFF Fuzzy states.

This forms a rule base of 3*3*3*2 =54 rules. However as the ISL learns only the rules required by the user we find that we don't need to learn the whole rule base, thus the rule base will be optimised.

Table 3. The learnt Rule Base for "User-1"

I1	I2	I3	I4	O1	O2	O3	O4	O5	O6	O7
XX	Med	Bright	Desk	Vhigh	Vhigh	OFF	OFF	Med	OFF	OFF
XX	High	Bright	Desk	Vhigh	Vhigh	OFF	OFF	Med	OFF	OFF
High	Low	Dim	Desk	Vhigh	Vhigh	OFF	OFF	Med	OFF	OFF
Med	Med	Bright	Bed	Vhigh	Vlow	ON	OFF	Vlow	OFF	OFF
Med	High	Bright	Bed	Vhigh	Vlow	ON	OFF	Vlow	OFF	OFF
High	Med	Bright	Bed	Vhigh	Vlow	ON	OFF	Vlow	OFF	OFF
High	High	Bright	Bed	Vhigh	Vlow	ON	OFF	Vlow	OFF	OFF
Med	Med	Dark	Desk	Med	Med	ON	ON	Med	OFF	OFF
Med	High	Dark	Desk	Med	Med	ON	ON	Med	OFF	OFF
High	Med	Dark	Desk	Vhigh	Vhigh	ON	OFF	Vhigh	ON	OFF
High	High	Dark	Desk	Vhigh	Vhigh	ON	OFF	Vhigh	ON	OFF
Low	Med	Dark	Bed	Vlow	Vlow	OFF	OFF	VLow	OFF	OFF
Low	High	Dark	Bed	Vlow	Vlow	OFF	OFF	VLow	OFF	OFF
Low	High	Dark	Desk	Med	Med	ON	ON	Med	OFF	OFF
Low	High	Dark	Desk	Med	Med	ON	ON	Med	OFF	OFF

Table 3 shows the learnt rule-base for "User-1" who occupied the room for more than two hours. In these experiments the users undertook many behaviours such as studying during the day (in which the lighting was bright) and studying in the evening (i.e. when external light was fading). This specific user preferred to use all the ceiling lights ON and the blind open to Medium (Med). Another behaviour was lying in bed reading with the blind adjusted to his convenience and the bedside light being sometimes used. He would also close the blinds in the evening using combinations of the ceiling lamps and desk lamp. The data also contains "going to sleep behaviours" including reading before sleeping and getting up spontaneously at night to work (they are students!).

The ISL learnt 15 rules of which the first 7 rules were learnt during the Initialisation phase. The next 4 rules were ported from similar users and were satisfactory to the user. The last 4 rules resulted from fine-tuning the ported rules, these rules dealt with darkness where the first two rules dealt with the user wanting to sleep and he wanted all lights off while the similar user slept with the desk lamp ON because he doesn't like darkness. The last two rules dealt with user returning to the desk to read as he couldn't sleep he switches all lights to Medium (Med) and switches ON the desk and the bedside lamp and the blind to Medium to allow more light, the similar user had the same behaviour but he was closing the blind. It is obvious that the room user and the similar user actions are very similar and we needed only a fine-tuning to satisfy the current user needs, this is an advantage of using the Experience Bank which reduces the life long learning time and satisfies the user. The XX in Table 3 indicate a No-Care situation, which resulted as the inside room lighting level was not important as when it was bright the user took always the same actions. This shows also that ISL besides optimising the rule base can help to identify the input parameters required by the user and hence giving more optimisation to the system.

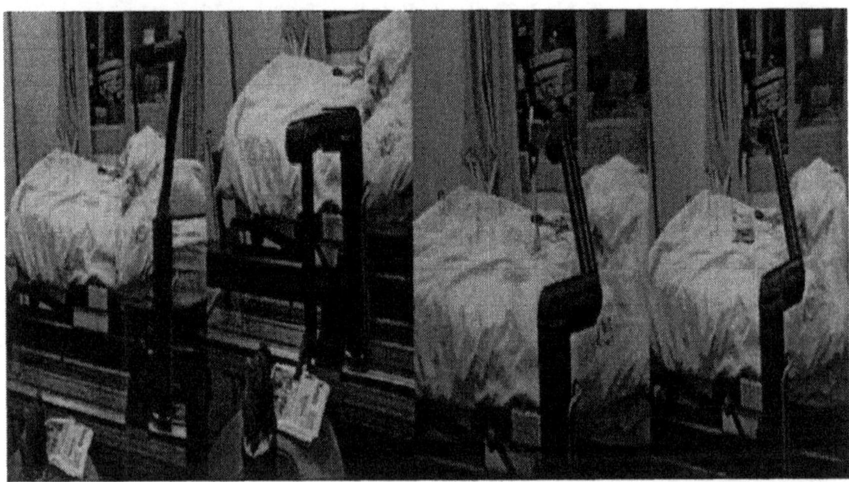

Fig. 9. The Building agent communicates with the Robotic agents to get the newspaper to the bed.

Fig. 9 shows the BA communicating with the RA agents in the form of a Manus arm to fetch a newspaper from the table and deliver it to the user in the bed. These experiments were conducted in cooperation with KAIST and in the Human-friendly Welfare Robotic system lab in Daejeon-Korea. The RA only receives a high level mission of getting the newspaper which is a high level output from the ISL.

7 Conclusions

In this chapter we have presented our learning and adaptation techniques applied in intelligent domestic environments. The system operated in both an interactive mode for Robotic agents and in a non-intrusive mode for Building agents. We have also illustrated how to form a heterogeneous multi embedded agent system in which the various member agents cooperate and communicate to satisfy the user desires and objectives. The intelligent and autonomous system we have described learns behaviours and adapts online and through interaction with the environment, including its occupants. We have also argued that embedded-intelligence can bring significant cost and effort savings over the evolving lifetime of product by avoiding expensive programming (and re-programming).

For our current and future work we have plans to conduct more and longer experiments with the iDorm (up to a year to get a full climate cycle), significantly expand the sensor-effector set and explore more fine-grained and course grained distributed embedded-agents (e.g. with agents within gadgets, or communicating rooms). We will try also to deal with dynamically varying I/O size. We are also currently conducting experiments which link agents in physically separated intelligent spaces together (e.g. UK to Korea) which are allowing us to explore the issues in virtual adjacency across time and culture zones.

Acknowledgement

We are pleased to acknowledge the funding support from the EU IST Disappearing Computer program (eGadgets) and the joint UK-Korean Scientific fund (cAgents), which has enabled much of this research to be undertaken. We especially acknowledge the work and effort from our Korean partners from KAIST who areco-operating with us in the Robotic and Building agent communication. In particular we would like to acknowledge the highly valuable advise from Professor Zenn Bien of KAIST and his team which includes Mr. Kim and Mr. Lee and Mr. Myung.

We are pleased to acknowledge the contribution of Malcolm Lear, Robin Dowling and Arran Holmes for their help building the intelligent dormitory and intelligent-artifacts. We would also like to thank, Anthony Pounds-Cornish, Sue Sharples, Gillian Kearney, Filiz Cayci, Adam King, and Fiayaz Doctor for their indirect contributions arising from many stimulating discussions on intelligent-artifact and embedded-agent issues.

References

1. Callaghan V, Clarke G, Colley M, Hagras H (2001) Embedding Intelligence: Research Issues for Ubiquitous Computing. Proceedings of the Ubiquitous Computing in Domestic Environments Conference, pp 13-14.
2. Colley M, Clarke G, Hagras H, Callaghan V (2001) Integrated Intelligent Environments: Cooperative Robotics & Buildings. Proceedings of the 32nd International Symposium on Robotics and Automation, pp 745-750.
3. Hagras H, Callaghan V, Colley M (2001a) Prototyping Design and Learning in Outdoor Mobile Robots operating in unstructured outdoor environments. IEEE International Robotics and Automation Magazine, 8: 53-69.
4. Saffiotti A (1997) Fuzzy Logic in Autonomous Robotics: Behaviour Co-ordination. Proceedings of the 6th IEEE International Conference on Fuzzy Systems, pp 573-578.
5. Tunstel E, Lippincott T, Jamshidi M (1997) Behaviour Hierarchy for Autonomous Mobile Robots: Fuzzy Behaviour Modulation and Evolution. International Journal of Intelligent Automation and soft computing, 3: 37-49.
6. Hagras H, Callaghan V, Colley M (2000a) On Line Calibration of the sensors Fuzzy Membership Functions in Autonomous Mobile Robots" Proceedings of the 2000 IEEE International Conference on Robotics and Automation, pp 3233-3238.
7. Hagras H, Callaghan V, Colley M (2000b) Learning Fuzzy Behaviour Co-ordination for Autonomous Multi-Agents Online using Genetic Algorithms & Real-Time Interaction with the Environment. Proceedings of the 2000 IEEE International Conference on Fuzzy Systems, pp 853-859.
8. Rocha M, Cortez P, Neves J (2000). The Relationship between Learning and Evolution in Static and Dynamic Environments. Proceedings of the Second ICSC Symposium on Engineering of Intelligent Systems, Paisley-UK, pp 500-506.
9. Nehmzow U (2000) Continuos Operation and perpetual Learning in Mobile Robots. Proceedings of the International Workshop on Recent Advances in Mobile Robots, Leicester, UK.
10. Bonarini A (1999) Comparing Reinforcement Learning Algorithms Applied to Crisp and Fuzzy Learning Classifier systems. Proceedings of the Genetic and Evolutionary Computation Conference, pp 52-60.
11. Dorigo M, Colombetti M (1995) Robot Shaping: Developing Autonomous agents through learning. Artificial Intelligence Journal, 71: 321-370.
12. Kasabov N, Watts M (1999) Neuro-Genetic Information Processing for Optimisation and Adaptation in Intelligent Systems. In: Neuro-Fuzzy Techniques for Intelligent Information Processing, N.Kasabov and R. Kozma, Eds. Heidelberg Physica Verlag, pp 97-110.
13. Srinivas M, Patnaik L (1996) Adaptation in Genetic Algorithms", Genetic Algorithms for pattern recognition. Eds, S.Pal, P.Wang, CRC press, pp. 45-64.
14. Hagras H, Callaghan V, Colley M, Clarke G (2000d) A Hierarchical Fuzzy Genetic Multi-Agent Architecture for Intelligent Buildings Sensing and Control. Proceedings of the International Conference on Recent Advances in Soft Computing 2000, pp 199-206.

Integration of Soft Computing Towards Autonomous Legged Robots

Anthony Wong and Marcelo H. Ang Jr.

Department of Mechanical and Production Engineering,
National University of Singapore,
Singapore 119260
{dtiwonga,mpeangh}@nus.edu.sg

Abstract. Mobile robots are extensively used in various terrains to handle situations inaccessible to man. Legged robots in particular are tasked to move in uneven terrain. Hence the primary problem of these robots is locomotion in an autonomous fashion. Autonomy is important as the tasks are plagued with many uncertainties. Some of these tasks include leg movement and coordination, navigation, localisation, and stability during movement, all operating in dynamic and unexplored territories.

Classical control and traditional programming methods provide stable and simple solutions in a known environment. The environment that legged robots work in is dynamic and unstructured, and such control methods are not always able to cope with. It is difficult to model the environment to provide the controller with the relevant data and program actions for all possible situations. Hence controllers with abilities to learn and to adapt are needed to solve this problem. Soft computing provides an attractive avenue to deal with these situations.

Soft computing methods are based on biological systems and they provide the following features: generalisation, adaptation and learning. As more is realised about the use and properties of soft computing methods, the development of controller is shifting towards using soft computing. They have properties that can be used to improve the stability, adaptability, and generalisation of the controllers. Some of the more popular methods used are fuzzy logic, artificial neural networks, reinforcement learning and genetic algorithms. They are commonly integrated with classical methods to enhance the features of classical controllers and vice versa. Each soft computing method serves a different purpose, with its advantages and disadvantages, and the methods are often used together to complement each other.

This chapter provides a survey on the different uses of soft computing methods in the different aspects of legged robotics. We see how soft computing methods and classical techniques compliment each other. Two areas of legged robotics are dealt with - control architecture and the problem of navigation. The Central Pattern Generator (CPG) controller and the behaviour-based controller are two architectures presented in this chapter. Various soft computing techniques are used to implement and improve these two controllers.

1 Introduction

Hard computing methods refer to traditional methods of programmed computing where the program provides all necessary actions for all possible conditions. They are used to control machines and they work well in a relatively controlled environment. They are preferred over soft computing methods as they are reliable and predictable in the task performed within the controlled environment. Once unpredicted changes occur to the environment, hard computing controllers lack the ability to adapt and this may lead to failure of the system. Soft computing is a relative new tool and is used to solve this problem. Soft computing systems are designed to mimic the performance of biological systems. It gives us new methods to better and optimize the control of machines both in industry and research.

Hard computing methods use established methods of mathematics, science and computing to solve engineering problems [1]. It consists of the following methods:

- binary logic,
- boolean logic
- analytical models,
- deterministic searches,
- crisp classifications, and
- numerical analysis.

Particularly with robot controllers, hard computing methods are characterised by rigidity and accurate tolerances. This means the controller performs deterministically pre-programmed moves which require some human intervention to work. It has the advantage of being precise and clear in the categorization of the system. The complexity of the mathematical description increases as the complexity of the control system increases. There are two choices to solve complex control problems. If an accurate model is preferred, it is difficult to identify the system and measure the parameters. Moreover the controller will be hard to design. On the other hand if a simple model is used, the identification and parameterisation is simple, and the controller is easy to implement. The problem with the simple model is that the results can be far from the specifications required. Soft computing tries to find a compromise. There are two situations that soft computing methods are implemented, i.e., if hard computing methods cannot be used to solve the problem or hard computing methods require improvement.

The following are architectures used to modularize the different components of the controllers. They are commonly fused with various techniques, including soft computing methods, to provide solutions. They are:

- schema-based methods [2],
- multi-agent methods [3],
- hierarchical methods,

and other forms of divide and conquer paradigms are used to divide the main task into manageable subtasks. These subtasks/modules are easier to solve and various techniques can be used in these subtasks to improve the system. Hence, they can be implemented either as hard or soft computing methods.

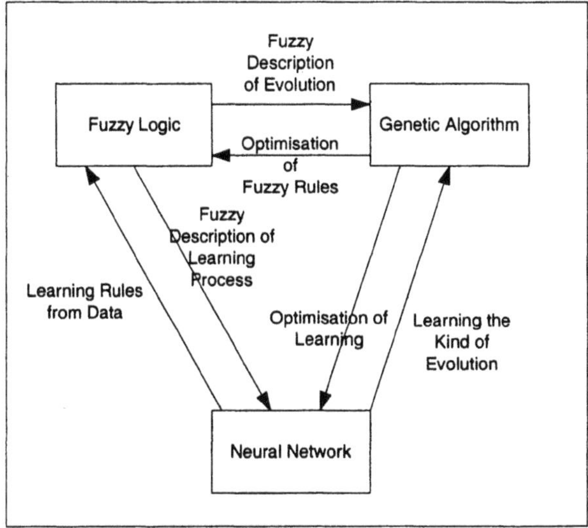

Fig. 1. Soft Computing Connections

Soft computing is relatively young in terms of control and hence there is still much to learn and improve on [1]. One of the advantages of soft computing is that it incorporates real life data to produce solutions that consider the environment when it works. To have a clearer picture of the various soft computing methods, the following is a list of the four common methods:

- fuzzy logic,
- artificial neural networks (ANNs),
- reinforcement learning, and
- genetic algorithms (GAs).

These methods differ in their use and each has its own advantages and disadvantages. Only fuzzy logic, artificial neural networks and generic algorithms are covered in this chapter as they are the more commonly used soft computing techniques in robotics. Also, implementation of combination of these various soft computing methods is popular as they capture the advantage of the different techniques and remove the disadvantages (Figure 1). Fuzzy logic allows the translation of human knowledge into machine representable and understandable form. Neural networks provide learning capabilities from data and can be used to learn behaviors and actions. Genetic algorithms provide optimisation algorithms applicable to hard problems with large search spaces [16]. They can be used to further optimise parameters.

2 Legged Robotics

There are two main types of mobile robots, namely wheeled robots and legged robots. This chapter deals with the difficulties of control of legged robots. Although this chapter deals only with legged robots, many tasks such as navigation, localisation and control architectures, are common to both legged and wheeled robots, and are therefore discussed.

The fundamental issue of legged robotics is the age old task of walking. Walking requires robots to have many actuators to execute their task of locomotion as each leg is usually manipulated by a few actuators. Also, legged robot locomotion requires information about the environment to locate itself, sense the ground for foot placement, and sense the surroundings for obstacles and danger. Therefore sensors play an important role in performing these tasks. An important aspect of mobile robots is sensor data interpretation. The robot uses sensors to "picture" the environment to make decisions. Sensors feed the system with quantitative values that must be interpreted by the controller for it to understand the situation or environment. The relation between sensory feedback and actuation is very complicated and require non-standard methods to comprehend and control, respectively. Soft computing methods, like fuzzy logic, are commonly employed to deal with the vagueness of the sensory data [5].

Legged robot control is a daunting task, hence it is ideal to spilt the task into various components. It usually consist of the overall control, the locomotion control, and other subsystems that aid it in its task as shown in Figure 2. Therefore scalability and modularity of the controller is important to the system to simplify the task of integrating the various components of the controller. Divide and conquer methods are applied and variation of these methods include the behaviour-based architecture [6].

2.1 Autonomy Requirements in Mobile Robotics

Autonomy refers to the robot's ability to perform a task robustly without human intervention, with the ability to recover when the system fails. At first glance, it might seem that intelligent control is not required for an action like walking since it is a cyclic action which can be pre-programmed into the controller. Walking translates to periodically lifting the leg, moving it forward and finally lowering it. Should soft computing techniques be used for such simple actions? This would work if the robot moves only on a flat piece of ground. But this is not practical, as legged robots are built to move in undulating unknown terrain. The ground might be soft, slippery and impassable at times, thus requiring the robot to have intelligence to deal with these situations.

The locomotion controller provides the robot with the ability to move from point to point. A user must be able to set goals of the robot, after which it is to act on its own without human supervision. Hence other functions have to

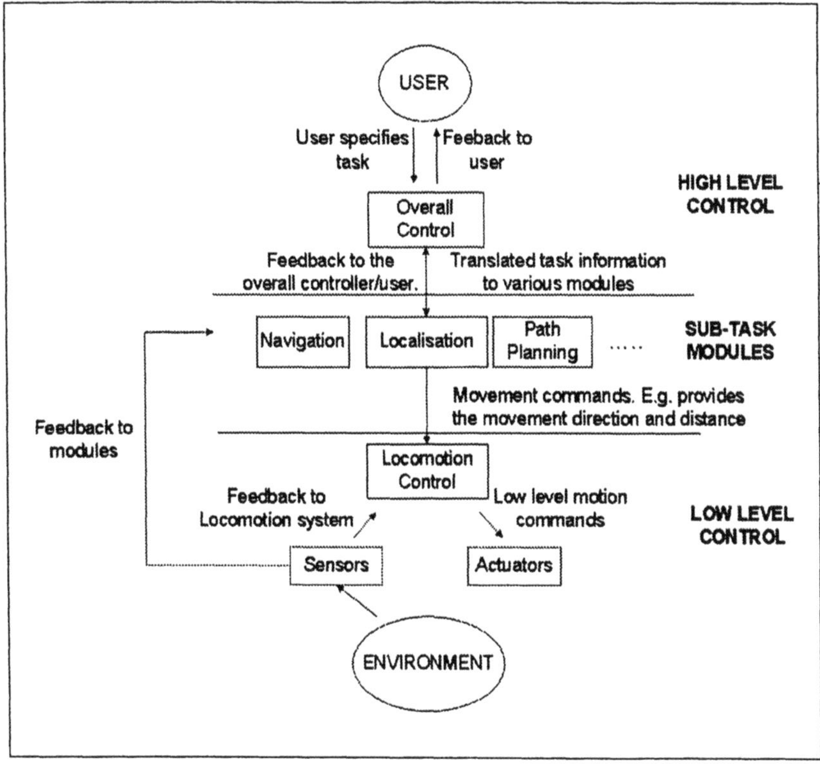

Fig. 2. Legged robot control structure

be considered to achieve autonomy. Legged locomotion is a task plagued with many uncertainties as it is dependent on the terrain and the way the robot interacts with the environment. If the robot is to move around an unknown environment, it should possess the following properties:

- coordination of joint movements in spite of perturbations,
- obstacle avoidance,
- navigation and localization,
- recovery from mistakes or mishaps,
- adaptation to the changing environment fast,
- avoidance of impassable routes,
- negotiation across unpredictable irregularities and rough terrain,
- stability, and
- adaptive reflex behavioural control.

These functions cannot be easily implemented with standard classical control methods and until this day, not many solutions can work well to deal with the above problems. Soft computing solutions available as yet are not as established compared with classical control methods and hence most mobile robots are slow, inefficient and unreliable. Due to the complexity of the various functions, however, soft computing techniques are employed as they provide various methods to simplify, optimise, and learn the various tasks to be performed. Soft computing methods are also used to store information about the robot movement and the environment which is manipulated by the controller in decision making. Some of these method will be presented in the following sections.

2.2 Bipedal Walking - Dealing with Dynamic Walking

Research in bipedal robots has a different focus. The research in the various areas mentioned can be used for bipeds but there is another area that is not easily dealt with. It is the area of dynamic walking.

The dream of all biped developers is to build a biped that walks like a human being. Bipedal walking, in the engineering sense, deals with the problem of dynamic walking (as compared to static walking) where stability and control is an issue. Similar to all multi-legged robotic walking, bipedal walking is also a rhythmic action that can vary in the speed of movement and the length of the stride. Biped locomotion control systems is plagued with the problem that the system is a high-order, highly coupled nonlinear dynamics due to the multiple degrees of freedom which is inherent to the structure [36]. Hence soft computing techniques are ideal to solve these problems. Some of the developments of biped research are presented. Although there are many problems associated with biped walking, research in this area is worth the effort as the world is evolving to be a more human-friendly environment (i.e. a place suitable for bipedal creatures and robots).

3 Soft Computing in Legged Robotics

Biological systems have developed over the centuries and the control mechanisms within these systems are optimised to perform the various tasks. Soft computing techniques mimic biological systems to take advantage of their various properties. They model biological systems that possess properties such as learning, generalisation, and adaptation [17]. But until today, soft computing methods are unable to perform as well as biological systems. Each soft computing technique captures the different aspects of biological systems and it is common to fuse various soft computing methods to capture the various advantages of these controllers [16]. Recent research in this area has given rise to many soft computing techniques, which will be presented in the following sections.

Although soft computing techniques have many advantages, they are not exclusively implemented in engineering systems. For a commercial point of view, these techniques are maturing and are being integrated into commercial products (fuzzy logic based systems in household appliances). They usually work together with classical hard computing techniques as there are many areas in control where hard computing methods excel.

Robot control architectures are large as they encompass many functions mentioned in the previous section. Some popular tools to organise large architectures are modular approach (subsumption approach [6]), hierarchical approach [13] and various similar approaches . Modular systems are popular with legged robot systems due to their complexities. Soft computing methods can be implemented as components within these modular systems. As an example, Atienza [14] developed a behaviour-based controller that took control of the robots actions in terms of moving the robot's feet based on the sensory data it obtained. At the lower control level, he implemented a fuzzy controller that dealt with actuator synchronization. Since his controller was based on Brooks' work on subsumption, other subsystems could be added to his work if required.

3.1 Advantages and Disadvantages of Soft and Hard Computing

Hard computing techniques are predictable and stable but are rigid and therefore cannot handle unanticipated changes well in the environment. Hence they are good for linear time invariant problem where changes in control environment are minimal. In other words, once a technique is developed for an environment, it only performs well under those conditions but might not work in other environments. These methods are established methods which have found their uses in various fields.

Different soft computing methods are used individually or together depending on the task. But the advantage of these controllers is that it is able to handle non-linear and time variant problems, improve robustness, and perform in non-ideal situations [4]. On top of that, it is able to improve generalisation and adaptiveness of systems thus enabling systems to be portable. Computational power also increases as some soft computing techniques provide parallel/distributed computing methods.

As with all systems, they are always trade-offs. Soft computing methods can be difficult to configure as there are many parameters and variable to deal with. This is true for both fuzzy logic and artificial neural networks as they are very variable systems that cannot be easily predicted. In terms of training and evolving, training of artificial neural networks and evolving of genetic algorithms can take a long time to attain the desired state. These processes are also computationally demanding as they require iterations and a significant amount of processing. But the advantages outweigh the disadvantages as these tools provide methods with abilities that surpasses that of current methods, capabilities that will be described later.

3.2 Soft Computing Techniques

Fuzzy Logic

Machines can improve their capabilities if they have human-like intelligence. Human knowledge and reasoning are qualitative rather than quantitative in nature, and are not representable nor comprehensible to machines. Fuzzy Logic is a method to translate qualitative human reasoning into machine representable and processable form. Fuzzy logic takes care of the uncertainties and vagueness present in describing a problem [15]. It is useful in solving the uncertainty problem but seldom used alone to solve engineering problems but rather as an entity within a system. It is easy to implement on software and embedded systems and is more cost effective than some traditional controllers for a wide range of applications. Hence it is popularly used to perform interpolation, pattern matching and decision making. These are some of the uses of Fuzzy Logic:

- advice providing expert systems,
- soft constraint propagation, and
- decision making systems

Fuzzy logic based systems are commonly used to deal with imprecision in data within these systems, especially in handling noisy signals like sensory readings of the environment, etc. [16]. Furthermore, it does not take up as much processing power as the other two methods discussed in this chapter.

Since the environment that the robot works in is a uncertain and unknown environment, it is difficult to get an accurate "picture" of the surrounding. That is not a problem for humans. Humans do not have to know the exact size or distance of the obstacle in front of us. We walk forward to clear the obstacle if we think we can clear it, and move around it if we think we cannot. Robots should be able to think in this fashion and fuzzy logic removes the need for exactness in the data that the robot acquires. Examples of how this is implemented in a mobile robotic system are described in the later sections.

Artificial Neural Networks (ANNs)

An ANN is a network structure of layers of neurons connected by weighted links. The weighted links, known as synaptic weights, and the layers of neurons define the network [17]. The values of the different weighted links are obtained through a learning process.

The advantage of ANN is not only that it derives its computational power from a massive parallel processing network (multiple layers of neurons working together), but also its ability to learn and generalize (to adapt to unforeseen situations) [17]. It is still not possible to build an entire engineering system that completely uses ANNs.

ANNs are useful in detecting trends or patterns from complicated or imprecise data, and therefore are used in tasks such as forecasting (recognition of trends), signal processing, control applications and pattern classification [17]. These tasks' operating environment tend to deviate from expected working conditions and therefore the adaptive property of ANN makes it the ideal candidate for the task. Some applications in the field of legged robotics are,

- in visual systems to recognise landmarks,
- to learn a complex control algorithm,
- to store information about location maps [18] or action maps, and
- to learn rhythmic actions (patterns) such as walking [7].

In relation to other soft computing methods, tuning of the parameters of a fuzzy system is a difficult task as there are many parameters to modify. This is an area where other soft computing techniques, such as ANNs can be fused to improve the performance of the fuzzy system. ANNs have been found to be useful especially pertaining to fuzzy controllers that have complex structures. ANNs add learning and generalization capabilities to fuzzy logic systems. The adaptability of neural networks however does not always improve the robustness of the system as the system might adapt to spurious disturbances which is detrimental to the system.

Legged robotics involve tasks that require adaptability, and knowledge storage and update where neural networks can provide. Human memory and experience matter when it comes to walking and moving within non-uniform terrain. The way that a baby adapts as he or she learns to walk is amazing. He or she does not forget the techniques that he or she employed in walking the day before. An ANN is used for this possibly to learn and store a pattern and implement it, with the ability to generalise in any given situation [7,8] . This allows robots to learn task within a controlled environment, and be able to perform as well in an unknown environment.

An ANN is a good tool but there are still areas in control that do not require its complexity to function. ANNs are commonly used in conjunction with hard computing methods when implemented in a robot controller.

Legged robotics is a field where there are multiple joints and sensors in the system. In terms of control, it is difficult to fully specify the required joint actuation and sensor relationships. Soft computing methods allow systems to be specified as long as corresponding output/input sets are known. The designer of the system treats an ANN controller as a black box. He or she trains the system with a set of standard outputs mapped to corresponding sets of inputs. Once trained, the system is able to work on its own, following the pattern if the input is "standard," and be able to adapt to "non-standard" inputs. This is a powerful tool when you have multiple variables to consider, hence an ANN can simplify certain aspects of legged robotics control.

Evolutionary Computation (EC)

EC can be considered as a method of searching for the optimum solution from a multi-dimensional solution space [19]. There are many variations to this method, such as Evolutionary Algorithm (EA), Evolutionary Strategies (ES), Evolutionary Programming (EP), Genetic Algorithm (GA), and Genetic Programming (GP).

Genetic Algorithm (GA) is the commonly used method among the different variations of EC. It is an iterative optimisation procedure. The algorithm selects the candidate solutions, based on performance, by testing different random solutions with a fitness function. Candidates that have good results survive and produce offsprings while bad solutions are discontinued. GA can be used as an evaluation function to tune rule sets or the membership values of a fuzzy controller. With regards to ANN systems, GA can be used to evolve the network topology, to find an optimal set of weights and to evolve the reward function of the network, thus making it adaptive.

GA has been used in optimisation of ANNs to work out various variables in the system including the weights, architecture and the number of neurons within the network. It has also been used with fuzzy logic to tune the membership functions and rule base. Examples of the use of GA are shown in later sections.

Reinforcement Learning (RL)

Reinforcement Learning uses expert knowledge in the area of exerting action selection and action evaluation, with information about the environment, to control systems which are dynamic in nature. Hence, it is a tool used to automate goal-directed learning and decision making [20]. A learning algorithm improves the controller as it interacts more with the environment, i.e., it is capable to learn and act based on experience which is what humans do. In short, RL provides the ability to act based on sensation to achieve a goal. The problem with the RL method is that it performs poorly in large state spaces with sparse reinforcements.

Robotic tasks are difficult to tune and implement and hence algorithms are used to learn the task given the aim or goal of the task. RL is one tool that mimics how biological systems learn using a reward and punishment system. Hence it is used in many legged robot systems, as they try to mimic biological systems in the way it evolves from an infant stage to a matured stage. More commonly, RL is used to teach other soft computing methods.

3.3 Uses of Soft Computing Techniques in Legged Robotics

The fact that legged robots move in a dynamical and unpredictable environment requires that the algorithms used be flexible and adaptable where the robot must "think." They should have the ability to learn and act from "experiences." Soft computing methods possess these capabilities and are used to implement them.

Soft computing techniques are used in the field of legged robots for

- coordination of leg movement [7,8],
- adapting to the environment,
- path planning,
- localisation [9],
- navigation[10–12], and
- sensor data fusion and interpretation [5].

When coordinating the legs of the robot, both the velocity and type of gaits are important as they determine the operating parameters of the different legs. These two factors are dictated by the terrain and the environment. For example, if the conditions of the ground deteriorate, the speed of the robot has to decrease to improve stability and if necessary, the robot has to alter it's course to avoid the impassable route. On top of that, other factors such as inter-leg collision and foot slippage must be considered during movement as these factors play an important role in the overall stability of the robot. Traditional mathematical and "hard computing" (i.e., "programmed") methods were used to solve this problem but it was found that there are too many factors to consider in the equations. It is impossible to anticipate all possible conditions that the robot may encounter, and program the response of the robot accordingly. Hence a better approach is to incorporate learning and adaptation techniques to the classical methods to improve the robustness of the controllers. For *sustainable* autonomy, it requires the ability to react fast before failure and recover upon failure. Classical methods fail to address these issues.

In path planning, accurate sensory feedback is important for the robot to make accurate judgements. This is especially true in a dynamic environment where the conditions change. Soft computing methods are employed in this area to help the robot in classifying and storing information about the environment using the available sensors. Furthermore methods such as landmark-based recognition are made feasibly usable with various soft computing methods with increased accuracy.

Sensory information is only useful if they are interpreted correctly. Some information require data from various sensors on the robot. Data fusion techniques integrate data from various different sensors, without having to interpret in detail all sensor signals, and are able to obtain useful information about the environment. It is possible to interpret the sensor data without full knowledge of the data with soft computing methods. It has been shown that soft computing methods are useful in producing adequate maps of the environment given the correct sensory information [4].

Calibration of sensors to the environment is very important as sensor noise is a major cause of discrepancy in data. Mobile robots depend heavily on sensors to maneuver and act correctly, hence if the sensors are not calibrated, environment conditions will be falsely interpreted. Most legged robots are designed to operate outdoors, and these conditions change drastically in that environment. Reliable sensors are required to operate in these environment but there are still variations in sensor data. These variations include - change in lighting (infrared sensors), noise from audio devices (ultrasound), magnetic field disturbance(electronic compass), and others depending on the sensors used. Soft computing methods have been found to be useful in adapting to imprecisions in the sensor readings [9].

4 Navigational Systems and Controllers in Legged Robotics

In this section, various representative work are reviewed to explore the areas for which soft computing is used in the different tasks of the mobile robots. The two important systems within a mobile system are navigation and control (both locomotion and overall control) architecture. Hence this section looks into the various fusion methods within these systems.

4.1 Navigational Systems

Navigational Systems are crucial to the autonomous mobile robotic systems as it is required for the robot to move from point to point, with the ability to avoid obstacles and locate itself in its environment. There are many problems within this task which has been solved with various soft computing methods. These include slippage of the leg which introduces positional error and solving certain imprecision in navigation that is not required by the controller.

Fuzzy Localisation Classical localisation methods, like Kalman filters, use position tracking to locate the robot, i.e. the current location is based on previous position, current movement and perception information. This is a relative position method. Although this has been applied to wheeled robots, the positional errors in legged robot are much more because of slippage and momentary lose of balance (the error tends to be relatively large). There are also absolute positional methods that keep a global view of a set of possible positions which enable recovery from positional error. The Markov localisation method uses both relative and absolute position and has proven to be highly accurate in locating the robot's position. But this method relies on the accuracy of the sophisticated sensors used. Legged robots' position in the 3D plane is unpredictable, hence degrading the accuracy of the sensor data.

Buschka [9] developed a fuzzy landmark-based self-localization method for Sony's AIBO robots. The algorithm uses a 3D fuzzy positional map to locate the robot in its surroundings. A low cost camera was used to local the landmarks in the environment so as to test the algorithm in a demanding situation. All tests were conducted during a Robocup (robot soccer) competition, where the robot had to react in a highly dynamic environment with the need for real-time reaction.

The landmark acquisition system used the AIBO camera to recognise the colour of an object and used a model-based approach to classify the various objects the robot "sees." Since the objects were known, the distance could be calculated from the captured picture of the object. The data obtained from the camera was not precise due to variation in the size of the image and also the angle that the camera takes the picture. Hence the distance calculation was not accurate. The data was not reliable as well since there were cases of false object recognition.

The distance and angle data was modelled using fuzzy sets to handle the unreliability in data. Their approach for self-location uses the fuzzy landmark-based localization (Saffiotti [10]), and the position probability grids (Burgard [21]). The position of the AIBO robot was represented by a possibility distribution over the set of possible positions, by a virtual 3D fuzzy position grid (FPG). The angle data gave the possible orientation of the robot. Hence the AIBO robot was able to locate itself without odometry information. The system not only reduced the cost of the sensors used, but also reduced the computational power required, needs an approximate sensor model, uses qualitative motion information and only requires sporadic observation data.

There are other fuzzy navigation systems which include Sanz's work on fuzzy logic based navigation systems [11]. They make use of fuzzy reasoning to reduce the precision required for sensory reading.

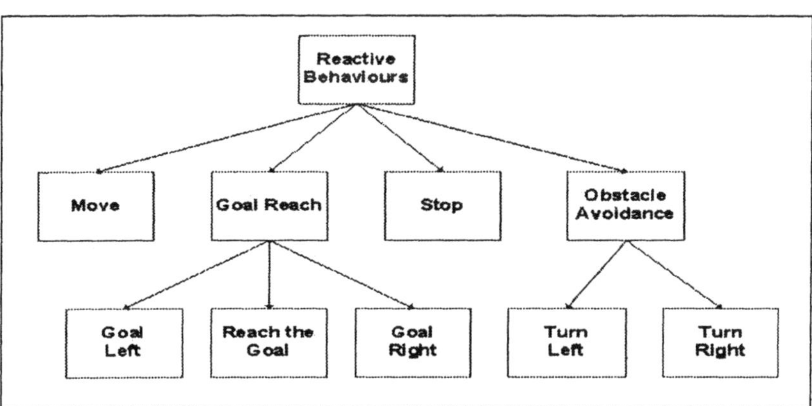

Fig. 3. This diagram was taken from Al-Jumaily [22]. The reactive behaviour of the system is shown in this tree diagram. The complexity of the system increases as more situations are discovered

Fuzzy Reactive Behaviour-Based Navigation System The work of Al-Jumaily [22] is another example of a fuzzy logic based navigation system. The base system is a combination of a reactive system and a behaviour-based system. Behaviour-based systems are plagued with the problem of having to deal with the multiple connections between the various modules within the system. Al-Jumaily used the different modules in the system to represent the various situations that the robot might face (Figure 3), hence the number of modules included in the system increased the complexity of the system.

The task of connecting the different module becomes difficult if it is analyzed using conventional methods as there are many ways to connect the different behaviours in the system. Whereas the problem can be easily resolved using fuzzy logic as it does not use a rigid mathematical description but instead, fuzzy rules and reasoning to resolve the conflicts and competition between the various behaviours. Hence hiding the rigid definition required by the system to perform the task. Furthermore a priority system was integrated into the fuzzy system, which improved the robustness of the system as it enforces the action selection of the behaviour. The system was able to produce promising results thus showing that fuzzy logic is capable of simplifying systems that are complex in nature. This reduces the computational effort and memory requirement of system.

A GA Behaviour-Based Navigational System Neural networks provide the adaptability and generalization but they require human experience, time and many trials before the network can be trained to deal with a real environment. EA solves this problem as it optimizes the network to perform its task in the environment.

Ram [12] uses unsupervised GA to learn a reactive control architecture for autonomous mobile robot navigation, which is able to manoeuvre in any environment configuration. This method is similar to Brooks' behaviour-based approach except the behaviour of the controller changes by altering a set of parameters that belong to the architecture. This method is known as Schema-based Reactive Control. Instead of using the data collected by the sensors to update a world model, this algorithm ties the sensor reading to the action taken by the robot according to a set of parameters. The algorithm was able to learn these various control parameters to suit the different environments. The simulation showed that it was able to navigate through a cluttered environment without hitting any obstacles. Not only was the algorithm able to react in real-time but it also performed well in dynamic and complex environments.

EA Optmised ANN Navigational Controller Han [23] describes an EA optimised modular neural network to develop a navigation controller. The author's aim is to develop a controller that mimics human response to the environment, and hence require the following capabilities -decision making skills, learning, and anticipation. The robot is a wheeled mobile robot with ultrasonic sensors that are used to sense objects around it. The controller was implemented using a ANN, which has proven successful in various robotic applications. The problem with ANN is that optimisation of the network is usually done by trial and error which is time consuming. On the other hand, the gradient descent method of optimisation is not suitable for mobile robots as the environment is too dynamic, which makes it difficult to model. EA learning methods do not require prior knowledge of the problem hence are ideal to solve this problem. It's disadvantage is that the search for the optimal solution is time consuming and the solution obtained for a particular environment is not necessarily applicable to other environments. Hence to improve in this aspect of EA, the network was split into a multiple module ANN. Each module was trained to handle the different localised environment that the robot might face within the working global environment. This reduced the search time and improved the generalisation of the EA.

The robot's task is to move from a start position to a goal position, without prior map or knowledge of the environment. The EA was used to train the weights of a multiple module steering network. The cost function of the EA was set to optimise the simplicity of the network and also the time taken for the robot to reach the goal. Environment classification was implemented to improve the network's ability to generalise. The robot had to decide on the local environment it was in to deal with the respective conditions. 3000 sets of sensory data collected from different local environments were classified using a clustering type ANN. Hence depending on the clustering network, the respective network module from the steering network would be singled out to deal with the situation. This is an example of two ANNs interacting to deal with different task faced in robot navigation. The robot was placed in an environment where it had to maneuver within has U turns, corners, walls and clear obstacles. The results of the exercise showed that a trained single network for one environment has limitations in the degree of generalisation and could not adjust to changes in the environment. Whereas the modular network working with the clustering network was able to increase the generalisation of the algorithm. The modular network was tested and the robot was able achieve the objective without collisions. This is an example of improvement made to a soft computing system with the aid of modular systems.

4.2 Control Architectures (including Locomotion Controllers)

Fuzzy Logic VLSI Controller in Robot Control from the Università di Catania ROBINSPEC (ROBot INSPECtion) [15] is a legged robot built to inspect areas inaccessible to man. It is a teleoperated robot that inspects the grounds with a set of sensors and a video camera. Each of the legs were fitted with a magnet for the robot to climb and move around on the ferromagnetic material of the environment it inspects. A fuzzy logic controller was implemented by using a WARP II fuzzy rule processor to control the trajectory planning of the moving legs. The robot was able to perform walking on arbitrary slopes and could handle changes in angle between the robot and the surface. On top of that, the stiffness of the leg is able to adapt to the kinds of surface that the robot walks on. This is an example of the use of a VLSI fuzzy logic controller chip on a legged robot. This shows the commercial viability of soft computing techniques.

Fuzzy Logic Used in Microrobot Control from the Università di Catania Another robot, the PLIF (Piezo Light Intelligent Flea) [15], is a three legged microrobot which moves by means of piezo ceramics bimorph actuators on two legs with the third as a passive leg. Each leg consists of two parts, the femur that moves the leg vertically and the tibia that moves the leg horizontally. An actuator moves each part of the leg by means of the voltage difference between the two surfaces of the piezo material in the actuator. As the actuator is able to move at very high frequencies, a speed of 18 cm/s is achieved. The robot did not move as expected in the beginning as one leg can be more active than the other at a certain frequency. Moreover, the interaction between the leg and the ground was not consistent, and this posed problems. Another robot with two transistors and a infrared emitting diode fixed on the robot provided feedback to the robot to aid in control of the robot. Soft computing methods were chosen as the controller for the robot with the following objective: to find the quickest way to reach the other robot using the relative position between the two robots. A neural network, that mapped the measurements of the phototransistor and the corresponding action, was trained using reinforcement learning techniques. Also a fuzzy controller was implemented to interpolate the actions from the map. The output from the neural network affects the fuzzy controller by altering the weights and changing erroneous rules of the fuzzy controller. The robot was able to achieve its objective with the ability to adapt to changes within the environment within minutes, as that was the time taken to train the algorithm.

Hierarchical Approach to Robot Control

The hierarchical approach is a structural architectural solution to robotic controllers. Its main purpose is to divide the complexity of the controller problem into modular manageable parts. Various soft computing based controllers use this method to divide the controller problem into many subtasks to reduce the complexity of the control problem. The problems faced within the subtask becomes more specific and therefore narrowing the scope of the problem at hand, hence simplifying to the task of dealing with these subtask using the available soft computing techniques.

Neural Schema for Autonomous Robots Autonomous robots exhibit complex behaviours in different environments. Hence a sophisticated software architecture is required to build a controller that can deal with these complex situations. Weitzenfeld [2] developed a hierarchical and distributed architecture schema modelling method, the Abstract Schema Language (ASL). It is used with neural network methods to build controllers for a group of different mobile robots. ASL is a unified schema computational model based on different robotic schema-based architectures. Neural networks is encapsulated into the ASL with the Neural Schema Language (NSL) to create the required neural-based schema architecture. Hence it has the ability to use neural networks to adapt and learn in developing various behavioural architectures for autonomous agents. All developed schemas work in parallel to produce an entity to operate the different robots. ANNs are not implemented in every schema in the architecture but only in schemas where modelling of the schemas require complex computation. Thus the resulting schema architecture is a fusion between traditional schemas and neural based schemas, improving the modelling, learning and adaptability of the architecture. This work gives researchers an opportunity to conduct biological behavioural experiments which gives them a better understanding of behavioural scenarios.

A Hierarchical Reinforcement Learning Gait Controller Kirchner [24] worked on a hierarchical Q-Learning technique (RL technique) that is used with a six-legged robot (SIR ARTHUR) to learn various gaits of the robot. As there are many dependable degrees of freedom involved in legged robots, it is difficult to control the coordination between the legs. Learning techniques require many trials before converging to an optimal solution, but this is difficult for legged robots as their actions are complicated and mechanical wear is a problem during the trials. Hence the problem was decompose into many sub-problems to reduce the complexity of the problem. It has been shown that the decomposition of a single complex problem can be learnt by learning the temporal sequences of the solutions to the sub problems, while at the same time learn the solutions to the sub-problems [29].

The different layers of the controller perform different roles, from the lowest layer that learns simple movement action (move leg up, down, left and right) to a middle layer that learn coordination of the various legs, and the highest layer that coordinated the middle layer to perform a task. The lowest level's task is to follow the trajectory of the actual leg movement whereas the task of the middle level is to find the correct activation sequence of the different legs to perform more complicated actions like move forward, left, right and back. The task of the highest layer is to learn a goal achieving task based on external stimuli.

This hierarchical structure is based on the fact that the top layers use the bottom layers to achieve their task. It has been found that dividing the task and learning the subtasks, as opposed to learning the entire task, reduces the overall learning time of the algorithm. Hence this approach uses a hierarchical method coupled with reinforcement learning to improve the learning rate of the system. The learning process produced stable movement of the leg after 200 trials.

Hexapod ("HUWE")- Hierarchical Soft Computing Architecture HUWE is a control architecture [25] for a walking robot that allows the robot to negotiate rough terrain environments with sparse footholds autonomously. This architecture builds on the hierarchical architecture and various soft computing techniques such as evolution strategies, cognitive maps, adaptive heuristic critics, temporal difference learning and adaptive artificial neural control in various areas to develop a robust, adaptive robot controller with the ability to map the environment. These are traits of human walking rely on the adaptiveness of our gait and knowledge of the environment to perform stable walking.

The controller is split into four hierarchical levels. The top level interprets motivation or goals into high level commands. This mechanism likens a behaviour selection mechanism for different situations that a robot faces. The second level converts the high level command/goal and plans the navigational path of the robot. It builds up a map of the environment based on the sensory interaction it has with the environment. The third level produces the kinematic plan (gait and leg trajectory planning) for the different actuators, and provides the actuation signals to minimize the error of the different actuator. The last level receives feedback for error detection of error caused either by the system or the environment.

The second layer to the system uses fuzzy-behavioural modules coupled with topological maps developed using neural networks trained by evolution strategies. Each behaviour module performs a particular task (e.g. obstacle avoidance, turn left, etc.) and is implemented in fuzzy logic. A cognitive map is developed under static conditions. A topological map is then developed by tracing based on the cognitive map. But this does not cater for local perturbations. Hence while the robot moves according to the traced path, the behaviour modules subsuming the controller when necessary for reactive action. For example, a behaviour module can adjust the bearings of the robot if the robot steers away from the actual bearing.

The third layer, the gait and leg trajectory planner, coordinates the timing and stability of the movement of the different legs. The leg trajectories were realised using two models, the minimum torque-change model and the minimum jerk model. The last layer is the dynamic compensation using an adaptive neural network controller, which compensates for any trajectory tracking errors that result from the implementation of the demand trajectory. The kinematic controller is essentially a PD controller implemented with a neural network to control the dynamics involved, i.e. the errors incurred due to the dynamics of the system. The system was trained using desired position and velocity values offline and then operated with online learning to reduce the tracking errors of the system.

Soft computing in this case was used to search the complex and large search solution space for an optimum solution with no a priori assumptions. The second layer uses soft computing to build the map of the environment through an evolution strategy method with a neural network to remember the map of the surroundings. The last layer uses a hybrid controller that performs classical kinematic control using PD with a neural network to handle the dynamics involved in the system. These examples show the success of implementation of soft computing in the various aspects of mobile robot control, from the highest decision making levels to the lowest actuator control level.

The Subsumption Architecture

Although this method of control is not a typical soft computing method, this tool is widely used in the field of mobile robotics and various people have worked on this architecture incorporating different techniques to improve the architecture or perform the different mobile robot tasks. The techniques used in this control method includes soft computing methods that improve on the performance and robustness of the control method.

Brooks' Subsumption architecture [6] is a modified reactive controller that builds on top of the advantage of a reactive controller. It is an incremental module-based network controller that uses a combination of simple modules to build a structure that achieves an overall goal. Each module independently is able to perform simple tasks and all task work in parallel to achieve the goal. Each modules is directly connected to actuators, sensors and its links to other modules, depending on the role of the module. There might be conflict of interest within the system if many modules want control over an actuator. Therefore Brooks developed the subsumption architecture where inhibitory and exhibitory signals were used to arbitrate between conflicting modules, allowing the developer to decide on the action that the module takes. Therefore these inhibitory and exhibitory signals determine the characteristic of the system. Another feature of the subsumption architecture is that different groups of modules can form layers within the system, each performing a unique task, integrated to form the entire control mechanism. Hence this method is hierarchical as well as modular in natural with a mechanism to resolve the parallelism of the architecture.

This method has found its uses with mobile robots and proved to be robust and useful in dynamic and complex situations. This architecture has provided the basis for control of mobile robots and hence many soft computing methods were used to add different features to mobile robot controllers. Michaud studied the interaction of behaviours with the environment based on observation of behaviours over time and adapting based on the observations using reinforcement learning techniques [40]. A soft computing technique is employed to teach and remember the history of the actions of the different behaviours. A neural based reinforcement learning technique is used in the process of design of the policies that are used in building a behaviour-based system [41]. This differs from normal methods where the control program decides on the desired behaviour. Soft computing methods are used in the design stage instead of the controller. Many examples of fusion between soft computing techniques and behaviour-based systems are mentioned Mataric's work [42].

4.3 Gait Generators and Balancing Control

ANN Gait and Balance Control for Biped Hu [39] developed a unsupervised learning, self-organising neural network mechanism with a CMAC (Cerebellar Model Articulation Controller) based adaptive control scheme. The CMAC is an ANN inspired by the cerebellum and is a good robotic motor controller. It is able to learn fast and has simple computation as compared to the traditional multi-layer perceptron with back-propagation learning. Hu used this technique to solve two particular problems in biped control - gait synthesis and lateral balance control. The algorithm was simulated on a six joint planar robot and the results showed that tracking (height and pitch) is well maintained even if the robot was experiencing external force impact. The algorithm was able to dynamically adapt to the changing situation and able to reject the external disturbances.

Neurofuzzy Biped Locomotion Controller Bipedal gait controllers are plagued with the problem of controlling a highly coupled nonlinear dynamic system. Zhou [43] developed a neurofuzzy controller that used the both ANN and fuzzy logic to deal with this problem. He uses fuzzy logic (a Sugeno fuzzy system) to characterize the dynamic model of the biped, hence eliminating the need to create a detailed dynamic or kinematic model of the biped. However the task of defining the rules of such a complex system is not simple. Hence ANN was used to tune the fuzzy rules so as to achieve dynamic balancing during walking. The ANN tool used is to model and fine tune the Adaptive-Network-based Fuzzy Inference System (ANFIS). ANFIS represents the fuzzy logic system as a set of ANN and tunes the fuzzy system using a back-propagation method. It has the ability to develop fuzzy rules with a set of desired input-output data, with self-learn capability that can fine tune and generate fuzzy rules when required.

Zhou used the ANFIS controller developed as part of a bipedal controller where it serves as a joint controller to the system (lowest level controller in the system). The higher level modules deal with path planning, obstacle crossing and gait selection aided by sensory information. They then provide the joint reference commands to the ANFIS controller, which the ANFIS controller use to process drive signals to the actuators of the system. The results of the simulation showed that walking with dynamic balance was achieved with the controller. It also showed that the gait produced was greatly improved with the on-line learning capability of the ANFIS controller.

Central Pattern Generator (CPG) - A Gait Controller

The CPG controller is a rhythmic generator system within the spinal system that is present in animals and humans. It controls legged locomotion in animals and has been found to be able to act without interaction with the brain. Since it is an efficient biological system, research in this area has been increasing to use the CPG to generate gaits for legged robots. Many methods have been used to develop CPG controllers, and they include ANNs, other classical methods (such as oscillatory ordinary differential equations) and the combination of the two methods. This method gives us a clearer picture of the biological mechanism of living beings, which provides us with a treasure chest of information. This not only benefits robot builders but also biologists who are trying to understand these working mechanism.

ANN CPG Controller Since the CPG is a biological system, it would be natural to consider using artificial neural networks to construct artificial CPG controllers. Micci-Barreca [26] implemented a CPG controller for a hexapod using a set of ANNs. The central controller is a single neural network with a control subnetwork attached to each leg. The central controller produces the rhythm for the gait and each subnetwork, which acts as sensory-motor controller, only acting based on the signal received from the central controller. These subnetworks receive feedback from the sensors on the leg and sends the corresponding motor signals to the joints. The equations in the neurons are defined by a group of Ordinary Differential Equations (ODEs) proposed by Ellias and Grossberg. The algorithm was able to handle changes in gait speed and gait type with the ability to adapt to the terrain. Franca [27] uses a Asymmetric Hopfield Network to build a CPG controller. Similarly, an ANN is used to generate and coordinate the rhythmic signal sent to the various legs of the robot.

Robust ANN CPG Controller A robust locomotion controller [28] was developed for a hexapod using a set of distributed neural networks. The distributed network is modelled after the neural control of insect locomotion [8]. Each leg consist of a network of three motor neurons, two sensor neurons and a pacemaker neuron. The inhibitory connections between the pacemaker neurons coordinate the leg movement and therefore determines the gait of the robot. The overall locomotion is achieved by coordinating the oscillation of the pacemaker neurons of all legs. The network was able to produce rhythmic pattern for legged movement. The controller was able to perform locomotion even with damage inflicted on itself, which included the loss of a sensor or of an effector. This distributed network is developed such that the failure of a network component or its connections does not affect the efficiency of the controller. The network was tested and was able to maintain static stability even if there were perturbations due to failure of different components in the system. Here a modular structure is used to isolate the different neurons in the network where they are dealt with individually. This allows the network to function when one module in the network is not functioning, thereby improving the robustness of the controller.

A modular ANN multi-link controller was developed at the university of Saskatchewan [30]. The heart of the controller is a CPG type controller that produces the oscillatory movement of legged locomotion. This controller controls movement trajectories instead of muscle activity, charateristic of a biological CPG , therefore it is termed as a "movement pattern generator". The ANN used is a modified Jordan's network which has the capability to store pattern sequences and in this case the network was used to store the sequence of various legged locomotion with the ability to evolve and learn. A set of gait is first preprogrammed into the network based on gait data obtained by observing animal gait patterns. Although the network can be configured to work with any number of limbs or links, it was tested on a three link model limbs. The network was able to learn and faithfully recall different cyclic temporal sequences amidst noise and erroneous inputs. From the preprogrammed gait, the network is able to evolve to other gaits that are found in animals. The function of the ANN used in the CPG controller is to learn a rhythmic pattern and secondly to produce the motor commands in accordance to the sensory data feedback and the rhythmic signal. The underlying mechanisms of these ANN is a set of equations (or Ordinary Differential Equations) that characterise the behaviour and the delay between the actuation of different legs. The adaptive and storage characteristics of the controller is shown in this example of a controller.

Figure 4 is a biologically inspired controller that uses soft computing as one of the modules that controls a quadruped named Meno. Each leg of the robot has a prismatic joint and two rotational joints. The rotational joints control horizontal motion, whereas the prismatic joint controls vertical motion. Hoff [31] built a controller for a simulation model of Meno. The control architecture of Meno is shown in Figure 4.2. The CPG controller generates the gait, while the gait planner controls the different movements of the robot. The gait optimiser takes care of deficiencies in the two previously mentioned controllers. A reinforcement learning schema training a neural network is incorporated into the gait optimiser. The learning problem for the optimiser was to correct the error for the trajectory based on foot slippage and postural instability by inferring from the sensory data.

Cruse's Artificial Neural Network Solution Cruse used a set of rules [32] to implement a CPG controller for a legged robot. This set of rules were created from observation of the movement of stick insect. There are two classes of these rules: ipsilateral (same side) leg relationships and contralateral (opposite side) leg relationships. There are a total of six simple relationship rules. He used a neural network to implement this rule base to

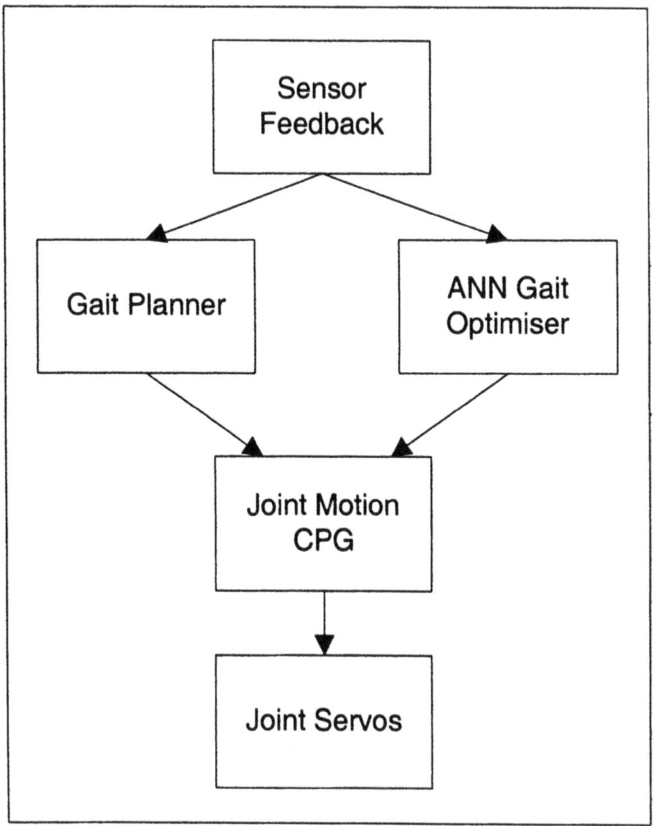

Fig. 4. A simple block diagram of Meno's control architecture

control a six legged robot [7]. Each leg is controlled by a local sub-network in his work. Each of these control sub-networks consists of three subnets. The first generates the return stroke while the other generates the power stroke, with the last subnet controlling the previous two subnets to ensure that the leg performs the correct action. The gait emerges from the interaction between the different control sub-networks of the different legs. The interaction between the different legs is governed by the six rules. His implementation concept was simple (i.e. computationally more efficient), yet was able to produce a robust gait controller [33]which was able to handle massive disturbances during walking. The gait was able to perform straight and turning gait with variation in gait speed. It was not only able to clear obstacle of a certain height, but also limited walking was possible with amputation.

Evolutionary Approach to CPG Based on Lamprey Ijspeert and Kodjabachians' study on the lamprey has led to the development of an evolutionary approach to ANN, to develop and study the CPG [34,35]. This method uses a genetic programming algorithm to evolve developmental programs which encode a growing dynamic neural network. The genetic programming approach is used with a developmental encoding known as Simple Geometry Oriented Cellular Encoding (SGOCE). This architecture evolves to control high level characteristics like the speed of locomotion and a change in direction and the aim of this approach is to obtain an automatic generation of a control mechanism for locomotion. This controller imitates the natural process of evolution, allowing the neural network to create its own synaptic connections and the amount of neuron in the network in a systematic way. It uses a control structure similar to Cruse's method [32] as each limp is controlled by a local controller. The coordination of all the controllers determines the gait of the system in control. This method has been used as a controller for a virtual six legged insect with behaviours such as gradient following and obstacle avoidance. This evolution system gives this implementation the ability to achieve its specified task without having to tune the controller. The only problem with this method is that it can take a long time to create the neural network (approximately 450 CPU hours on an Ultra Sun station).

Rhythmic Action - Neural Oscillator for Bipedal Walking Cao [37] proposed two methods of developing a neural oscillator for rhythmic walking in a 3D plane. But before she started work on the model, an investigation and analysis was conducted on human walking. The investigation gave her insight on human walk, hence she was able to develop a model that mimic human walking closely. The human model chosen was a eight joint human model with two joints at the arms, two at the knees, two at the hip and the last two are the lateral joints at the hip. She then defined the gait with a set of parameters and an ANN. The outputs to the ANN was the joint trajectories for the human model. After running simulations, it was found that the connection weights of the ANN determined the walking patterns and the parameters affected the frequency and the amplitude of the oscillation. The first method that was implemented was to use observed mechanisms in human walking to reduce the system dimension. After which to solve for suitable connection weights for the system.

The second method uses GA to calculate a number of connection weights for the system. To define the problem (and also reduce the search space for the GA), the following characteristics were considered:

- Knee joints are bent forward.
- Phase difference between right and left leg is π.
- Periodical movement.
- Bilaterally symmetric.
- The walking pattern parameters determine the posture.

The parameters for the walking gait is fixed and the weights of the ANN are the variables to the system. Two fitness functions are used with the above restrictions to form the objective function for the walking gait. GA is used to select the optimum set of weight connections that performed the desired walking gait.

Both methods proved to be effective to produce human-like rhythmic walking. These methods are similar to the CPG method mentioned in the previous section. ANN proved effective in producing the rhythmic signals required for walking. Additionally, GA was used to optimize the weights of the ANN.

Adaptive Neural Oscillator for Bipedal Locomotion Zheng [38] categorizes walking into two categories -voluntary and involuntary walking. Voluntary walking takes place when there are no obstacles or irregularities in the path and hence walking is a rhythmic action that requires no sensory feedback as there are no surprises. On the other hand, when there are irregularities, reflexive motion, hence sensory feedback, is required for a human being to maneurvere in a rhythmic motion. This is referred to as involuntary walking. But the human is able to learn to perform this involuntary action which after practice becomes a voluntary action. Hence Zheng states that controlling walking involves voluntary motion, involuntary motion and learning.

Zheng describes a gait controller which consists of the following:

- CPG : generates the patterns for both voluntary and involuntary motions.
- Adaptive neural network : responsible for generating reflexive action. It receives signal from both the CPG and external sensors to generate reflexive motion. It has the capability to learn various motion patterns.
- Switching unit : to decide to use voluntary and involuntary motion in real-time.
- Knowledge base : store parameters which represent gait patterns that map to a particular terrain. This information is fed to the CPG to generate the correct motion.
- Learning unit : builds up the library of parameters that are mapped to terrain conditions.

The CPG is implemented with a set of decoupled differential equations (van der Pol oscillators). Zheng used the set of equations to generate the joint trajectories for different joints on the biped. He thus simplified the CPG problem so that real-time computation was possible. A multi-layer neural network with back-propagation was used to implement the adaptive unit. This network was used to learn and map sensory data to the desired motion patterns. The unit uses the information it acquires from the sensors to modify the gait pattern to suit the terrain. The switching mechanism, knowledge base and learning mechanism are used to enhance the features of the CPG and the adaptive unit. The gait controller is trained to handle locomotion on basic terrain and use the adaptive unit to learn to move within other terrains.

Once again, neural network has proven its usefulness in learning and storing patterns used in legged locomotion.

5 Conclusion and Trends

Legged robots are built to move autonomously in undulating terrain which requires the robot to be intelligent and have the ability to act robustly and recover from failure. Therefore the development of an autonomous legged robotic system involves many factors and has to take the environment into consideration. These tasks include navigation, localisation, locomotion, and sensor interpretation. Various methods involving soft computing, hard computing, and organisational architectures have been employed to implement these tasks. The four common soft computing methods used in robotics are fuzzy logic, artificial neural networks, genetic algorithm and reinforcement learning. These methods provide the ability to learn, adapt, and generalise; these are important requirements in an autonomous system. Hard computing methods are still important in the area of robots and are implemented if they serve the purpose due to their stability. If hard computing methods cannot fulfill the role. Soft computing methods are fused with hard computing methods to improve the capability of the systems. The resultant system combines the stability of hard computing methods, and flexibility and learning capability of soft computing methods.

Two main areas were reviewed in this chapter - navigation and control architecture of mobile robots. Various fusion techniques were reviewed. In general the robots were able to learn without prior information, adapt to different erroneous situations, improve in stability and perform more complex tasks. A few organisational architecture techniques were reviewed as this is widely used in development of controllers in the field of legged robotics. Two examples of biologically inspired controllers were presented - the CPG, a locomotion controller and the behaviour-based controller, a modular robotic controller architecture. These were used as examples of how soft computing techniques can be used to enhance and realise robust and adaptive controllers.

Legged robots are unlike industrial robots where motion is more predictable and the environment is known. The environment and situation in which legged robots operate in is unknown. On top of that, legged robotics usually involves control of many actuators and interpretation of many sensors. Hence classical methods are difficult to implement with legged robotics. Human reasoning and decision making is far more superior in this aspect. Hence these soft computing techniques are used to mimic human reasoning and thought. From the various examples of soft computing in legged robotics, we learn that

- fuzzy logic is able to reduce the precision needed in sensory information, hence improving the exactness of the data obtain. ([9]). Fuzzy reduces the complexity of the problem as the parameters do not have to be clearly defined. This is especially useful in the area of legged robotics where there are many variables and parameters to control ([15], [22]).

- ANN is a tool that is seldom used alone as it is not an easy tool to train. It has proven its use in the area mapping in navigation ([23],[25]) and also in the area of gait controller ([7], [26], [27], [28], [30], [34], [35], [37], [38], [39], [43]) where the action of the legs are a pattern action. ANN is able to generalise and cater for adaptability. This is required for movement in uneven terrain where every step that the robot take is not the same. Its ability to learn patterns (e.g. maps and gait pattern) by training mimics human memory, which simplifies the way pattern recognition is perform.
- RL and GA are powerful tools used in legged robotics where models of the environment or control situation is difficult to build. A problem faced in the area of robot control learning is that the task is too surmountable for unmodified learning techniques. Hence a controller has to be subdivided into more manageable modules to be handled([24]). This is typical of most implementation of robot controllers as the size of these controller increase due to the many functions that is required to perform its task([25]). Soft computing methods hence complement modular techniques([2],[3],[6]) in providing beautiful solutions to legged robotic control.
- fusion of the above techniques strengthen each other and rid one another of their weaknesses. Many controllers use the advantages of the different techniques to develop robust and adaptive algorithms in the two areas of legged robotic research -control and navigation. Human capabilities such as decision making, learning and anticipation are possible with soft computing methods which improve the autonomy and stability of the robot ([25]). This improvement of stability and robustness is not easily implemented without soft computing methods.

The importance of autonomy in legged robots justifies the need to work at creating techniques to solve the problems. Soft computing provides many methods to change the way the different systems function. The different concoctions of various soft computing and hard computing techniques are able to deal with many problems associated with legged robotics. They allow man to create a system that is human-like, with the ability to think, understand and act like a human. But in this sense, development of autonomous control of legged robots is still in its infancy. Humans and animals are able to perform locomotion and navigation with relative ease as we are able to interpret the various information available to us and use this information to produce the correct action within a short period of time. Robotics is lacking in this area but as soft computing research continues, a better system that mimics human thought and action can be achieved.

References

1. Kaynak, O. and Rudas, I. (1995): Soft computing methodologies and their fusion in mechatronic products. *Computing and Control Engineering Journal*,

6(2):68–72.
2. Cervantes, F. Olivares, R., Weitzenfeld, A., Arkin, R. and Corbacho, F. (1998): A neural schema architecture for autonomous robots. *Proc. of 1998 International Symposium on Robotics and Automation*, pages 245–252. Saltillo, Mexico.
3. Kolushev, F.A., Timofeev, A.V. and Bogdanov, A.A. (1999): Hybrid algorithms of multi-agent control of mobile robots. *International Joint Conference on Neural Networks*, 6:4115–4118.
4. Pasparakis, G., Luk, B.L., Galt, S., Kalyvas, T. and Virk, G.S. (1996): A.I. solutions for semi-autonomous legged robots. *IEE Colloquium on Information Technology for Climbing and Walking Robots*, 1996/167:9/1–9/4.
5. Benediktsson, H., Benediktsson, J.A., and Arnason, K. (2000): Absolute neuro-fuzzy classification of remote sensing data. *Geoscience and Remote Sensing Symposium Proceedings. IEEE 2000 International.*, 3:969–971.
6. Brooks, R.A. (1985): A robust layered control system for a mobile robot. Massachusetts Institute of Technology Artificial Intelligence Laboratory.
7. Cymbalyuk, G., Dean, J., Cruse, H., Bartling, C.H., and Dreifert, M. (1994): A neural net controller for six-legged walking system. *From Perception to Action Conference IEEE Computer Society Press*, pages 55–65. Edited by P. Gaussier, J.-D. Nicoud. Los Alamitos, California.
8. Beer, R.D., Chiel, H.J. (1989): A lesion study of a heterogeneous artificial neural network for hexapod locomotion. *International Joint Conference on Neural Networks*, 1:407–414.
9. Saffiotti, A., Buschka, P., and Wasik, Z. (2000): Fuzzy landmark-based localization for a legged robot. *IEEE/RSJ International Conference on Intelligent Robots and Systems*, pages 1205–1210. Takamatsu, Japan.
10. Saffiotti, A. (1997): The use of fuzzy logic for autonomous robot navigation. *Soft Computing*, 1(4):180–197.
11. Sanz, A. (1997): The uses of fuzzy logic in autonomous robot navigation. *6th IEEE International Conference on Fuzzy Systems*, 2:1089–1093. Barcelona, Spain.
12. Boone, G., Ram, A., Arkin, R. and Pearce, M. (1994): Using genetic algorithms to learn reactive control parameters for autonomous robotic navigation. *Adaptive Behavior*, 2:277–304.
13. Bruce, J., Lenser, S., and Veloso, M. (2002): A modular hierarchical behavior-based architecture. In *Birk, A., Coradeschi, S. and and Tadokoro, S., editors, RoboCup-2001: The Fifth RoboCup Competitions and Conferences.* Springer Verlag, Berlin, 2002, forthcoming.
14. Atienza, R.O. (1998): *An AI-Enhanced Control System for a Four-legged Robot.* National University of Singapore.
15. Muscato,G. (1998): Soft computing techniques for the control of walking robots. *Computing and Control Engineering Journal*, 9(4):193–200, August 1998.
16. Bonissone, P.P. (1997): Soft computing: The convergence of emerging reasoning technologies. *Soft Computing*, 1:6–18.
17. Haykin, S. (1999): *Neural Networks - A Comprehensive Foundation.* McMaster University, Prentice-Hall Inc., 2nd edition.
18. Tomas, L.M., Zamora, M.A., Toledo, F.J., Luis, J.D., and Martinez, H. (2000): Map building with ultrasonic sensors of indoor environments using neural networks. *IEEE International Conference on Systems, Man, and Cybernetics*, 2:3334–3339.

19. Whitley, D. (1994): A genetic algorithm tutorial. *Statistics and Computing*, 4:65–85.
20. Sutton, R.S. and Barto, A.G. (1998): Reinforcement learning : An introduction, http://www-anw.cs.umass.edu/ rich/book/the-book.html.
21. Hennig, D., Burgard, W., Fox, D. and Schmidt, T. (1996): Position tracking with position probability grids. *Proceedings of the First Euromicro Workshop on Advanced Mobile Robot*, pages 2–9.
22. Al-Jumaily, A.A.S. and Amin, S.H.M. (1999): Fuzzy logic based behaviors blending for intelligent reactive navigation of walking robot. *Proceedings of the Fifth International Symposium on Signal Processing and Its Applications*, 1:155–158.
23. Oh, S.Y. and Han, S.J. (2001): Evolutionary algorithm based neural network controller with selective sensor usage for autonomous mobile robot navigation, INNS-IEEE International Joint Conference on Neural Networks, 3:2194–2199, Washington. DC, USA, July 2001.
24. Kirchner, F. (1997): Q-learning of complex behaviours on a six-legged walking machine. *Proceedings of the second Euromicro Workshop on Advanced Mobile Robots IEEE*, pages 51–59. Brescia, Italy.
25. Randell, M.J. and Pipe, A.G. (2000): A novel soft computing architecture for the control of autonomous walking robots. *Soft Computing*, 4:165–185.
26. Micci-Barreca, D., Ogmen, H. (1994): A central pattern generator for insect gait production. *From Perception to Action Conference, IEEE*, pages 348–351.
27. Felipe, M.G., Yang, F. and Yang, Z. (2000): Building artificial CPGs with asymmetric hopfield networks. *of the IEEE-INNS-ENNS International Joint Conference on Neural Networks*, 4:290–295.
28. Quinn, R.D., Espenschied, K.S., Chiel, H.J., Beer, R.D. (1992): Robustness of a distributed neural network controller for locomotion in a hexapod. *IEEE Transactions on Robotics and Automation*, 8(3):293–303.
29. Singh, S.P.(1992): Transfer of training by composing solutions for elemental sequential tasks. *Machine Learning*, 8(3/4):323–339.
30. Gander, R.E., Srinivasan, S. and Wood, H.C. (1992): A movement pattern generator model using artificial neural networks. *IEEE Transactions on Biomedical Engineering*, 39(7):716–722.
31. Hoff, J. and Bekey, G.A. (1997): A cerebellar approach to adaptive locomotion for legged robots. *IEEE International Symposium on Computational Intelligence in Robotics and Automation*, pages 94–100. Monterey, California.
32. Muller, U., Cruse, H., Dean, J. and Schmitz, J. (1991): The stick insect as a walking robot. *Proceedings of the Fifth International Conference on Advanced Robotics IEEE*, 2:936–940.
33. Cruse, H., Kindermann, T. and Dean, J. (1998): A biologically motivated controller for a six-legged walking system. *Proceedings of the 24th Annual Conference of the IEEE Industrial Electronics Society*, 4:2168–2173.
34. Ijspeert, A.J. (2001): A connectionist central pattern generator for the aquatic and terrestrial gaits of a simulated salamander. *Biological Cybernetics*, 84:331–348.
35. Ijspeert, A.J. and Kodjabachian, J. (1999): Evolution and development of a central pattern generator for the swimming of a lamprey. *Artificial Life*, 5:247–269.
36. Yang, J. (1993): Adaptive control for a biped locomotion system. *Proceedings of the 36th Midwest Symposium on Circuits and Systems*, 1:657 – 660.

37. Cao, M. and Kawaniura, A. (1998): A design method of neural oscillatory networks for generation of humanoid biped walking patterns. *Proceedings of IEEE International Conference on Robotics and Automation*, pages 2357–2362.
38. Zheng, Y.F. (1990): A neural synthesizer for autonomous biped robots. *Proceedings of IEEE International*, pages 657–660.
39. Hu, J. and Pratt, G. (1999): Self-organising cmac neural networks and adaptive dynamic control. *Proceedings of IEEE International Symposium on Intelligent Control/Intelligent Systems and Semiotics*, pages 259–265.
40. Michaud, F. and Matarić, M.J. (1998): A history-based approach for adaptive robot behaviour in dynamic environments. *Proceedings, Autonomous Agents*, pages 422–429.
41. Bekey, G.A., Fagg, A.H. and Lotspeich, D. (1994): A reinforcement-learning approach to reactive control policy design for autonomous robots. *Proceedings of the 1994 IEEE International Conference on Robotics and Automation*, pages 39–44.
42. Matarić, M.J. (1998): Behavior-based robotics as a tool for synthesis of artificial behavior and analysis of natural behavior. *Trends in Cognitive Science*, 2(3):82–87.
43. Zhou, C. and Jagannathan, K. (1996): Adaptive network based fuzzy control of a dynamic biped walking robot. *Proceedings of the IEEE International Joint Symposia on Intelligence and Systems*, pages 109–116.

Grasp Learning by Active Experimentation Using Continuous B-Spline Model

Jianwei Zhang and Bernd Rössler

Faculty of Technology, University of Bielefeld
Bielefeld 33501, Germany
{zhang, broessle}@techfak.uni-bielefeld.de

Abstract. In this chapter we present a self-valuing learning system based on continuous B-spline model which is capable of learning how to grasp unfamiliar objects and generalize the learned abilities. The learning system consists of two learners which distinguish between local and global grasping criteria. The local criteria are not object specific while the global criteria cover physical properties of each object. The system is self-valuing, i.e. rates its actions by evaluating sensory information and the use of image processing techniques. An experimental setup consisting of a PUMA-260 manipulator, equipped with hand-camera and force/torque sensor, was used to test this scheme. The system has shown the ability to grasp a wide range of objects and to apply previously learned knowledge to new objects.

1 Introduction

In a wide range of robotic systems grasping is a basic skill that is crucial for manipulation tasks and interaction with an environment. In most industrial applications the problem of grasping is solved via *teaching-by-doing* or static programs, that is, a well defined sequence of actions which the robot simply reproduces. However, sensor-based motions like *visual servoing* are not considered in these industrial robotic systems. But when thinking of recent research fields, e.g. service robots or humanoids, aspects of sensor based grasping will play a very important role. New techniques must be developed for the robots to operate in uncharted and unknown territory. They should consider elements of human learning abilities when constructing a robotic grasping systems. Such an approach is presented in this chapter.

When an infant learns how to grasp different objects he needs no help from a teacher. It learns from interaction with the object, i.e. success and failure of stable grips. For example, when he tries to get hold of an object which slips out of his fingers, at the next moment, the infant would grasp the object at a different and more promising position until it finally reaches a stable grasp. Next time the infant can employ the learned faculty when faced to the same or a similar object. It *generalizes* between different objects. This principle of learning, i.e. learning by interaction with the environment and generalization of learned abilities, is applied in our system.

A robotic system, based on a self-valuing learning technique, is developed, which is capable of learning how to grasp unfamiliar objects and generalize the learned abilities to new ones. For the design of the system supposed basic human grasping abilities were considered. A graphic simulation tool was developed for testing and evaluating the implemented learning algorithms and finally a setup of one PUMA-260 manipulator was used to evaluate the system under real conditions.

1.1 Grasping

As mentioned above, grasping is a crucial skill for many robotic systems. On the other hand, it is very difficult to let a robot perform skilled grasping operations without predefined sequences.

The action of grasping an object can be defined as the placement of the gripper relative to the object and is characterized by a set of contacts called a *grasp point*. Stable grasp points must be found where the robot is able to get hold of an object. This decision depends on the object itself, the gripper, the workspace, and the task to be performed while holding the object. However, the last two topics are beyond the focus of grasp point selection for only stable grips. When choosing stable grips mainly the following two requirements have to be met:

1. Robustness against friction, i.e. the object must not slip between the fingers of the used gripper.
2. The grasp point should be as close as possible to the center of mass of the object to reduce resulting torques.

Especially, it is clear that the second requirement cannot be fulfilled only by analytical approaches if no complete model of the object is available. In fact, the robot has to "interact" with the object, i.e. measure the physical properties and evaluate the goodness of a grasp.

1.2 Related Research

A lot of work has been done in the field of robot grasping. [1] gives a brief overview of the field over the last two decades. Most work deals with analytical approaches that try to compute optimal grips according to special heuristics (e.g. in [2] and [3]). In these cases one has either a fully specified model of the object and its mass distribution or has to use the center of area of the object, extracted via image processing, to approximate the real center of gravity. The first case is very difficult to obtain via external sensors and without any previous knowledge. One would have to gain a complete 3D representation of the object via image processing and additionally try to examine factors like the material of the object. However, hidden internal inhomogeneous mass distribution can never be found with such an approach. The latter case of using the center of the object's area is certainly only a kind of approximation. This works fine if the center of gravity coincides with the object's center of area, but this approach cannot deal with inhomogeneity, too. However, relatively few efforts handle the problem of learning how to grasp. In [4] a system is presented that learns how to grasp objects with a parallel-jaw gripper. Two main subproblems are learned: to choose grasping points and to predict the quality of a given grasp. The disadvantage of this system is that only *local* criteria are used to store grasping configurations. Without *global* criteria it is for example impossible to learn how to grasp objects where the center of gravity does not coincide with that of the object's image area. Without self-valuing learning techniques it is not possible to handle the real physical properties of an object. [5] presented a learning system for visual guided grasping, constructed by two learners. This system is not self-valuing, i.e. the optimal grasp point has to be given initially to the learner. Therefore, the two learners are also not generalizable to new objects. In [6] an uncalibrated vision-guided system was developed for manipulating objects that may be placed anywhere in the robot's 3-D workspace even though not visible in the initial fields of view of the cameras. More recent work was done on grasping with multi-finger hands. Such robotic hands are so complex that many problems arise before one can focus on learning techniques.

2 Learning Scheme

To construct a robotic learning system, it is useful to investigate elements of human learning abilities. No enlightening work exists that deals with the learning theory of human grasping. When discussing this problem, of course, the human hand with its five fingers is considered which is much more complex than a parallel-jaw gripper as used in this setup. Therefore, in this chapter an approach is suggested that is based on our *supposed* human learning abilities when grasping an object. Although no well studied human learning abilities are taken to construct a robotic learning system, the system is used to show that the proposed learning abilities in the field of grasping could in fact be similar to what we supposed for a human.

2.1 Local and Global Grasp Criteria

Our work is based on the observation that when a human wants to grasp an unfamiliar object, he mainly considers two criteria on how to choose optimal grasp points. These two criteria are further referred to as *local grasp criteria* and *global grasp criteria*. These two criteria form the basis for the underlying learning system design.

Local Grasp Criteria: A local grasp criterion is mostly independent from a special shape and therefore from global aspects like the distribution of mass of an object. Therefore, it can be applied in the same way to any kind of object. Local criteria are considered first when one decides to grasp an unfamiliar object. Such a criterion is for example to choose a grasp point at two opposite parallel edges.

Global Grasp Criteria: Global grasp criteria, in contrast to the local ones, are strongly interconnected with a special object and therefore seldom to be applied to different kinds. They are considered right after the local criteria to find the optimal grasp point. These criteria consider aspects like the distribution of mass of an object, e.g. grasping an object near its center of gravity.

The terms *local* and *global* need some more specifications. The local criteria refer to *local* environmental features near the grasp point whereas the global criteria describe the *global* properties of the position of a grasp point within an object. Therefore, it can be shown that the local criteria are universally valid and the global ones are mostly restricted to a special object.

Technically speaking, the local criteria define an axis on which the grasp point can be searched to further fulfill the global criteria. For example the grasp configuration in Figure 1(b) could be gained from the search direction proposed by grasp point 3 in Figure 1(a). In the learning process these criteria are repeatedly considered one after another in a finite number of steps until a good grasp point is found. The number of steps varies with the skill of the learner and the shape of the object. For a familiar type of object the global and local criteria are mostly considered in only one step.

In fact, since the same local criteria can be applied to any kind of object, as mentioned above, they are fully learned precociously. Thenceforward, only the global criteria will be learned to grasp unfamiliar objects.

2.2 Optimality

A grasp point is optimal according to the local criteria, if

1. the fingers can cover the object at this grasp point, and
2. no friction occurs between the fingers and the object.

It is considered to be optimal according to the global criteria, if

1. no torque occurs between the fingers grasping the object, and
2. the object does not slip out of the fingers, and
3. the grasp is stable, i.e. the object does not slip between the fingers.

Some sample grasp configurations are shown in Figure 1.

2.3 Higher Level Criteria

An additional and *higher level criterion* for human grasping is the role of the grip, i.e. the role it fulfills to perform further operations, e.g. grasping a cup at

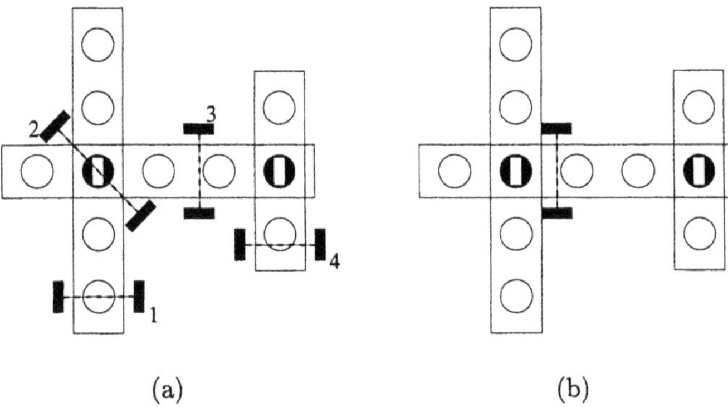

Fig. 1. Local and global grasp criteria. 1(a): some sample grasp configurations which are optimal according to the local grasp criteria. 1(b): the grasp point is optimal according to both criteria

its bail in order to drink something or a sledge at its handle to bang a nail into the wall[1]. Other higher level criteria are for example the material or surface of an object. To consider these criteria additional sensors or sophisticated image processing techniques ought to be integrated. However, this lies not in the scope of this work. Our objective is to emulate the abilities of an infant who just intends to get hold of an object as good as possible.

3 Two-Learner System

The criteria mentioned above advise a system consisting of two learners, one for the local and the other for the global grasp criteria. The states for the first learner provide only the local features $s = (f_{l_1}, \ldots, f_{l_m})$. The learner tries to map them to actions consisting of a rotational component $a = (\phi)$. The second learner tries to map states of global features $s = (f_{g_1}, \ldots, f_{g_n})$ to actions of translational components: $a = (x, y)$. Because the local criteria are covered mainly from the relative orientation of the gripper, the responsible learner is called *orientation learner*. The global criteria are determined through the position of the grasp point in the object and therefore the proper learner is further referred as *position learner*. These two learners operate right after each other (Algorithm 1), as a human being is supposed to. The local and global features used in our system are shown in Figure 2. The first component of the state vector of the orientation learner is the length L of the grasp-line. With this feature, good grasp points are distinguished from those which are

[1] For this higher level criteria, aspects of optimality like reducing torque must possibly be shelved.

Algorithm 1 Algorithm for learning an optimal grasp point

choose an initial grasp point configuration
steps $\leftarrow 0$
repeat
  steps $\leftarrow$ *steps* $+ 1$
  repeat
    learn with the orientation learner
  until [the grasp point is optimal according to orientation OR number of episodes exceeds a given value]
  repeat
    learn with the position learner
  until [the grasp point is optimal according to the position in the object OR number of episodes exceeds a given value]
until [the optimal grasp point is found OR $steps > steps_{max}$]

not adequate, because the gripper cannot cover the object. The remaining features are the corresponding angles $\Theta_1, \ldots, \Theta_4$ between the grasp-line and the flanking straight line segments gained by a simple contour tracking process. The features for the position learner are the distance D between the center of the grasp-line and the center of area of the object's image and the torque T around the normal vector $\boldsymbol{n}$ of the gripper. Due to the learner sep-

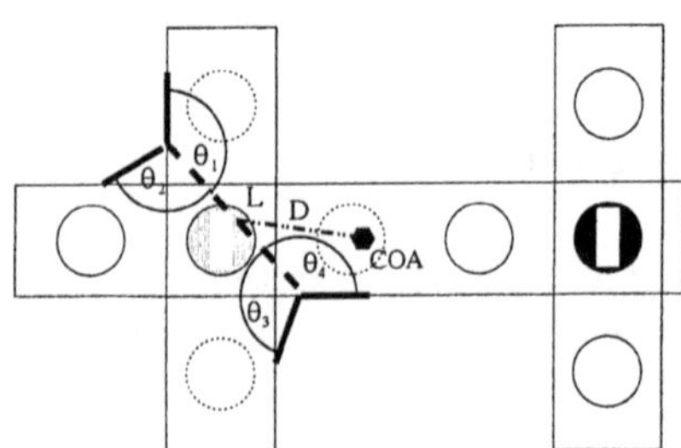

Fig. 2. State coding for the learners. The orientation learner uses length L and angles $\Theta_1, \ldots, \Theta_4$ while the position learner integrates the distance D between the center of the grasp-line and the center of area of the object's image

aration the local criteria have not to be learned for every object anew. The orientation learner is a universal learner which means that the same learner can be used for every object. So this learner will for example learn to grasp objects at opposite parallel or concave edges. The new aspect of this work is the intention for the use of these two learners. As described above this design was chosen in respect to the local and global criteria and their generalization properties.

4 Self-Valuation by Active Experimentation

The presented system is self-valuing, a method to gain estimation for the learning algorithms. Self-valuation is realized via a force/torque sensor and several image processing techniques. It is important to mention that no optimal grasp point is pre-known. The system finds its own grasp points by taking into account the optimality conditions.

4.1 Hardware Configuration

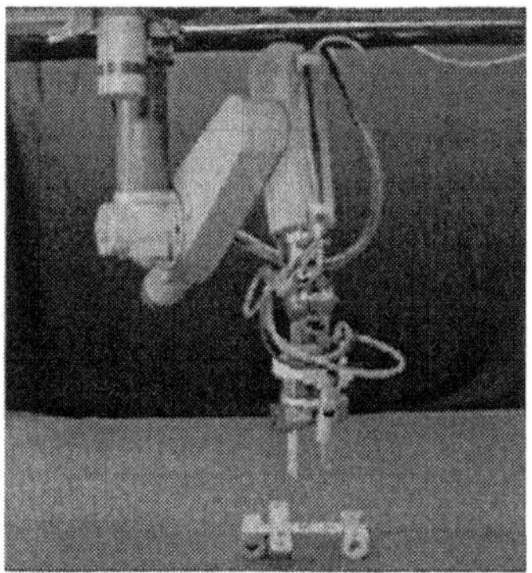

Fig. 3. The hardware setup

The physical set-up of this system consists of the following components:

Main actuator: One 6 d.o.f. PUMA-260 manipulator is installed overhead in a stationary assembly cell. On the wrist of the manipulator, a pneumatic jaw-gripper with integrated force/torque sensor and "self-viewing" hand-eye system (local sensors) is mounted. The robot is controlled by RCCL (*Robot Control C Library*).

Objects: The most kind of objects are constructed from *Baufix* elements, wooden toys for children containing parts like screws, ledges and cubes. Therefore, these objects are also referred to as *aggregates*. An advantage of these parts is that one can construct very quickly new aggregates that can be tested with the system.

4.2 Orientation Learner

A good grasp, determined by the orientation learner, can be best estimated by the second optimality condition, i.e. no friction at the fingers of the parallel-jaw gripper. When an object slips between or out of the fingers at the moment of closing the gripper, the selected grasp configuration was not optimal according to the local criteria. Some existing systems (e.g. [2]) try to determine the friction occurring within the gripper analytically, i.e. by computing the friction cone via geometrical features. Here, several grasp configurations are tried out with the real robot that values the success or failure of the performed grasp thereafter - like a human who does not analytically compute its optimal grips, but learns by success and failure. Because a parallel-jaw gripper, as used in this work, is very rigid and does not slip like human fingers at the object's surface, friction appears either as a rotation or as a displacement of the object itself. So the valuation signal for the orientation learner is basically observed by image processing. A penalty for self valuation is computed as follows:

$$P = \begin{cases} -(\Theta_{\text{diff}} + \mathcal{D}_{\text{diff}}) & \text{if grip was successful} \\ -\mathcal{P}_{\text{const}} & \text{otherwise} \end{cases}$$

where Θ_{diff} is the angle between the initial and the least inertia axis after the performed grasp, D_{diff} the displacement of the center of area and P_{const} a high constant penalty. Figure 4 shows a grasp configuration which results in a rotation of the object itself. If a grasp has totally failed and so the

Fig. 4. Friction of the fingers result in rotation of the object

first optimality criteria cannot be fulfilled[2], a predefined penalty is given. Determining if something is in the gripper after a performed grasp is done via the force sensor. Figure 5 shows that the force in direction of the approach vector raises suddenly while lifting up the object.

4.3 Position Learner

While the valuation technique for the orientation learner is primarily based on processing images from camera sensors, the self-valuation of the position

[2] This either occurs when the orientation of the gripper does not permit to cover the object or the object slips out of the fingers while closing them.

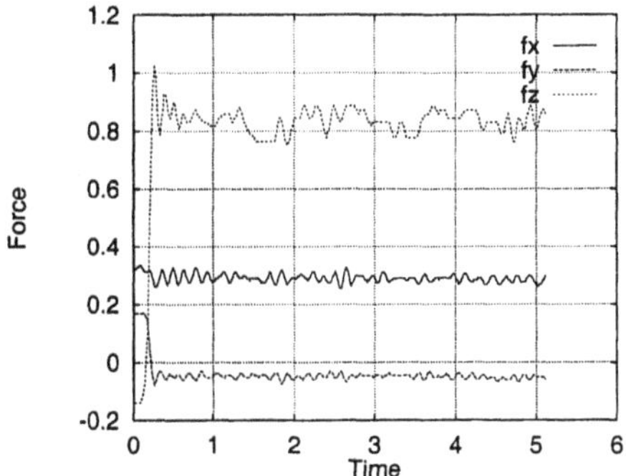

Fig. 5. Force profiles during the lift-up process of an object as in Figure 6(b)

learner is primarily gained via the force/torque sensor. The three optimality conditions of the global grasp criteria, expressed above, are taken into account for self-valuation as follows:

Stable Grasp: The grip is stable, according to the optimality conditions, if the grasped object does not move between the fingers of the gripper. This occurs especially when a heavy object is grasped far away from its center of gravity. The gripper then is perhaps not strong enough to fix the object at this position. Such a situation is shown in Figure 6(a). This lift-up movement of the manipulator results in forces shown in Figure 7. Nearly during the whole lift-up movement, the force in direction of the approach vector $\boldsymbol{a}$ is approximatively constant. At the moment when the object looses contact with the table (in this example at $4s$) the force raises to a higher value. This attitude can be evaluated and used within the learner, e.g. this situation is valued with a predefined high penalty to express that such grips are not desired.

Slipping: Slipping of an object out of the fingers of the gripper is undesirable. This effect mostly occurs as consequence of an unstable grasp as described above. In such a situation the force in direction of $\boldsymbol{a}$ sinks suddenly to zero and the grasp can be thought to have failed. Therefore, a constant penalty is given to prevent the system from using this grip in the future.

Reducing torque: Another goal of the position learner is to reduce torque within the fingers of the gripper. Figure 6(b) shows an example of a grip that produces large torque. The torque profiles are shown in Figure 8. Immediately after the beginning of the lift-up process, the torque around the normal vector $\boldsymbol{n}$ raises to a value which is different from zero and stays constant while holding the object. This torque is computed, negated and directly used with the position learner. Here, no constant penalty is given, because a grip with large torque is not necessarily bad. The system must

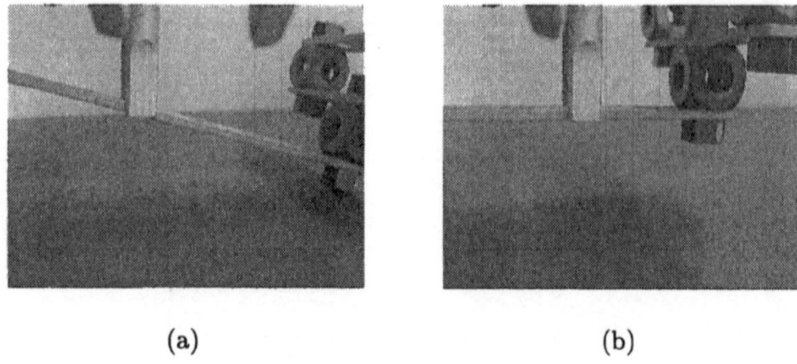

Fig. 6. Grips that are suboptimal according to the optimality conditions

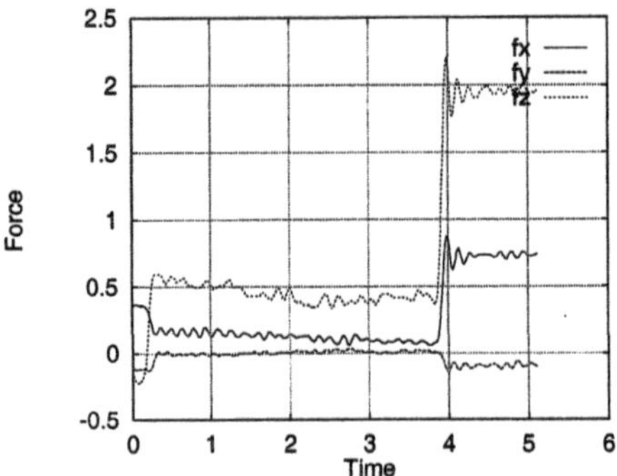

Fig. 7. Force profiles when the grip is not stable as shown in Figure 6(a)

have the possibility to distinguish between grasp points with different torques and choose the best among them.

5 Generalization

The orientation learner is fully applicable to any kind of object, i.e. it provides a total generalization. In other work, where the learning process is not divided into two separate learners, the generalization is only partial. The result could be the same, i.e. these systems also learn to grasp, but they have to reconsider the local criteria every time for grasping a new object. This results in slower learning phases for new objects. Propositions like "grasping at parallel edges is always good" cannot be made by such systems at all. Here, once the orientation learner has learned several grasping situations it can be used with any kind of new object the robot is faced to.

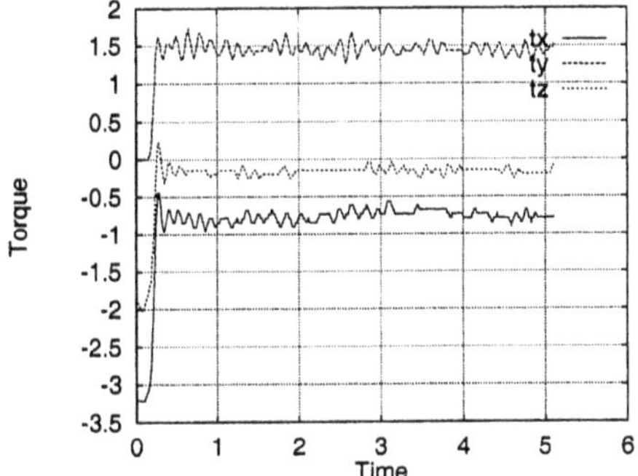

Fig. 8. Torque profiles of the grip in Figure 6(b)

The case of the position learner is more complicated. Because of different shapes of objects the global positions of the learned grasp points cannot be applied to every object. One has to consider if a new position learner must be initialized or can be adopted from a previously learned object. A good technique for generalizing the position learner is to specify simple *sub-objects* which position learners can be used as the basis for any new object. These are for example *basic parts* of aggregates which are not constructed from smaller objects, e.g. a ledge. Not only basic parts, but also small objects which are parts from bigger ones, can be used to provide initial position learners. Three main subproblems must be solved:

1. In which situation can a previously learned position learner fully be adopted to a new object?
2. When can a previously learned position learner be used as a basis for a new object?
3. When must a completely new position learner be initiated?

The idea to be studied is the object hierarchy. This is a challenging task, but can improve generalization functionality among different kind of objects. What we are examining is, if one has found suitable features for describing sub-objects, a complex aggregate can be coded in a tree consisting of these sub-objects. Then, a tree distance model, as for example proposed in [7] and [8], can be used to compare several objects. In this manner one can determine the "most similar" object out of the set of previously learned aggregates for a new object. The three sub-problems mentioned above could then be solved for example as follows:
Let $\mathcal{L}_s = \{(o_i, l_i) | i = 1 \ldots n\}$ be the set of tuples of n previously stored objects o_i in tree notation, together with their stored position learners l_i, and $dist(o_i, o_k)$ the distance of the trees according to a distance measure. Then,

1. a previously learned position learner l' of an object o' can be fully adopted to a new object o, if $\forall(o_i, l_i) \in \mathcal{L}_s \backslash (o', l')$:

$$dist(o', o) \leq dist(o_i, o) \leq D_{min};$$

2. a previously learned position learner l' of an object o' can be used as basis for a new object o , if $\forall(o_i, l_i) \in \mathcal{L}_s \backslash (o', l')$:

$$D_{max} \geq dist(o', o) \leq dist(o_i, o) > D_{min};$$

3. a completely new position learner is initiated for a new object o, if $\forall(o_i, l_i) \in \mathcal{L}_s$

$$dist(o_i, o) > D_{max},$$

where D_{max} and D_{min} are adequate thresholds for accepting and refusing an object to be equal, respectively.

6 Implementation of Learning

6.1 Temporal Difference Learning

The overall objective of our learning system is to maximize the following well-known function in reinforcement learning:

$$Q(s_t, a_t) = r_{t+1} + \gamma Q^{\pi}(s_{t+1}, a_{t+1}) \tag{1}$$

where s_t, a_t are the states and actions at time step t as described in Section 3 and r_t is the penalty after each performed grasp as expressed in Section 4. This function, called Q-function by Watkins [9], measures the expected cumulative reward of executing action a_t at state s_t and thereafter following a policy π, i.e. a strategy of selecting an action in a certain state of the environment. For learning of Q in our system the method of temporal difference learning [10] is employed. The general update formula computes the difference between the current and the next prediction of cumulative reward and updates Q by a fraction of this difference as follows:

$$Q(s_t, a_t) = Q(s_t, a_t) + \alpha \left[r_{t+1} + \gamma Q(s_{t+1}, a_{t+1}) - Q(s_t, a_t)\right] \tag{2}$$

As seen above, our system must handle continuous states and actions, i.e. angles, torques, etc. In such a situation we cannot provide a single value Q for every state-action pair but rather have to use a function approximator. Such a function approximator is of the form $Q_{\omega}(s, a)$, where $\omega = (\omega(1), \omega(2), \ldots, \omega(n))^T$ is a set of adjustable weights. The update of the current estimate of Q is performed by modifying the weights according to the following rule:

$$\Delta\omega_t = \alpha \left[r_{t+1} + \gamma Q_{t+1} - Q_t\right] \nabla_{\omega_t} Q_t \tag{3}$$

A general advantage of function approximators is that they are able to *generalize.* The system is able to estimate the expected return of state-action pairs that where never visited before. Although a function approximator can deal with continuous state and action spaces, it may not be able to accurately represent Q for the entire state and action space due to its finite resources.

We employ the B-spline function approximator for the Q-function which is a natural generalization of coarse coding to continuously-valued features.

6.2 The B-Spline Model

To solve the problem of numerical approximation for smoothing statistical data, "basis splines" (B-splines) were introduced by Schoenberg [11]. B-splines were used later by Riesenfeld [12] and Gordon [13] in *Computer Aided Geometric Design* (CAGD) for curve and surface representation. Because of their versatility based on only low-order polynomials and their straightforward computation, B-splines have become more and more popular. Nowadays, B-spline techniques represent one of the most important trends in CAD/CAM; they have been extensively applied in modeling free shape curves and surfaces. Recently, splines have also been proposed for neural network modeling and control [14,15].

Assume x is a general input variable of a control system that is defined on the universe of discourse $[x_1, x_m]$. Given a sequence of ordered parameters (knots): $x_1, x_2, \ldots$, the ith B-spline $N_{i,k}$ of order k (degree $k-1$) is recursively defined as follows (Figure 9):

$$N_{i,k}(x) = \begin{cases} \begin{cases} 1 & \text{for } x \in [x_i, x_{i+1}) \\ 0 & \text{otherwise} \end{cases} & \text{if } k = 1 \\ \frac{x - x_i}{x_{i+k-1} - x_i} N_{i,k-1}(x) + \frac{x_{i+k} - x}{x_{i+k} - x_{i+1}} N_{i+1,k-1}(x) & \text{if } k > 1 \end{cases} \quad (4)$$

with $i = 1, \ldots, m-k$. Therefore m knots $x_i (i = 1, \ldots, m)$ form $l = m - k$ B-splines.

Figure 10 illustrates the partition of a two-dimensional B-spline model with 8 B-splines on each uniformly subdivided input interval and the activated ones (slightly shaded) for a given input. Since learning one new part of the input space affects only a given number of controller response values (darkly shaded area of Figure 10), fast on-line learning can be devised. By using the B-spline model the approximation ability is only limited by the number of knots distributed over the input intervals. Regarding that most observed data are disturbed to a certain degree, the over-fitting problem may occur. Genetic algorithm (GA) optimized B-spline models are a promising approach to find sparse models, which are able to bridge the gap between high bias and high variance of a model.

The B-spline model provides an ideal implementation of the CMAC. The CMAC model provides a neurophysiological interpretation of the B-spline model.

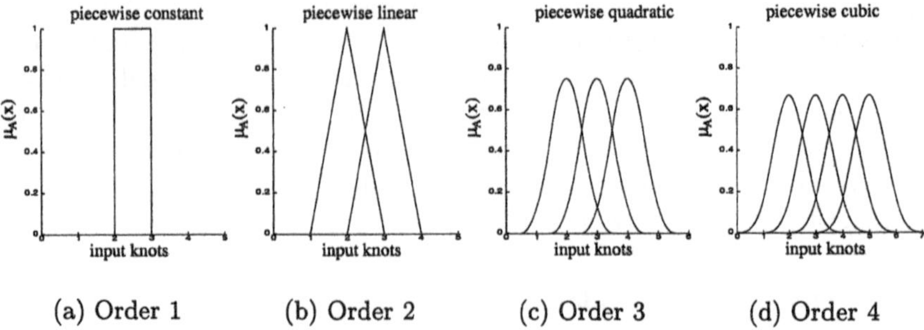

(a) Order 1 (b) Order 2 (c) Order 3 (d) Order 4

Fig. 9. Univariate B-splines of order 1–4

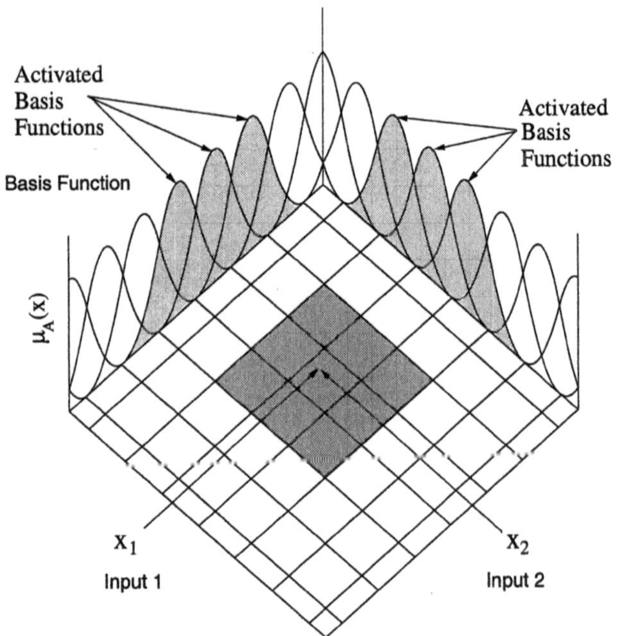

Fig. 10. The B-spline model – a two-dimensional illustration

6.3 Approximating Q

In the following we define $\boldsymbol{x}$ as the concatenation of the current state $s = (s_1, \ldots, s_n)$ and the taken action $a = (a_1, \ldots, a_m)$, that is: $\boldsymbol{x} = (s_1, \ldots, s_n, a_1, \ldots, a$ The output for the B-spline function approximator which is the prediction of

Q is computed by:

$$Q(\boldsymbol{x}) = \frac{\sum_{i_1=1}^{l_1} \cdots \sum_{i_n=1}^{l_n} \left(c_{i_1,\ldots,i_n} \prod_{j=1}^{n} N_{i_j,k_j}^{j}(x_j) \right)}{\sum_{i_1=1}^{l_1} \cdots \sum_{i_n=1}^{l_n} \prod_{j=1}^{n} N_{i_j,k_j}^{j}(x_j)} \tag{5}$$

$$= \sum_{i_1=1}^{l_1} \cdots \sum_{i_m=1}^{l_n} \left(c_{i_1,\ldots,i_n} \prod_{j=1}^{n} N_{i_j,k_j}^{j}(x_j) \right) \tag{6}$$

where:

- x_j: the jth input ($j = 1, \ldots, n$),
- k_j: the order of the B-splines used for x_j,
- N_{i_j,k_j}^{j}: the ith B-spline of x_j,
- $i_j = 1, \ldots, l_j$: represents the index of the B-spline of x_j,
- l denotes the number of B-splines and
- $c_{i_1,i_2,\ldots,i_n}$: the *control vertices*[3].

This is called a *general NUBS hypersurface* , which possesses the following properties:

- If the B-splines of order $k_1, k_2, \ldots, k_n$ are employed to cover the spaces of the input variables $x_1, x_2, \ldots, x_n$, it can be guaranteed that the output variable y is $(k_j - 2)$ times continuously differentiable with respect to the input variables $x_j, j = 1, \ldots, n$.
- If the input space is partitioned fine enough and at the correct positions, the interpolation with the B-spline hypersurface can reach a given precision.

Because the introduced weights ω of Q here correspond to the control vertices $c_{i_1,i_2,\ldots,i_n}$ of the B-spline function approximator, the gradient of Q with respect ω from Equation (3) is:

$$\nabla_{\omega} Q = \nabla_{c_{i_1,i_2,\ldots,i_n}} Q$$

Now the learning update from Equation (3) turns into the following formula:

$$\Delta c_{i_1,i_2,\ldots,i_n} = \alpha \left[r_{t+1} + \gamma Q_{t+1} - Q_t \right] \prod_{j=1}^{n} N_{i_j,k_j}^{j}(x_j) \tag{7}$$

Based on this, the control vertices are updated online after each grasping trial of the system.

[3] Corresponding to *de Boor points* in CAGD.

6.4 Accumulating Trails

A practical problem that arises is that the system will learn a path from an initial state, i.e initial grasping configuration, upto a final state, i.e. successful grip. To overcome this side effect in systems where the goal state is the most important outcome and not the path there, we propose an easy and slightly new approach for increasing performance of such a learning system, called *accumulating trails.* When a learning systems learns a type of path from an initial state to a final state, i.e. by applying a set of actions $a_0, \ldots a_n$ to s and its successors, it is sometimes possible to get to the same goal state if applying a set of actions $a'_0 \ldots a'_m$ to the state s and its successors, where $m < n$. That is to say, that one would reach the goal state $n - m$ steps earlier.

Let ψ denote the function applying an action a to a state s, denoted $\psi : \mathcal{A} \rightarrow (\mathcal{S} \rightarrow \mathcal{S})$, where $\mathcal{A}, \mathcal{S}$ are the total sets of actions and states, respectively. The outcome of this function, applying it to an action, is a function on the state space $\mathcal{S}$ called *action execution function.*

Using the definition above, each learning episode can be considered as a composition of functions $(A : \mathcal{S} \rightarrow \mathcal{S})$

$$A(s) = \psi(a_n) \circ \psi(a_{n-1}) \circ \cdots \circ \psi(a_0)(s)$$

where s is the starting state of the episode and a_i is the action applied in time step i. This function composition is further referred as *sequence.*

A sequence B of action executions $\psi(b_m) \circ \cdots \circ \psi(b_0)$ is called a *sub-sequence* of sequence $A = \psi(a_n) \circ \cdots \circ \psi(a_0)$, if $A(s) = B(s)$:

$$\psi(a_n) \circ \cdots \circ \psi(a_0)(s) = \psi(b_m) \circ \cdots \circ \psi(b_0)(s), \; m \leq n$$

where s is the starting state. Then, sequence A is called *substitutable* through B. The sub-sequence B always produces the same resulting state as the sequence A. That means, if starting in state s it makes no difference whether to "follow" sequence B or sequence A. The state at the end of the sequence is always the same. If a sequence A is not substitutable through any other sequence B, it is called *final.* When the agents intention is to reach the goal states as soon as possible, as for example in this work[4], the learning algorithm should converge to a situation of only final sequences.

An accumulation function on action executions is defined as

$$\oplus : (\mathcal{S} \rightarrow \mathcal{S}) \times (\mathcal{S} \rightarrow \mathcal{S}) \rightarrow (\mathcal{S} \rightarrow \mathcal{S}).$$

A sequence $A = \psi(a_n) \circ \cdots \circ \psi(a_0)$ of actions executions is *accumulatable*, if

$$\psi(a_n) \oplus \psi(a_{n-1}) \oplus \cdots \oplus \psi(a_0) = B,$$

[4] It is desirable to find an optimal grasp point as soon as possible.

where B is subsequence of A. The accumulation function describes how to combine action executions to produce shorter sequences. This function has to be defined according to the learning system one wants to develop. The accumulation is defined on action executions and not solely on actions, because it depends on the states if such an accumulation can be performed. In some situation the accumulation function is defined as follows:

$$\psi(a_k) \oplus \psi(a_l) = \psi(a_k \diamond a_l), \tag{8}$$

where $\diamond$ is a function $\diamond : \mathcal{A} \times \mathcal{A} \rightarrow \mathcal{A}$.

In most situations, the accumulation function must include a kind of model of the environment and this is only possible by taking into account also the states rather than only the actions as supposed by Equation (8). The agent must "know" in which situations it is possible to accumulate action executions and in which situation it is not. However, for some tasks Equation (8) is an easy and sufficient definition.

As an example, for application within the orientation learner the accumulation function $\diamond$ is defined as:

$$a_k \diamond a_l = \begin{cases} a_k + a_l & \text{if} \quad -90 \leq a_k + a_l \leq 90 \\ a_k + a_l + 180 & \text{if} \quad a_k + a_l < -90 \\ a_k + a_l - 180 & \text{if} \quad a_k + a_l > 90 \end{cases}$$

assuming that the actions of the orientation learner are rotational movements from the interval $[-90, \ldots, 90]$.

7 Experimental Results

To get a uniform and matchable view of the objects, the system learns to grasp, the manipulator initially moves itself over the object so that the x-axis of the cameras coordinate system appears parallel to the axis of least inertia of the object and the center of area in the right side of the image. The center of the object's bounding box coincides with the center of the image. An additional tool-transformation is performed, so that the camera is moved in direction to the working surface.

Several objects were used to test the performance of the whole system. Some of them are shown in Figure 11. The robot has found a good and stable grasp point for each object that fulfills the optimality conditions given above, most times near the object's center of gravity. Two special results of a grasping operation are shown in Figure 12. In Figure 12(a) the manipulator grasped the object at a point different from the center of area but near the center of mass of the object. Figure 12(b) shows a successful grasp at a convex edge of a different object. To show the generalization ability of the orientation learner, it was first trained on a new object until a defined number of epoches. Thereafter, the same learner was used on a different object to show that the

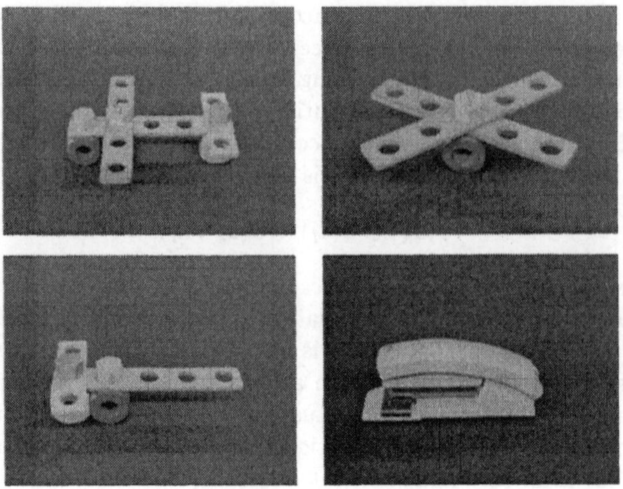

Fig. 11. Sample objects

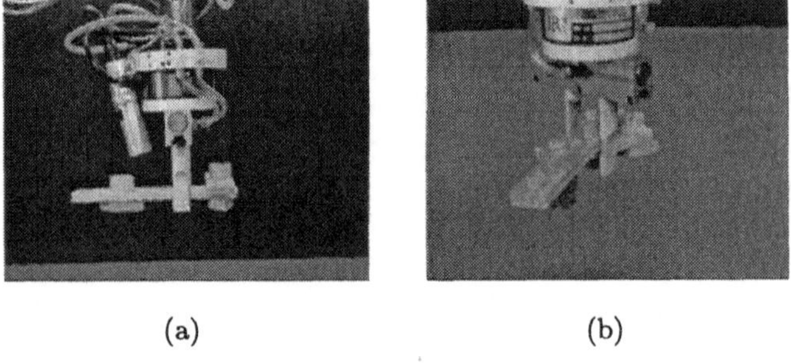

(a) (b)

Fig. 12. Successfully performed grasping operations

average steps until the goal state decrease much faster. The result is shown in Figure 13. In the second part of the experiment the orientation learner did not start at the average steps of 3 where the initially performed orientation learner ended. This is due to the fact that in the first cycle a simple ledge was used and the learner still was not fully trained while in the second cycle a more complex object was used. However, one can see that in the second cycle the orientation learner was quicker. Only new states that do not occur on the simple ledge has to be learned additionally.

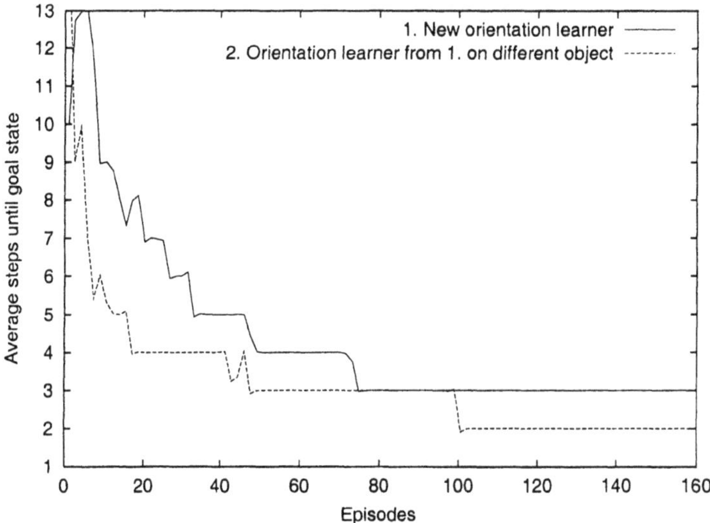

Fig. 13. Generalization of the orientation learner

8 Discussion and Future Work

We present a self-valuing learning system that is capable of grasping various kind of objects. Our system consists of two learners based on local and global grasping criteria. These criteria are supposed human learning abilities in the field of grasping. While the orientation learner is applicable to any kind of object and therefore fully generalizes between them, the position learner is mostly dependent on a special object and its physical properties. The system shows the ability to grasp several kind of objects and to generalize the learned faculties to new ones.

We used the B-spline model to implement the function approximator of the learning system. This turns out to be a good technique, because if the input space is partitioned fine enough and at the correct positions, the interpolation with the B-spline hypersurface can reach a given precision.

An interesting future work is to adopt the presented system to a multi-fingered robot hand. With such a hand a single grasp point is much more complex than with a parallel-jaw gripper. Also the possible actions of the learners are more challenging. However, the basic principle of two learners, based on local and global criteria, and the self-valuing approach could be maintained.

References

1. A. Bicchi and V. Kumar. Robotic grasping and contact: A review. In *Proceedings of the IEEE International Conference on Robotics and Automation*, 2000.
2. G. Smith, E. Lee, K. Goldberg, K. Boehringer, and J. Craig. Computing parallel-jaw grips. In *Proceedings of the IEEE International Conference on Robotics and Automation*, pages 1897–1903, 1999.
3. A. Morales, P. J. Sanz G. Recatalá, and Ángel P. del Pobil. Heuristic vision-based computation of planar antipodal grasps on unknown objects. In *Proceedings of the IEEE International Conference on Robotics and Automation*, 2001.
4. I. Kamon, T. Flash, and S. Edelman. Learning to grasp using visual information. In *Proceedings of the IEEE International Conference on Robotics and Automation*, pages 2470–2476, 1996.
5. J. Zhang, G. Brinkschröder, and A. Knoll. Visuelles Reinforcement-Lernen zur Feinpositionierung eines Roboterarms über kompakte Zustandskodierung. In *Tagungsband Autonome Mobile Robotersysteme*, München, 1999.
6. M.-C. Nguyen and V. Graefe. Self-learning vision-guided robots for searching and grasping objects. In *Proceedings of the IEEE International Conference on Robotics and Automation*, pages 1633–1638, 2000.
7. K. C. Tai. The tree-to-tree correction problem. *J. Assoc. Comput. Mach.*, 26(3):422–433, 1979.
8. T. Jiang, L. Wang, and K. Zhang. Alignment of Trees - An Alternative to Tree Edit. *Theoretical Computer Science*, 143(1):137–148, 1995.
9. Ch. J. C. H. Watkins. *Learning from Delayed Rewards.* PhD thesis, King's College, Cambridge, England, 1989.
10. R. S. Sutton. Learning to predict by the method of temporal differences. *Machine Learning,3:9-44*, 1988.
11. I. J. Schoenberg. Contributions to the problem of approximation of equidistant data by analytic functions. *Quarterly of Applied Mathematics*, 4:45–99, 112–141, 1946.
12. R. F. Riesenfeld. *Applications of B-Spline approximation to geometric problems of computer-aided design.* PhD thesis, Syracuse University, 1973.
13. W. J. Gordon and R. F. Riesenfeld. B-spline curves and surfaces. In R. E. Barnhill and R. F. Riesenfeld, editors, *Computer Aided Geometric Design.* Academic Press, 1974.
14. M. Brown and C. J. Harris. *Neural networks for modelling and control*, chapter I, pages 17–55. In "Advances in Intelligent Control", Ed., C. J. Harris, Taylor & Francis, London, 1994.
15. H. W. Werntges. Partition of unity improve neural function approximators. In *Proceedings of IEEE International Conference on Neural Networks, San Francisco*, volume 2, pages 914–918, 1993.

Online Adaptive Fuzzy Neural Identification and Control of Nonlinear Dynamic Systems

Meng Joo Er and Yang Gao

School of Electrical and Electronic Engineering
Nanyang Technological University
Nanyang Avenue, Singapore 639798 (Republic of Singapore)
emjer@ntu.edu.sg, ygao@pmail.ntu.edu.sg

Abstract. This chapter presents a robust Adaptive Fuzzy Neural Controller (AFN C) suitable for identification and control of uncertain Multi-Input-Multi-Output (MIMO) nonlinear systems. The proposed controller has the following salient features: (1) Self-organizing fuzzy neural structure, i.e. fuzzy control rules can be generated or deleted automatically; (2) Online learning ability of uncertain MIMO nonlinear systems; (3) Fast learning speed; (4) Fast convergence of tracking errors; (5) Adaptive control, where structure and parameters of the AFNC can be self-adaptive in the presence of disturbances to maintain high control performance; (6) Robust control, where global stability of the system is established using the Lyapunov approach. Two simulation examples are used to demonstrate excellent performance of the proposed controller.

1 Introduction

Design of robust adaptive controllers suitable for real-time control of MIMO nonlinear systems is one of the most challenging tasks for many control engineers, especially when the nonlinear system is required to manoeuvre very quickly under external disturbances. In the late 1980's to early 1990's, adaptive control has undergone rapid development leading to global stability and excellent tracking results for reasonably large classes of nonlinear systems [1,2].

Conventional adaptive controllers based on nonlinear control laws can achieve fine control and compensate partially unknown system dynamics. However, they often suffer from heavy computational burden and this hinders their real-time applications [2]. Although variable-structure control strategy using sliding mode is an effective way to deal with uncertainties in nonlinear systems, the chattering phenomenon due to switching operation greatly affects the accuracy of tracking performance. Moreover, in the design of a sliding-mode controller, mathematical models of the system and the bound of uncertainties need to be known in advance [3]. Hence, there is a need for model-free control strategies with learning and adaptive ability.

In the last few decades, much research effort has been put into the design of intelligent controllers using fuzzy logic and neural networks. Fuzzy logic provides human reasoning capabilities to capture uncertainties, which cannot be described by precise mathematical models. Neural networks offer exciting advantages such as adaptive learning, parallelism, fault tolerance and generalization. They have been proven to be very powerful techniques in the discipline of system control, especially when the controlled system is hard to be modeled mathematically, or when the controlled system has large uncertainties and strong nonlinearities. Therefore, fuzzy logic and neural networks have been greatly adopted in model-free adaptive control of nonlinear systems [3–7]. Furthermore, a few hybrid techniques were applied to adaptation of parameters in fuzzy and/or neural controllers, like sliding mode control [3,8–10], Bayesian probability [11,12], genetic algorithms [13], neuron-like structure [14], hybrid pi-sigma network [15] and RBF neural networks [7,16]. However, it turns out that only adjustment of parameters will not be sufficient in many cases. For example, if the number of fuzzy rules or number of hidden layers and neurons is very large, real-time implementation will be difficult or impossible. More importantly, this reduces the flexibility and numerical processing capability of the controller and results in redundant or inefficient computation. Therefore, the controller structure needs to be adaptive so that a compact fuzzy or neural control system can be obtained. In [17], the structure of fuzzy rules was optimized by genetic algorithms. In [18], evolution strategies were proposed to optimize fuzzy control structure and its associated parameters simultaneously. These methods are found to be quite useful in system control. Unfortunately, offline learning is required and there are still difficulties in the initialization of control structure and the associated parameters. In [19], BackPropagation (BP) network was utilized to determine the number of fuzzy rules for each input variable. In [20], a dynamically structured neural network using wavelets was proposed and the idea was extended to adaptive control of robot manipulators in [21]. These methods are successful in that it is not necessary to determine exactly the structure and parameters of a fuzzy or neural controller in advance. However, the problem with either BP or wavelets algorithm is that the learning and adaptation speed are slow. As a consequence, control system response deteriorates.

This motivates us to investigate adaptive learning algorithms for constructing a fuzzy and/or neural control system systematically and automatically. The resulting intelligent controller must have fast online adaptability to guarantee good real-time control performance. In line with this objective, a new Adaptive Fuzzy Neural Controller (AFNC), which is built based on a Generalized Fuzzy Neural Network (G-FNN) controller employing the G-FNN learning algorithm, is proposed. The G-FNN controller possesses both the advantages of fuzzy logic, such as human-like thinking and ease of incorporating expert knowledge, and neural networks, such as learning abilities and connectionist structures. By virtue of this, low-level learning and computational power of neural networks can be incorporated into the fuzzy logic system on one hand and high-level human-like thinking and reasoning of fuzzy logic systems can be incorporated into neural networks on the other hand. The G-FNN algorithm offers a fast online learning algorithm, which can recruit or delete fuzzy control rules or neurons dynamically without pre-definition of the structure. Its outstanding computational efficiency in terms of learning speed, adaptability and generalization has been verified in some of our latest work [22,23]. In essence, the G-FNN algorithm enables the G-FNN controller to successfully model the nonlinear system dynamics and its uncertainties online.

The rest of the chapter is organized as follows. Section 2 introduces the dynamic model of the MIMO nonlinear systems under consideration. Section 3 presents nonlinear systems' identification using the G-FNN controller. This is followed by Section 4 that describes the design procedure of the AFNC in details. Convergence of the G-FNN controller and global stability of the closed-loop control system are proven using the Lyapunov theory. Section 5 presents simulation results and discussions on an inverted pendulum system and a two-link robot manipulator. Finally, Section 6 concludes the chapter. Detailed descriptions of the G-FNN architecture, learning algorithm and modelling method are shown in the Appendix.

2 MIMO Nonlinear System Dynamics

The class of nth-order MIMO nonlinear systems considered in this chapter, termed *companion form* or *controllability canonical form*, is given by [1]:

$$\mathbf{z}^{(n)} = \mathbf{F}(\underline{\mathbf{z}}) + \mathbf{G}(\underline{\mathbf{z}})\mathbf{u} + \mathbf{D} \tag{1}$$

where

- $\mathbf{u} \in \Re^{n_i}$ and $\mathbf{z} \in \Re^{n_o}$ are the input and output vectors of the MIMO nonlinear system respectively, with n_i and n_o being the total number of system inputs and outputs respectively.
- $\underline{\mathbf{z}} = [\mathbf{z}^T \ \dot{\mathbf{z}}^T \ldots \mathbf{z}^{(n-1)T}]^T \in \Re^{n_o n}$ is the state vector of the system.
- $\mathbf{F}(\underline{\mathbf{z}}) \in \Re^{n_o}$ and $\mathbf{G}(\underline{\mathbf{z}}) \in \Re^{n_o \times n_i}$ represent smooth nonlinearities of the dynamic system.
- $\mathbf{D} \in \Re^{n_o}$ is an unknown function representing system uncertainties and external disturbances.

Since we require that the MIMO nonlinear system (1) is controllable, the input gain $\mathbf{G}(\underline{\mathbf{z}})$ needs to be invertible for all $\underline{\mathbf{z}} \in U_c$. The function $\mathbf{G}$ is assumed to be known and bounded. Furthermore, $\mathbf{F}$, and $\mathbf{D}$ are assumed to be bounded.

3 Adaptive Fuzzy Neural Identification of Nonlinear Systems

In the context of using the G-FNN directly for nonlinear control, G-FNN is viewed as a means of system identification, or even a framework for knowledge representation. The knowledge about system dynamics and mapping characteristics are implicitly stored within the network. Therefore, training a G-FNN using input-output data from a nonlinear system becomes a central issue to its use in control. The adaptive modelling capability of the G-FNN is summarized in Appendix A.3. In this section, adaptive fuzzy neural identification of MIMO nonlinear systems using G-FNN will be elaborated.

Inverse modelling of dynamical systems plays a crucial role in a range of control problems, which will become apparent in the next section. In this chapter, direct modelling of system's inverse dynamics by the G-FNN as illustrated in Fig. 1, is attempted. It can be easily derived from Eq. (1) that the inverse dynamics of the nonlinear system is given by:

$$\mathbf{u}(\overline{\mathbf{z}}) = \mathbf{G}(\underline{\mathbf{z}})^{\dagger}[\mathbf{z}^{(n)} - \mathbf{F}(\underline{\mathbf{z}}) - \mathbf{D}] = \mathbf{\Omega}(\overline{\mathbf{z}}) \tag{2}$$

where $\mathbf{G}(\underline{\mathbf{z}})^{\dagger}$ is the pseudoinverse of $\mathbf{G}$ if $n_i \neq n_o$ or the inverse of $\mathbf{G}$ if $n_i = n_o$, and $\overline{\mathbf{z}}_r = [\underline{\mathbf{z}}^T \;\; \mathbf{z}^{(n)T}]^T$. The G-FNN is trained to obtain an estimate of the inverse dynamics, $\hat{\mathbf{\Omega}}$, i.e. the one-to-one mapping relationship from $\overline{\mathbf{z}}$ to $\mathbf{u}$. This is achieved by applying the G-FNN learning algorithm, which is capable of estimating the mapping relationship by determining the appropriate structure and parameters of a fuzzy neural system.

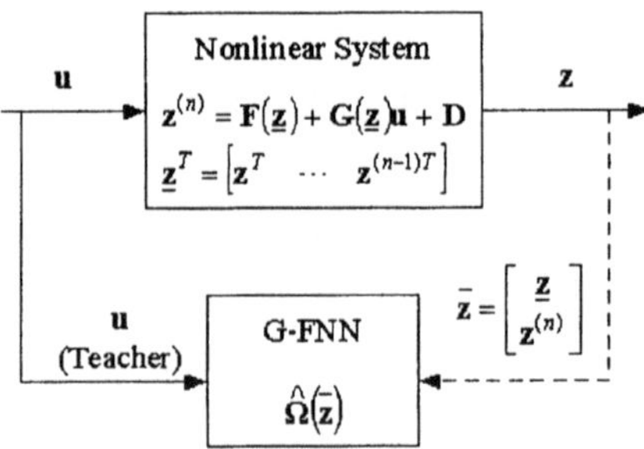

Fig. 1. G-FNN inverse modelling of a nonlinear system

The resulting G-FNN can therefore be used to estimate the input signal $\mathbf{u}_{G-FNN}$ of the nonlinear system given the desired output vector $\overline{\mathbf{z}}$. From Eqs. (A-5) and (2), the output of the G-FNN can be shown to be

$$\mathbf{u}_{G-FNN}(\overline{\mathbf{z}}|\mathbf{W}) = \hat{\mathbf{\Omega}}(\overline{\mathbf{z}}|\mathbf{W}) = \mathbf{W}^T\mathbf{\Phi}(\overline{\mathbf{z}}) \tag{3}$$

where $\mathbf{W} \in \Re^{N_r(N_i+1)\times N_o}$ is the weight matrix of the G-FNN, $\mathbf{\Phi} \in \Re^{N_r(N_i+1)}$ is the regressor vector associated with $\overline{\mathbf{z}}$, $N_i = n_o(n+1)$ and $N_o = n_i$ are the number of inputs and outputs of the G-FNN respectively, and N_r is the total number of fuzzy control rules inside G-FNN.

Define the optimal parameter $\mathbf{W}^*$ for a given regressor vector $\mathbf{\Phi}(\overline{\mathbf{z}})$ as

$$\mathbf{W}^* \triangleq \arg\min[\sup_{\overline{\mathbf{z}} \in U_c} ||\mathbf{\Omega}(\overline{\mathbf{z}}) - \hat{\mathbf{\Omega}}(\overline{\mathbf{z}}|\mathbf{W})||] \tag{4}$$

and the *minimum approximation error vector*, $\varepsilon \in \Re^{n_i}$ as

$$\varepsilon \triangleq \mathbf{\Omega}(\overline{\mathbf{z}}) - \hat{\mathbf{\Omega}}(\overline{\mathbf{z}}|\mathbf{W}^*) \tag{5}$$

In this chapter, we assume that ε is upper bounded by $|\varepsilon| \leq \varepsilon_N$, where ε_N is a known vector containing positive values. As a consequence, Eq. (2) can be established by the G-FNN as follows:

$$\mathbf{u}(\overline{\mathbf{z}}) = \hat{\mathbf{\Omega}}(\overline{\mathbf{z}}|\mathbf{W}^*) + \varepsilon \tag{6}$$

$$= \mathbf{W}^{*T}\mathbf{\Phi}(\overline{\mathbf{z}}) + \varepsilon \tag{7}$$

4 Adaptive Fuzzy Neural Control of Nonlinear Systems

4.1 Adaptive Fuzzy Neural Control Structure

The objective of this work is to design a robust Adaptive Fuzzy Neural Controller (AFNC) for a class of MIMO nonlinear systems in companion form given by Eq. (1), which guarantees boundedness of all closed-loop variables and tracking of a given desired signal $\mathbf{z}_d(t)$. We define the tracking error $\mathbf{e}$ as follows:

$$\mathbf{e} \triangleq \mathbf{z}_d - \mathbf{z} \in \Re^{n_o} \tag{8}$$

To facilitate the following development, we define the dynamic tracking error $\mathbf{s}$ as follows:

$$\mathbf{s} \triangleq \mathbf{z}_r^{(n-1)} - \mathbf{z}^{(n-1)} \in \Re^{n_o} \tag{9}$$

where $\mathbf{z}_r^{(n-1)}$ is the so-called "reference" value of $\mathbf{z}^{(n-1)}$ obtained by modifying $\mathbf{z}_d^{(n-1)}$ according to the tracking errors as follows:

$$\mathbf{z}_r^{(n-1)} = \mathbf{z}_d^{(n-1)} + \mathbf{\Upsilon}_{n-2}\mathbf{e}^{(n-2)} + \ldots + \mathbf{\Upsilon}_0\mathbf{e} \tag{10}$$

Here, $\mathbf{\Upsilon}_i \in \Re^{n_o \times n_o}$, $i = 0, 1, \ldots, n-2$ is a diagonal matrix whose diagonal entries are positive values. Substituting (10) into (9), we have

$$\mathbf{s} = \mathbf{e}^{(n-1)} + \mathbf{\Upsilon}_{n-2}\mathbf{e}^{(n-2)} + \ldots + \mathbf{\Upsilon}_0\mathbf{e} \tag{11}$$

To achieve asymptotic perfect tracking of the desired configuration $\mathbf{z}_d$ with all $\dot{\mathbf{z}}_d \ldots \mathbf{z}_d^{(n)}$ known and bounded, the perfect control law $\mathbf{u}^*$ can be designed as follows:

$$\mathbf{u}^* = \mathbf{G}(\underline{\mathbf{z}})^\dagger[\mathbf{z}_r^{(n)} - \mathbf{F}(\underline{\mathbf{z}}) - \mathbf{D}] + \mathbf{K_D}\mathbf{s} \tag{12}$$

$$= \mathbf{\Omega}(\overline{\mathbf{z}}_r) + \mathbf{K_D}\mathbf{s} \tag{13}$$

where the function $\mathbf{\Omega}(.)$ represents the inverse dynamics of the nonlinear system, $\overline{\mathbf{z}}_r = [\underline{\mathbf{z}}^T \ \mathbf{z}_r^{(n)T}]^T \in \Re^{n_o(n+1)}$, and $\mathbf{K_D} \in \Re^{n_i \times n_o}$.

If the parameters of the system are all known, this control law leads to the following tracking error dynamics:

$$\mathbf{G}(\underline{\mathbf{z}})^{\dagger}\dot{\mathbf{s}} + \mathbf{K_D}\mathbf{s} = 0 \tag{14}$$

which gives exponential convergence of $\mathbf{s}$ by proper choice of $\mathbf{K_D}$, which in turn guarantees convergence of $\mathbf{e}$.

However, external disturbances and unmodeled dynamics represented by $\mathbf{D}$ are unknown in practice. Therefore, we cannot implement the perfect control law. To circumvent this problem, the G-FNN is proposed to generate the optimal torque to approximate the perfect control law. By substituting Eqs. (6) and (7) into Eq. (13), the perfect control law can be executed using the G-FNN as follows:

$$\mathbf{u}^* = \hat{\mathbf{\Omega}}(\overline{\mathbf{z}}_r|\mathbf{W}^*) + \varepsilon + \mathbf{K_D}\mathbf{s} \tag{15}$$

$$= \mathbf{W}^{*T}\mathbf{\Phi}(\overline{\mathbf{z}}_r) + \varepsilon + \mathbf{K_D}\mathbf{s} \tag{16}$$

The proposed robust AFNC is therefore designed to mimic the perfect control law given by Eq. (16). As depicted in Fig. 2, the G-FNN controller is connected in parallel with a linear controller to generate a compensated control signal. The control law of the AFNC is given by

$$\mathbf{u}_c = \hat{\mathbf{\Omega}}(\overline{\mathbf{z}}_r|\mathbf{W}) + \varepsilon_N \cdot \operatorname{sgn}(\mathbf{G}^T\mathbf{P}\mathbf{s}) + \mathbf{K_D}\mathbf{s} \tag{17}$$

$$= \mathbf{W}^T\mathbf{\Phi}(\overline{\mathbf{z}}_r) + \varepsilon_N \cdot \operatorname{sgn}(\mathbf{G}^T\mathbf{P}\mathbf{s}) + \mathbf{K_D}\mathbf{s} \tag{18}$$

where the operator "$\cdot$" denotes element-by-element multiplication of the vector ε_N and $\operatorname{sgn}(\mathbf{G}^T\mathbf{P}\mathbf{s})$, and $\mathbf{P} \in \Re^{n_o \times n_o}$ is a symmetric positive definite matrix.

In other words, the G-FNN controller is formed in such a way that it captures the inverse dynamic of the controlled system, i.e. the mapping relationship from $\overline{\mathbf{z}}$ to $\mathbf{u}_c$. This is achieved by applying the G-FNN learning algorithm, which is able to determine the appropriate structure and parameters of a fuzzy neural system in order to estimate the desired mapping relationship. Its online learning ability enables parameters of the G-FNN, such as c_{ij}, σ_{ij} and $\mathbf{W}(0)$, to be obtained during real-time control of the plant. Detailed explanations of the G-FNN architecture and the G-FNN learning algorithm are shown in Appendix A.1 and A.2 respectively. After the initial value of the weight matrix $\mathbf{W}(0)$ is obtained from the G-FNN learning algorithm, $\mathbf{W}$ is further adjusted and obtained by the adaptive law derived in Section 4.2. This is to compensate for modeling errors of the G-FNN learning algorithm and fulfill stability requirement of the entire control system. The stability of the AFNC system is proven in Section 4.3.

Remark 1. It should be noted that the G-FNN learning algorithm is operating in discrete time to produce the parameters (i.e. c_{ij}, σ_{ij} and $\mathbf{W}(0)$) for

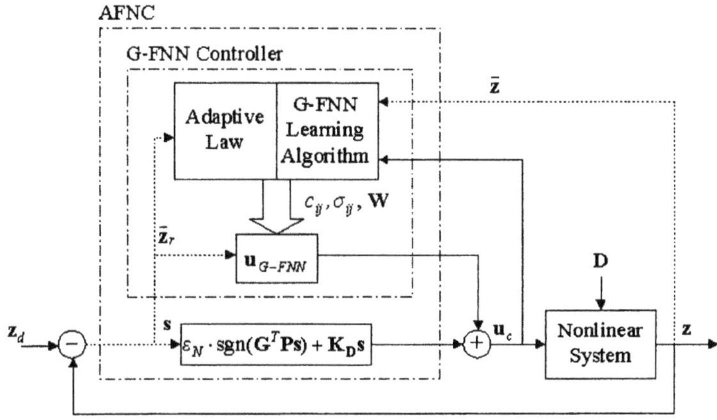

Fig. 2. Adaptive fuzzy neural control structure

the G-FNN controller. The parameter $\mathbf{W}(0)$ is the initial value of the weight matrix $\mathbf{W}$, which will be further adjusted in continuous time according to the adaptive law derived in Section 4.2.

4.2 Convergence Analysis of AFNC

From Eqs. (1), (15) and (17), the system tracking error equation can be shown to be

$$\dot{\mathbf{s}} = \mathbf{As} + \mathbf{G}(\underline{\mathbf{z}})(\mathbf{u}^* - \mathbf{u}_c) \tag{19}$$

where $\mathbf{A} = -\mathbf{G}(\underline{\mathbf{z}})\mathbf{K_D} \in \Re^{n_o \times n_o}$ is designed to be a Hurwitz matrix.

Substituting Eqs. (16) and (18) into Eq. (19), we have

$$\dot{\mathbf{s}} = \mathbf{As} + \mathbf{G}(\underline{\mathbf{z}})[(\mathbf{W}^* - \mathbf{W})^T \mathbf{\Phi}(\bar{\mathbf{z}}_r) + \varepsilon - \varepsilon_N \cdot \mathrm{sgn}(\mathbf{G}^T\mathbf{Ps})] \tag{20}$$

Eq. (20) shows that only $\mathbf{W}$ need to be further adjusted to minimize the tracking error. The adaptive law of $\mathbf{W}$ is designed as follows:

$$\dot{\mathbf{w}}_{jk} = a_j \lambda_j \mathbf{\Phi}_j \mathbf{s}^T \mathbf{P} \mathbf{g}_k - (1 - a_j)\eta_{jk} \mathrm{sgn}(\mathbf{w}_{jk})$$
$$j = 1 \dots N_r, \quad k = 1 \dots N_o \tag{21}$$

where $\mathbf{w}_{jk} \in \Re^{N_i+1}$ is the weight vector associated with the jth rule for the kth output variable, i.e. $\mathbf{W} = \begin{bmatrix} \mathbf{w}_{11} & \mathbf{w}_{12} & \cdots & \mathbf{w}_{1N_o} \\ \mathbf{w}_{21} & \mathbf{w}_{22} & \cdots & \mathbf{w}_{2N_o} \\ \vdots & \vdots & \cdots & \vdots \\ \mathbf{w}_{N_r1} & \mathbf{w}_{N_r2} & \cdots & \mathbf{w}_{N_rN_o} \end{bmatrix}$ (refer to Eq. (A-6)), $\mathbf{\Phi} = [\mathbf{\Phi}_1^T \; \mathbf{\Phi}_2^T \dots \mathbf{\Phi}_{N_r}^T]^T$ (refer to Eq. (A-7)), and $\mathbf{G} = [\mathbf{g}_1 \; \mathbf{g}_1 \dots \mathbf{g}_{N_o}]$.

The term $\lambda_j > 0$ is the learning rate for the weight vector associated with the jth rule. The term $\eta_{jk} = \lambda_j \hat{\eta}_{jk} > 0$ is the time-varying decay rate, which is chosen so that

$$\hat{\eta}_{jk} \geq \text{sum}|\mathbf{\Phi}_j \mathbf{s}^T \mathbf{P} \mathbf{g}_k| \tag{22}$$

where the operator "sum" denotes summation of all elements of the vector $|\mathbf{\Phi}_j \mathbf{s}^T \mathbf{P} \mathbf{g}_k|$.

The variable $a_j \in [0,\ 1]$ is a *structural adaptation parameter*, which is obtained from the G-FNN learning algorithm. It reflects the generation and deletion of fuzzy control rules associated with the G-FNN in the following manner. The jth rule is introduced into the G-FNN at time t by setting $a_j(t) = 1$, and the corresponding weight vector is initialized to $\mathbf{w}_{jk}(0)$ acquired by the G-FNN learning algorithm. The jth rule is targeted for deletion from the G-FNN at time t by setting $a_j(t) = 0$, which causes the adaptation mechanism to drive each of the output weights associated with the jth rule node to zero in a time bounded by $\max(\mathbf{w}_{jk})/\eta$. When $a_j(t) = 0$ and weight vectors of the jth rule node have reached zero, the rule is considered deleted from the G-FNN, and its associated memory and computational resources may be freed for reuse.

The symmetric positive definite matrix $\mathbf{P}$ is designed to satisfy the following relationship:

$$\mathbf{PA} + \mathbf{A}^T\mathbf{P} = -\mathbf{Q} \tag{23}$$

where $\mathbf{Q}$ is a positive definite matrix and is selected by the user.

To guarantee stability of the control system, the G-FNN must converge, which requires the parameters of the G-FNN to be bounded. Eq. (A-5) shows that the outputs of the G-FNN are bounded if the weights $\mathbf{W}$ are bounded. Define the constraint set $\mathbf{\Gamma}$ for $\mathbf{W}$ as follows:

$$\mathbf{\Gamma} \triangleq \{||\mathbf{w}_{jk}|| \leq ||\mathbf{w}_{jk}(0)||\} \ : \ j = 1 \ldots N_r, \ \ k = 1 \ldots N_o \tag{24}$$

where $||.||$ denotes the two-norm of a vector. According to the projection algorithm [6], the adaptation law given by Eq. (21) can be modified as follows:

$$\dot{\mathbf{w}}_{jk} = \begin{cases} a_j \lambda_j \mathbf{\Phi}_j \mathbf{s}^T \mathbf{P} \mathbf{g}_k - (1 - a_j)\eta_{jk}\text{sgn}(\mathbf{w}_{jk}) \\ \text{if } (||\mathbf{w}_{jk}|| < ||\mathbf{w}_{jk}(0)||) \text{ or} \\ (||\mathbf{w}_{jk}|| = ||\mathbf{w}_{jk}(0)|| \text{ and } \mathbf{w}_{jk}^T \mathbf{\Phi}_j \mathbf{s}^T \mathbf{P} \mathbf{g}_k \leq 0) \\ \\ a_j \lambda_j (\mathbf{I} - \frac{\mathbf{w}_{jk}\mathbf{w}_{jk}^T}{||\mathbf{w}_{jk}||^2}) \mathbf{\Phi}_j \mathbf{s}^T \mathbf{P} \mathbf{g}_k - (1 - a_j)\eta_{jk}\text{sgn}(\mathbf{w}_{jk}) \\ \text{if } (||\mathbf{w}_{jk}|| = ||\mathbf{w}_{jk}(0)|| \text{ and } \mathbf{w}_{jk}^T \mathbf{\Phi}_j \mathbf{s}^T \mathbf{P} \mathbf{g}_k > 0) \end{cases} \tag{25}$$

Concerning the boundedness of the weights of the G-FNN, we have

Theorem 1. If the initial values of the weights $\mathbf{w}_{jk}(0) \in \mathbf{\Gamma}$, the adaptation law given by Eq. (25) guarantees $\mathbf{w}_{jk}(t) \in \mathbf{\Gamma}$, $\forall t > 0$.

Proof: Consider the following Lyapunov function

$$V_{jk} = \frac{1}{2}\mathbf{w}_{jk}^T \mathbf{w}_{jk} \tag{26}$$

Taking the derivative of the Lyapunov function with respect to time, we have

$$\dot{V}_{jk} = \mathbf{w}_{jk}^T \dot{\mathbf{w}}_{jk} \tag{27}$$

When ($||\mathbf{w}_{jk}|| = ||\mathbf{w}_{jk}(0)||$ and $\mathbf{w}_{jk}^T \mathbf{\Phi}_j \mathbf{s}^T \mathbf{P} \mathbf{g}_k \leq 0$), $\dot{V}_{jk} \leq 0$. Thus, it can be guaranteed that $||\mathbf{w}_{jk}|| \leq ||\mathbf{w}_{jk}(0)||$. When ($||\mathbf{w}_{jk}|| = ||\mathbf{w}_{jk}(0)||$ and $\mathbf{w}_{jk}^T \mathbf{\Phi}_j \mathbf{s}^T \mathbf{P} \mathbf{g}_k > 0$), $\dot{V}_{jk} \leq 0$. Thus, $||\mathbf{w}_{jk}|| \leq ||\mathbf{w}_{jk}(0)||$ is also guaranteed. Since the initial value $\mathbf{w}_{jk}(0) \in \mathbf{\Gamma}$, $\mathbf{w}_{jk}$ is bounded by the constraint set $\mathbf{\Gamma}$ for all $t \geq 0$. □

Remark 2. As shown in Appendix A.2, the initial weight vector $\mathbf{w}_{jk}(0)$ is obtained using the LLS method incorporating the SVD technique to minimize the error energy of the modeling process. Using the LLS method, the obtained vector norm $||\mathbf{w}_{jk}(0)||$ is guaranteed to be the shortest amongst a set of nearly degenerate solutions. The derivation of the adaptive law of $\mathbf{w}_{jk}$ in order to fulfill the stability requirement of the control system is shown in Section 4.2. Theorem 1 further proves that $\mathbf{w}_{jk}$ will remain within its boundary $||\mathbf{w}_{jk}(0)||$.

4.3 Stability Analysis of AFNC

Concerning the stability of the closed-loop system, we have the following theorem.

Theorem 2. Consider the MIMO nonlinear dynamic system represented by Eq. (1). If the robust control law of Eq. (17) and the adaptive law given by Eq. (25) are applied, asymptotic stability is guaranteed.

Proof: We consider the following Lyapunov function candidate, which is based on Eq. (20),

$$V(t) = \frac{1}{2}\mathbf{s}^T \mathbf{P} \mathbf{s} + \frac{1}{2} tr[(\mathbf{W}^* - \mathbf{W})^T \mathbf{\Lambda}^{-1} (\mathbf{W}^* - \mathbf{W})] \tag{28}$$

where $\mathbf{\Lambda} \in \Re^{N_r(N_i+1) \times N_r(N_i+1)}$ is a diagonal matrix whose jth (N_i+1) entries are the learning rate λ_j.

Taking the derivative of the Lyapunov function in Eq. (28) and using Eqs. (20) and (23), we have

$$\begin{aligned} \dot{V}(t) &= \frac{1}{2}\dot{\mathbf{s}}^T \mathbf{P} \mathbf{s} + \frac{1}{2}\mathbf{s}^T \mathbf{P} \dot{\mathbf{s}} - tr[(\mathbf{W}^* - \mathbf{W})^T \mathbf{\Lambda}^{-1} \dot{\mathbf{W}}] \\ &= \frac{1}{2}\mathbf{s}^T (\mathbf{P}\mathbf{A} + \mathbf{A}^T \mathbf{P}) \mathbf{s} + \mathbf{s}^T \mathbf{P} \mathbf{G} [\varepsilon - \varepsilon_N \cdot \mathrm{sgn}(\mathbf{G}^T \mathbf{P} \mathbf{s})] \end{aligned}$$

$$
\begin{aligned}
&+\mathbf{s}^T\mathbf{PG}(\mathbf{W}^* - \mathbf{W})^T\mathbf{\Phi} - \sum_{k=1}^{N_o}\sum_{j=1}^{N_r}(\mathbf{w}_{jk}^* - \mathbf{w}_{jk})^T\lambda_j^{-1}\dot{\mathbf{w}}_{jk} \\
&= -\frac{1}{2}\mathbf{s}^T\mathbf{Qs} - \mathbf{s}^T\mathbf{PG}[(\varepsilon_N \pm |\varepsilon|)\cdot \mathrm{sgn}(\mathbf{G}^T\mathbf{Ps})] \\
&\quad + \sum_{k=1}^{N_o}\mathbf{s}^T\mathbf{Pg}_k\sum_{j=1}^{N_r}(\mathbf{w}_{jk}^* - \mathbf{w}_{jk})^T\mathbf{\Phi}_j - \sum_{k=1}^{N_o}\sum_{j=1}^{N_r}(\mathbf{w}_{jk}^* - \mathbf{w}_{jk})^T\lambda_j^{-1}\dot{\mathbf{w}}_{jk}
\end{aligned} \tag{29}
$$

Under condition 1 of Eq. (25), Eq. (29) becomes

$$
\begin{aligned}
\dot{V}(t) &= -\frac{1}{2}\mathbf{s}^T\mathbf{Qs} - \mathbf{s}^T\mathbf{P}[(\varepsilon_N \pm |\varepsilon|)\cdot \mathrm{sgn}(\mathbf{G}^T\mathbf{Ps})] + \sum_{k=1}^{N_o}\mathbf{s}^T\mathbf{Pg}_k \\
&\quad \sum_{j=1}^{N_r}(\mathbf{w}_{jk}^* - \mathbf{w}_{jk})^T\mathbf{\Phi}_j - \sum_{k=1}^{N_o}\sum_{j=1}^{N_r}(\mathbf{w}_{jk}^* - \mathbf{w}_{jk\lambda^{-1}})^T\lambda_j^{-1} \\
&\quad [a_j\lambda_j\mathbf{\Phi}_j\mathbf{s}^T\mathbf{Pg}_k - (1-a_j)\eta_{jk}\mathrm{sgn}(\mathbf{w}_{jk})] \\
&= -\frac{1}{2}\mathbf{s}^T\mathbf{Qs} - \mathbf{s}^T\mathbf{PG}[(\varepsilon_N \pm |\varepsilon|)\cdot \mathrm{sgn}(\mathbf{G}^T\mathbf{Ps})] \\
&\quad + \sum_{k=1}^{N_o}\sum_{j=1}^{N_r}(\mathbf{w}_{jk}^* - \mathbf{w}_{jk})^T\mathbf{\Phi}_j\mathbf{s}^T\mathbf{Pg}_k - \sum_{k=1}^{N_o}\sum_{j=1}^{N_r}a_j(\mathbf{w}_{jk}^* - \mathbf{w}_{jk})^T\mathbf{\Phi}_j\mathbf{s}^T\mathbf{Pg}_k \\
&\quad + \sum_{k=1}^{N_o}\sum_{j=1}^{N_r}(1-a_j)\hat{\eta}_{jk}(\mathbf{w}_{jk}^* - \mathbf{w}_{jk})^T\mathrm{sgn}(\mathbf{w}_{jk}) \\
&= -\frac{1}{2}\mathbf{s}^T\mathbf{Qs} - \mathbf{s}^T\mathbf{PG}[(\varepsilon_N \pm |\varepsilon|)\cdot \mathrm{sgn}(\mathbf{G}^T\mathbf{Ps})] \\
&\quad + \sum_{k=1}^{N_o}\sum_{j=1}^{N_r}(1-a_j)(\mathbf{w}_{jk}^* - \mathbf{w}_{jk})^T[\mathbf{\Phi}_j\mathbf{s}^T\mathbf{Pg}_k + \hat{\eta}_{jk}\mathrm{sgn}(\mathbf{w}_{jk})] \\
&= -\frac{1}{2}\mathbf{s}^T\mathbf{Qs} - \mathbf{s}^T\mathbf{PG}[(\varepsilon_N \pm |\varepsilon|)\cdot \mathrm{sgn}(\mathbf{G}^T\mathbf{Ps})] + \sum_{k=1}^{N_o}\sum_{j=1}^{N_r}v_{jk}
\end{aligned} \tag{30}
$$

When $a_j = 1$, i.e. the jth rule is generated, $v_{jk} = 0$. When $a_j = 0$, i.e. the jth rule is going to be deleted and $\mathbf{w}_{jk}^* = 0$, $v_{jk} = -\mathbf{w}_{jk}^T[\mathbf{\Phi}_j\mathbf{s}^T\mathbf{Pg}_k + \hat{\eta}_{jk}\mathrm{sgn}(\mathbf{w}_{jk})] \leq 0$ using Eq. (22). Therefore, Eq. (30) becomes

$$
\dot{V}(t) \leq -\frac{1}{2}\mathbf{s}^T\mathbf{Qs} - \mathbf{s}^T\mathbf{PG}[(\varepsilon_N \pm |\varepsilon|)\cdot \mathrm{sgn}(\mathbf{G}^T\mathbf{Ps})] \leq 0 \tag{31}
$$

Under condition 2 of Eq. (25), Eq. (29) becomes

$$
\dot{V}(t) = -\frac{1}{2}\mathbf{s}^T\mathbf{Qs} - \mathbf{s}^T\mathbf{PG}[(\varepsilon_N \pm |\varepsilon|)\cdot \mathrm{sgn}(\mathbf{G}^T\mathbf{Ps})] + \sum_{k=1}^{N_o}\mathbf{s}^T\mathbf{Pg}_k
$$

$$\sum_{j=1}^{N_r}(\mathbf{w}_{jk}^* - \mathbf{w}_{jk})^T\mathbf{\Phi}_j - \sum_{k=1}^{N_o}\sum_{j=1}^{N_r}(\mathbf{w}_{jk}^* - \mathbf{w}_{jk})^T\lambda_j^{-1}$$

$$[a_j\lambda_j(\mathbf{I} - \frac{\mathbf{w}_{jk}\mathbf{w}_{jk}^T}{||\mathbf{w}_{jk}||^2})\mathbf{\Phi}_j\mathbf{s}^T\mathbf{P}\mathbf{g}_k - (1-a_j)\eta_{jk}\mathrm{sgn}(\mathbf{w}_{jk})]$$

$$= -\frac{1}{2}\mathbf{s}^T\mathbf{Q}\mathbf{s} - \mathbf{s}^T\mathbf{P}\mathbf{G}[(\varepsilon_N \pm |\varepsilon|)\cdot \mathrm{sgn}(\mathbf{G}^T\mathbf{P}\mathbf{s})]$$

$$+\sum_{k=1}^{N_o}\sum_{j=1}^{N_r}(\mathbf{w}_{jk}^* - \mathbf{w}_{jk})^T\mathbf{\Phi}_j\mathbf{s}^T\mathbf{P}\mathbf{g}_k - \sum_{k=1}^{N_o}\sum_{j=1}^{N_r}a_j(\mathbf{w}_{jk}^* - \mathbf{w}_{jk})^T\mathbf{\Phi}_j\mathbf{s}^T\mathbf{P}\mathbf{g}_k$$

$$+\sum_{k=1}^{N_o}\sum_{j=1}^{N_r}a_j(\mathbf{w}_{jk}^* - \mathbf{w}_{jk})^T\frac{\mathbf{w}_{jk}\mathbf{w}_{jk}^T}{||\mathbf{w}_{jk}||^2}\mathbf{\Phi}_j\mathbf{s}^T\mathbf{P}\mathbf{g}_k$$

$$+\sum_{k=1}^{N_o}\sum_{j=1}^{N_r}(1-a_j)\hat{\eta}_{jk}(\mathbf{w}_{jk}^* - \mathbf{w}_{jk})^T\mathrm{sgn}(\mathbf{w}_{jk})$$

$$= -\frac{1}{2}\mathbf{s}^T\mathbf{Q}\mathbf{s} - \mathbf{s}^T\mathbf{P}\mathbf{G}[(\varepsilon_N \pm |\varepsilon|)\cdot \mathrm{sgn}(\mathbf{G}^T\mathbf{P}\mathbf{s})]$$

$$+\sum_{k=1}^{N_o}\sum_{j=1}^{N_r}(1-a_j)(\mathbf{w}_{jk}^* - \mathbf{w}_{jk})^T[\mathbf{\Phi}_j\mathbf{s}^T\mathbf{P}\mathbf{g}_k + \hat{\eta}_{jk}\mathrm{sgn}(\mathbf{w}_{jk})]$$

$$-\sum_{k=1}^{N_o}\sum_{j=1}^{N_r}a_j(1 - \frac{\mathbf{w}_{jk}^{*T}\mathbf{w}_{jk}}{||\mathbf{w}_{jk}||^2})\mathbf{w}_{jk}^T\mathbf{\Phi}_j\mathbf{s}^T\mathbf{P}\mathbf{g}_k$$

$$= -\frac{1}{2}\mathbf{s}^T\mathbf{Q}\mathbf{s} - \mathbf{s}^T\mathbf{P}\mathbf{G}[(\varepsilon_N \pm |\varepsilon|)\cdot \mathrm{sgn}(\mathbf{G}^T\mathbf{P}\mathbf{s})] + \sum_{k=1}^{N_o}\sum_{j=1}^{N_r}v_{jk} - \sum_{k=1}^{N_o}\sum_{j=1}^{N_r}u_{jk} \tag{32}$$

Since $\mathbf{w}_i^* \in \mathbf{\Gamma}$ and $1 - \frac{(\mathbf{w}_i^*)^T\mathbf{w}_i}{||\mathbf{w}_i||^2} \geq 0$, when $a_j = 1$, i.e. the jth rule is generated, $v_{jk} = 0$ and $u_{jk} \geq 0$. When $a_j = 0$, i.e. the jth rule is going to be deleted and $\mathbf{w}_{jk}^* = 0$, $v_{jk} \leq 0$ and $u_{jk} = 0$. Therefore, Eq. (32) becomes

$$\dot{V}(t) \leq -\frac{1}{2}\mathbf{s}^T\mathbf{Q}\mathbf{s} - \mathbf{s}^T\mathbf{P}\mathbf{G}[(\varepsilon_N \pm |\varepsilon|)\cdot \mathrm{sgn}(\mathbf{G}^T\mathbf{P}\mathbf{s})] \leq 0 \tag{33}$$

Eqs. (28), (31) and (33) show that $V(t) \geq 0$ and $\dot{V}(t) \leq 0$. Furthermore, Eqs. (31) and (33) imply that $\dot{V}(t) = 0$ if and only if $V(t) = 0$. Therefore, global stability is guaranteed by the Lyapunov theorem. By using Barbalat's lemma [1,2], it can be shown that $\mathbf{s}(t) \to 0$ as $t \to \infty$. As a result, the control system is asymptotically stable. Moreover, the tracking error of the system will converge to zero. □

5 Simulation Studies

Example 1: In this example, we illustrate the effectiveness of the proposed AFNC design in tracking control of an inverted pendulum system used in [1,6] whose dynamic equation is given by

$$\ddot{\theta} = \frac{g\sin\theta - \frac{ml\dot{\theta}\cos\theta\sin\theta}{m_c+m}}{l(\frac{4}{3} - \frac{m\cos^2\theta}{m_c+m})} + \frac{\frac{\cos\theta}{m_c+m}}{l(\frac{4}{3} - \frac{m\cos^2\theta}{m_c+m})}u + d \tag{34}$$

Here, u represents the input force, θ represents the angle of the pendulum with respect to the vertical line and d represents plant uncertainties. The inverted pendulum system is depicted in Fig. 3.

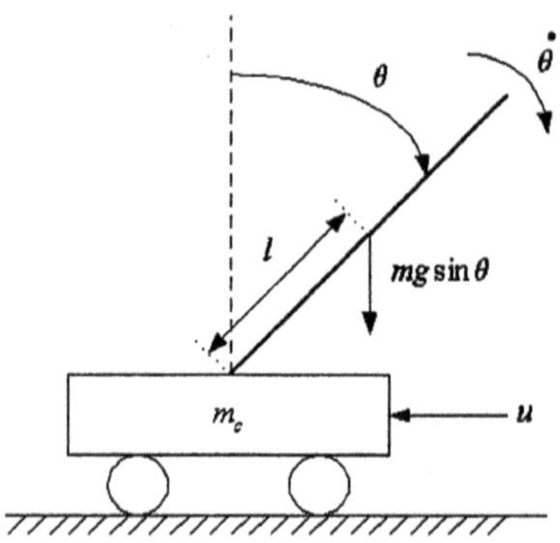

Fig. 3. Inverted pendulum system

In this simulation, the system parameters are chosen as: mass of the cart, $m_c = 1kg$; mass of the pole, $m = 0.1kg$; half-length of the pole, $l = 0.5m$ and acceleration due to gravity, $g = 9.8m/s^2$. The initial position is chosen as $\theta = \dot{\theta} = \ddot{\theta} = 0$. The desired angle trajectory is assumed to be $\theta_d = 0.1\sin(2\pi t)rad$ and plant uncertainties are given by $d = 0.2\sin(4\pi t)rad/s^2$.

The parameters of the continuous AFNC are designed as: $\boldsymbol{\Upsilon}_0 = 2$, $\mathbf{K_D} = 8$, $\mathbf{P} = 5$, and $\lambda_j = 0.5$. The G-FNN learning algorithm operates in discrete time at a sampling rate of 50 *sample/s*. Other parameters of the learning algorithm are designed as: $e_{\max} = 0.1$, $e_{\min} = 0.005$, $d_{\max} = \sqrt{\ln(\frac{1}{0.5})}$, $d_{\min} = \sqrt{\ln(\frac{1}{0.8})}$, $K_{s,\min} = 0.9$, $K_{mf} = 0.5$, $K_{err} = 0.01$, and $N_d = 100$.

Using the G-FNN learning algorithm, the fuzzy neural structure and parameters are generated simultaneously and automatically. In this simulation, a total of three fuzzy rules are generated online as shown in Fig. 4. The corresponding Gaussian fuzzy membership functions were obtained with respect to the input training variable as shown in Fig. 5. It can be seen that the membership functions are evenly distributed over the input training interval. This is in line with the aspiration of "local representation" in fuzzy logic. Fig. 6 shows the tracking result of the control system. The results clearly demonstrate that the pole angle was able to track the desired trajectories from second trial onwards.

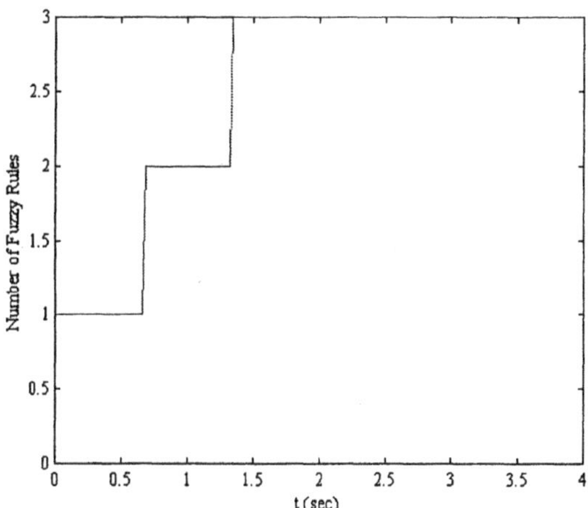

Fig. 4. Number of fuzzy rules generated (example 1)

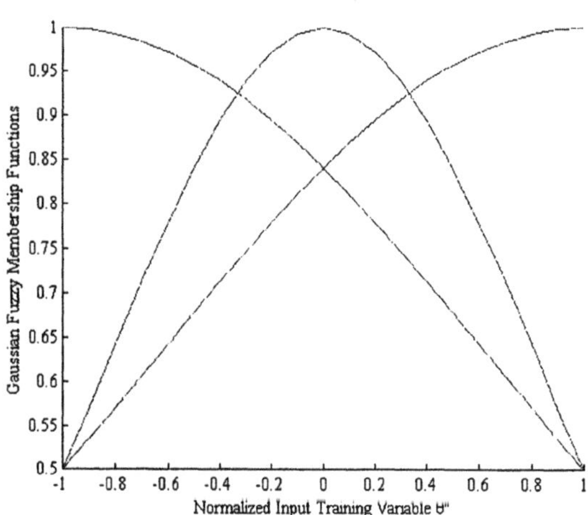

Fig. 5. Gaussian membership functions w.r.t input training variable (example 1)

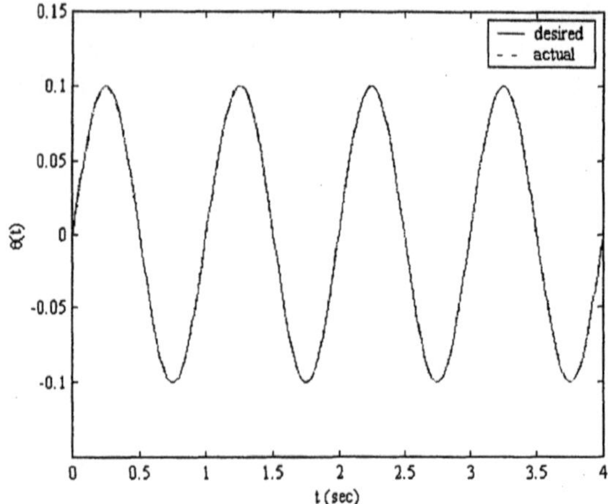

Fig. 6. Tracking response using AFNC (example 1)

Example 2: In this simulation example, we test the proposed AFNC in tracking control of a two-link robot manipulator depicted in Fig. 7. The dynamic equation of the manipulator used in [1] is given by

$$\ddot{\mathbf{q}} = -\mathbf{M}(\mathbf{q})^{-1}\mathbf{C}(\mathbf{q},\dot{\mathbf{q}}) + \mathbf{M}(\mathbf{q})^{-1}\tau - \mathbf{M}(\mathbf{q})^{-1}\tau_d \tag{35}$$

where

$$\begin{aligned}
\mathbf{M}(\mathbf{q}) &= \begin{bmatrix} a_1 + 2a_3\cos q_2 + 2a_4 \sin q_2 & a_2 + a_3\cos q_2 + a_4\sin q_2 \\ a_2 + a_3\cos q_2 + a_4 \sin q_2 & a_2 \end{bmatrix} \\
\mathbf{C}(\mathbf{q},\dot{\mathbf{q}}) &= \begin{bmatrix} -(a_3\sin q_2 - a_4\cos q_2)\dot{q}_1\dot{q}_2 - (a_3\sin q_2 - a_4\cos q_2)(\dot{q}_1+\dot{q}_2)\dot{q}_2 \\ (a_3\sin q_2 - a_4\cos q_2)\dot{q}_1^2 \end{bmatrix} \\
a_1 &= I_1 + m_1 l_{c1}^2 + I_e + m_e l_{ce}^2 + m_e l_1^2 \\
a_2 &= I_e + m_e l_{ce}^2 \\
a_3 &= m_e l_1 l_{ce}\cos\delta_e \\
a_4 &= m_e l_1 l_{ce}\sin\delta_e
\end{aligned} \tag{36}$$

and $\mathbf{M}$ is the 2×2 inertia matrix of the manipulator, $\mathbf{C}$ is the 2×1 vector of centrifugal, Coriolis, friction forces and gravity, τ is the 2×1 vector of input torque generated by the joint motor, $\ddot{\mathbf{q}}$, $\dot{\mathbf{q}}$ and $\mathbf{q}$ are 2×1 vectors of output link acceleration, velocity and position respectively, and τ_d is the 2×1 vector of unknown terms arising from unmodeled dynamics and external disturbances.

Parameters of the two-link planar manipulator used for the simulation are: mass of link 1, $m_1 = 1kg$; length of link 1, $l_1 = 1m$; mass of link 2 and payload, $m_e = 2kg$; angle of payload with respect to link 2, $\delta_e = 30°$; centroidal moment of inertia of link 1, $I_1 = 0.12kgm^2$; length of CG of link

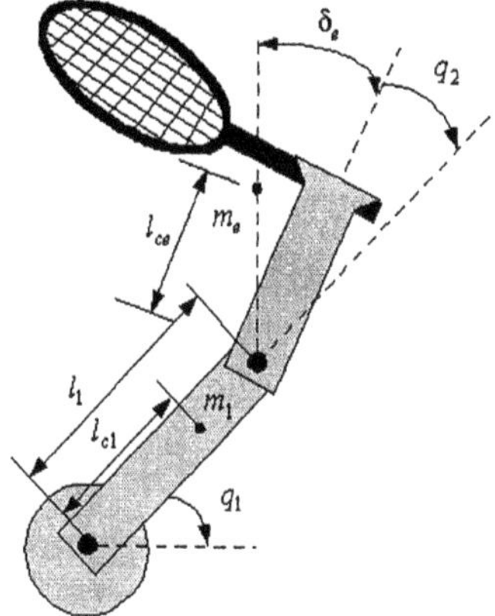

Fig. 7. Articulated two-link robot manipulator [1]

1 from the axis of rotation, $l_{c1} = 0.5m$; centroidal moment of inertia of link 2 and payload, $I_e = 0.25kgm^2$; length of CG of link 2 and payload from the axis of rotation, $l_{ce} = 0.6m$. Initial conditions are chosen as $q_1 = q_2 = 0\ rad$, $\dot{q}_1 = \dot{q}_2 = 0\ rad/sec$ and $\ddot{q}_1 = \ddot{q}_2 = 0\ rad/sec^2$. The desired trajectory is an ellipse in the $q_1 - q_2$ plane not starting from the initial position with $q_{d1} = 1.5\sin(2\pi t)\ rad$ and $q_{d2} = 0.7\cos(2\pi t)\ rad$. In a real system, there always exist uncertainties and disturbances. Hence, we deliberately introduce disturbances $\tau_{d1} = 100\sin(2\pi t)\ Nm$ and $\tau_{d2} = 50\sin(2\pi t)\ Nm$, which are comparable to the control torques of the robot manipulator.

The parameters of the continuous AFNC are designed as follows: $\mathbf{\Upsilon}_0 = diag[5,5]$, $\mathbf{K_D} = diag[7,7]$, $\mathbf{P} = diag[7,7]$, and $\lambda_j = 0.4$. The same set of parameters (as in Example 1) are used for the G-FNN learning algorithm.

Due to the complexity of the system dynamics, more fuzzy control rules are generated compared with the first example. Figs. 8 and 9 illustrate the generation of fuzzy rules and Gaussian fuzzy membership functions respectively. Fig. 10 shows control responses of the two links. The results clearly demonstrate that the two links were able to track the desired trajectories from second trial onwards. It should be highlighted that for q_2, whose initial position was set to be 100% error from the desired initial link position, i.e. $q_2(0) = 0$, the link was still able to catch up with the desired trajectory very fast without any overshoots.

The proposed adaptive fuzzy neural control scheme has some advantages over normal fuzzy control and neural networks control schemes. For example, in normal fuzzy controller design, certain controller parameters must be tuned by trial and error. Such parameters include scalars (or gains) for its input data and output data, the number of membership functions, the width of a single membership function, and the number of control rules. On the con-

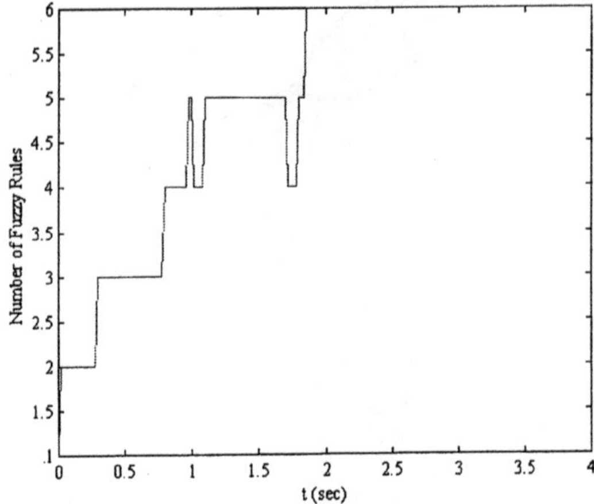

Fig. 8. Number of fuzzy rules generated (example 2)

trary, the adaptive fuzzy neural controller based on estimation of the dynamic model will allow the parameters to be tuned automatically. In addition, the performance of the AFNC is also considered to be very good.

6 Conclusions

In this chapter, an adaptive fuzzy neural control scheme for a class of nonlinear systems was proposed and its adaptive capability to handle modeling errors and external disturbances was demonstrated. The error convergence rate with the AFNC was found to be fast. Asymptotic stability of the control system is established using Lyapunov approach. Computer simulation studies of an inverted pendulum system and a two-link robot manipulator demonstrate the flexibility, adaptation and tracking performance of the proposed AFNC. The major contributions of the chapter are:

- Successful development of a robust adaptive control scheme where the proposed AFNC is employed to compensate for modeling errors and to handle external disturbances of nonlinear systems.
- Successful development of an online parameter training and adjustment algorithm, which is developed based on the G-FNN learning algorithm and the gradient descent method, to enhance learning and adaptation capabilities of the AFNC.

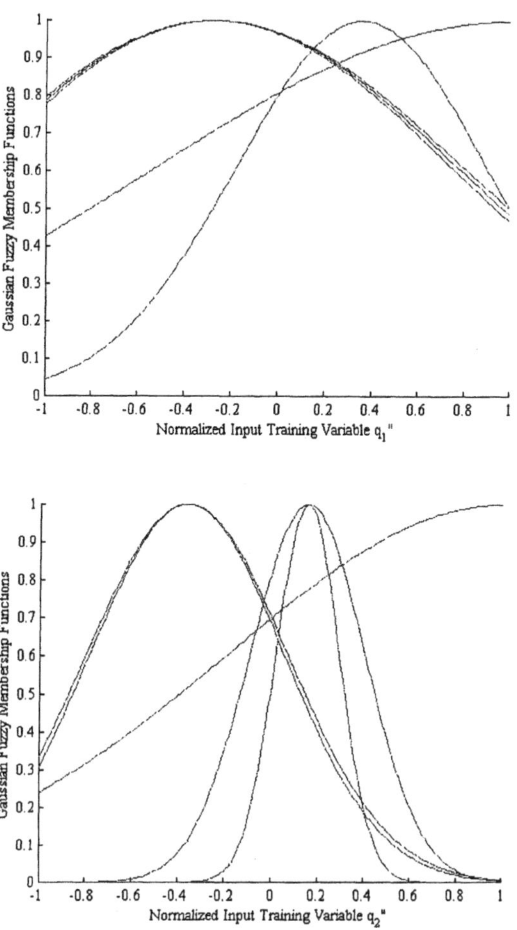

Fig. 9. Gaussian membership functions w.r.t input training variable (example 2)

APPENDIX Generalized Fuzzy Neural Network (G-FNN)

G-FNN is a newly developed neural-network-based fuzzy logic control/decision system. It is an improved version of the Generalized Dynamic Fuzzy Neural Network (GD-FNN) proposed in our previous work [23]. The G-FNN is a multi-layer feedforward network which integrates the TSK-type Fuzzy Inference System (FIS) and the RBF Neural Network (RBF-NN) into a connectionist structure. In addition, a special designed online supervised learning algorithm, namely G-FNN learning algorithm, provides an efficient way for

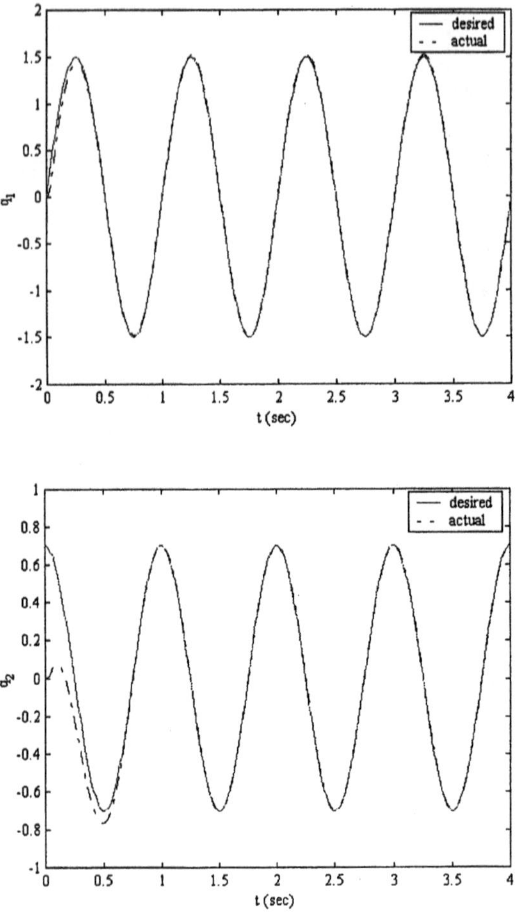

Fig. 10. Tracking responses using AFNC (example 2)

constructing the G-FNN in real time and introduces a novel scheme to combine structure learning and parameter learning simultaneously within the FNN. The salient characteristics of the G-FNN are: (1) Hierarchical online learning ability, where not only parameters can be adjusted, but also structure can be self-adaptive without partitioning the input space a priori; (2) Fast learning speed, which is suitable for real-time implementation. The G-FNN can serve as a generalized FNN for system modeling and control.

A.1 G-FNN Architecture

The G-FNN is constructed based on the extended RBF-NN, which is functionally equivalent to the TSK-type FIS. The G-FNN (see Fig. A-1) has a total of four layers. Nodes in layer one are input nodes which represent input linguistic variables. Layer four is the output layer. Nodes in layer two act as membership functions to represent the terms of the respective linguistic variables. Each node in layer three is a rule node which represents one fuzzy rule. Thus, all the layer-three nodes form a fuzzy *rule set*. Links in layers three and four function as a connectionist *inference mechanism*, which avoids the rule-matching process. Layer-three links define the preconditions of the rule nodes, and layer-four links define the consequences of the rule nodes. Therefore, for each rule node, there is at most one link from an input node. The links in layers two and four are fully connected with linguistic nodes. Nodes and links in layers one and two act as the *fuzzifier*, and those in layer four as *defuzzifier*. The arrow on the link indicates the signal flow direction when this network is in use after it has been built and trained. In the following context, we shall indicate the signal propagation and the basic function in each layer of the G-FNN.

- Layer one transmits values of the input linguistic variables $x_i : i = 1 \ldots N_i$ to the next layer directly.
- Layer two performs membership functions to the input variables. The membership function is chosen as the Gaussian function of the following form:

$$mf_{ij}(x_i) = \exp\left[-\frac{(x_i - c_{ij})^2}{\sigma_{ij}^2}\right] \tag{A-1}$$

  where $c_{ij}, i = 1 \ldots N_i, j = 1 \ldots N_r$ and $\sigma_{ij}, i = 1 \ldots N_i, j = 1 \ldots N_r$ are, respectively, the center (or mean) and the width (or variance) of the Gaussian function of jth term of the ith input linguistic variable x_i.
- Layer three is rule layer. The number of nodes (or neurons) in this layer indicates the number of fuzzy rules. The outputs are given by

$$\phi_j = \prod_{i=1}^{N_i} mf_{ij} \tag{A-2}$$

- Layer-four nodes define output linguistic variables. Each output variable $y_k, k = 1 \ldots N_o$ is the weighted sum of incoming signals, which performs defuzzification of the TSK-type FIS, i.e.

$$y_k = \sum_{j-1}^{N_r} \phi_j w_{jk} \tag{A-3}$$

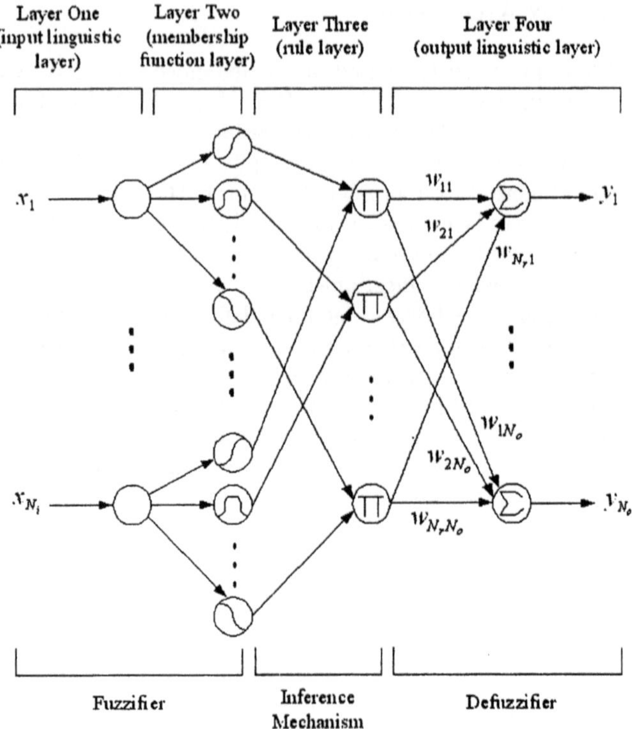

Fig. A-1. Architecture of G-FNN

and the weight or the consequence is

$$w_{jk} = K_0^{jk} + K_1^{jk} x_1 + \ldots + K_{N_i}^{jk} x_{N_i} \tag{A-4}$$

where K_i's are real-valued parameters. It is not difficult to see that

$$\mathbf{y} = [y_1 \;\; y_2 \;\; \cdots \;\; y_{N_o}]^T = \mathbf{W}^T \mathbf{\Phi} \tag{A-5}$$

where

$$\mathbf{W} = \begin{bmatrix} K_0^{11} \; K_1^{11} \ldots K_{N_i}^{11} & \ldots\ldots & K_0^{N_r 1} \; K_1^{N_r 1} \ldots K_{N_i}^{N_r 1} \\ \vdots & \vdots & \vdots \\ K_0^{1N_o} \; K_1^{1N_o} \ldots K_{N_i}^{1N_o} & \ldots\ldots & K_0^{N_r N_o} \; K_1^{N_r N_o} \ldots K_{N_i}^{N_r N_o} \end{bmatrix}^T \tag{A-6}$$

$$\mathbf{\Phi} = \left[\phi_1 \;\; \phi_1 x_1 \ldots \phi_1 x_{N_i} \;\; \ldots\ldots \;\; \phi_{N_r} \;\; \phi_{N_r} x_1 \ldots \phi_{N_r} x_{N_i}\right]^T \tag{A-7}$$

A.2 G-FNN Learning Algorithm

This section proposes an online supervised structure and parameter learning algorithm for constructing the G-FNN automatically and dynamically. The learning of G-FNN structure includes determining the proper number of membership function nodes in layers two and three, i.e. the coarse of input fuzzy partitions and the finding of correct fuzzy logic rules. The learning of network parameters includes the adjustment of the node parameters in layer two and link parameters in layer four. This corresponds to the learning of the premise and consequent parameters of the G-FNN. Premise parameters include membership function parameters $mf_{ij}(c_{ij}, \sigma_{ij})$, and consequent parameter refers to the weight w_{jk} of the G-FNN.

Parameter learning is performed by combining the semi-closed fuzzy set concept for the membership learning and the Linear Least Square (LLS) method for weight learning. For structure learning, two criteria are proposed for fuzzy rule generation and an Error Reduction Ratio (ERR) concept is proposed for rule pruning. Given the supervised training data, the proposed learning algorithm first decides whether or not to perform the fuzzy rule generation based on two proposed criteria. If structure learning is necessary, premise parameters of a new fuzzy rule will be obtained using semi-closed fuzzy set concept. The G-FNN will further decide whether there are redundant rules to be deleted based on the ERR, and it will also change the consequences of all the fuzzy rules properly. If no structure learning is necessary, the parameter learning will be performed to adjust the current premise and consequent parameters. This structure/parameter learning will be repeated for each online incoming training input/output data pair.

With G-FNN learning algorithm, the network starts with no fuzzy rules. Fuzzy logic rules can be recruited and deleted automatically and dynamically according to their significance to the system performance and the complexity of the mapped system. Therefore, not only the parameters can be adjusted, but also the structure can be self-adaptive.

A.2.1 Two Criteria of Fuzzy Rule Generation

1. System Error
   The output error of G-FNN with respect to the teaching signal is an important factor in determining whether a new rule should be added. For each training data pair $[\mathbf{x}(m), \mathbf{t}(m)] : m = 1 \dots N_d$, where $\mathbf{x}(m) = [x_1(m) \ x_2(m) \ \dots \ x_{N_i}(m)]^T$ is the mth input vector, $\mathbf{t}(m) = [t_1(m) \ t_2 (m) \ \dots \ t_{N_o}(m)]^T$ is the mth desired output or the supervised teaching signal and N_d is the total number of training data, compute the G-FNN output $\mathbf{y}(m)$ for the existing structure using (A-1)-(A-7). Define the mth system error $e(m)$ as the Euclidean distance between vectors $\mathbf{t}(m)$ and $\mathbf{y}(m)$, i.e.

$$e(m) = ||\mathbf{e}(m)|| = ||\mathbf{t}(m) - \mathbf{y}(m)|| \tag{A-8}$$

If

$$e(m) > K_e \tag{A-9}$$

a new fuzzy rule should be considered. K_e is designed based on the desired accuracy of the G-FNN.

2. ϵ-Completeness of Fuzzy Rules

Definition A-1. ϵ-Completeness of Fuzzy Rule For any input within the operating range, there exists at least one fuzzy rule such that the match degree (or firing strength) is not less than ϵ [6]. The minimum value of ϵ is usually selected as $\epsilon_{\min} = 0.5$.

A fuzzy rule is a local representation over the region of input space. If a new pattern falls within the accommodation boundary of the existing rules, i.e., it satisfies ϵ-completeness, G-FNN will not generate a new rule but update parameters of existing rules, and vice versa. According to ϵ-completeness, when the mth observation $[\mathbf{x}(m), \mathbf{t}(m)]$ enters the system, we calculate regularized Mahalanobis distance $md_j(m), j = 1 \ldots N_r$ between $\mathbf{x}(m)$ and the jth existing hyper ellipsoidal region as follows:

$$md_j(m) = \sqrt{[\mathbf{x}(m) - \mathbf{c}_j]^T \sum_j [\mathbf{x}(m) - \mathbf{c}_j]} \tag{A-10}$$

where $\mathbf{c}_j = [c_{1j} \ c_{2j} \ \ldots \ c_{N_ij}]^T$, $\sum_j = diag(\frac{1}{\sigma_{1j}^2} \ \frac{1}{\sigma_{2j}^2} \ \cdots \ \frac{1}{\sigma_{N_ij}^2})$ and N_r is the number of existing rules. Find

$$\bar{j} = \arg\min md_j(m) \tag{A-11}$$

If

$$d(m) = md_{\bar{j}}(m) > K_d \tag{A-12}$$

where $K_d = \sqrt{\ln \frac{1}{\epsilon}}$, a new rule should be considered because the existing system is not satisfied by ϵ-completeness. Otherwise, the new input training data can be represented by the nearest existing rule.

Remark A-1. An idea of hierarchical learning is adopted, where the thresholds K_e and K_d decay during the online learning process as follows:

$$K_e = \begin{cases} e_{\max} & 1 < m < \frac{N_d}{3} \\ \max[e_{\max}(\frac{e_{\min}}{e_{\max}})^{\frac{3m}{N_d}-1}, \ e_{\min}] & \frac{N_d}{3} \le m \le \frac{2N_d}{3} \\ e_{\min} & \frac{2N_d}{3} \le m \le N_d \end{cases} \tag{A-13}$$

$$K_d = \begin{cases} d_{\max} = \sqrt{\ln \frac{1}{\epsilon_{\min}}} & 1 < m < \frac{N_d}{3} \\ \max[d_{\max}(\frac{d_{\min}}{d_{\max}})^{\frac{3m}{N_d}-1}, \ d_{\min}] & \frac{N_d}{3} \le m \le \frac{2N_d}{3} \\ d_{\min} = \sqrt{\ln \frac{1}{\epsilon_{\max}}} & \frac{2N_d}{3} \le m \le N_d \end{cases} \tag{A-14}$$

The key idea of hierarchical learning is that K_e and K_d are set to be $e_{\max}$ and $d_{\max}$ initially. This is known as coarse learning. They will gradually reach $e_{\min}$ and $d_{\min}$ respectively to achieve fine learning.

A.2.2 Pruning of Fuzzy Rules If inactive hidden units can be deleted as learning progresses, a more parsimonious network topology can be achieved. Therefore, pruning becomes imperative for identification of non-linear systems with time-varying dynamics. In the G-FNN learning algorithm, the Error Reduction Ratio (ERR) method is proposed for fuzzy rule pruning.

As the mth observation $[\mathbf{x}(m), \mathbf{t}(m)]$ enters the G-FNN, all the existing m data are memorized by the G-FNN as $(\mathbf{X}, \mathbf{T})$ where $\mathbf{X} = [\mathbf{x}(1)\ \ \mathbf{x}(2)\ \ \ldots\ \ \mathbf{x}(m)]$ and $\mathbf{T} = [\mathbf{t}(1)\ \ \mathbf{t}(2)\ \ \ldots\ \ \mathbf{t}(m)]$. From (A-5)-(A-8), we have

$$\mathbf{T} = \mathbf{W}^T\mathbf{\Theta} + \mathbf{E} \tag{A-15}$$

and

$$\begin{aligned}\mathbf{\Theta} &= \begin{bmatrix} \mathbf{\Phi}(1)^T \\ \vdots \\ \mathbf{\Phi}(m)^T \end{bmatrix}^T \\ &= \begin{bmatrix} \phi_1(1)\ \ \phi_1 x_1(1)\ldots\phi_1 x_{N_i}(1) & \ldots\ldots & \phi_{N_r}(1)\ \ \phi_{N_r} x_1(1)\ldots\phi_{N_r} x_{N_i}(1) \\ \vdots & \vdots & \vdots \\ \phi_1(m)\ \ \phi_1 x_1(m)\ldots\phi_1 x_{N_i}(m) & \ldots\ldots & \phi_{N_r}(m)\ \ \phi_{N_r} x_1(m)\ldots\phi_{N_r} x_{N_i}(m) \end{bmatrix}^T\end{aligned}$$

where $\mathbf{T} = [\mathbf{t}_1\ \ \mathbf{t}_2\ \ \ldots\ \ \mathbf{t}_{N_o}]^T \in \Re^{N_o\times m}$ is the teaching signal, $\mathbf{W} = [\mathbf{w}_1\ \ \mathbf{w}_2\ \ \ldots\ \ \mathbf{w}_{N_o}] \in \Re^{v\times N_o}$ are real parameters, $\mathbf{\Theta} = [\mathbf{\Phi}(1)\ \ \mathbf{\Phi}(2)\ \ \ldots\ \ \mathbf{\Phi}(m)] \in \Re^{v\times m}$ is known as the regressor, $\mathbf{E} = [\mathbf{e}_1\ \ \mathbf{e}_2\ \ \ldots\ \ \mathbf{e}_{N_o}]^T \in \Re^{N_o\times m}$ is the system error which is assumed to be uncorrelated with the regressor $\mathbf{\Theta}$, and $v = N_r(N_i + 1)$.

For simplicity, we use only the single-output system to illustrate the ERR method. Therefore, Eq. (A-15) for the kth output variable becomes

$$\mathbf{t}_k = \mathbf{w}_k^T\mathbf{\Theta} + \mathbf{e}_k : k = 1\ldots N_o \tag{A-16}$$

Taking transpose of both sides, we have

$$\mathbf{t}_k^T = \mathbf{\Theta}^T\mathbf{w}_k + \mathbf{e}_k^T \tag{A-17}$$

where $\mathbf{t}_k^T, \mathbf{e}_k^T \in \Re^{m\times 1}, \Theta^T \in \Re^{m\times v}$ and $\mathbf{w}_k \in \Re^{v\times 1}$. For any matrix $\mathbf{\Theta}^T$, if its row number is larger than the column number, i.e. $m \geq v$, it can be transformed into a set of orthogonal basis vectors by QR decomposition [24]. Thus

$$\mathbf{\Theta}^T = \mathbf{QR} \tag{A-18}$$

where $\mathbf{Q} = [\mathbf{q}_1 \ \mathbf{q}_2 \ \ldots \ \mathbf{q}_v] \in \Re^{m \times v}$ has orthogonal columns, and $\mathbf{R} \in \Re^{v \times v}$ is upper triangular matrix. Therefore, Eq. (A-17) can then be written as

$$\mathbf{t}_k^T = \mathbf{Q}\mathbf{R}\mathbf{w}_k + \mathbf{e}_k^T = \mathbf{Q}\mathbf{g} + \mathbf{e}_k^T \tag{A-19}$$

where $\mathbf{g} = [g_1 \ g_2 \ \ldots \ g_v]^T$ and the LLS method gives

$$\mathbf{g} = \mathbf{Q}^T(\mathbf{Q}\mathbf{Q}^T)^{-1}\mathbf{t}_k^T \quad or \quad g_\gamma = \mathbf{q}_\gamma^T(\mathbf{q}_\gamma \mathbf{q}_\gamma^T)^{-1}\mathbf{t}_k^T : \gamma = 1 \ldots v \tag{A-20}$$

As $\mathbf{Q}$ has orthogonal columns, the energy function of $\mathbf{t}_k^T$ can be represented as [25]

$$\mathbf{t}_k \mathbf{t}_k^T = \sum_{\gamma=1}^{v} g_\gamma^2 \mathbf{q}_\gamma^T \mathbf{q}_\gamma + \mathbf{e}_k \mathbf{e}_k^T \tag{A-21}$$

If $\mathbf{t}_k^T$ is the desired output after its mean has been removed, the variance of $\mathbf{t}_k^T$ would be

$$m^{-1}\mathbf{t}_k \mathbf{t}_k^T = m^{-1}\sum_{\gamma=1}^{v} g_\gamma^2 \mathbf{q}_\gamma^T \mathbf{q}_\gamma + m^{-1}\mathbf{e}_k \mathbf{e}_k^T \tag{A-22}$$

where $m^{-1}\sum g_\gamma^2 \mathbf{q}_\gamma^T \mathbf{q}_\gamma$ is the variance that can be explained by the regressor $\mathbf{q}_\gamma$, while $m^{-1}\mathbf{e}_k \mathbf{e}_k^T$ is the unexplained variance of $\mathbf{t}_k^T$. Thus, $m^{-1}\sum g_\gamma^2 \mathbf{q}_\gamma^T \mathbf{q}_\gamma$ is the increment to the explained desired output variance introduced by $\mathbf{q}_\gamma$. Thus, an error reduction ratio (ERR) due to $\mathbf{q}_\gamma$ on the kth output variable can be defined as [25]

$$err_\gamma^k = \frac{g_\gamma^2 \mathbf{q}_\gamma^T \mathbf{q}_\gamma}{\mathbf{t}_k \mathbf{t}_k^T} \tag{A-23}$$

Substituting g_γ by Eq. (A-20), we have

$$err_\gamma^k = \frac{[\mathbf{q}_\gamma^T(\mathbf{q}_\gamma \mathbf{q}_\gamma^T)^{-1}\mathbf{t}_k^T]^2 \mathbf{q}_\gamma^T \mathbf{q}_\gamma}{\mathbf{t}_k \mathbf{t}_k^T} \tag{A-24}$$

The ERR offers a simple and effective means of seeking a subset of significant regressors. The term err_γ in (A-24) reflects the similarity of $\mathbf{q}_\gamma$ and $\mathbf{t}_k^T$ or the inner product of $\mathbf{q}_\gamma$ and $\mathbf{t}_k^T$. Larger err_γ implies that $\mathbf{q}_\gamma$ is more significant to the output $\mathbf{t}_k^T$.

For a MIMO system with N_i inputs and N_o outputs, the ERR matrix is defined as

$$\mathbf{ERR} = \left[\begin{array}{cccc} err_1^1 & err_2^1 & \cdots & err_{N_r}^1 \\ err_{N_r+1}^1 & err_{N_r+2}^1 & \cdots & err_{N_r+N_r}^1 \\ \vdots & \vdots & \cdots & \vdots \\ err_{N_i \times N_r+1}^1 & err_{N_i \times N_r+2}^1 & \cdots & err_{N_i \times N_r+N_r}^1 \\ \hline \vdots & \vdots & \cdots & \vdots \\ \hline err_1^{N_o} & err_2^{N_o} & \cdots & err_{N_r}^{N_o} \\ err_{N_r+1}^{N_o} & err_{N_r+2}^{N_o} & \cdots & err_{N_r+N_r}^{N_o} \\ \vdots & \vdots & \cdots & \vdots \\ err_{N_i \times N_r+1}^{N_o} & err_{N_i \times N_r+2}^{N_o} & \cdots & err_{N_i \times N_r+N_r}^{N_o} \end{array}\right] \tag{A-25}$$

$$= \left[\mathbf{err}_1 \quad \mathbf{err}_2 \quad \ldots \quad \mathbf{err}_{N_r}\right] \tag{A-26}$$

We can further define the total error reduction ratio $Terr_j, j = 1 \ldots N_r$ corresponding to the jth rule as

$$Terr_j = \sqrt{\frac{(\mathbf{err}_j)^T \mathbf{err}_j}{(N_i + 1)N_o}} \tag{A-27}$$

If

$$Terr_j < K_{err} \tag{A-28}$$

where K_{err} is a predefined threshold, the jth fuzzy rule should be deleted, and vice versa.

A.2.3 Determination of Premise Parameters

The concept of semi-closed fuzzy set is used to allocate premise parameters of the G-FNN.

Definition A-2. Semi-Closed Fuzzy Set For input variable $x_i, i = 1 \ldots N_i$, if each fuzzy Gaussian membership function $mf_{ij}(c_{ij}, \sigma_{ij})$ is in the operating interval $[x_{i,\min}, x_{i,\max}]$ and satisfies the following boundary conditions:

$$\begin{array}{ll} mf_{ij}\left(c_{ij} = x_{i,\min}, \sigma_{ij} = \frac{|x_{i,\max} - x_{i,\min}|}{\sqrt{\ln \frac{1}{\epsilon}}}\right) & |x_i - x_{i,\min}| \le \delta \\ mf_{ij}\left(c_{ij} = x_{i,\max}, \sigma_{ij} = \frac{|x_{i,\max} - x_{i,\min}|}{\sqrt{\ln \frac{1}{\epsilon}}}\right) & |x_i - x_{i,\max}| \le \delta \\ mf_{ij}\left(c_{ij} = x_{i,\max}, \sigma_{ij} = \frac{\max\left(|c_{ij} - c_{ij_a}|, |c_{ij} - c_{ij_b}|\right)}{\sqrt{\ln \frac{1}{\epsilon}}}\right) & |x_i - x_{i,\min}| \ge \delta \text{ or} \\ & |x_i - x_{i,\max}| \ge \delta \end{array} \tag{A-29}$$

where δ is a tolerable small value, c_{ij_a} and c_{ij_b} are centers of two most neighboring membership functions $mf_{ij}(c_{ij}, \sigma_{ij})$, mf_{ij} is a semi-closed fuzzy set which satisfies the ϵ completeness of fuzzy rules.

Using concept of the semi-closed fuzzy set, the premise parameters are then allocated under the following three cases:

1. $e(m) > K_e$ and $d(m) > K_d$
   Multidimensional input vector $\mathbf{x}(m)$ is first projected to the ith axis as $x_i(m), i = 1 \ldots N_i$. Compute the Euclidean distance between $x_i(m)$ and the boundary point $b_{ij_n}, i = 1 \ldots N_i, j_n = 1, 2, \ldots N_r + 2$:

$$ed_{ij_n}(m) = ||x_i(m) - b_{ij_n}|| \tag{A-30}$$

   where $b_{ij_n} \in \{c_{i1}, c_{i2}, \ldots, c_{iN_r}, x_{i,\min}, x_{i,\max}\}$ and N_r is the number of existing rules. Find

$$\bar{j}_n = \arg\min ed_{ij_n}(m) \tag{A-31}$$

   If

$$ed_{i\bar{j}_n}(m) < K_{mf} \tag{A-32}$$

   where K_{mf} is a predetermined threshold which implies the similarity of neighboring membership function. Incorporated with the first two conditions in (A-29), we have

$$c_{i(N_r+1)} = b_{i\bar{j}_n}, \sigma_{i(N_r+1)} = \sigma_{i\bar{j}_n} \tag{A-33}$$

   Otherwise, we choose to incorporate with the third condition in (A-29)

$$c_{i(N_r+1)} = x_i(m) \tag{A-34}$$

$$\sigma_{i(N_r+1)} = \frac{\max\left(|c_{i(N_r+1)} - c_{i(N_r+1)_a}|, |c_{i(N_r+1)} - c_{i(N_r+1)_b}|\right)}{\sqrt{\ln \frac{1}{\epsilon}}} \tag{A-35}$$

2. $e(m) > K_e$ but $d(m) \leq K_d$
   $\mathbf{x}(m)$ can be clustered by the adjacent fuzzy rule, however, the rule is not significant enough to accommodate all the patterns covered by its ellipsoidal field. Therefore, the ellipsoidal field needs to be decreased to obtain a better local approximation. A simple method to reduce the Gaussian width is as follows

$$\sigma_{i\bar{j}\,new} = K_s \times \sigma_{i\bar{j}\,old} \tag{A-36}$$

   where K_s is a reduction factor which depends on the sensitivity of the input variables. The sensitivity measure of the ith input variable in the jth fuzzy rule is denoted as s_{ij}, which is calculated using the ERR matrix of (A-25), i.e.

$$s_{ij} = \frac{\sum_{k=1}^{N_o} err^k_{i \times N_r + j}}{\sum_{i=1}^{N_i} \sum_{k=1}^{N_o} err^k_{i \times N_r + j}} \tag{A-37}$$

The threshold K_s could be designed with a minimum value of $K_{s,\min}$ when $s_{ij} = 0$, and a maximum value of 1 when s_{ij} is greater than or equal to the average sensitivity level of the input variables in the jth rule, i.e.

$$K_s = \begin{cases} \frac{1}{1+\frac{(1-k_{s,min})N_r{}^2}{k_{s,min}}(s_{ij}-\frac{1}{N_i})^2} & s_{ij} < \frac{1}{N_i} \\ 1 & \frac{1}{N_i} \leq s_{ij} \end{cases} \tag{A-38}$$

3. $e(m) \leq K_e$ but $d(m) > K_d$ or $e(m) \leq K_e$ and $d(m) \leq K_d$
The system has good generalization and nothing need to be done except adjusting weight.

A.2.4 Determination of Consequent Parameters TSK-type consequence for each rule is determined using the Linear Least Squared (LLS) method. The LLS method is employed to find the weight matrix $\mathbf{W}$ such that the error energy $trace(\mathbf{E}^T\mathbf{E})$ is minimized in (A-15) [26]. Furthermore, the LLS method provides a computationally simple but efficient procedure for determining the weight so that it can be computed very quickly and used for real-time control. The weight matrix is calculated as

$$\mathbf{W} = (\mathbf{T}\mathbf{\Theta}^\dagger)^T \tag{A-39}$$

where $\mathbf{\Theta}^\dagger$ is the pseudoinverse of $\mathbf{\Theta}$

$$\mathbf{\Theta}^\dagger = (\mathbf{\Theta}^T\mathbf{\Theta})^{-1}\mathbf{\Theta}^T \tag{A-40}$$

In practice, direct solution of (A-40) can lead to numerical difficulties due to the possibility of $\mathbf{\Theta}^T\mathbf{\Theta}$ being singular or nearly singular. This problem is best resolved by using singular value decomposition (SVD) technique [24] in our G-FNN learning algorithm. SVD approach avoids problems due to the accumulation of numerical roundoff errors, and automatically selects (from amongst a set of nearly degenerate solutions) the one for which the weight vector length $||\mathbf{w}_k||$ of the kth output variable is the shortest.

A.2.5 Normalization of Learning Data Set While G-FNN is working as a MIMO system, different input/output variables usually have different ranges. This costs difficulties and complexities in parameter determinations as shown in previous sections. To solve this problem, a data preprocessing method is introduced to normalize input and output variables into a uniform map $[-1, 1]$. We denote each input variable as $x_i \in [x_{i,\min}, x_{i,\max}]$ and each desired output as $t_k \in [t_{k,\min}, t_{k,\max}]$. The normalized results are defined as

$$x_i(m) = \frac{x_i(m)}{\max(|x_{i,\min}|, |x_{i,\max}|)} \tag{A-41}$$

$$t_k(m) = \frac{t_k(m)}{\max(|t_{k,\min}|, |t_{k,\max}|)} \tag{A-42}$$

Advantages of this preprocessing are: (1) Predefined parameters of the G-FNN can be unified, such as the error factor K_e in Eq. (A-9) and the E-distance factor K_{mf} in Eq. (A-32); (2) Online learning speed is improved because the mapping is restricted to a uniform small region; (3) The ERR sensitivity analysis is considered to be more accurate because thresholds K_{err} in Eq. (A-28) and K_s in Eq. (A-36) are predefined based on the average level of input/output variables.

A.3 G-FNN Direct Modelling

As an integration of the FIS and ANN, G-FNN is capable of system identification and can, as a consequence, be used for control. Functionally, the G-FNN can be described as a function approximator, i.e. it aims to approximate a MIMO function $\mathbf{\Omega}$: $\Re^{N_i} \rightarrow \Re^{N_o}$ from sample patterns $(\mathbf{x}, \mathbf{t})$ drawn from $\mathbf{\Omega}$. Conceptually, the simplest approach is the direct modelling [5], where a synthetic training signal $\mathbf{u}$ is introduced to the modelled system to generate the system output signal $\mathbf{z}$. If the forward dynamics of the system is to be modelled, the sample pattern is defined as $(\mathbf{x}, \mathbf{t}) \triangleq (\mathbf{u}, \mathbf{z})$. For inverse dynamics modelling, the system output $\mathbf{z}$ is then used as the input to the G-FNN and the synthetic training signal $\mathbf{u}$ is the teaching signal, i.e. $(\mathbf{x}, \mathbf{t}) \triangleq (\mathbf{z}, \mathbf{u})$. An example of inverse dynamics modelling using G-FNN is depicted in Fig. A-2. This structure will clearly result in a good representation of the plant inverse dynamics using the G-FNN. However, there are drawbacks in the direct modelling approach:

- If the actual inputs of the G-FNN are outside its input training space, direct modelling performance may be poor.
- If the nonlinear system's forward or inverse mapping is not one-to-one, an incorrect inverse may result.

Therefore, before using the direct modelling method, a few assumptions have to be made in this chapter:

- As long as the direct modelling of a G-FNN is used for control purpose, its actual inputs will be within its input training space. To ensure that this assumption is valid, the synthetic training signal should be properly chosen during learning.
- The dynamic mapping is chosen to be one-to-one.

References

1. Slotine, J.J.E., Li, W. (1991): Applied Nonlinear Control, Prentice Hall, New Jersey

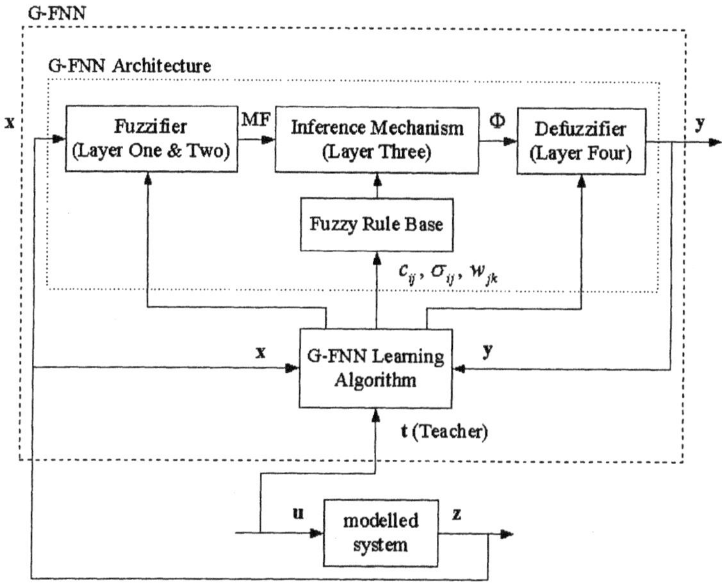

Fig. A-2. G-FNN direct modelling of system inverse dynamics

2. Åström, K.J., Wittenmark, B. (1995): Adaptive Control, Addison-Wesley, New York
3. Ge, S.S., Lee, T.H., Harris, C.J. (1998): Adaptive Neural Network Control of Robotic Manipulators, World Scientific Publishing, New Jersey
4. Brown, M., Harris, C. (1994): Neurofuzzy Adaptive Modelling and Control, Prentice Hall, London
5. Hunt, K.J., Sbarbaro, D., Zbikowski, R., Gawthrop, P.J. (1992): Neural Networks for Control Systems - a Survey. Automation **28**, 1083-1112
6. Wang, L.X. (1997): A Course in Fuzzy Systems and Control, Prentice Hall, New Jersey
7. Lin, F.J., Hwang, W.J., Wai, R.J. (1999): A Supervisory Fuzzy Neural Network Control System for Tracking Periodic Input. IEEE Trans. on Fuzzy Systems **7**, 41-52
8. Wu, J.C., Liu, T.S. (1996): A Sliding Mode Approach to Fuzzy Control Design. IEEE Trans. on Control Systems Technology **4**, 459-468
9. Ohtani, Y., Yoshimura, T. (1996): Fuzzy Control of a Manipulator Using the Concept of Sliding Mode. Int. J. of Systems Science **27**, 1727-1731
10. Chang, Y.C. (2001) Adaptive Fuzzy-Based Tracking Control for Nonlinear SISO Systems via VSS and H^{∞} Approaches. IEEE Trans. on Fuzzy Systems **9**, 278-292
11. Pan, Y., Klir, G.J., Yuan, B. (1996): Bayesian Inference Based on Fuzzy Probabilities. Proc. 5th IEEE Int. Conf. on Fuzzy Systems **3**, New Orleans, Louisiana, USA, 1693-1699

12. MacKay, D.J.C. (1995): Bayesian Neural Networks and Density Networks. Nuclear Instruments and Methods in Physics Research **354**, pp. 73-80
13. Park, D., Kandel, A., Langholz, G. (1994): Genetic-Based New Fuzzy reasoning models with Application to Fuzzy Control. IEEE Trans. on System, Man, and Cybernetics **24**, 39-47
14. Berenji, H.R., Khedkar, P. (1992): Learning and Tuning Fuzzy Controllers Through Reinforcement. IEEE Trans. on Neural Network **3**, 724-739
15. Jin, Y., Jiang, J., Zhu, J. (1995): Neural Network Based Fuzzy Identification with Application to Modelling and Control of Complex Systems. IEEE Trans. on System, Man, and Cybernetics **25**, 990-997
16. Jang, J.S.R., Sun, C.T., Mizutani, E. (1997): Neuro-Fuzzy and Soft Computing, Prentice Hall, New Jersey
17. Jin, Y. (1998): Decentralized Adaptive Fuzzy Control of Robot Manipulators. IEEE Trans. on System, Man, and Cybernetics **28**, 47-58
18. Jin, Y., Seelen, W.V., Sendhoff, B. (1999): On Generating FC^3 Fuzzy Rule Systems from Data Using Evolution Strategies. IEEE Trans. on System, Man, and Cybernetics **29**, 829-845
19. Nie, J., Linkens, D. (1995) Fuzzy-Neural Control: Principles, Algorithms and Applications, Prentice Hall, New Jersey
20. Cannon, M. R., Slotine, J.J.E. (1995): Space-Frequency Localized Basis Function Networks for Nonlinear System Estimation and Control. Neurocomputing **9**, 293-342
21. Sanner, R.M., Slotine, J.J.E. (1998): Structurally Dynamic Wavelet Networks for Adaptive Control of Robotic Systems. Int. J. Control **70**, 405-421
22. Gao, Y., Er, M.J., Yang, S. (2001): Adaptive Fuzzy Neural Control of Robot Manipulators. IEEE Trans. on Industrial Electronics **48**, 1274-1278
23. Wu, S.Q., Er, M.J., Gao, Y. (2001): A Fast Approach for Automatic Generation of Fuzzy Rules by Generalized Dynamic Fuzzy Neural Networks. IEEE Trans. on Fuzzy Systems **9**, 578-594
24. Press, W.H., Teukolsky, S.A., Vetterling, W.T., Flannery, B.P. (1992): Numerical Recipes in C: The Art of Scientific Computing, Cambridge University Press
25. Chen, S., Cowan, C.F.N., Grant, P.M. (1991): Orthogonal Least Squares Learning Algorithm for Radial Basis Function Network. IEEE Trans. Neural Networks **2**, 302-309
26. Bishop, C.M. (1995): Neural Networks for Pattern Recognition, Oxford University Press

Hybrid Fuzzy Proportional-Integral plus Conventional Derivative Control of Robotics Systems

Meng Joo Er and Ya Lei Sun

School of Electrical and Electronic Engineering
Nanyang Technological University, Blk S1
Nanyang Avenue, Singapore 639798
emjer@ntu.edu.sg, eylsun@ntu.edu.sg

Abstract. This chapter presents a new approach towards optimal design of a hybrid fuzzy proportional-integral-derivative (PID) controller for robotics systems using genetic algorithm (GA). The proposed hybrid fuzzy PID controller is derived by replacing the conventional PI controller by a two-input normalized linear fuzzy logic controller and implementing the conventional D controller in an incremental form. The salient features of the proposed controller are: (1) Gain scheduling method is incorporated in the design to linearize the robotics system for a given reference trajectory. (2) Only one well-defined linear fuzzy control space is required for multiple local linearized systems. (3) The linearly defined fuzzy logic controller can generate sector bounded nonlinear outputs so that the closed-loop system is stable and has better performance. (4) Optimal tuning of controller gains is carried out by using GA. (5) It is simple and easy to implement. Simulation studies on a pole balancing robot and a multi-link robot manipulator demonstrate the effectiveness and robustness of the proposed controller. Comprehensive comparisons with other latest approaches show that the proposed approach is superior in terms of tracking performance and robustness.

1 Introduction

A fuzzy logic controller (FLC) makes control decisions by its well-known fuzzy IF-THEN rules. In the antecedence of the fuzzy rules (i.e. *the IF part*), the control space is partitioned into small regions with respect to different input conditions. In each region, there is a corresponding membership function (MF) to fuzzify each of the input variables. In the center of the region approach, *values of fuzziness* of all the input variables will be set to 1. Using the convex MF's, fuzziness values at the edge of each region will be set to 0 or a very small positive value. For continuity of the fuzzy space, the regions are usually overlapped by their neighbors. By manipulating all the input fuzziness values in the fuzzy rule base, an output will be given in the subsequence (i.e. *the THEN part*), where FLC's can be classified into two major categories: Mamdani type and Takagi-Sugeno (TS) type. A Mamdani type FLC uses fuzzy numbers to make decisions [7] and a TS type FLC generates control actions by linear functions of the input variables [12].

In the early years, most FLC's were designed by trial and error. Since the complexity of an FLC will increase exponentially when it is used to control complex systems, it is tedious to design and tune FLC's manually for most industrial problems like robotics systems. This is why the conventional nonlinear design method [11] was adopted in the fuzzy control area, such as fuzzy sliding control, fuzzy gain scheduling (GS) [10], and adaptive fuzzy control [15], in order to alleviate difficulties in constructing the fuzzy rule base. For most control systems, state feedback and its derivatives are assumed to be available. The reference input is assumed to be *piecewise continuous* so that error signals and at least their first derivative are available to the controller. Analytical calculations show that a two-input FLC employing proportional error signal and velocity error signal is a nonlinear proportional-integral (PI) or proportional-derivative (PD) controller [1],[17]. Due to the popularity of PID controllers in industrial applications, most of the development of fuzzy controllers revolved around fuzzy PID controllers in the past decade [2],[6],[8],[19]. To distinguish themselves from *conventional stand-alone FLC's*, these controllers are usually called *fuzzy controllers*. To emphasize the existence of conventional controllers in the overall control structure, they are called *hybrid fuzzy controllers.*

There are some difficulties that prevent the design of hybrid fuzzy controllers from being systematic:

1. The choice of the overall control structure is the first problem faced by many designers. Each conventional nonlinear design method has its own merits and drawbacks. The design of hybrid fuzzy controllers can be viewed as, from one point of view, the use of conventional control methods to facilitate the FLC's design; and, from the other point of view, to incorporate the FLC's into the conventional control structure so as to enhance the power of the conventional design method. In either point of view, the contradiction between conventional functions and fuzzy systems has to be solved in order to integrate the design.
2. Partitioning the control space or the fuzzification problem consists of two aspects, namely what kind of MF's should be used and where the MF's should be located. To the first question, although many types of MF's can be used, for simplicity and without loss of generality, triangular or Gaussian MF's are most commonly used. There is no standard answer to the second question. However, for control problems in a robotics system, the reference trajectory may guide us to a fairly good fuzzification of system output signals. Nevertheless, the question of fuzzification of error signals is not answered yet.
3. It is common sense that the more fuzzy rules are used, a better controller will be obtained. Theoretically, by firing a very large number of fuzzy rules, a fuzzy system can be viewed as an universal approximator on a compact set of arbitrary accuracy [16] and a FLC can generate perfect control actions for any dynamic systems. Obviously, such kind of design

is impractical due to either the tuning problem of design parameters or limitations of hardware implementation. Construction of a fuzzy rule base by using limited number of rules to approximate the input-output relationship of the FLC with a high accuracy is always a challenge for many designers of fuzzy controllers.

4. In a hybrid fuzzy controller, not only parameters of the FLC need to be designed, but also the gains of the conventional controller need to be tuned. Because of their complicated cross-effects, the analytical tuning algorithm for these parameters is not available.

In this chapter, a novel design method of hybrid fuzzy control is proposed to solve the aforementioned problems. The key idea of the proposed method is as follows: First, gain scheduling is applied to linearize the robotics system at *frozen time*. For each frozen system, a fuzzy PID controller is designed by replacing the conventional PI controller by an incremental FLC. All the fuzzy PID controllers will have the same structure and fuzzy rule base, but different gains. Second, fuzzification of the reference input is performed by general knowledge of the system, while the control space of error signals is linearly partitioned after normalization. Third, the linear fuzzy rule base is constructed in an incremental manner to obtain better fuzzy PID control of the frozen system. Finally, to facilitate optimal tuning of the design parameters, genetic algorithm (GA) is applied to both the frozen system and the overall closed-loop system.

In the next section, the robotics systems are reviewed as a general class of nonlinear systems. The idea of gain scheduling is introduced to transform the trajectory tracking problem into the local stabilization problem. In Section 3, a novel fuzzy PID controller is proposed. The incremental design of the fuzzy rule base guarantees the stability of local closed-loop systems. In Section 4, optimal tuning of parameters by using GA with analytical constraints is discussed. In Section 5, control of a pole-balancing robot and a multi-link robot manipulator illustrate how the proposed design method can be easily applied to a robotics system. Concluding remarks are given in the last section.

2 The Control Problem

Tracking control of a given reference trajectory is a common task in controlling a robotics system. Since the force generated by motors can be explicitly measured or calculated from sensor feedback data, most robotics systems can be expressed by the following nonlinear autonomous system

$$\dot{\mathbf{x}} = \mathbf{f}(\mathbf{x}) + \mathbf{g}(\mathbf{x})\mathbf{u} , \tag{1}$$

where $\mathbf{x} = [x_1, x_2, \ldots, x_n]^T \in R^{n\times 1}$ is the state vector, $\mathbf{u} = [u_1, u_2, \ldots, u_m]^T \in R^{m\times 1}$ is the control input vector, $\mathbf{f}(\mathbf{x}) \in R^{n\times 1}$ and $\mathbf{g}(\mathbf{x}) \in R^{n\times 1}$ are vector functions of states.

Assume $\mathbf{x^d}(t) \in R^{n\times 1}$ is the given reference trajectory whose corresponding reference input is $\mathbf{u^d}(t)$:

$$\dot{\mathbf{x}}^{\mathbf{d}} = \mathbf{f}(\mathbf{x^d}) + \mathbf{g}(\mathbf{x^d})\mathbf{u^d} \,. \tag{2}$$

Taking Lyapunov-linearization around the operating points $(\mathbf{x^d}, \mathbf{u^d})$, we have

$$\dot{\mathbf{x}} = \dot{\mathbf{x}}^{\mathbf{d}} + \mathbf{A}(\mathbf{x^d})(\mathbf{x} - \mathbf{x^d}) + \mathbf{B}(\mathbf{x^d})(\mathbf{u} - \mathbf{u^d}) \,, \tag{3}$$

where

$$\begin{aligned} \mathbf{A}(\mathbf{x^d}) &= \frac{d\mathbf{f}}{d\mathbf{x}}\bigg|_{\mathbf{x}=\mathbf{x^d}} \,, \\ \mathbf{B}(\mathbf{x^d}) &= \mathbf{g}(\mathbf{x^d}) \,. \end{aligned} \tag{4}$$

Let $\mathbf{e} = \mathbf{x} - \mathbf{x^d} \in R^{n\times 1}$, $\dot{\mathbf{e}} = \dot{\mathbf{x}} - \dot{\mathbf{x}}^{\mathbf{d}} \in R^{n\times 1}$, and $\mathbf{u^e} = \mathbf{u} - \mathbf{u^d} \in R^{m\times 1}$. System (3) is equivalent to

$$\dot{\mathbf{e}} = \mathbf{A^d}\mathbf{e} + \mathbf{B^d}\mathbf{u^e} \,, \tag{5}$$

where $\mathbf{A^d}$ and $\mathbf{B^d}$ are assumed to be in the controllability canonical form (CCF) or at least can be transformed into diagonal CCF, which is valid for many robotics systems. Because the reference trajectory $\mathbf{x^d}$ is a function of time t, we can linearize the nonlinear autonomous system (1) at frozen time τ so that the tracking problem of the nonlinear system is transformed into the stabilization problem of the linear system (5) in error state space. The equilibrium points are shifted from the reference trajectory points $\mathbf{x^d}(\tau)$ to the origin $\mathbf{0}$. By the Lyapunov theorem, if the linearized system is strictly stable, then the equilibrium point is asymptotically stable for the actual nonlinear system [11].

3 Design of Hybrid Fuzzy PID Controller

In this section, a fuzzy PID controller with fuzzy PI plus conventional derivative controller is proposed. The fuzzy PI controller is derived by discretizing the conventional PI controller and constructed with simple linear fuzzy rules in an incremental way, which guarantees the sector condition of the output. Local stability analysis also explores the relationship between the conventional derivative gain and the fuzzy gain. Although the proposed controller is developed as a hybrid fuzzy controller, the overall structure shows its potential to be a new form of stand-alone FLC, that is, the combination of the Mamdani type FLC and the TS type FLC. The feedback connection of a linear system and a nonlinear element is shown in Fig. 1.

3.1 Design of Fuzzy PI Controller

The most widely adopted conventional PID controller structure used in industrial applications is the following structure:

$$U_{PID}(s) = \left(K_P^c + \frac{K_I^c}{s} + K_D^c s \right) E(s) \,, \tag{6}$$

where K_P^c, K_I^c, and K_D^c are the conventional proportional, integral and derivation gains respectively, $U_{PID}(s)$ is the controller output and $E(s)$ is the error signal. To replace the PI portion, we discretize (6) via bilinear transformations. It can be shown that for a sampling time, T, the conventional PI controller is governed by

$$U_{PI}(z) = \left(K_P^c + \frac{K_I^c T}{2} - \frac{K_I^c T}{1 - z^{-1}} \right) E(z) \,, \tag{7}$$

or equivalently

$$u_{PI}(kT) - u_{PI}(kT - T) = K_P[e(kT) - e(kT - T)] + K_I T e(kT) \,, \tag{8}$$

where $K_P = K_P^c - K_I^c T/2$ and $K_I = K_I^c$. Hence, in incremental form, the conventional PI controller can be expressed as

$$u_{PI}(kT) = u_{PI}(kT - T) + T \Delta u_{PI}(kT) \,, \tag{9}$$

where

$$\Delta u_{PI}(kT) = K_I e_p(kT) + K_P e_v(kT) \,, \tag{10a}$$

$$e_p(kT) = e(kT) = y_r(kT) - y(kT) \,, \tag{10b}$$

$$e_v(kT) = [e(kT) - e(kT - T)] \,/T \,. \tag{10c}$$

Here, $\Delta u_{PI}(kT)$ is the incremental control output of the PI controller, $e_p(kT)$ is the error signal or proportional error signal, $e_v(kT)$ is the change of error signal or velocity error signal, $y(kT)$ is the output feedback, and $y_r(kT)$ is the reference input signal. In the design of the fuzzy PI controller, the term $T \Delta u_{PI}(kT)$ in Eq. (9) will be replaced by a fuzzy control action, denoted by $K_F \Delta u$, where K_F is the output scaling factor of the FLC. Since the output of the FLC is nonlinear, Eq. (9) can be rewritten as

$$u_{PI} = \phi + K_F \Delta u \,, \tag{11}$$

where ϕ is the nonlinear output of the fuzzy PI controller at time step, kT and u_{PI} is the output we want to obtain from the fuzzy PI controller at the next time step $(k+1)T$. The conventional PID controller only employs the proportional error signal, e_p, and the velocity error signal, e_v. However, it is clear from Eq. (11) that the proposed fuzzy PI controller employs both error signals and the control input signal at the previous step.

The fuzzy rule base is constructed by linear fuzzy rules in an incremental form. Without loss of generality, only the design of the single-out FLC is illustrated. First, the fuzzy rules about u_{PI} are linearly defined by

$$\begin{aligned} \mathrm{R}^j &: \text{IF } e_1 \text{ is } LE_{1j} \text{ AND } e_2 \text{ is } LE_{2j} \ldots \text{ AND } e_n \text{ is } LE_{nj}\,, \\ &\text{THEN } u_{PI} \text{ is } LU_j = LE_{1j} + LE_{2j} + \ldots + LE_{nj}\,, \end{aligned} \tag{12}$$

where $LE_{1j}, \ldots, LE_{nj}$ are the linguistic values of error signals of the jth fuzzy rule and LU_j is the linguistic value for the output u_{PI}. It should be emphasized that the summation of linguistic variables is carried by the method of computing with words (CW) [18] rather than the conventional arithmetic operation. For example, if $LE_1 = positive$, $LE_2 = positive$ and $LU = LE_1 + LE_2$, we may have $LU = positive\ medium$ or *positive big* depending on the number of output MF's. It is easy to see that the linguistic value of u_{PI} is the linear combination of the linguistic values of $\mathbf{e}$ and so is their crisp values. If we define $y = \mathbf{Ce}$, where $\mathbf{C} \in R^{1\times n}$, the u_{PI} can be viewed as a function of y. Since in time domain, $\phi(kT) = u_{PI}(kT - T)$, the output of the fuzzy PI controller, ϕ, can be viewed as a nonlinear function of y (as depicted in Fig. 2).

The linguistic values of u_{PI} in (12) can be manipulated by CW to generate finer control actions without increasing the number of output MF's. For example, if $LU_j = positive$, by introducing more singleton values of Δu, we can have control actions like *postive medium* and *positve big*. Using the output scaling factor K_F in (11), these singleton values can be easily defined as equally distributed normalized values. Hence, the incremental fuzzy PI controller will have the following fuzzy rules:

$$\begin{aligned} \mathrm{R}^j &: \text{IF } e_1 \text{ is } LE_{1j} \text{ AND } e_2 \text{ is } LE_{2j} \ldots \text{ AND } e_n \text{ is } LE_{nj}\,, \\ &\text{AND } u_{PI} \text{ is } LU_j\,, \\ &\text{THEN } \Delta u \text{ is } LD_j = LE_{1j} + LE_{2j} + \ldots + LE_{nj} - LU_j\,. \end{aligned} \tag{13}$$

3.2 Stability Analysis of Local Closed-loop Systems

By introducing conventional derivative control, the open-loop system (5) can be transformed into the following closed-loop system, which can be regarded as a feedback connection of a linear system and a nonlinear element (depicted in Fig. 1),

$$\begin{aligned} \dot{\mathbf{e}} &= \hat{\mathbf{A}}^{\mathbf{d}}\mathbf{e} - \hat{\mathbf{B}}^{\mathbf{d}}\phi\,, \\ y &= \mathbf{Ce}\,, \end{aligned} \tag{14}$$

where ϕ is the nonlinear output of the fuzzy PI controller and y is the linear combination of error signals. Stabilizing a linear system with error signals $\mathbf{e}$ is equivalent to minimizing z as close to the origin as possible. The conventional gain K_D^c in (6) can be chosen to be within a range such that $\hat{\mathbf{A}}^{\mathbf{d}}$ is stable.

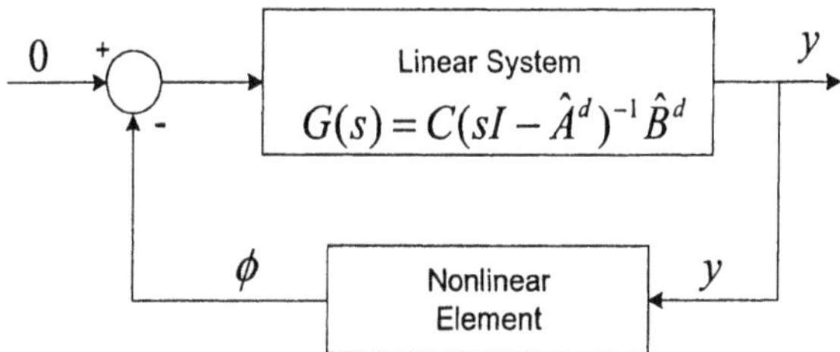

Fig. 1. Feedback Connection of a Linear System and a Nonlinear Element

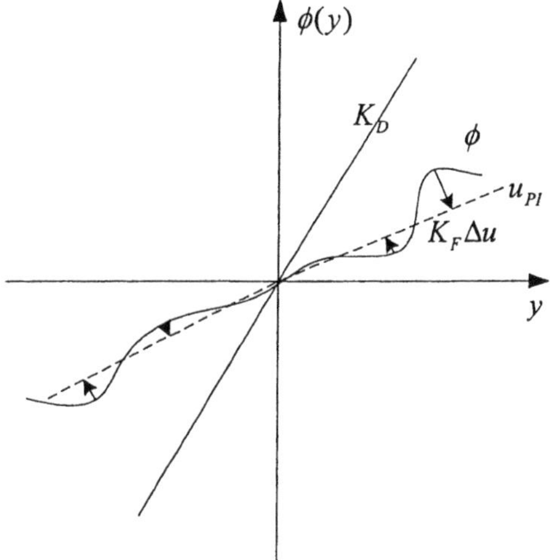

Fig. 2. Sector condition of the output of the fuzzy PI controller

The sector condition of the nonlinear output ϕ is depicted in Fig. 2. The line u_{PI} represents the output of the fuzzy rule base given by Eq. (12), which can be designed to have a very weak nonlinearity by using simple MF's of error signals and have the sector condition of $[0, K_D]$. The slope K_D will be affected by both input fuzzy gains and the conventional derivative gain. By manipulating K_F and K_D, the sector condition of ϕ can be obtained as follows:

$$\phi[\phi - K_D y] \leq 0 \,. \tag{15}$$

For each stable $\hat{\mathbf{A}}^{\mathbf{d}}$, there exists a symmetric positive definite matrix $\mathbf{P} \in R^{n \times n}$, a vector $\mathbf{q} \in R^{n \times 1}$ and a constant $\epsilon > 0$ such that [4]

$$\begin{aligned} \mathbf{P}(\hat{\mathbf{A}}^{\mathbf{d}}) + (\hat{\mathbf{A}}^{\mathbf{d}})^T \mathbf{P} &= -\mathbf{q}\mathbf{q}^T - \epsilon \mathbf{P} \,, \\ \mathbf{P}\mathbf{B}^{\mathbf{d}} &= K_D \mathbf{C} - \sqrt{2}\mathbf{q} \,. \end{aligned} \tag{16}$$

Consider the following Lyapunov function candidate in quadratic form

$$V(\mathbf{e}) = \mathbf{e}^T\mathbf{P}\mathbf{e}\,. \tag{17}$$

The time derivative along the states of system (14) is given by

$$\dot{V}(\mathbf{e}) = \mathbf{e}^T[\mathbf{P}(\hat{\mathbf{A}}^{\mathbf{d}}) + (\hat{\mathbf{A}}^{\mathbf{d}})^T\mathbf{P}]\mathbf{e} - 2\mathbf{e}^T\mathbf{P}\hat{\mathbf{B}}^{\mathbf{d}}\phi\,. \tag{18}$$

By subtracting the non-positive term in Eq. (15), we have

$$\begin{aligned}\dot{V}(\mathbf{e}) &\leq \mathbf{e}^T[\mathbf{P}(\hat{\mathbf{A}}^{\mathbf{d}}) + (\hat{\mathbf{A}}^{\mathbf{d}})^T\mathbf{P}]\mathbf{e} - 2\mathbf{e}^T\mathbf{P}\hat{\mathbf{B}}^{\mathbf{d}}\phi - 2\phi[\phi - K_D\mathbf{C}\mathbf{e}]\,,\\ &= \mathbf{e}^T[\mathbf{P}(\hat{\mathbf{A}}^{\mathbf{d}}) + (\hat{\mathbf{A}}^{\mathbf{d}})^T\mathbf{P}]\mathbf{e} - 2\mathbf{e}^T[K_D\mathbf{C} - \mathbf{P}\hat{\mathbf{B}}^{\mathbf{d}}]\phi - 2\phi^2\,.\end{aligned} \tag{19}$$

Note that $\mathbf{Ce} = \mathbf{e^T C}$ is a scalar. Substituting Eq. (16) into Eq. (19), we have

$$\begin{aligned}\dot{V}(\mathbf{e}) &\leq -\epsilon\mathbf{e}^T\mathbf{P}\mathbf{e} - \mathbf{e}^T\mathbf{q}\mathbf{q}^T\mathbf{e} + 2\sqrt{2}\mathbf{e}^T\mathbf{q}\phi - 2\phi^2\,,\\ &= -\epsilon\mathbf{e}^T\mathbf{P}\mathbf{e} - [\mathbf{e}^T\mathbf{q} - \sqrt{2}\phi][\mathbf{q}^T\mathbf{e} - \sqrt{2}\phi]\,,\\ &\leq -\epsilon\mathbf{e}^T\mathbf{P}\mathbf{e}\,.\end{aligned} \tag{20}$$

Since $\dot{V}(\mathbf{e})$ is negative definite, the local closed-loop system is stable by the Lyapunov theorem.

3.3 The Overall Control Structure

The proposed controller can be designed step by step so that it is simple and easy to be implemented on the actual robotics system. In Step 1, for a given reference trajectory, $\mathbf{x^d}(t)$, at some frozen times τ_i, the corresponding control input can be approximated by Eq. (2), which is $\mathbf{x^d}(\tau_i)$ or $\mathbf{x}^i$ in short. In the partitioning of the state space, these $\mathbf{x}^i$ will be the centers of MF's, LX^i. The nonlinear system Eq. (1) can be transformed into several local linear systems:

$$\mathrm{R}^i : \mathrm{IF}\ \mathbf{x^d}\ \mathrm{is}\ LX^i\ \mathrm{THEN}\ \dot{\mathbf{e}} = \mathbf{Ae} + \mathbf{Bu^e}\,. \tag{21}$$

In Step 2, the combination of fuzzy PI control plus conventional derivative control is given by

$$\mathrm{R}^j_m : \mathrm{IF}\ \mathbf{e}\ \mathrm{is}\ LE^j\ \mathrm{THEN}\ u^e_m = d(\mathbf{x} - \mathbf{x^d}) + LU^j_m\,, \tag{22}$$

where the function $d(\cdot)$ calculates the conventional derivative control output and LU^j_m is the linguistic value defined in Eq. (12) for the single control input u_m. Since the linguistic value of error signals, LE, can be obtained by computing with the linguistic values, LX, the proposed hybrid fuzzy controller will have an equivalent overall control structure given by

$$\begin{aligned}\mathrm{R}^{ij}_m : &\mathrm{IF}\ \mathbf{x^d}\ \mathrm{is}\ LX^i\ \mathrm{AND}\ \mathbf{x}\ \mathrm{is}\ LX^j\,,\\ &\mathrm{THEN}\ u_m = K_m(\mathbf{x}, \mathbf{x^d}) + LU_m\,.\end{aligned} \tag{23}$$

This form shows that the proposed hybrid fuzzy controller uses both linear functions of the input variables, $K(\mathbf{x^d}, \mathbf{x})$, and the fuzzy number, LU, as the output of fuzzy rules. Hence, it can be viewed as the combination of the Mamdani type FLC and the TS type FLC (depicted in Fig. 3).

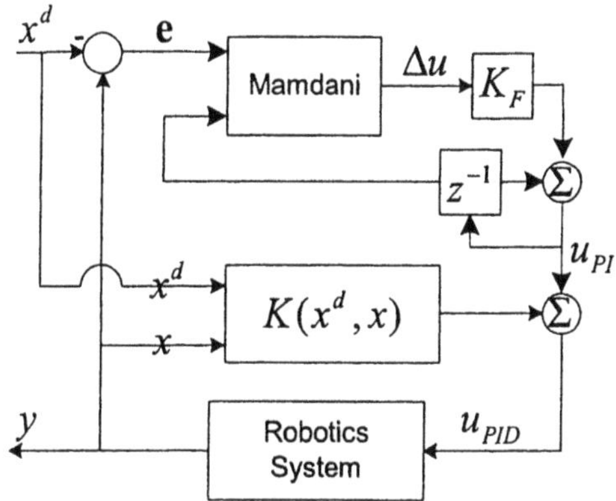

Fig. 3. The overall control structure

The controller gains in Step 2 are only tuned for local closed-loop systems. In order to compensate the linearization error and fine tune the controller globally, an additional step, Step 3, is needed to add an compensator, δu_m, to each output, u_m. Obviously, there are no analytical calculations available for the compensators and it is not possible to tune them manually. Some external tuning mechanisms have to be sought if we want to apply Step 3.

4 Optimal Tuning Using GA

GA is biologically inspired multi-parameters search algorithm that has proven to be effective in solving a variety of complex problems [3]. The majority of GA developed for fuzzy controllers in the literature are for offline optimization [5]. In this chapter, offline GA will be used to tune all the controller gains, namely the conventional derivative gain K_D^c, the input scaling factor K_E, the output scaling factor K_F, and the output compensators (if any) optimally. Much research effort has been directed towards variations of genetic operators, such as coding, selection, crossover and mutation. However, the variables to be optimized here are already real values, so coding is not a problem and simple GA will work [14]. Several improved techniques have been used in this chapter to speed up the tuning process. First of all, defining good fitness functions is much more important as the entire GA process relies on it. Since optimization of static function is not adequate for many cases of control problems, we proposed different cost functions from one example to another. Cost functions are similar to fitness functions, which will be minimized by GA to obtain optimal performance of the control system.

Some commonly used control criteria, such as rise time, t_r, settling time, t_s, and percent maximum overshoot, M_p, are used as basic components of cost functions. The system output y and the controller output u are often of interest since they are the two most important signals in any control systems. To monitor the system output, the error signal is usually evaluated by Integral Absolute Error (IAE), Integral Square Error (ISE), and Integral Time Absolute Error (ITAE). The ISE is a good numerical criterion for many control systems. Absolute Control Effort (ACE), the summation of all absolute control input signals, is used to monitor the energy pumped into the system. Weightings of linear combinations of the criteria are part of the design if a single cost function is to be used. The standard form of a cost function for control problems can be governed by

$$J = \sum_{i=1}^{N} (w_i T_i + P_i) \ , \tag{24}$$

where w_i are the weightings, T_i are the aforementioned performance criteria, and P_i is the extra penalty term defined for some constraints of specific problems. For example, if a control system is required to have a ceiling value of percentage maximum overshoot, then a penalty term P_1 can be set for the overshoot constraint. During offline tuning, whenever the system output exceeds the limitation, a very large value will be assigned to P_1 so that the current set of parameters will be redundant. If multiple constraints are required, several distinct penalty terms can be used to assist the analysis of the closed-loop system. The performance index J is actually a penalty function or cost function. A small value of J indicates that the system has good performance.

Initial population and searching range play important roles in the convergence of GA. The initial population is the guess on where the search should start, and the searching range defines range of the search space. In this chapter, based on the controller structure, the search space is divided into two layers. First, GA is used for local closed-loop systems, which means only the conventional gain and the fuzzy scaling factors will be optimized locally. For local linearized systems, some conventional PID tuning methods like Ziegler-Nichols (ZN) tuning can help in the initial guess of the parameters [13]. The search space is defined to contain any possible successful initial guesses and may be bounded by stability regions of the parameters. Then, GA is carried out to optimize the global system so as to compensate modeling errors in Eq. (3). A numerical term, δu^i can be added to the antecedent part of each fuzzy rule in Eq. (23). Summations of absolute values of the tracking error could be the criterion of GA. However, if the number of the frozen systems is big, the computation load of GA will be heavy.

Some GA settings used are: number of generations = 30, population size = 30, tournament selection with niching technique, two points crossover with $p_c = 0.7$ and a high mutation rate with $p_m = 0.2$ to speed up the search process.

5 Illustrative Examples

5.1 A Pole Balancing Robot - SISO Case

In this example, the proposed controller is used to control a pole balancing robot (also well known as the inverted pendulum system or the cart-pole system) depicted in Fig. 5.1. The cart is allowed to move horizontally in order to keep the pole at the upright position from different initial angles or to ensure that the angular position of the pole follows certain trajectories. The dynamics of the system is given by

$$\ddot{\theta} = \frac{(m_p + m_c)g\sin\theta - m_p\dot{\theta}^2 l\cos\theta\sin\theta - F\cos\theta}{(m_p + m_c)l(4/3 - m_p\cos^2\theta)} , \tag{25}$$

where θ is the angular position, $\dot{\theta}$ is the angular velocity and F is the external force applied to the cart. The gravity acceleration, g is $9.8m/s^2$, the mass of the cart, m_c is $1.0kg$, the mass of the pole, m_p is $0.1kg$ and the half-length of the pole, l is $0.5m$.

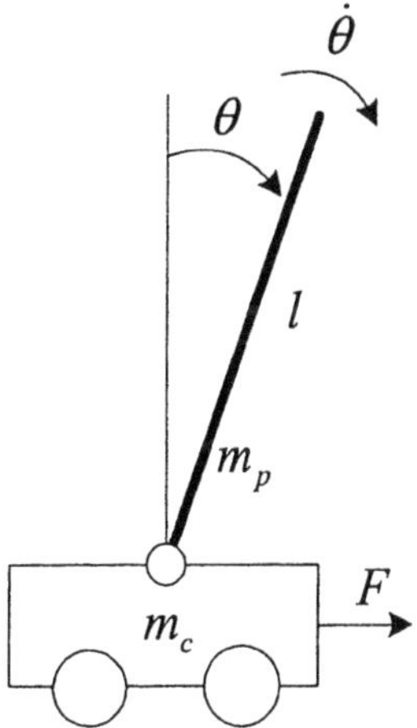

Fig. 4. An inverted pendulum system

Let $\mathbf{x} = [\theta \ \dot{\theta}]^T$ and $u = F$, the system equation can be written as

$$\dot{\mathbf{x}} = \mathbf{f}(\mathbf{x}) + \mathbf{g}(\mathbf{x})u , \tag{26}$$

where

$$\mathbf{f}(\mathbf{x}) = \begin{bmatrix} \dot{\theta} & \dfrac{(m_p + m_c)g\sin\theta - m_p\dot{\theta}^2 l\cos\theta\sin\theta}{(m_p + m_c)l(4/3 - m_p\cos^2\theta)} \end{bmatrix}^T ,$$

$$\mathbf{g}(\mathbf{x}) = \begin{bmatrix} 0 & \dfrac{-\cos\theta}{(m_p + m_c)l(4/3 - m_p\cos^2\theta)} \end{bmatrix}^T .$$

Assume that the pole angle is required to follow a particular trajectory, θ^d , we can calculate the corresponding control input, u^d, approximately at frozen times. The system can be linearized to

$$\dot{\mathbf{x}} = \dot{\mathbf{x}}^{\mathbf{d}} + \mathbf{A}^{\mathbf{d}}(\mathbf{x} - \mathbf{x}^{\mathbf{d}}) + \mathbf{B}^{\mathbf{d}}(u - u^d) , \tag{27}$$

where $\mathbf{x}^{\mathbf{d}} = [\theta^d \ \dot{\theta}^d]^T$, $\mathbf{A}^{\mathbf{d}} = d\mathbf{f}/d\mathbf{x}|_{\mathbf{x}=\mathbf{x}^{\mathbf{d}}}$, and $\mathbf{B}^{\mathbf{d}} = \mathbf{g}(\mathbf{x}^{\mathbf{d}})$.

Consider the problem of keeping the pole at the upright position, or stabilizing the system (26) at the origin $\mathbf{0}$. Since the reference point is the origin $\mathbf{0}$, the numerical values of the absolute signals and the error signals are identical. For notation simplicity, we use θ and $\dot{\theta}$ to represent error signals of the pole angle and the angular velocity respectively.

Using the general knowledge of the pole balancing system, the fuzzy rule base for u_{PI} can be constructed intuitively as

$$\begin{aligned} &R^1 : \text{IF } \theta \text{ is } \textit{positive} \text{ AND } \dot{\theta} \text{ is } \textit{positive} \text{ THEN } u_{PI} \text{ is } \textit{positive} ,\\ &R^2 : \text{IF } \theta \text{ is } \textit{positive} \text{ AND } \dot{\theta} \text{ is } \textit{negative} \text{ THEN } u_{PI} \text{ is } \textit{zero} ,\\ &R^3 : \text{IF } \theta \text{ is } \textit{negative} \text{ AND } \dot{\theta} \text{ is } \textit{positive} \text{ THEN } u_{PI} \text{ is } \textit{zero} ,\\ &R^4 : \text{IF } \theta \text{ is } \textit{negative} \text{ AND } \dot{\theta} \text{ is } \textit{negative} \text{ THEN } u_{PI} \text{ is } \textit{negative} . \end{aligned} \tag{28}$$

We choose the Sigmoid function and the Gaussian function as MF's:

$$\begin{aligned} \mu_{positive}(\theta) &= \frac{1}{1 + \exp(-K_\theta\theta)} ,\\ \mu_{negative}(\theta) &= \frac{1}{1 + \exp(+K_\theta\theta)} ,\\ \mu_{positive}(\dot{\theta}) &= \frac{1}{1 + \exp(-K_{\dot{\theta}}\dot{\theta})} ,\\ \mu_{negative}(\dot{\theta}) &= \frac{1}{1 + \exp(+K_{\dot{\theta}}\dot{\theta})} ,\\ \mu_{positive}(\phi) &= \frac{1}{1 + \exp[-A_u(\phi - C_u)]} ,\\ \mu_{zero}(\phi) &= \exp(-K_u\phi^2) ,\\ \mu_{negative}(\phi) &= \frac{1}{1 + \exp[+A_u(\phi + C_u)]} , \end{aligned} \tag{29}$$

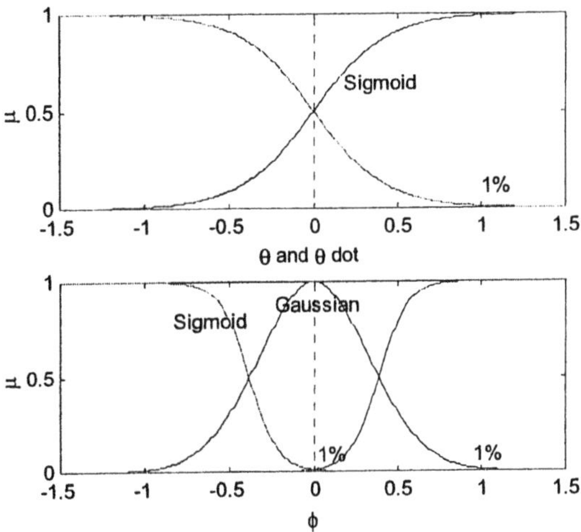

Fig. 5. Membership functions of the error signals and the controller output

where K_θ and $K_{\dot{\theta}}$ are input scaling factors for θ and $\dot{\theta}$ respectively and K_u is the output gain. The parameters of the output Sigmoid MF's, A_u and C_u, can be calculated by using 50% over-lapping and 1% low-bound fuzziness value (as depicted in Fig. 5). In this chapter, we will use the Sigmoid-Sigmoid (SS) function (the upper subplot of Fig. 5) and Sigmoid-Gaussian-Sigmoid (SGS) function (the lower subplot of Fig. 5) for the variables that have two and three MF's respectively.

From Eq. (11) and Eq. (13), the fuzzy rule base can be reconstructed recursively. For example, for the first rule in Eq. (28), the output is required to be *positive* at the next step, while at the current step, it could have values of *positive*, *zero* and *negative*. Without extra knowledge, the incremental fuzzy output of R^1 in Eq. (28) can be obtained as follows:

$$\begin{aligned} R^1_1 &: \text{IF } \theta \text{ is } \textit{positive} \text{ AND } \dot{\theta} \text{ is } \textit{positive} \text{ AND } \phi \text{ is } \textit{positive}\ , \\ &\quad \text{THEN } \Delta u \text{ is } \textit{zero}\ , \\ R^1_2 &: \text{IF } \theta \text{ is } \textit{positive} \text{ AND } \dot{\theta} \text{ is } \textit{positive} \text{ AND } \phi \text{ is } \textit{zero}\ , \\ &\quad \text{THEN } \Delta u \text{ is } \textit{positive}\ , \\ R^1_3 &: \text{IF } \theta \text{ is } \textit{positive} \text{ AND } \dot{\theta} \text{ is } \textit{positive} \text{ AND } \phi \text{ is } \textit{negative}\ , \\ &\quad \text{THEN } \Delta u \text{ is } \textit{positive big}\ . \end{aligned} \tag{30}$$

By using the CA defuzzification method, normalized singleton values can be used for the incremental output, Δu, such as *zero* $= 0$, *positive* $= 1$ and *positive big* $= 2$. Other linguistic values of Δu can be calculated linearly by simple arithmetic operations.

To compare with the fuzzy Lyapunov control method of [9], the equivalent set of parameters $K_\theta = 30$, $K_{\dot{\theta}} = 30$, $K_u = 0.18$, $A_u = 2.37$ and $C_u = 1.94$ are used.

Figs. 6 and 7 depict stabilization by the fuzzy Lyapunov controller and the proposed controller for different initial conditions, $(\theta(0), 0)$. The proposed controller not only can stabilize the system much faster than the fuzzy Lyapunov controller (e.g. for the initial pole angles $0.06rad$, $0.16rad$ and $0.21rad$), but also it has a much wider stable range (marginal stable angle is $0.73rad$ compared to $0.23rad$ of the fuzzy Lyapunov controller).

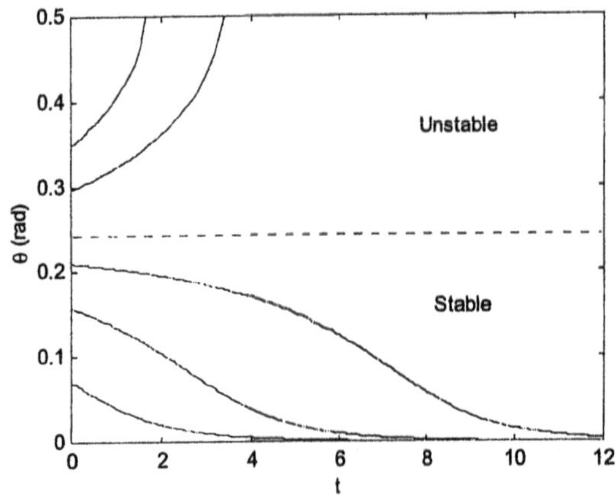

Fig. 6. Stabilization control by the fuzzy Lyapunov controller for five different initial conditions

Fig. 8 compares stabilization control by the proposed controller for the initial condition $(0.15rad, 0.15rad/s)$ under different incremental bias conditions, which are numerically the singleton values of the incremental output, Δu. Without changing the MF's of the control input ϕ, we can shift the output of the fuzzy PI controller towards controlling θ or $\dot{\theta}$ more effectively. The solid phase-plane trajectory is the control result that θ has more credit in the fuzzy rule base. It shows that the controller will bring the pole angle quickly ($t = 2s$) to the origin, while sacrificing the performance of the angle velocity. If the fuzzy rule base does not give credit to either variable (the dotted line), $\dot{\theta}$ will be stabilized prior to θ. This is because in Eq. (25), $\dot{\theta}$ has less effect on the system than θ, which automatically gives some credit to $\dot{\theta}$ in the fuzzy rule base. However, it should be emphasized that there are three factors that can affect the bias conditions, namely the symmetric design of the fuzzy rules, the bias of the incremental output and the input scaling factors of the variables.

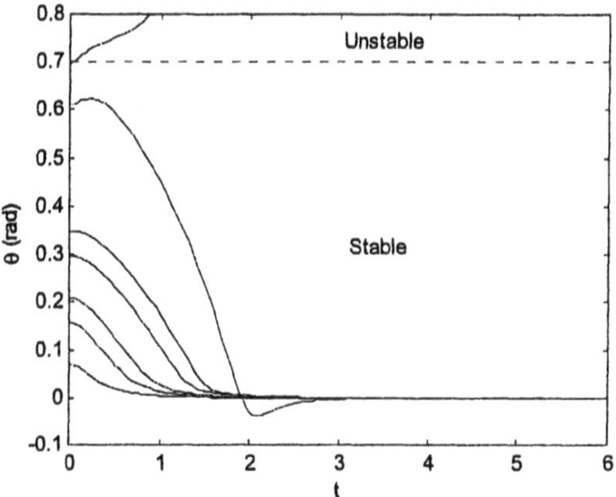

Fig. 7. Stabilization control by the proposed controller for seven different initial conditions

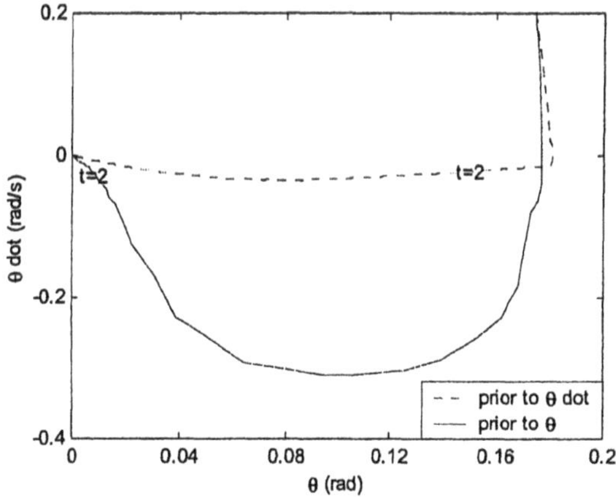

Fig. 8. Stabilization control using different incremental bias conditions

As depicted in Fig. 2, stability range of the nonlinear output, ϕ depends on two factors: the sector condition, K_D, and the output scaling factor, K_F. The sector condition K_D is proportional to the conventional derivative gain, K_D^c. Fig. 9 shows stabilization control using different conventional derivative gains. The initial condition is $(0.15rad,\ 0.15rad)$. Set $K_F = 3$, the solid line shows a marginal stable case for $K_D^c = 1.98$. Any conventional derivative gain that is larger than this value will stabilize the system (for example, $K_D^c = 2$ for the dotted line). But, if K_D^c is too big, the closed-loop system will have overdamped response with a very large rise time. Figs. 10 and 11 depict the stabilization result and the corresponding control input using different fuzzy output scaling factors. Let $K_D^c = 1.5$, the unstable result for $K_F = 3$ verifies the previous conclusion. In fact, in this case, the marginal stable value of the fuzzy output scaling factor is $K_F = 4.79$. However, the solid line depicts a chattering phenomenon with $K_F = 10$, which shows that if K_D^c is too small, the stability range of the output scaling factor will be very limited. The dotted lines in these two figures depict the control performance and the control input with $K_F = 4.8$. Although it stabilizes the system after some time, the transient response is not satisfactory.

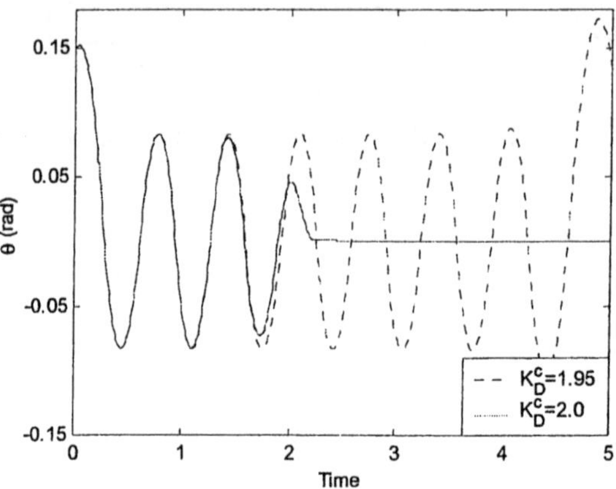

Fig. 9. Stabilization control using different conventional derivative gains

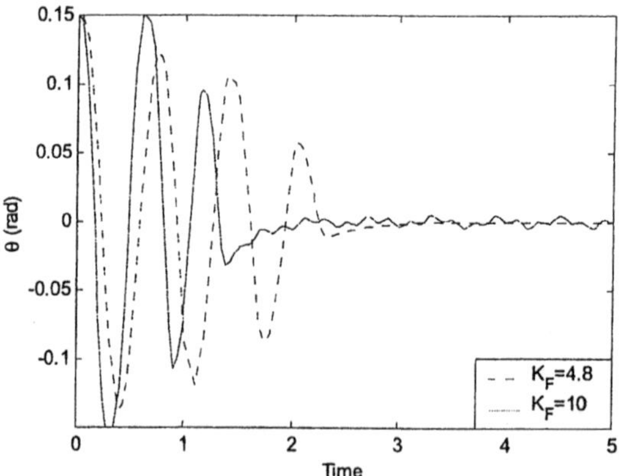

Fig. 10. Stabilization control using different fuzzy output scaling factors

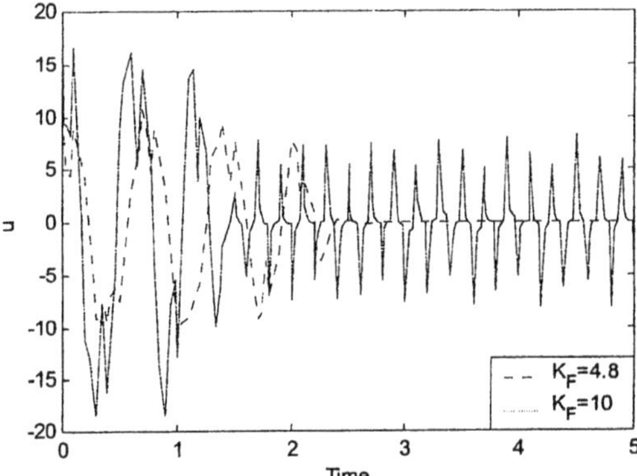

Fig. 11. Control input for stabilization using different fuzzy output scaling factors

Fig. 12 depicts the control performance of the controller for tracking a given trajectory $\theta^d = 0.1 \sin t$ with initial conditions $(-0.05rad,\ 0rad/s)$. Fig. 13 compares the tracking error for the fuzzy Lyapunov controller (the dotted line) and the proposed controller (the solid line). Since the mass of the pole is small, external disturbances may be significant, when distributed white noises are present at the inputs for both controllers, the control performances of both controllers are shown in Fig. 14.

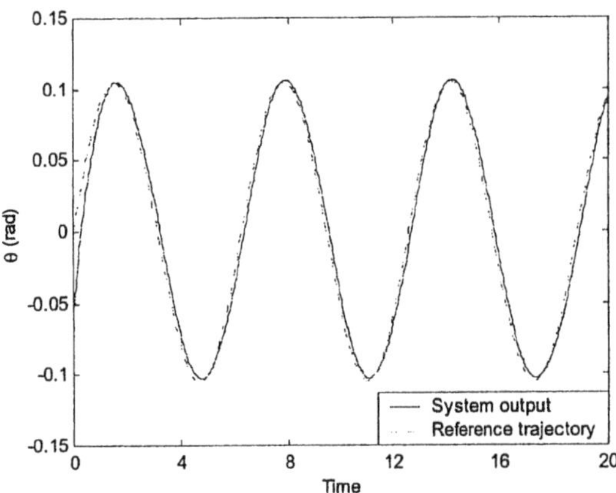

Fig. 12. Tracking control by the proposed controller

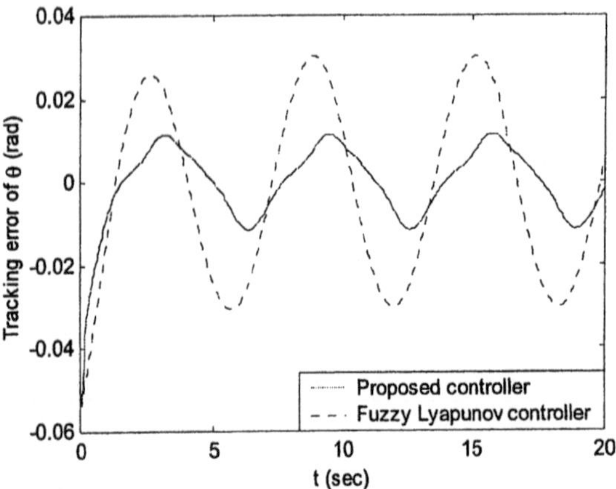

Fig. 13. Tracking control by the proposed controller

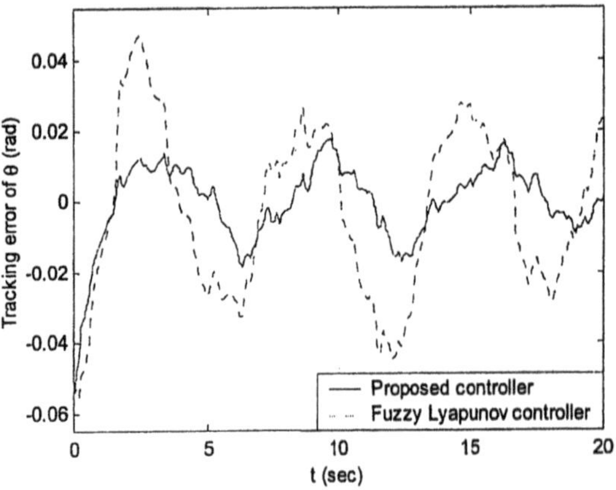

Fig. 14. Tracking control by the proposed controller

5.2 A Multi-Link Robot Manipulator - MIMO Case

Multi-link robot manipulators are familiar examples of trajectory-controllable mechanical systems. Their dynamics are usually strongly nonlinear and can be expressed as

$$\mathbf{H}(\mathbf{q})\ddot{\mathbf{q}} + \mathbf{C}(\mathbf{q}, \dot{\mathbf{q}}) = \mathbf{u} \,, \tag{31}$$

where $\mathbf{q} \in R^{n\times 1}$ is the position vector indicating joint angles, $\dot{\mathbf{q}} \in R^{n\times 1}$ is the velocity vector and $\ddot{\mathbf{q}} \in R^{n\times 1}$ is the acceleration vector. The term $\mathbf{H}(\mathbf{q}) \in R^{n\times n}$ is the manipulator inertia matrix, which is symmetric positive definite, $\mathbf{C}(\mathbf{q}, \dot{\mathbf{q}}) \in R^{n\times n}$ is the matrix of damping, centrifugal, coriolis, gravitational forces, $\mathbf{u} \in R^{n\times 1}$ represents the generalized torques including the input torques generated by joint motors and some unknown terms arising from model uncertainties and external disturbances.

Since the inertia matrix $\mathbf{H}(\mathbf{q})$ is invertible, the dynamics of the robot manipulator in Eq. (31) can be transformed into

$$\ddot{\mathbf{q}} = -\mathbf{H}(\mathbf{q})^{-1}\mathbf{C}(\mathbf{q},\dot{\mathbf{q}}) + \mathbf{H}(\mathbf{q})^{-1}\mathbf{u}\,. \tag{32}$$

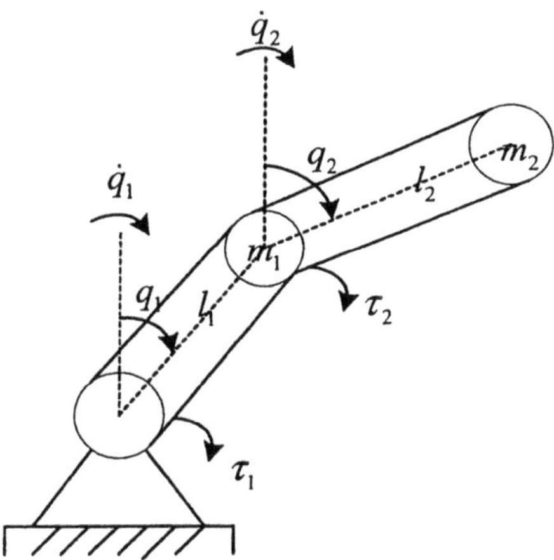

Fig. 15. A two-link robot manipulator

Consider a two-link robot manipulator with its masses concentrated at the ends of each link and where the motor inertia are neglected (as depicted in Fig. 15). Hence, $\mathbf{q} = [q_1\ q_2]^T$, $\mathbf{u} = [u_1\ u_2]^T$, and

$$\mathbf{H}(\mathbf{q}) = \begin{bmatrix} h_{11} & h_{12} \\ h_{21} & h_{22} \end{bmatrix},$$
$$\mathbf{C}(\mathbf{q},\dot{\mathbf{q}}) = \begin{bmatrix} c_1 \\ c_2 \end{bmatrix}, \tag{33}$$

where

$$\begin{aligned}
h_{11} &= (m_1+m_2)l_1^2\,,\\
h_{12} &= h_{21} = m_2 l_1 l_2 \cos(q_1-q_2)\,,\\
h_{22} &= m_2 l_2^2\,,\\
c_1 &= -\dot{q}_2^2 m_2 l_1 l_2 \sin(q_2-q_1) - (m_1+m_2)gl_1 \sin q_1 + K_{q1}\dot{q}_1\,,\\
c_2 &= -\dot{q}_1^2 m_2 l_1 l_2 \sin(q_1-q_2) - m_2 g l_2 \sin q_2 + K_{q2}\dot{q}_2\,.
\end{aligned}$$

with parameters: the gravitational acceleration, g is $9.8m/s^2$, the mass of link 1, m_1 is $1.5kg$, the length of link 1, l_1 is $0.2m$, the mass of link 2 (including payload), m_2 is $1kg$, the length of link 2, l_2 is $0.2m$, and both damping coefficients, K_{q1} and K_{q2} are $10kgm^2/s$.

Let $\mathbf{x} = [q_1\ \dot{q}_1\ q_2\ \dot{q}_2]^T$ and $\mathbf{u} = [u_1\ u_2]^T$, Eq. (32) can be written as

$$\begin{aligned}\dot{\mathbf{x}} &= \mathbf{f}(\mathbf{x}) + \mathbf{g}(\mathbf{x})\mathbf{u} \,, \\ &= \begin{bmatrix} \dot{q}_1 \\ f_1 \\ \dot{q}_2 \\ f_2 \end{bmatrix} + \begin{bmatrix} 0 & 0 \\ g_{11} & g_{12} \\ 0 & 0 \\ g_{21} & g_{22} \end{bmatrix} \begin{bmatrix} u_1 \\ u_2 \end{bmatrix} \,,\end{aligned} \tag{34}$$

where

$$\begin{aligned}\det(H) &= m_2 l_1^2 l_2^2 [m_1 + m_2 \sin^2(q_1 - q_2)] \,, \\ g_{11} &= h_{22}/\det(H) \,, \\ g_{12} = g_{21} &= -h_{12}/\det(H) \,, \\ g_{22} &= h_{11}/\det(H) \,, \\ f_1 &= g_{11}c_1 + g_{12}c_2 \,, \\ f_2 &= g_{21}c_1 + g_{22}c_2 \,.\end{aligned}$$

Assume that the trajectories of this robot manipulator are always bounded by

$$q_1,\ q_2 \in [-\pi/3, \pi/3] \,, \quad \text{and} \quad \dot{q}_1,\ \dot{q}_2 \in [-\pi/2, \pi/2] \,,$$

so that the partition of the state space is shown in Fig. 16, where the MF's used for $\mathbf{q}$ are SGS functions and the MF's used for $\dot{\mathbf{q}}$ are SS functions. The definition of these MF's can be found in the previous example.

The desired control input, $\mathbf{u^d}$ corresponding to the operation points of the reference trajectory, $\mathbf{x^d}$ can be calculated by using Eq. (31)

$$\mathbf{u^d} = \hat{\mathbf{H}}(\mathbf{q^d})\ddot{\mathbf{q}} + \hat{\mathbf{C}}(\mathbf{q^d}, \dot{\mathbf{q}}^\mathbf{d}) \,, \tag{35}$$

where $\hat{H}(q^d)$ and $\hat{C}(q^d, \dot{q}^d)$ are the approximate coefficient matrices. By using the desired control input, $\mathbf{u^d}$, system (34) can be linearized into the form of Eq. (27), where $\mathbf{q^e} = \mathbf{q} - \mathbf{q^d} = [q_1^e\ \dot{q}_1^e\ q_2^e\ \dot{q}_2^e]^T$.

For the local FLC's, the fuzzy rules are

$$\begin{aligned}&R: \text{IF } q_1^e \text{ is } LE1 \text{ AND } \dot{q}_1^e \text{ is } LDE1 \text{ AND } q_2^e \text{ is } LE2 \text{ AND } \dot{q}_2^e \text{ is } LDE2 \,, \\ &\quad \text{THEN } u_{PI1} \text{ is } LU1, \quad u_{PI2} \text{ is } LU2 \;.\end{aligned} \tag{36}$$

Even if the simple SS MF's are used for $\mathbf{q^e}$ and SGS MF's are used for $\mathbf{u}_{PI}$, the incremental fuzzy rule base will contain $2^4 \times 3^2 = 144$ fuzzy rules. Knowledge from the human experts is needed to diminish any repeated or unnecessary rules. By observations, u_{PI1} is mostly decided by q_1^e and $\dot{q}_1^e$ and so is u_{PI2}. Hence, the fuzzy rule base in Eq. (36) can be decoupled into two

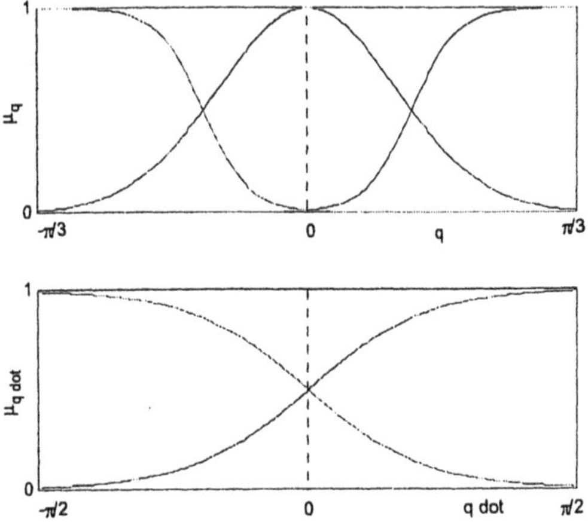

Fig. 16. Membership functions of the states, $\mathbf{q}$ and $\dot{\mathbf{q}}$

parallel fuzzy rule bases, i.e.

$$\begin{aligned} R_{u1} &: \text{IF } q_1^e \text{ is } LE1 \text{ AND } \dot{q}_1^e \text{ is } LDE1 \ , \\ &\quad \text{THEN } u_{PI1} \text{ is } LU1 \ , \end{aligned} \tag{37}$$

$$\begin{aligned} R_{u2} &: \text{IF } q_2^e \text{ is } LE2 \text{ AND } \dot{q}_2^e \text{ is } LDE2 \ , \\ &\quad \text{THEN } u_{PI2} \text{ is } LU2 \ , \end{aligned} \tag{38}$$

where each fuzzy rule base only contains $2 \times 2 \times 3 = 12$ fuzzy rules and can be constructed linearly. To compensate the cross-effect of u_{PI1} and u_{PI2}, we can define a fuzzy rule base as follows:

$$\begin{aligned} R_{\delta u} &: \text{IF } u_{PI1} \text{ is } LU1 \text{ AND } u_{PI2} \text{ is } LU2 \ , \\ &\quad \text{THEN } \delta u_1 \text{ is } LD1, \quad \delta u_2 \text{ is } LD2 \ . \end{aligned} \tag{39}$$

Although this fuzzy rule base only contains $3 \times 3 = 9$ rules, it is not straight forward to defined them manually. Since the cross-effects between u_{PI1} and u_{PI2} are relatively small (less than 0.5%) in comparison to the control signals themselves, we would use the direct compensators (See Sec. 3.3, Step 3) instead of constructing the fuzzy rule base in Eq. (39).

The trajectory tracking control performance of the proposed controller is compared to the fuzzy GS method in [10], where the linearization technique used is the same as what has been used in this chapter. The difference of their control performance is mainly caused by different local controllers used. In [10], conventional pole placement design is applied, while in this chapter, the fuzzy PID controller is used. In the following figures, the solid line and the dotted line refers to the control performance of the proposed controller and the fuzzy GS controller respectively.

Figs. 17 and 18 depict the tracking errors of the angle and the angular velocity respectively, where the reference trajectory is given by

$$\mathbf{x^d} = [-\frac{\pi}{3}\sin(\frac{3}{2}t), \quad -\frac{\pi}{2}\cos(\frac{3}{2}t), \quad \frac{\pi}{2}+\frac{\pi}{3}\sin(\frac{3}{2}t), \quad \frac{\pi}{2}\cos(\frac{3}{2}t)]^T .$$

Both figures show that the proposed controller performs as well as the fuzzy GS controller to link 1 and has superior control for link 2. This is because in the proposed controller, control actions to link 2 are mainly generated by the fuzzy rule base in Eq. (38). The variation of the error signals in link 1, if not very large, will not affect the control input to link 2. In contrast, the output MF's of u_2 of the fuzzy GS controller contains a big portion of link 1 error signal q_1^e, which helps to stabilize the local closed-loop system but sacrifices the control performance of link 2.

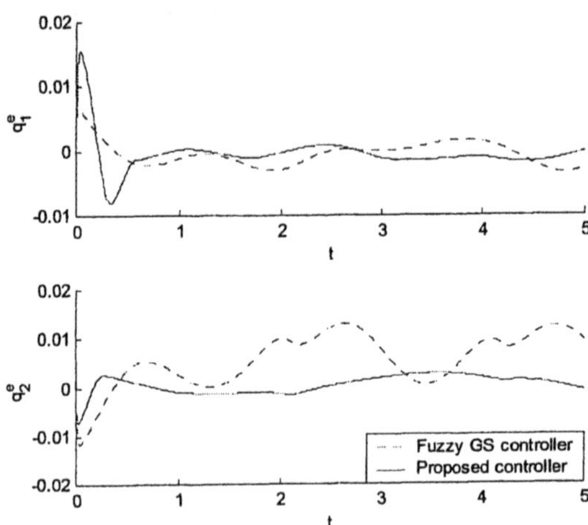

Fig. 17. Comparison on the tracking errors of the angle

Fig. 19 depicts the scenario where the payload m_2 is increased to $2.5kg$. It is evident from the plot that the proposed controller has excellent control of link 2.

The reference trajectory used in the example depicted in Fig. 20 is given by

$$\mathbf{x^d} = [-\frac{\pi}{6}\sin(3t), \quad -\frac{\pi}{2}\cos(3t), \quad \frac{\pi}{2}+\frac{\pi}{6}\sin(3t), \quad \frac{\pi}{2}\cos(3t)]^T .$$

It should be noted that since both controllers interpolate the points in between of the operation points $(\mathbf{x^d}, \mathbf{u^d})$ by their fuzzy rules, the slow varying trajectory is still an assumption for both of the design approaches. However, it shows that the performance of the proposed controller at high frequency is also superior.

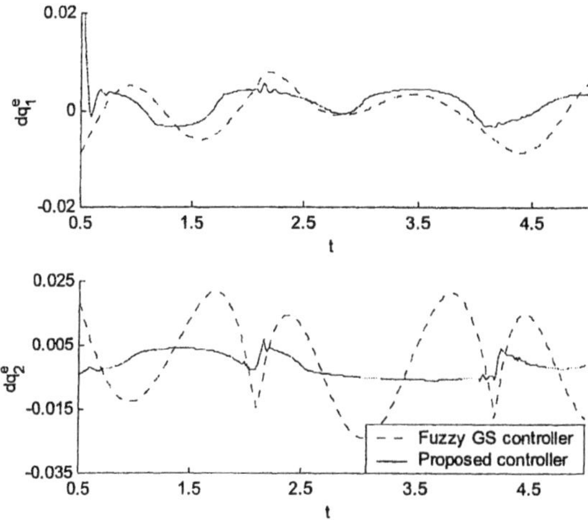

Fig. 18. Comparison on the tracking errors of the angular velocity

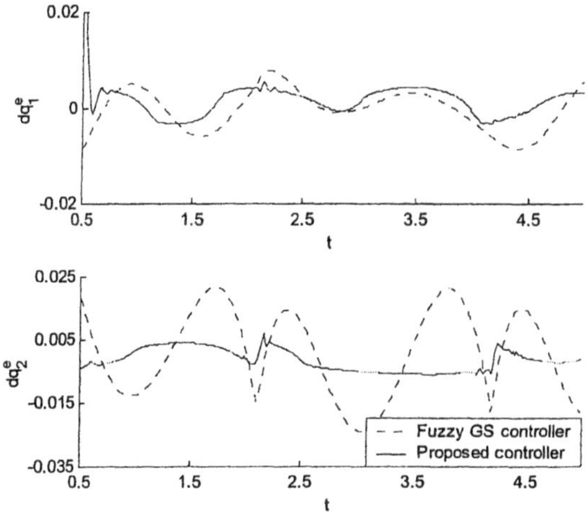

Fig. 19. Comparison on the tracking errors of the angle with heavy payload

6 Conclusions

In this chapter, a novel approach towards optimal design of a hybrid fuzzy PID controller using GA is proposed. Instead of analyzing the fuzzy controller by numerical calculations, the proposed design method focuses on constructing the fuzzy rule base by using linguistic language or words. The discretization of the conventional PID controller directs to an incremental form of the FLC. Stability analysis of the closed-loop system will be guaranteed by the

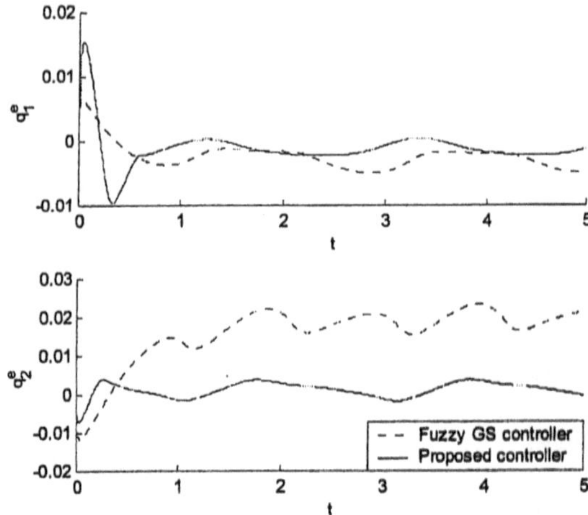

Fig. 20. Comparison on the tracking errors of the angle with fast changing trajectory

sector condition of the nonlinear output of the FLC. The proposed controller not only demonstrates excellent control performance for a pole balancing robot and a two-link robot manipulator, but also, it shows its potential to be a new form of stand-alone FLC.

References

1. Astrom, K.J., Hagglund, T. (1995): PID Controllers : Theory, Design, and Tuning, Instrument Society of America, Research Triangle Park, N.C.
2. Er, M.J., Y.L., Sun. (2000): Design of a Hybrid Fuzzy Proportional-Integral plus Conventional Derivative Controller, Asian Control Conference, 1155-1159
3. Goldberg, D.E. (1989): Genetic Algorithms in Search, Optimization, and Machine Learning, Addison-Wesley
4. Khalil, H.K. (1992): Nonlinear Systems, Macmillan, New York
5. Linkens, D.A., Nyongesa, H.O. (1995): Genetic Algorithms for Fuzzy Control: Offline System Development and Application, IEE Proceeding Control Theory Applications, **142**, 161-176.
6. Li, W. (1998): Design of a Hybrid Fuzzy Logic Proportional plus Conventional Integral-Derivative Controller, IEEE Trans. on Fuzzy Systems, **6**, 449-463
7. Mamdani, E.H., Assilian, S. (1975): An experiment with in linguistic synthesis with a fuzzy logic controller, International Journal Man-Machine Studies, **7**, 1-13
8. Mann, G.K.I., Hu, B.G., Gosine, R.G. (1999): Analysis of Direct Action Fuzzy PID Controller Structures, IEEE Trans. on Fuzzy Systems, **29**, 371-388
9. Margaliot, M., Langholz, G. (1999): Fuzzy Lyapunov-Based Approach to the Design of Fuzzy controllers, Fuzzy Sets and Systems, **106**, 49-59

10. Palm R., Driankov D., Hellendoorn H. (1997) Model Based Fuzzy Control, Springer, Berlin New York Heidelberg
11. Slotine J.J.E., Li W.P. (1991): Applied Nonlinear Control, Prentice Hall, Englewood Cliffs, New Jersey
12. Takagi, T., Sugeno, M. (1985): Fuzzy Identification of Systems and its Applications to Modeling and Control, IEEE Trans. on System Man Cybernetics, **15**, 116-132
13. Tan, K.K., Wang, Q.G., Hang, C.C. (1999): Advances in PID Control, Springer, London Berlin Heidelberg
14. Vose, M.D. (1999): The Simple Genetic Algorithm : Foundations and Theory, MIT Press, Cambridge
15. Wang, L.X. (1994): Adaptive Fuzzy Systems and Control: Design and Stability Analysis, Prentice-Hall, Englewood Cliffs, New Jersey
16. Wang, L.X., Mendel, J.M. (1992): Fuzzy Basis Functions, Universal Approximation, and Orthogonal Least-Squares Learning, IEEE Trans. on Neural Networks, **3**, 807-814
17. Ying, H. (1993): A Nonlinear Fuzzy Controller with Linear Control Rules is the Sum of a Global Two-Dimensional Multilevel Relay and a Local Nonlinear Proportional-Integral Controller, Automatica, **29**, 499-505
18. Zadeh, L.A. (1996): Fuzzy Logic = Computing with Words, IEEE Trans. on Fuzzy Systems, **4**, 103-111
19. Zhao, Z.Y., Tomizuka, M., Isaka, S. (1993): Fuzzy Gain Scheduling of PID Controllers, IEEE Trans. on Systems Man Cybernetics, **23**, 1392-1398

Part 4

VISION AND PERCEPTION

Robot Vision Using Cellular Neural Networks

Marco Balsi[1] and Xavier Vilasís-Cardona[2]

[1] Dipartimento di Ingegneria Elettronica, Università "La Sapienza", via Eudossiana 18, 00184 Rome, Italy
balsi@uniroma1.it

[2] Universitat "Ramon Llull", Enginyeria "La Salle", Departament d'Electrònica, Pg. Bonanova 8, 08022 Barcelona, Spain
xvilasis@salleURL.edu

Abstract. We show how Cellular Neural Networks (CNNs) can provide the necessary image processing to guide an autonomous mobile robot in a maze made of black lines on a light surface. The system consists of a fuzzy controller performing the elementary navigation tasks fed by the result of processing the image only by CNN techniques. We use this solution to make some considerations on more difficult problems such as curved or dashed line following and obstacle avoidance.

1 Introduction

Vision-based navigation for autonomous mobile robots in an unstructured environment still remains a challenge despite a large effort is being poured into this problem. Maybe we, humans, feel that a clever verbalisation of our strategies for interacting with our environment should lead to a simple and clear solution. The most sophisticated image processing techniques, ranging from mathematical morphology to artificial neural networks, have been applied to achieve partial solutions. Algorithms exist to solve obstacle avoidance, cliff detection, autonomous highway driving or active stereo vision for tracking. Still, many of these achievements require large amounts of computing time and jeopardy real time operation. Our proposal is to put forward a new candidate for robot vision processing: Cellular Neural Networks (CNNs) [1–3]. These are arrays of cells or neurons locally-connected with two precise strong-points: direct VLSI implementation, allowing for real time parallel processing and a natural suitability for image processing. Our point is to prove that CNNs can be used to solve the robot vision problem, at least in some simple case, namely, line following and landmark recognition in a maze [4,5]. This should be a first step toward tackling more complex problems such as curved line following or obstacle avoidance.

The chapter is organised as follows. First of all, we give some basic considerations on robot vision. Next, we make a fast review of Cellular Neural Networks and previous uses in robotics. Then, we present our solution for guidance in a maze: splitting the navigation and image processing problem, describing the image processing in CNN terms and designing the fuzzy controller. This section is ended by some experimental results. Finally, we study how can we use the maze solution to study curved line tracking and obstacle avoidance.

2 Robot Vision

The standard robot architecture divides the machine structure into three parts following the standard control theory scheme: sensors, controller and actuators [6]. The first provide the system with the necessary observability and their task is to observe and retrieve information from the environment. The controller or central unit decides the robot actions according to the sensor observations. Finally, actuators allow for the robot actions in the environment, thus ensuring the system controlability. Yet a control system, a robot is a complex one. Depending on the task to be achieved, the sensors must provide enough information to depict the environment accurately. Also, decision taking may require arbitrarily complex algorithms involving learning or evolution. For this reason robotics is an excellent test-bed for artificial intelligence techniques, particularly for studying hard and soft computing solutions.

In the particular case of mobile robots, we find among the tasks to be accomplished, landmark recognition, navigation, path planning and obstacle avoidance (e.g. [7]). The accuracy in performing these tasks depends on a precise description of the environment. For this purpose, we find a number of sensory solutions in the literature, ranging from infrared, light or contact sensors, to ultrasound devices, radars, etc. Robots usually combine a number of such sensors in order to obtain better representations of the environment and improve their navigation capabilities. This combination is known as multi-modal processing or data fusion [8]. Sensor data, single or fused, requires preprocessing in order to be useful to the decision taking algorithm. In the same way a controller shall not use information that cannot possibly be provided by the sensors. Finally, control action should meet the actuators requirements and limitations. This interrelation proves that one cannot devise independently any of the three parts of the robot, namely sensors, controllers and actuators.

Computer vision is believed to be a powerful solution to the sensory problem, perhaps because human interaction with the environment happens mainly through vision. Actually, a careful interpretation of many navigation algorithms with their visual requirements shows that those arise from the verbalisation of human strategies. For instance, the location and interpretation of objects and landmarks in an image are much enhanced by the programmers' experience. In this sense, we can think of visual feedback as a particular case of an expert system in which the human designer incorporates his knowledge and experience into the robot behaviour. A large number of sophisticated image processing algorithms help in this task. Moreover, advances in CCD and CMOS light sensing technology have increased the availability of cameras to be mounted on mobile robots. Still image processing algorithms are usually complex and costly. So in order to meet the robotics requirement for real time solutions either one reverts to very simple processing, or large computing power must be called for, either on board, or externally, causing the robot not to be fully autonomous.

The robot task is sometimes simplified by reducing the type and complexity of landmarks. When only a finite number of landmarks is considered, we speak of structured environment, in contradistinction to the general case called so forth unstructured environment. Line following is one of the reference problems connected with navigation in a structured environment. It is a relatively simple, yet non-trivial, problem that is relevant to real-life situations, such as road navigation, or motion through a maze (e.g. industrial environments) and contains the essential features such as navigation or landmark recognition. Line (or equivalent marking) following has been considered elsewhere. Solutions proposed in the literature may be classified into two main categories: simple solutions with very limited capability, or solutions for difficult, realistic problems obtained by use of rather complicated hardware.

An example of simple system is the ARGO partially autonomous vehicle [9]. It is a normal passenger car fitted with an automatic steering mechanism, controlled by a 486PC. The PC processes the images taken by two B/W cameras with the aim of finding the right line of the road, and keeping it in the appropriate position in the field of view. The processing is very simple, yet effective for the purpose.

The more complex systems are generally based on a computer which is either on board (for large vehicles, e.g. [10,11]), or remotely connected (e.g. [12]). In the quoted systems, images are processed by an additional unit, due to the necessity of high computing power. Of course, these vehicles can generally tackle fairly complex problems, such as unstructured road navigation, target following, obstacle avoidance, besides line following.

3 Cellular Neural Networks

3.1 The CNN Paradigm

Cellular Neural Networks (CNNs) [1–3] are arrays of dynamical artificial neurons (cells) that are only locally interconnected. This essential characteristic of CNNs has made hardware implementation of large networks possible on a single VLSI chip [13,14].

Cells are typically organised in a fairly large two-dimensional grid, so that CNNs can be used as massively parallel image processing tools, by associating each pixel of an image with one cell. CNN chips can include on board image sensing. This makes them capable of focal-plane processing, avoiding a bottleneck caused by electronic image loading. Moreover, distributed memory and a global control module can be present on chip. In this way the

CNN becomes a stored-program massively parallel processor known as the *CNN Universal Machine* - (CNN-UM) [15], which is capable of executing a complex sequence of operations (*analogic* programs) in real time. Therefore, CNNs are very good candidates as hardware platforms for robot vision.

In this chapter, we shall refer to the Discrete-Time CNN model (DTCNN hereafter) [3], however all operations described in the following can as well be done using continuous-time CNNs. DTCNN core operation is described by the following system of iterative equations:

$$x_{ij}(n+1) = \left(\sum_{kl \in N(ij)} A_{k-i,l-j} \operatorname{sign} x_{ij}(n) + \sum_{kl \in N(ij)} B_{k-i,l-j} u_{ij}(n) + I \right) \tag{1}$$

where:

x_{ij} is the state of the cell (neuron) in position ij, that corresponds to the image pixel in the same position;

u_{ij} is the input to the same cell, representing the luminosity of the corresponding image pixel, suitably normalised;

A is a matrix representing the interaction between cells, which is local (as specified by the fact that summations are taken over the set N of indexes neighbour cells) and space-invariant (as implied by the fact that weights depend on the difference between cell indexes, rather than their absolute values);

B is a matrix representing forward connections issuing from a neighbourhood of inputs.

I is a bias.

$N(i,j)$ is the set of indexes corresponding to cell ij itself and a small neighbourhood (e.g. cell ij and its 8 nearest neighbours). Due to the locality of the computation, implied by the summation over this neighbourhood, and space-invariance, implied by the differences of indexes in matrices A and B in equation (1), it is sufficient to define A and B for a few instances of the indexes, so that they may be represented by small matrices.

The operation performed by the network is fully defined by the so-called *cloning template* $\{A, B, I\}$ (see examples in Table 1). Moreover, under suitable conditions, and with time-invariant input u, a steady state is reached. This steady state depends, in general, on initial state values and input u. Images to be processed are fed to the network as initial state and/or input, and the result taken as steady state value, which realistically means a state value after some time steps (ranging normally from 10 to 100 according to the task).

Equation (1) can be generalised by including nonlinear and delayed interactions [15]. While non-linearity in the feed-forward (connection to input) path can be exchanged for a suitable linear-connection-based algorithm [16], nonlinear feedback connections allow for real additional functionality.

Unlike most neural networks, CNN functionality is determined by a very small number of parameters, so that it is possible to explicitly design [17] a cloning template for a specific task (besides learning it, e.g. [18], and to broadcast such parameters to all neurons so as to use them as *analogic instructions* for the CNN. *Analogic* instructions can then be combined into *analogic* (analog/logic) programs, using the global controller and local memory. In this way, sophisticated image processing can be implemented in real time by analog software in a very efficient way. Many cloning templates and analog algorithms have been designed for the most diverse tasks [19].

3.2 CNN in Robotics : Previous Approaches

Cellular Neural Networks have been applied to machine vision tasks by several authors. Main results available in literature shall be briefly described in the following subsections.

Optical Flow and Time-to-Contact Estimation Optical flow estimation is instrumental in assessing ego-motion with respect to the surrounding environment and in moving object tracking. The complete optical flow gives very rich information at the cost of rather intensive computation. A regularisation-based approach to optical flow permits to cast the problem in a form that lends itself to local parallel computation [20] by a CNN-like architecture. For obstacle avoidance purposes, estimation of velocity field source (focus of expansion) and magnitude permits evaluation of direction and distance of obstacles. In this case, it is sufficient to use one or few one-dimensional arrays, that can be efficiently integrated in VLSI [21]. It is to be noted that the CNN model used for this task is not standard, so that general purpose CNN-UM chips cannot accommodate such kind of processing.

Depth Estimation by Stereo Vision Stereo vision is widely used for global estimation of depth information. This is also a computationally-intensive task that is very demanding for traditional architectures. Cellular Neural Networks can be applied to real-time solution of the stereo-matching problem [22]. Hardware implementation of such algorithms [23] is very efficient, but it can be necessary to resort to a multi-chip assembly to deal with high-definition images [24].

Line Following A simple line-following task has been addressed in reference [25]. Even if the task is not complex, simplicity and robustness of the solution is remarkable. Very limited CNN-based hardware resources are proved to perform efficiently in real-time.

4 Autonomous Robot Architecture

4.1 Statement of the Problem

Our goal is to study up to what point CNNs are a good solution for the robot visual feedback problem. As it has been already discussed in Section 3, they provide a number of interesting characteristics. Among the most important, we find VLSI hardware implementation giving the possibility of very fast image processing in parallel with the robot main controller. However, the particular locality features of CNNs limit the number of operations on images that can be envisaged. Therefore, the robot vision problem has to be cast adequately to be decomposed in terms of the action of a succession of templates. In CNNs terms, a CNN Universal Machine programme.

To start with, we chose to deal with a simple problem such as the guidance of a robot in a maze made of black lines on a light background. Despite its simplicity, the problem involves the key features such as navigation (line following) and landmark recognition (crossings, turns). The maze structure also provides an orientation problem although we shall leave it out of the scope of this chapter since it can be considered a task of the main controller once landmark recognition is correctly performed. As we already mentioned in Section 2, this problem can also have industrial applications such as motion in a warehouse. Although this problem can be solved using other techniques, even simpler, we consider this problem as a significant benchmark to prove that CNNS can provide an efficient solution.

4.2 Modular Structure

The image processing to obtain the data necessary to guide the robot depends critically on the navigation algorithm. We therefore need to define first how are we going to guide the robot to follow the line. For this purpose, we resort to a variation of a well established fuzzy control system used for car parking purposes, but that can be immediately translated into line following [26]. This fuzzy controller delivers the steering angle given the distance X_c of the centre of the steering wheel from the line to be followed and the angle A_{th} made by the axis of the robot and the line. This is portrayed in Figure 1, where A_{st} is the steering angle.

Then, the robot guidance problem is split into two processes, namely, the image processing to extract the mathematical features of the lines and the navigation of the robot to follow them. From the first, we expect to obtain the relative position of the mobile robot with respect to the line being followed, in order to correct possible deviations owing to misalignment of the wheels or mechanical or electrical fluctuations. This correction is to be performed by the navigation module. From the image processing we also need to extract information on the forthcoming crossings or bends, so as to decide what direction is to be taken and what is the correct moment to start turning. The

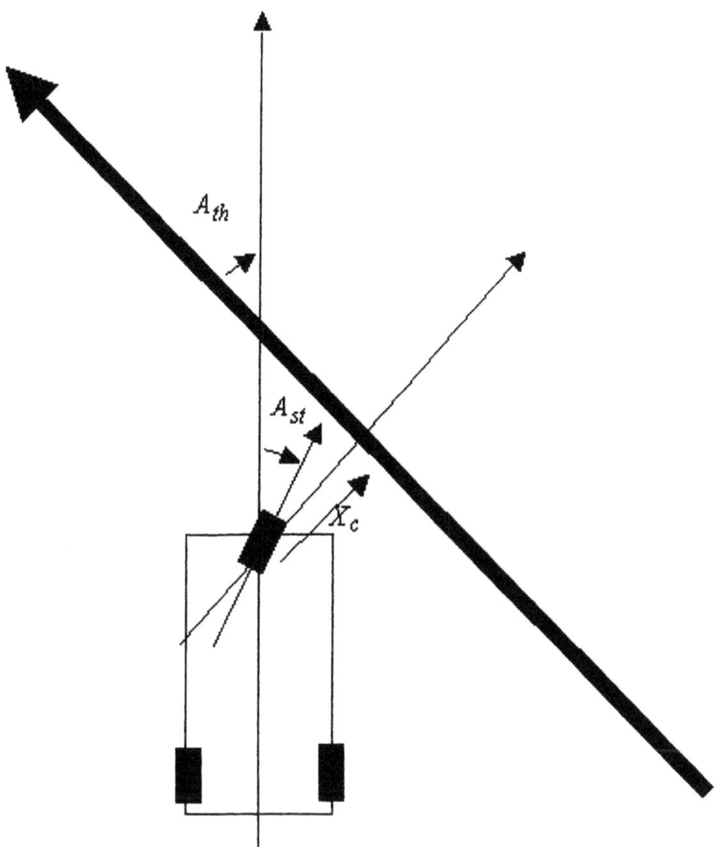

Fig. 1. Line parameters. In the situation portrayed A_{st} is positive, A_{th} is positive and X_c is negative (*i.e.* left)

mathematical information needed to control the robot (A_{th}, X_c) has to be extracted from a CNN operation. The steering angle of the front wheel A_{st} shall be the result of the fuzzy controller.

Hardware Implementation We have built a small autonomous robot to be guided by the algorithms proposed in this chapter (Figure 2). It is a three-wheeled cart driven by a motor mounted on the single front wheel, that also steers by means of a second motor. The cart is approximately 27cm long, 18cm wide, 16cm tall, and weights about 4kg. A PAL camera is fitted on the front of the robot, oriented downwards, 30° from the horizontal. Images are grabbed and digitised by dedicated circuitry implemented in a CPLD (ALTERA EPM7128STC100), which also performs image decimation to reduce unnecessary definition. Image processing (CNN simulation) is performed in a

DSP (TMS320c32), and the control system (fuzzy rule base) is implemented in a 386-microprocessor-based microcontroller. Image size is 60×45 pixels and lines on the floor are obtained using a 2cm wide black tape. A power board feeds the motors, and batteries are carried on board. In this way the robot it is completely autonomous. The image processing stage can currently process one image per second. We estimated by simulation that this allows the robot to move smoothly at a speed of 2.5 cm/s. Of course if a CNN chip were used instead of the DSP, it would be possible to reach a much higher speed.

Fig. 2. A picture of the actual robot

4.3 Image Processing

The image processing relies on the following stages: acquisition, preprocessing (cleaning and contrast enhancement), line recognition, and extraction of line parameters (angle, position). Acquisition of course depends on the actual implementation and in our case it is performed by the camera and the digitalisation system. For the preprocessing we used a so-called *small-object-killer* template (see Table 1) to make a preliminary cleaning and binarisation of the acquired image (Figure 3 (a) and (b)).

The calculation of the parameters of lines visible in an image is universally performed by the Hough transform [27]. Such approach is a very effective technique, yet computationally intensive, for obtaining direction and

Table 1. Templates used in the processing stage.

	A	B	I
small object killer	$\begin{pmatrix} 1 & 1 & 1 \\ 1 & 2 & 1 \\ 1 & 1 & 1 \end{pmatrix}$	0	0
left edge	2	$\begin{pmatrix} 0 & 0 & 0 \\ -1 & 0 & 0 \\ 0 & 0 & 0 \end{pmatrix}$	-1
vertically-tuned filter	2	$\begin{pmatrix} -1 & 0.5 & 1 & 0.5 & -1 \\ -1 & 1 & 1 & 1 & -1 \\ -1 & -1 & 5 & -1 & -1 \\ -1 & 1 & 1 & 1 & -1 \\ -1 & 0.5 & 1 & 0.5 & -1 \end{pmatrix}$	-13
connected-component detector	$\begin{pmatrix} 0 & 1 & 0 \\ 0 & 2 & 0 \\ 0 & -1 & 0 \end{pmatrix}$	0	0

positions of all lines existing in an image. However, it becomes unnecessarily complicated when only few lines are present in the image.

Our particular case demands a reasonably fast response with our limited hardware. For this reason, together with the fact that the Hough transform is a global operation that does not lend itself to efficient implementation on the CNN, we chose to devise a different approach based on CNNs.

In order to perform direction and position evaluation, we need to get a thin line first. A *skeletonisation* algorithm, would be rather slow because it involves many iterations of an eight-step routine (*i.e.* cyclical application of eight cloning templates). For this reason, we resorted to the design [17] of a cloning template that extracts only one of the two edges of a stripe (it actually extracts only those parts of edges of black-and-white objects that lie on the left of the object itself). This template is given in Table 1, and its effect is shown in Figure 3 (c). Of course, when dealing with approximately horizontal lines, we use a rotated version of this template, which extracts the upper edge.

The line position and orientation computation is based on two steps. We assume that a maximum of two, approximately orthogonal lines are within sight of the camera. This is not very restrictive, because we chose a setup in which the camera is oriented at an angle that was chosen as a compromise between looking a reasonably long distance forward, and avoiding a large deformation due to perspective.

The first step is a directional filtering that extracts lines approximately oriented along two orthogonal directions. During normal operation these directions are the vertical and horizontal ones, corresponding to the line being followed and a possible orthogonal line following a bend or crossing. However, when the robot is turning or is largely displaced from all lines, it is necessary to switch to diagonal directions. As this situation is known to the controller, it will signal to the image processing stage which filter should be used.

Operation of a vertically-tuned filter (Table 1) is depicted in Figures 4 and 5. Other direction-selective filters are obtained by rotation of this cloning template. After the tuned filters have been applied, we get two images containing at most one line. Direction and position of the line is then extracted by performing a horizontal and vertical projection in order to read the positions of the first and last black pixels of the two projections. These four numbers, together with the information about which extreme of the line is closer to one of the borders of the image, are enough to compute direction and position of the line to the maximum precision allowed by image definition.

Extraction of the needed projections can be done by means of the so-called *connected-component-detector* cloning template (Table 1). Operation of such template at an intermediate and final stage of processing is depicted in Figure 3(d,e). It is apparent that besides obtaining the desired projections, also the information about which extreme is closer to the border can be obtained from examination of intermediate results.

Using the CNN operation described above, by simple trigonometry, it is easy to obtain the parameters that are necessary for navigation, namely, angle and distance. Such parameters are passed to the controller.

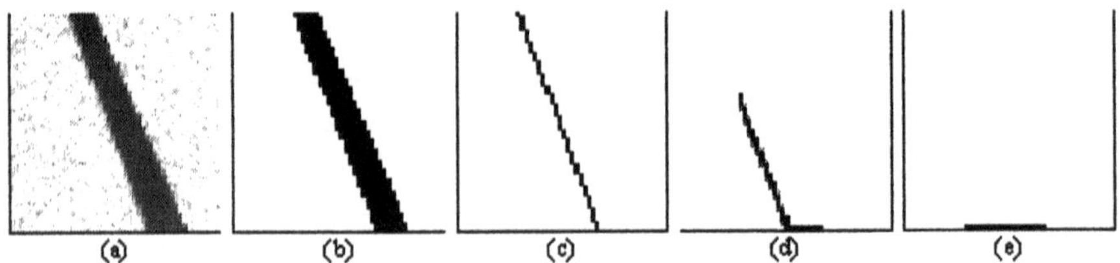

Fig. 3. Original image, taken from the actual camera with size 60 × 45 pixels (a), cleaning and binarisation (b), left edge (c), connected-component detector - intermediate result (d), final result (e).

A strong-point of CNN processing is its capability of dealing with low quality or ill defined images, as shown in Figure 6.

4.4 Fuzzy Navigation

For the sake of simplicity, we have chosen to implement a quite standard fuzzy controller for the line following task. It is inspired by the solution to the car backer-upper problem which can be recast into a line following problem. The fuzzy rule base is built out of driving experience inspired in reference [26]. The consequent sets are simplified to singletons. This scheme allows for a

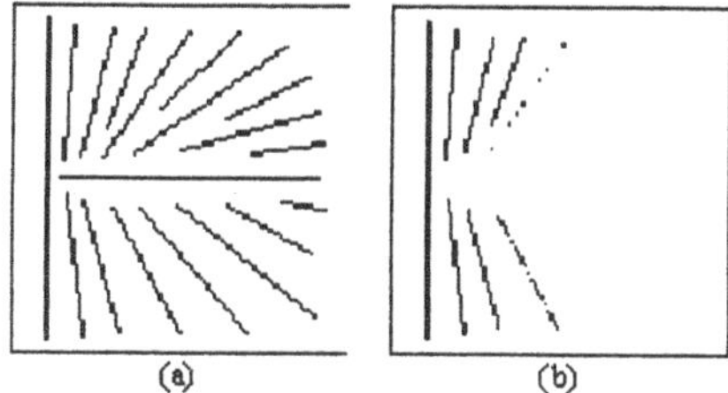

Fig. 4. Vertical line extracting filter: original image (a), result (b).

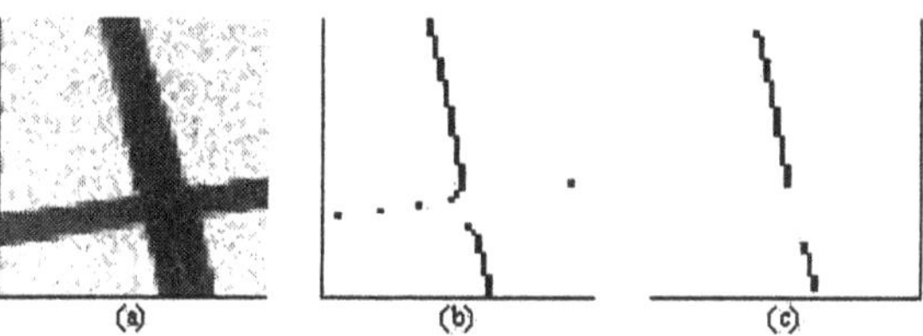

Fig. 5. Vertical line extracting filter applied to the image of a crossing (a) and after edge extraction (b): result (c).

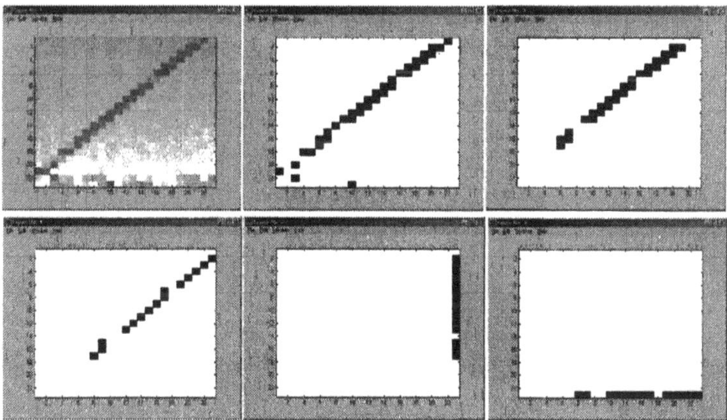

Fig. 6. Successful processing of a low quality image with low contrast.

simpler on-line implementation because it requires fewer calculations. The values for the singletons are adjusted according to the robot platform steering capabilities. The controller, fed with distance X_c and angle A_{th} to the line uses 15 rules to deliver the steering angle A_{st}. They are listed in Table 2. Membership functions for X_c and A_{th} are shown in Figure 7. Consequents for

Table 2. Fuzzy control rule set for A_{st}.

		A_{th}				
		Positive Large	Positive Small	Zero	Negative Small	Negative Large
X_c	Left	Medium Left	Small Right	Medium Right	Medium Right	Large Right
	Centre	Medium Left	Small Left	Zero	Small Right	Medium Right
	Right	Large Left	Medium Left	Medium Left	Small Left	Medium Right

A_{st} are 0 for Zero and ±20, ±40, ±80 for Small, Medium, Large Left/Right respectively.

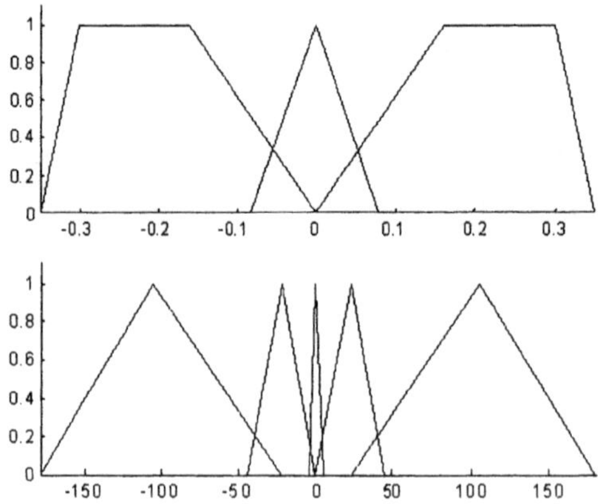

Fig. 7. Membership functions for X_c and A_{th}.

The global control strategy works as follows. With a single line in sight, the robot tries to align itself as fast as possible on it. When a second line gets into the view, the type of bend or crossing is assessed, by examining the relative positions of the two lines. The possible actions are, then, evaluated. If more than one is possible (e.g. *go straight on* or *turn right*), then a command from a higher order level is called for (in our experiments we just implemented a pre-defined list of actions). If a turn is called for, the robot continues on the current line until the crossing is at a pre-defined optimal distance (determined by the maximum steering angle). Then, the algorithm just switches the line to be followed. This involves also switching the directional filter that is selected for line following, and correctly interpreting angular displacement in order to choose the correct turn (left or right).

4.5 Experimental Results

The robot layout and the algorithms have been thoroughly tested by employing a realistic simulation of the vehicle realised in Working Model, interfaced with a simulation of the fuzzy control and image processing system realised in Matlab. All mechanical and physical parameters of the actual robot were taken into account (going from size and weight to wheel and floor materials), as well as actual processing times of the mounted board, that has already been successfully tested. Mechanical mount and electrical testing of the robot is currently being completed, and we expect the robot to be fully functional soon. Meanwhile, Figure 8 shows a simulation of the path followed by the robot, starting a little displaced ($A_{th} = 10°$,$X_c = 4$cm) and turning at a crossing. For each position of the robot, a camera shot is assumed and the geometrical parameters of the robot position are calculated and passed to the controller. Some snapshots of the model of the robot on its path are displayed in figure 9.

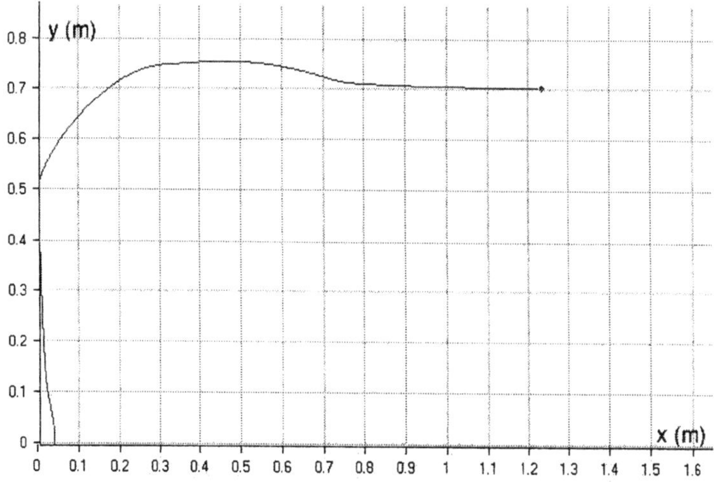

Fig. 8. Path followed by the front wheel of the robot. Axis marking in meters.

5 Further Perspectives

Based on the results described in the previous section, we are currently addressing more difficult problems, in order to devise solutions for real-life application by a step-by-step approach.

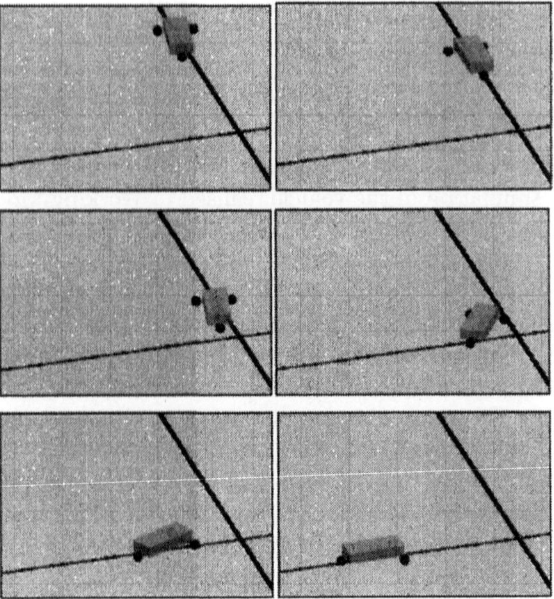

Fig. 9. Snapshots of the robot simulation at time $t = 0, 8, 25, 32, 41, 53$s.

5.1 Curved Line Following

Curved lines can be addressed by modifying the algorithms used for continuous straight lines. One just needs to find a suitable straight line to model the curved one locally and send the parameters to the fuzzy controller. An accurate driving strategy has to take into account the following requirements:

- when the line curvature is small, the robot should follow the line as close as possible,
- when the line curvature is too large to be followed exactly, it should depart from the line and perform a manoeuvre that is globally optimised. This case is similar to turning at a crossing as shown in Section 4.

For this purpose, we estimate a *short-term* approximation to the curved line which is a the tangent line at the closest end (line a in Figure 10), and a *long-term* one which is the secant line joining farthest ends of curve in the image (line b in Figure 10).

When the long-term approximation is similar to the short-term one, close following is possible using of the tangent line.

When the angle difference between the two approximations is large, then substantial correction of the steering angle is necessary. This can be performed by adding a supervisory fuzzy control algorithm using the information of both the tangent and the secant. Rules are of the form

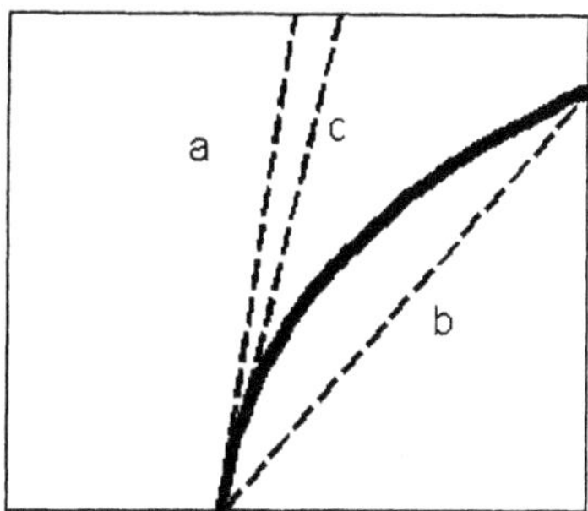

Fig. 10. Tangent and secant to the curve.

> *if secant is more to the right than tangent, then correct steering angle towards right,*

or else

> *if secant is more to the left than tangent, then correct steering angle towards left.*

This correction can also be applied in two other cases: when coarse sampling in time is being used or when speed is relatively high so that the robot has moved a long path before the evaluation of the following image. The tangent line can then be estimated by the projection approach described in Section 4, but just projecting an area of few lines near the bottom of the image instead of the whole image, so as to obtain a short-term secant, resembling the tangent (line *c* in Figure 10). This solution, however, is very sensitive to the image noise and quantisation.

An alternative approach is based on computing one or more intermediate projections. It is equivalent to making a piece-wise-linear approximation of the curve. The bottom projection delivers the short-term approximate line, while other projections can be used for correction.

5.2 Obstacle Avoidance

Obstacle avoidance involves two main tasks:

- recognising obstacles as such;
- driving to avoid collision.

Obstacle recognition depends strongly on the characteristics of the environment. In a very structured environment, where the appearance of boundaries and object is known and has little variance, it is sufficient to look for unexpected situations such as edges, changes of colour or texture, to recognise an obstacle.

In more realistic situation the formalisation of an obstacle is much more difficult. For our purposes, we consider an obstacle anything that develops in the vertical dimension, departing from the floor level. This means that we shall not consider colour, luminosity, texture or shape as indicators. These could just be patterns on the floor and make no obstacle for the robot.

When binocular vision is used, this task is tackled by stereo-matching-based range-finding (see for instance references [22,23]). For high-definition or foveated imaging we can also use an optical-flow-based strategy (e.g. [20,21]). Still we look for solutions using our current hardware based on monocular low-definition vision. We believe that working with such limitations will prompt us to look for solutions that optimise simplicity and cost.

We are currently considering two main approaches: one based on sensor fusion, and the other on vision only.

The sensor fusion approach assumes that the obstacle is always visible. This means that a feature appears in the image, and we have to assess whether it has a vertical dimension or not (which is the case e.g. of shadows and floor patterns). In order to do this, we can use a very simple range-finder, such as an infrared of ultrasonic sensor, to give use a coarse information about distances of objects ahead. When the range finder alerts about presence of something that is closer than a pre-defined distance, we can roughly mark the position of the object on the image. With this information, it is possible to apply CNNs to reconstruct exact shape of the object (see for instance the *recall* template in reference [19].

The vision-only approach is based on projecting a suitable light pattern ahead, and checking how it is deformed. For instance, if we project a straight line of light it will remain straight and in in a known position. When an obstacle (or level change, such as a step) is present, the line is deformed and displaced (Figure 11). It is easy to recognise such situation by CNNs, for instance by projecting shadows of perceived objects on the expected line (Figure 12).

Since obstacle recognition takes a relatively large amount of resources, it may be convenient to track objects once they are recognised rather than performing recognition from scratch at every step. Efficient CNN-based tracking strategies can be based on active contours [28,29].

An obstacle-avoiding driving strategy can be built on the line-following approach as follows. Suppose the robot is driving along a corridor with an obstacle in its way (Figure 13). Once the obstacle is recognised, we can use its contours and boundaries of the hallway as limits for the path. We apply then a CNN template to fill all available area by propagation (Figure 14). We can then take the skeleton of the shape obtained (Figure 15), or just draw a line pointing to the centre of the farthest and widest end of the shape obtained.

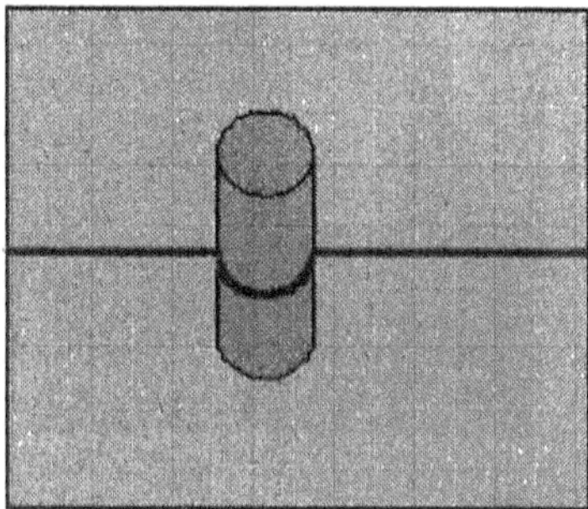

Fig. 11. Line deformation in the presence of an object.

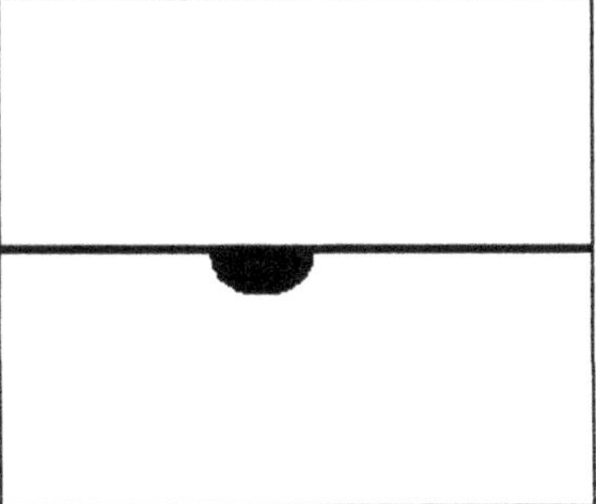

Fig. 12. Projection of the deformed line.

Fig. 13. A corridor with an obstacle.

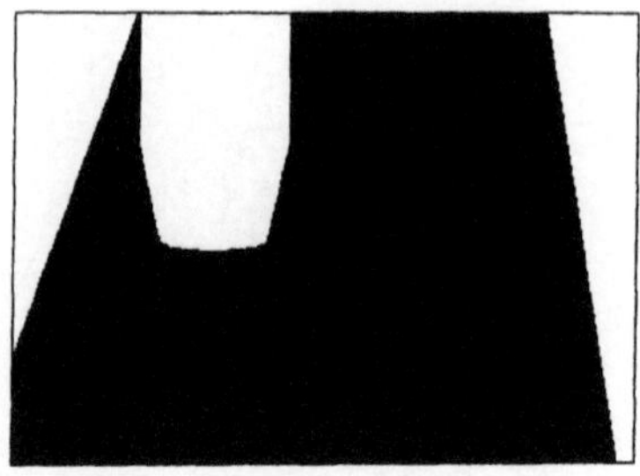

Fig. 14. A CNN template can fill the area available to move the robot.

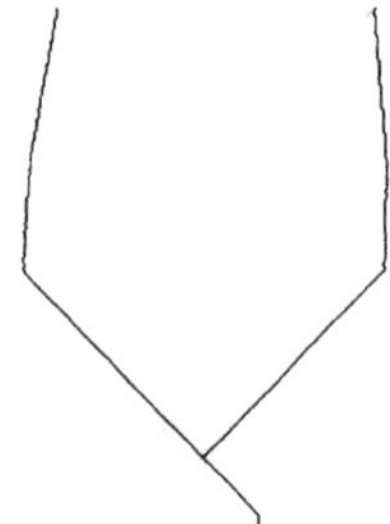

Fig. 15. A skeletonisation of the available area.

6 Summary and Conclusions

We have shown how the combination of Cellular Neural Network image processing and a fuzzy control can solve the robot vision problem in a simple case such as the motion in a maze. The architecture proposed solves the problem by a combination of hard and soft computing tools. The CNN itself is in some sense halfway between a properly soft-computing tool (neural network) and a hard computing one (non-linear image filter).

While the current implementation is based on software realisation of building blocks on programmable hardware, the target implementation is based on special-purpose VLSI CNN hardware, yielding substantially enhanced performance.

The solution for this problem gives a hint for studying more complicated problems such as curved line following or obstacle avoidance. Still, this is a first step towards a wider use of CNNs in robotics for visual feedback. Further contributions should stress the use of CNN chips on a hardware implementation and the formulation of the image processing steps in CNN terms.

Acknowledgements This work is partially supported by Funitec (PGR2001-05). Authors would like to thank Jordi Solsona, Sonia Luengo, Alessandro Maraschini, Giada Apicella, Valentina Giovenale and Ruben Funosas for their collaboration in the project.

References

1. Chua, L.O., Yang, L. (1988) Cellular Neural Networks: Theory. IEEE Trans. Circ. Syst. **CAS-35** 1257-1272
2. Chua, L.O., Roska, T. (1993) The CNN Paradigm. IEEE Trans. Circ. Syst. **CAS-I-40** 147-156
3. Harrer, H., Nossek, J.A. (1992) Discrete-time Cellular Neural Networks. Int. J. of Circ. Th. Appl.**20** 453-467
4. Balsi, M., Maraschini, A., Apicella, G., Luengo, S., Solsona, J., Vilasís-Cardona, X. (2001) Cellular Neural Networks for mobile robot vision. *Mira, J., Prieto, A. Eds. Proceedings of IWANN 2001*, Lecture Notes in Computer Science **2085** 484-491
5. Vilasís-Cardona, X., Luengo, S., Solsona, J., Apicella, G., Maraschini, A., Balsi, M. (2001) Mobile robot vision using Cellular Neural Networks. *Tan, K.K., Lim, S.Y. Eds. Proceedings of CIRAS 2001* ISSN 0219-6131 367-372
6. Latombe, J.C. (1991) Robot Motion Planning. Kluwer Academic Publishers Boston, MA
7. Singhal, A. (1997) Issues in Autonomous Mobile Robot Navigation. University of Rochester Computer Science Department Technical Report
8. Luo, R.C., Kay, M.G. (1992) Data Fusion and Sensor Integration: State-of-Art in the 1990s. *Data Fusion in Robotics and Machine Intelligence.* Academic Press Ltd San Diego CA
9. ARGO Project (2000), http://nannetta.ce.unipr.it/ÃRGO/english/index.html (as of December 2000)
10. Chen, K.H., Tsai, W.H. (1997) Vision-Based Autonomous Land Vehicle Guidance in Outdoor Road Environments using Combined Line and Road Following Techniques. Journal of Robotic Systems **14** (10) 711-728
11. Cheok, K.C., Smid, G.E., Kobayashi, K., Overholt, J.L., Lescoe, P. (1997) A Fuzzy Logic Intelligent Control System Paradigm for an In-Line-of-Sight Leader-Following HMMWV. Journal of Robotic Systems **14**(6) 407-420
12. Maeda, M., Shimakawa, M., Murakami, S. (1995) Predictive Fuzzy Control of an Autonomous Mobile Robot with Forecast Learning Function. Fuzzy Sets and Systems **72** 51-60
13. Liñan, L., Espejo, S., Domínguez-Castro, R., Rodríguez-Vázquez, A. (2002) ACE4K: An Analog VO 64x64 Visual MicroProcessor Chip with 7-bit Analog Accuracy. Int. J. of Circ. Th. Appl., in press
14. Kananen, A., Paasio, A., Laiho, M., Halonen, K. (2002) CNN Applications from the Hardware Point of View: Video Sequence Segmentation. Int. J. of Circ. Th. Appl., in press
15. Roska, T., Chua, L.O. (1993) The CNN Universal Machine: an Analogic Array Computer . IEEE Trans. Circ. Syst. **CAS-I-40** 163-173
16. Kék, L., Zarándy, A. (1998) Implementation of large-neighbourhood non-linear templates on the CNN universal machine. Int. J. of Circ. Th. Appl. **26** 551-566

17. Zaràndy, À. (1999) The Art of CNN Template Design. Int. J. of Circ. Th. Appl. **27** 5-23
18. Kozek, T., Roska, T., Chua, L.O. (1993) Genetic algorithm for CNN Template Learning. IEEE Trans. Circ. Syst., **CAS-I-40** 392-402
19. Roska, T., Kék, L., Nemes, L., Zaràndy, À., Brendel M. (2000) CSL-CNN Software Library. Report of the Analogical and Neural Computing Laboratory, Computer and Automation Institute, Hungarian Academy of Sciences, Budapest, Hungary
20. Balsi, M. (1998), Focal-Plane Optical Flow Computation by Foveated CNNs. Proc. of Fifth IEEE Int. Workshop on Cellular Neural Networks and their Applications (CNNA-98), London, UK, Apr. 14-17, 1998, 149-154
21. Shi, B.E. (1999) A one-dimensional CMOS focal plane array for Gabor-type image filtering. IEEE Trans. Circ. Syst. **CAS-I-46** 323-327
22. Zanela, A., Taraglio, S. (1998) Seeing the Third Dimension in Stereo Vision Systems: a Cellular Neural Network Approach. Eng. Appl. Artif. Intel. **11** 203-213
23. Sargeni, F., Bonaiuto, V. (2001) CNN Cell for Computing Disparity Map. Elec. Lett. **37** 682-683
24. Titus, A.H., Drabik, T.J. (2000) Analog VLSI Implementation of the Help if Needed Stereopsis Algorithm. IEEE Trans. Circ. Syst. **CAS-II-47** 1328-1337
25. Szolgay, P., Katona, A., Eröss, Gy., Kiss, Á. (1994) An Experimental System for Path Tracking of a Robot using a 16*16 Connected Component Detector CNN Chip with Direct Optical Input. Proc. of Third IEEE Int. Workshop on Cellular Neural Networks and their Applications (CNNA-94), Rome, Italy, Dec. 18-21, 1994, 261-266
26. Kosko, B. (1992) Neural Fuzzy Systems, Prentice Hall Englewood Cliffs, NJ
27. Hough, P.V.C. (1962) Method and means of recognizing complex patterns. U.S.Patent 3,069,654. *See also* Jain, A.K. (1989) Fundamentals of digital Image Processing, Prentice Hall Englewood Cliffs NJ.
28. Vilariño, D.L., Brea, V.M., Cabello, D., Pardo, J.M. (1998) Discrete-time CNN for image segmentation by active contours. Patt. Rec. Lett, **19** 721-734
29. Vilariño, D.L. (2001) Contornos activos a nivel de pixel: diseño e implementación sobre arquitecturas de redes no lineales celulares, Ph.D. thesis, Univ. of Santiago de Compostela, Spain

Multiresolution Vision in Autonomous Systems

Pelegrín Camacho, Fabián Arrebola, and Francisco Sandoval

Dpto. Tecnología Electrónica, ETSI Telecomunicación,
Universidad de Málaga, 29071 Málaga-Spain
pelegrin@dte.uma.es

Abstract. The performance of many autonomous systems based on artificial vision depends mainly on the speed of response and the field of view of the vision systems. The many tasks to be carried out, such as object detection, recognition, tracking, etc., the complexity of reliable algorithms and tasks to be done in real time, and the huge data volumes involved with stereo vision systems, imply processing times and resources that, in some cases, are incompatible with or unsuitable for acceptable system operation. Multiresolution systems are one alternative to cover wide fields of view without involving high data volumes and, therefore, considerably reduce the constraints imposed by off-the-shelf uniresolution vision systems.

Our work is related to adaptive space-variant sensors, able to supply any number of resolution levels with reconfigurable resolution profiles around regions or objects of interest, and to the specific algorithms and hierarchical data structures related to processing multiresolution data involved in tasks of image segmentation, object detection, etc. required for operation in dynamic environments.

1 Introduction

Most of the computer vision systems found today in a wide range of industrial, medical and military applications have two common characteristics: one is that they use a constant spatial resolution throughout the field of view, and the other is that their functionality is inspired in the Reconstructive Vision paradigm or Marr paradigm [1], which consists in trying to discover *what* is present in the scenes and *where* it is located by means of images. This concept or understanding of the vision process does not include the purposive aspects of vision used by animates to guide their activities, i.e. the processes of acquisition and extraction of visual information are deeply related to the tasks being performed and they serve to interact with the environment in an *intentional* and *active* manner: In the animal kingdom, the gaze is driven and events are attended to in a selective manner, according to the tasks requirements and contextual needs. The observation of biological vision systems shows that, in most vertebrates, the environment is perceived with a variable acuity. The retina is a layer of photoreceptors, cones and rods, whose function is to encode the light information entering through the pupil and transmit it to the brain via the ganglion cells and the optical nerve. Unlike the uniform resolution of a typical computer vision system, most retinae usually have one small region, sometimes two, where the cones are concentrated with a very high spatial resolution. This region, the *fovea* or *area centralis*, covering roughly 0.05% of the visual field but accounting for about 10% of the channels available in the optical nerve, allows a high acuity in the region of the scene projected onto it. Meanwhile, the density of rods in the retina decreases the further away from the fovea, forming a kind of elliptical rings with decreasing radial resolution. This variable spatial resolution implies a progressive acuity loss or a reduction of the ability to perceive details at the periphery of the visual field. Nevertheless, the acuity loss is compensated for by the ability of look, a complex purposive function, related to cognition and implying a brain activity translated into saccadic eye movements to fix the gaze.

Thus, in relation to many of the visual systems found in nature and approaching vision functions in a distinctive manner, the Selective Vision paradigm involves precisely detecting only the conditions or patterns concerning tasks under execution, without requiring a full understanding of the scene. As a consequence, when this paradigm is applied to vision systems, only selected regions are processed, applying the resources and operations required to the scene context.

This chapter briefly reviews the concepts of Active Vision in Section 2, introducing foveal vision in Section 3. Multiresolution structures for foveal vision are presented in Sections 4 and 5 and an imager implementation based on them is explained in Section 6. Finally, Section 7 presents the algorithms for processing multiresolution images from the imager, and is followed by the Conclusions.

2 Active Vision Systems

The selective perception concept falls within the Active Vision, a computer vision field whose main objectives include the study and development of perception systems operating in real time, for integration in systems acting in dynamic environments [2]. The concept of Active Vision [3,4] is inspired in biological visual systems and proposes active interactions with the perception process, consisting of determining and controlling appropriate camera parameters, as well as operating the processing modules in order to simplify scene comprehension.

At the development phase of an active vision system, several strategies could be adopted:

- *Control of camera parameters* as a function of the tasks being done and external stimuli, emulating the ability of many vertebrates to adapt their visual systems to current activities and scenes lighting conditions. Active vision systems, while controlling parameters such as pan and tilt, vergence, lens aperture, zoom, etc., process the obtained images in a relatively easy manner, extracting a few characteristics of the scenes after applying simple algorithms. This technique has proven to be more efficient and robust than performing a single observation followed by an exhaustive analysis of the image. However, this operating mode [5] requires some kind of observations planning: a gaze control subsystem must be in charge of the implementation of perception strategies, normally integrating the function of fixations or gaze stabilization and the function of gaze changes or saccades, to emulate rapid eye movements to attend to events captured by the retina in regions not fully covered by or outside of the fovea.

- *Attention mechanisms to obtain selective data reduction*, based on the fact that only a small fraction of the information contained in the scenes is relevant to the visual tasks. Relevant data is determined according to different criteria, such as *depth* or distance, measured by mean of stereo or focusing algorithms, able to segment areas [6] as a function of distance to the camera, or based on *motion*, extracting regions of interest for tracking purposes and obstacle avoidance. After extracting the useful information, the system must concentrate the computational resources on the regions of interest and therefore, a module would be needed [7] to control the attention mechanisms, providing feedback loops, as well as alert mechanisms to decide where and what to look at next.
- *Selective data reduction by foveal sensing*, emulating the spatial distribution of photoreceptors in the retina with specific devices [8] or synthesizing retinal structures in real time [9], to transform the uniresolution images obtained with off-the-shelf sensors. This has been our strategy for selective data reduction and will be presented with further detail in the following sections.

3 Foveal Vision

The behavior of an artificial vision system will be deeply conditioned by three factors: i) required spatial resolution, ii) amplitude of the field of view and iii) required speed of response. Thus, systems for medical applications, where the detection range or the classifying precision must be optimum, are mainly conditioned by the spatial resolution, while systems related to mobiles detection and tracking have a wide field of view constraint in order to reduce the probability of loosing a target and to increase the ability of detecting short-term events. By contrast, vision systems for robot guidance and interaction in dynamic environments, will need real-time response, wide field of view and high resolution. The last two conditions could lead to a high data volume, involving computationally intensive tasks, which, in turn, could degrade the desired time response. An alternative for overcoming this tradeoff is to apply multiresolution techniques emulating the space variant resolution adopted in biological systems. Such multiresolution images, called foveal or retinal images, combined with the ability to move the camera in a controlled manner, make it possible to acquire the information required to carry out certain tasks without considering all the data contained in uniresolution images, but covering the same field of view [10]. Thus, in foveal vision systems, two basic tasks have to be carried out after saccades: i) the analysis of the information corresponding to the fovea, the *what*, and ii) the selection within the field of view of the next fixation point, the *where*. This is what distinguishes foveal vision from other types of computer vision [11]. The emulation of receptive field distribution and photoreceptors size in the retina [12] has been approached by several research groups, applying different techniques:

- Optical methods: Using specific lenses to obtain space variant projection of scenes light onto uniform arrays [13].
- VLSI techniques: Using specific non-uniform arrays of photosensors with a fixed topology [8].
- Hardware synthesis: Using standard off-the-shelf devices and uniresolution sensors to obtain real-time foveal imagers [14].
- Software synthesis: Applying algorithms to uniresolution images to transform them into foveal images [15].

According to the space variant strategy applied, the obtained retinotopologies vary by either changing the shape and distribution of photoreceptors or the sampling methodology applied on uniform sensor arrays. Within the first strategy, the most common have been the log-polar [8] and the exponential Cartesian [16] topologies. Concerning photoreceptors shape, different geometries have been adopted, from hexagonal [17], exponential-Cartesian [16] to log-polar sectors [18].

3.1 Log-Polar Geometries

The receptive fields used in these structures are located in concentric circles, whose sizes increase linearly from the innermost circle to the periphery of the sensor. The position of each sensor element is specified by the radius and the orientation angle, as shown in Fig. 1. The parameters defining the sensor structure are the number of elements in each circle, N_a, the number of concentric circles, N_c, and the scale factor k, determining the size ratio between elements of the consecutive circles. Normally, the complex transform $W{=}log_k(Z)$ is applied to operate with log-polar images, implementing a mapping of the cells in the retinal or polar plane $Z(\rho,\theta)$ to the cortical or Cartesian plane $W(\xi,\eta)$, as shown in Fig. 1. The main advantage of this transform is that a radial expansion or compression of an object in the polar plane Z implies a shift on the ξ axis in plane W, while a rotation of the object in relation to the sensor center is equivalent to a shift on the η axis. The shifts mean an invariant transformation on scale and rotations referred

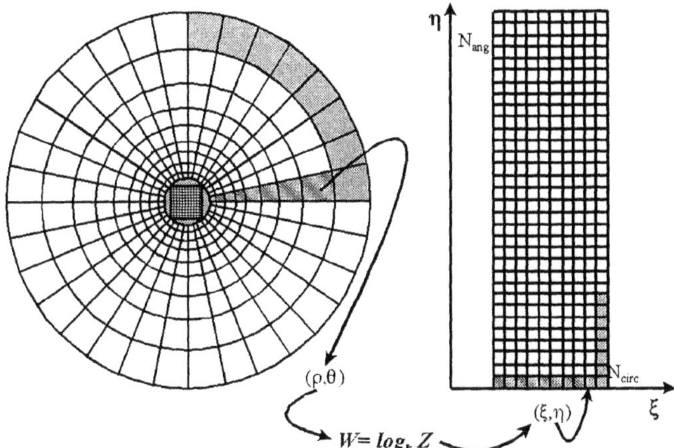

Fig. 1. Complex transformation of the log-polar geometry.

to sensor origin. This retinotopology has the advantage of low computational costs of the motion equations for estimating the optical flow and robust time to impact calculation [19]. Also, applying correlation techniques, it has been shown that an efficient vergence control could be achieved in stereo systems for active vision [20].

3.2 Exponential-Cartesian Geometries

In these geometries, whose lattice is shown in Fig. 2a, the size and dimensions of the peripheral elements or *rexels* in the sensor rings, are integer numbers expressing relations to fovea pixels. The retinal images corresponding to these structures are defined by two parameters: m, number of rings surrounding the fovea, and d, number of subrings in each ring.

Considering the storage and handling of this type of images, it is clear that a bidimensional array is not an adequate data structure, since to identify an element three parameters are required: the ring in which it is located and two coordinates within that ring. A multilevel data structure is required, as indicated in Fig. 2c, with two types of cells: measured cells, in gray, and computed cells, in white. Measured cells come directly from the sensor, while computed cells, holes in Fig. 2b, are averages of cell groups in the level immediately below. Above level m, or waist level, further levels can be obtained using the same 2^N law. The multilevel structure has the following properties:

- Each layer or level k can be considered as a uniresolution image, whose cells have a receptive field of $2^k \times 2^k$ pixels of the original image
- The data cube up to level m is a subset of the vision pyramid, as seen in Fig. 2d.
- The field of view at waist level, m, is the same as in the original image.

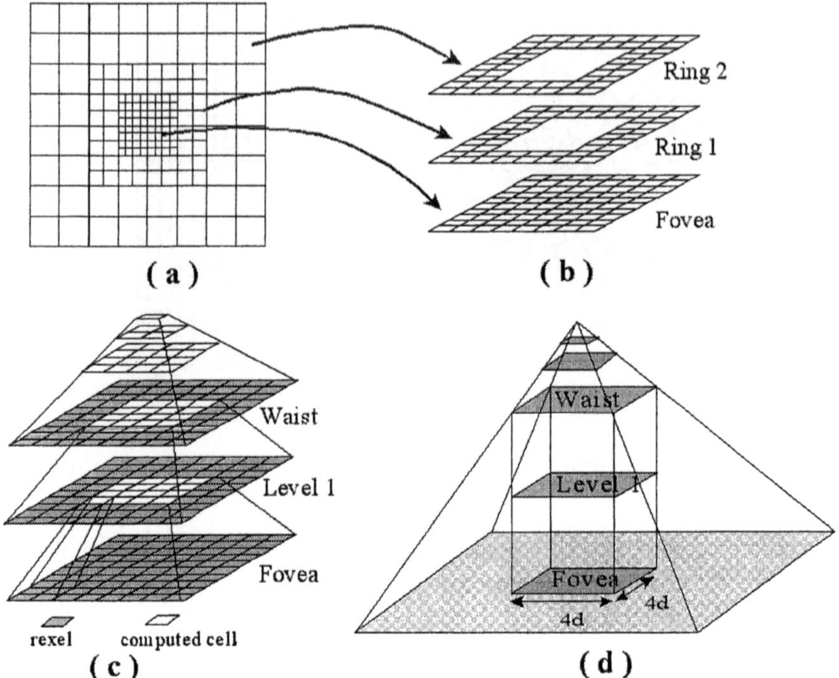

Fig. 2. Exponential-Cartesian geometry. a) Lattice b) Cell distribution in layers. c) Data structure d) Hierarchical data structures within the vision pyramid.

- Above m level, the data are the same as in the vision pyramid.

Exponential-Cartesian geometries have some advantages over the log-polar retinotopologies:

- Most vision devices on the market (cameras, *frame-grabbers*, etc.), as well as the techniques and algorithms used for image processing, operate with data sampled in Cartesian coordinates.
- Sensor implementation, either in silicon or synthesized, is simpler than the log polar implementation.
- Any algorithm developed for vision pyramids or uniresolution images is easily adapted to exponential-Cartesian structures.
- Although the implementation in silicon implies a fixed configuration, the synthesized versions are easily reconfigured to different retinotopologies, which is equivalent to controlling some camera parameters mentioned in relation to the strategies to be followed in Active Vision. This important advantage of synthesized sensors will be discussed in the following sections.

4 Multiresolution Geometries with Shifted Fovea

An essential requirement of an active vision system based on foveal multiresolution is the ability to perform fovea fixations, or foveations, once the system has decided where to look next. Thus, whenever the sensors have a fixed structure, like those implemented in silicon, a mechanical system will have to be used to carry out the pointing. The camera or sensor motion could be controlled by a mechanical subsystem, whose precision and speed will partially determine the effectiveness of the system. Although convenient pan-tilt pointing motors have been developed [21], there are specific alternatives for synthesized sensors when the incremental angles for pointing at a target fall within the solid angle being covered by the optics, making the high precision motors and the controller unnecessary. This is an important consideration, no matter the sensor, since any camera motion implies background changes, which could lead to serious reference loss and additional computations and time to recover the reference. Therefore, the lower the number of mechanical saccades, the higher the efficiency of the overall system.

The alternative of placing the fovea at different locations in the field of view without moving the camera is electronic foveation, which can be easily achieved with either shifted or adaptive fovea structures [9]. These structures could be classified into four different types [22], depending on the number of cells, C, in each resolution level and the relative shift, S, between two consecutive levels. If both, C and S are constant, as in Fig. 3a, only two parameters are needed to define the fovea location: the parameters, given in maxels, refer to the horizontal and vertical shifts sh, sv of any ring i within the ring $i+1$. In shifted fovea types, the associated structure has a constant number of elements, $4h \times 4v$, in the fovea and levels up to the waist, and the same relative position of cells in all resolution levels, as seen in Fig. 5a., where$h$ and v are related to m and the field of view dimensions, H and V:

$$\mathrm{h} = \frac{H}{2^{m+2}} \qquad \mathrm{v} = \frac{V}{2^{m+2}} \tag{1}$$

A constant S forces a minimum jump Δ_0 to move foveae in any direction. The jump, being $N = 0$ for fovea level, is given by the expression:

$$\Delta_N = 2^{N+1} + 2^{N+2} + \ldots + 2^m = \sum_{i=N}^{m-1} 2^{i+1} \tag{2}$$

which will also be used to define region jumps at the Nth level of pyramids.

To reduce Δ_0, making it independent of m, ring i is allowed to move within ring $i+1$. This second type corresponds to a variable S, allowing a minimum Δ_0=2, leading to different relative shifts in the resolution levels, as shown in Fig. 3b. The third and fourth structure types are described in the next section.

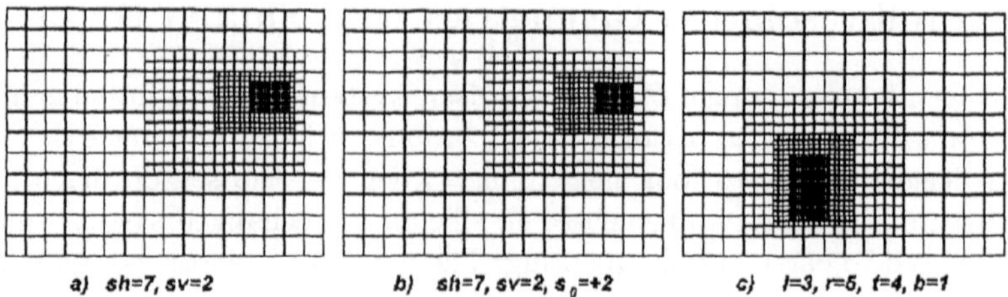

Fig. 3. Parameterized retinal structures: a) Constant C and S, b) Constant C, variable S, c) C variable and S constant.

5 Multiresolution Geometries with Adaptive Fovea

Shifted fovea retinotopologies have fixed size foveae, which may not be ideal for capturing regions much smaller than fovea, since it would imply unnecessary data processing, or objects bigger than fovea, requiring more than one foveation. To adapt fovea sizes to target sizes, we define adaptive retinal structures -ARS- [23], where both fovea dimensions and position are determined by a set of four parameters in relation to field of view dimensions, $H\times V$, and the coordinates of two opposite bounding box corners X_{min}, Y_{min} and X_{max}, Y_{max}:

$$\begin{array}{llll} \textit{Left factor} \quad l = \left\lfloor \dfrac{X_{min}}{\Delta_0} \right\rfloor & \textit{Top factor} \quad t = \left\lfloor \dfrac{Y_{min}}{\Delta_0} \right\rfloor \\ \textit{Right factor} \quad r = \left\lfloor \dfrac{H - X_{max}}{\Delta_0} \right\rfloor & \textit{Bottom factor} \quad b = \left\lfloor \dfrac{V - Y_{max}}{\Delta_0} \right\rfloor \end{array} \tag{3}$$

Shifted fovea structures are a particular case of ARSs defined by (3) and shown in Fig. 3c and Fig. 4, encompassing all exponential-Cartesian retinotopologies with constant S. There is a close relation between ARSs and vision pyramids, which means that the hierarchical structure of the ARSs, shown in Fig. 5b, can be defined as a particular case of the pyramidal data structure. Regions of interest, or ROIs, at each pyramid level have dimensions related to factors *Ld, Rd, Td, Bd*, also named *l, r, t, b* for short. Fig. 6 shows the ROIs for the data structures of Fig. 5. Note that the *Nth* level has its own Δ_N as defined in Eq.(2), which expresses the coordinates of ROIs within the *Nth* level as:

$$\begin{array}{ll} X_{m(N)} = \dfrac{1}{2^N}\left(\Delta_N \cdot l\right) & Y_{m(N)} = \dfrac{1}{2^N}\left(\Delta_N \cdot t\right) \\ X_{M(N)} = \dfrac{1}{2^N}\left(H - \Delta_N \cdot r\right) & Y_{M(N)} = \dfrac{1}{2^N}\left(V - \Delta_N \cdot b\right) \end{array} \tag{4}$$

Fig. 4. Left: Original image (320×256) Right: Adaptive retinal image (*l,r,t,b=11,10,12,4*)

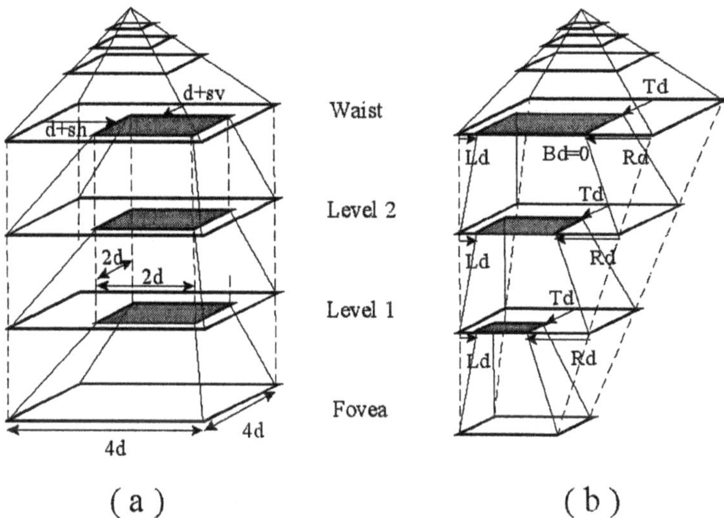

Fig. 5. Hierarchical data structures. a) Constant C and S. b) Variable C and constant S.

where $m(N)$ and $M(N)$ stand for the minimum and maximum of the respective coordinates at the *Nth* level, and the 2^{-N} factor indicates that coordinates are given in maxels of the respective level.

A fourth type of structure, derived from type three, could be defined with a variable S for precise fovea pointing of ± 2 pixels.

As can be derived from (4), the width and height of any upper ROI below the waist depend on fovea dimensions alone, while the coordinates depend on fovea position. These relations will be used appropriately for the hardware implementation of an imager obeying the ARS laws.

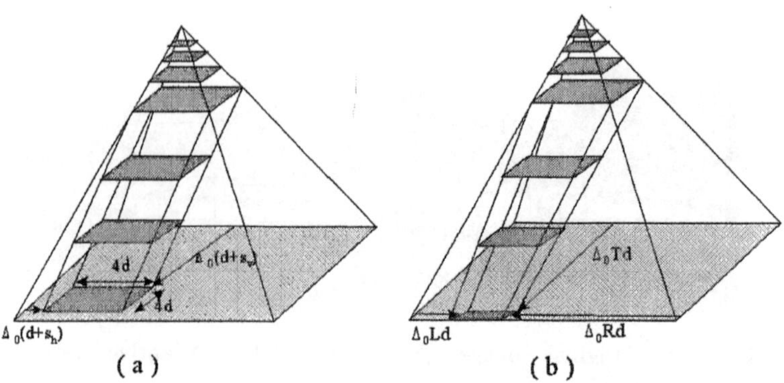

Fig. 6. Mapping of hierarchical structures in Fig. 5 into the vision pyramid.

6 Multiresolution Imager Implementation

The imager core is a FPGA implementing the interfaces with CMOS uniresolution sensors and the vision system, generating a pyramid and reducing data selectively according to the instructions received from the vision system to locate foveae, adjust sensor parameters or modify the frame rate, initially set at 25 frames per second. These functions are some of the considerations involved in active vision and controlled at system level but implemented in the imager to speed up system performance. In-system programmability of FPGAs makes them a versatile alternative for adapting the imager core to different interfaces, sensors and image formats according to vision applications. The selected FPGA family is the Altera's 10KE, with over 50Kgates and 40K SRAM bits. This enables the configuration of specific storage devices without the use of external components, reducing the imager to the CMOS progressive scan sensors, the FPGA and a few discrete components. The diagram in Fig. 7 shows the imager blocks and its interconnections.

6.1 Pyramid Generator and Regions Extractor

Although there are many pyramid-generating algorithms [24], some involve high computational complexity without offering clear advantages in terms of processing speed when applying some of the high level algorithms used for segmentation and block matching. Thus, we construct pyramid levels by successive 4-to-1 averaging:

$$g_L(i,j) = \left\lfloor \frac{1}{4} \sum_{u=0}^{1} \sum_{v=0}^{1} g_{L-1}(2i+u,\ 2j+v) \right\rfloor \qquad (0 \leq L \leq 3) \qquad (5)$$

where $g_L(i,j)$ is the gray level of maxel (i,j) of the Lth level, and $g_0(i,j)$ represents the pixels of the input image or pyramid base.

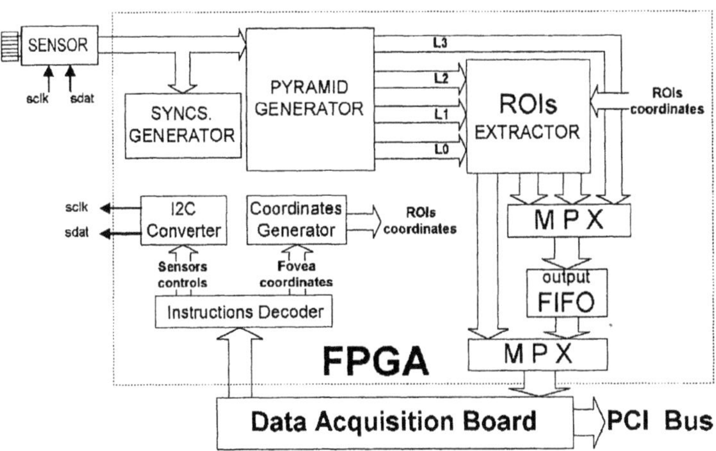

Fig. 7. Imager block diagram.

The Pyramid Generator, whose details are shown in Fig. 8, is a 3-step pipeline to recursively processing pixels entering from the progressive scan CMOS sensors. The first pipeline processes pairs of adjacent image lines, adding two consecutive pixels in the first adder. During odd lines, the sums are stored in a FIFO whose content is read during even lines and added with the corresponding sums of the same pairs of pixels in even lines. The second adder sums, right-shifted two bits, are the gray levels of maxels pertaining to level L1, which are temporarily stored in the L1 register. Simultaneously, pairs of L1 maxels are added in the second pipeline in a process identical to the one described for pixels entering the first pipeline. The results, the maxels of the second pyramid level, enter the L2 register and the third pipeline, which outputs the maxels of the level L3. The total delay between the last pixel of a frame and the last maxel is under 400 ns, which is to say that all the pyramid levels are obtained almost simultaneously with the entrance of pixels from the sensors. As a consequence of this parallel generation of maxels, some storage device is needed to buffer pyramid speed and interface banwidth. An alternative would be to use the internal SRAMs, but write and read addressing is rather complex, apart from calling for excessive FPGA resources, while external SRAMs would considerably increase FPGA pin out. Therefore, a FIFO alternative was chosen, using the SRAM bits of the FPGA.

The function of the ROIs extractor block on the right in Fig. 8 is to extract fovea pixels and ROIs maxels corresponding to the hierarchical structure mentioned in Section 5. To reliably implement this function in real time, both the Pyramid Generator and ROIs extractor are synchronized with the pixel rate and blanking coding supplied by the sensors and managed at the imager Synchronism block. ROIs extractor operation starts by loading the

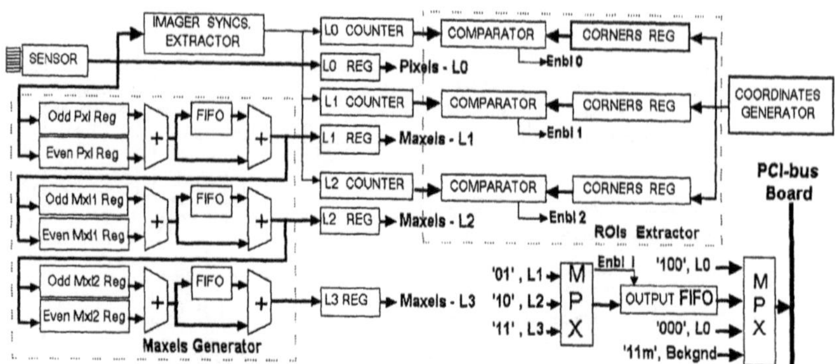

Fig. 8. Pyramid Generator and Regions of Interest Extractor.

ROIs coordinates into the Corners registers. Fovea coordinates are obtained using (3), after receiving l, r, t, b as a part of processor instructions.

Then, applying (4) recursively and using the relation:

$$\Delta_{N+1} = \Delta_N - 2^{N+1} \tag{6}$$

derived from Eq.(2), the upper ROIs coordinates are obtained as:

$$\begin{aligned} X_{m(N+1)} &= \frac{X_{m(N)}}{2} - l \qquad & Y_{m(N+1)} &= \frac{Y_{m(N)}}{2} - t \\ X_{M(N+1)} &= \frac{X_{m(N)}}{2} + r \qquad & Y_{M(N+1)} &= \frac{Y_{M(N)}}{2} + b \end{aligned} \tag{7}$$

easily implemented in FPGAs, without the multiplications implied in (4).

Although fovea coordinates are sent by the application processor at any time, coordinates loading occurs during the vertical blanking periods or inter-frames times, ensuring that only full frames are processed and full sets of ROIs are extracted. A set of counters keep the current coordinates of elements entering from the sensor, L_0, or from the maxel generator for levels up to L_{m-1}. Their counts are compared with those in the Corners registers and, if the counts are within the range of the ROIs coordinates, time-window signals are sent as write enables to the output FIFO, buffering the multiplexed outputs L_1 to L_m from the Pyramid Generator. Since the whole top level L_m is always sent to the processor, L_3 maxels do not enter the ROIs extractor, but go directly to the FIFO.

Unlike maxels processing, fovea pixels entering with a 24 MHz clock rate are directly sent to the processor through the interface multiplexer with a negligible delay. MPXs controls and FIFO clocks are synchronized with the generation of L_1 to L_m, allowing to store the L_i gray values into the FIFO.

A 320×256 image and the set of ROIs supplied by the imager, related to the pyramid levels and the fixation point are shown in Fig. 9.

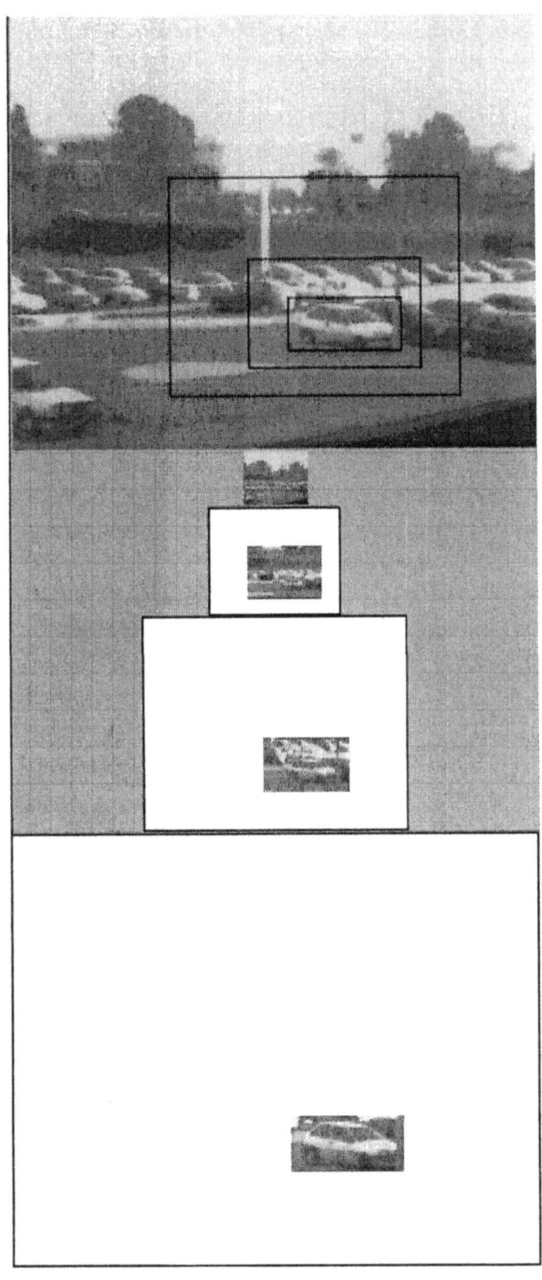

Fig. 9. Pyramidal ROIs from the retinal imager (Base image 320×256).

6.2 Imager Interface

As shown in (5), m=3. This implies a peripheral maxels field of 8x8 pixels, acceptable for detection purposes even with 320×256 images, as shown in Figs. 6 and 9. Using 640×480 VGA images, with foveae limited to 164×158 pixels, a rather high figure for multiresolution, generating ROIs up to 116×102, 92×74 and 80×60, the required bandwidth is under 1.25 MB at 25 fps. A PCI data acquisition board with 16 I/O lines and 50MB bandwidth, shown in Fig. 7, is the imager interface in an 800 MHz PC. Fovea pixels are sent to the interface at the sensor rate. However, the pipelined parallel generation interleaves maxels and levels in transmission. Since maxels are generated orderly within the levels, if headers are used to code maxels levels, ROIs can be extracted at the PC by headers filtering. Maxels stored in the 10-bit wide FIFO during image pixels time, are read during sensors horizontal blanking times. FIFO depth is 288 words, i.e. the number of maxels that could be generated during rows which are multiple of 8, where all maxels are possible. It is worth to note that in case of foveae 164×158, the total number of bytes sent is 49352 per frame, i.e. the data compression is at least 84%.

Instructions are sent to the imager during the vertical blanking times of sensors. FPGA commands are sent to fix fovea corners, with sets of *l, r, t, b* parameters, or to fix imager frame rate, selecting 1-out-of-N frames. Sensors commands to fix light conversion gain, thresholds, etc. can be received at any time, being converted to an I2C protocol, using the serial lines *sclk* and *sdat*. These types of commands are part of the camera control strategies in active vision. Image format, blanking times, ADC offset and other sensor parameters are fixed at application start-up.

7 Image Processing of Adaptive Retinal Structures

Depending on the results obtained after processing data from previous fixations and the task being performed, the highest modules of the active vision system using the ROIs received from the ARS imager will process the data associated with fovea or any level or will apply an attention mechanism using the hierarchical data structure to determine new targets, sending new instructions to the imager to perform new fixations. Concerning the first alternative, since fovea and ROIs are uniresolution images, conventional image processing algorithms can be applied to any of them. The steps are as follows:

- Extraction of cells integrating the edges of the chosen target.
- Determination and description of the regions boundaries using contour chain codes.
- Target characterization, using a contour corner detector or calculating the curvature function associated with the contour [25].
- Target identification or recognition by means of curvature classification patterns [26].

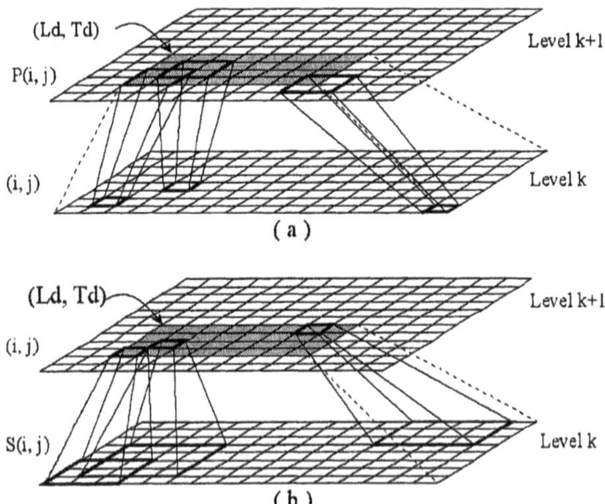

Fig. 10. Links stabilization process. a) Potential fathers cells. b) Potential son cells.

For attention mechanisms activation, there are two phases. In the first one, a hierarchical multiclass segmentation algorithm is applied, adapting a multiresolution algorithm initially developed [27] for vision pyramids to our data structures. During this phase, besides dividing the scene into a controllable number of classes, a stable father-sons linking structure is generated between cells of the different levels. The second phase function is to detect new regions of interest in the scene, using the above fathers-sons links. Both phases are described in further detail in the following sections.

7.1 Hierarchical Multiclass Segmentation

The first step of the attention mechanism involves a hierarchical technique named multiresolution pixel linking, allowing segmentation of the exponential-Cartesian retinal images. The technique exploits the principle of adaptive linking between cells of adjacent levels and a similarity criterion among fathers and sons. Departing from the ROIs supplied by the multiresolution imager, the following steps are applied [28], in reference to Fig. 10:

- Initialization of the areas of all cells in the hierarchical data structure. Areas of maxels of the *kth* level are initialized to $2^k \times 2^k$ and areas of computed cells are set to zero.
- Stabilization of structure links. Starting at fovea level, two processes are iteratively performed at each level, ending when links are stabilized. The two processes are:
  - Set up links between cells of consecutive levels. For each cell in level k, the potential father cells in level $k+1$ are searched for the most

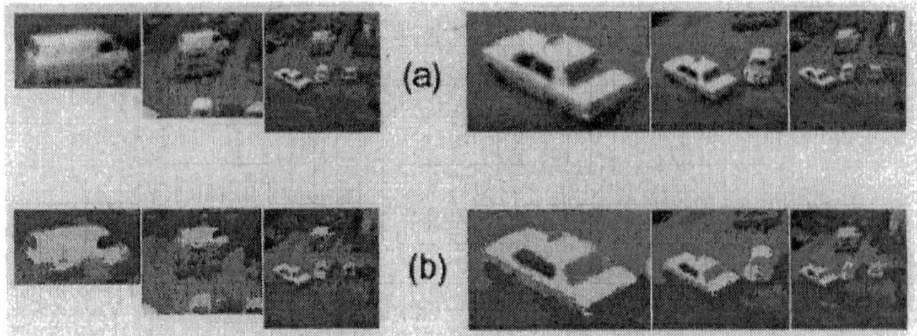

Fig. 11. Hierarchical segmentation results. a) Pyramid levels. b) Segmented pyramid levels.

similar cell. As shown in Fig. 10a, within an ARS, each cell (i,j) has fathers cells, whose coordinates $P(i,j)$ are:

$$P(i,j) = \left(\left\lfloor \frac{i-1}{2} \right\rfloor + Ld + n,\ \left\lfloor \frac{j-1}{2} \right\rfloor + Td + m \right) \quad 0 \le n,m \le 1 \tag{8}$$

– Recalculate gray level and area of computed cells at level $k+1$, considering the links calculated in the previous process. The area will be the sum of sons areas and the contribution of each son to its fathers gray level will be proportional to the area of the son. Each father cell could have a variable number of linked son cells, ranging from 0 to 16. Within an ARS, each cell (i,j) in a region of computed cells could have up to 16 sons, as can shown in Fig. 10b, whose coordinates $S(i,j)$ are:

$$S(i,j) = (2(i - Ld) + n\ ,\ 2(j - Td) + m) \qquad -1 \le n,m \le 2 \tag{9}$$

- Class generation. Once all links between cells of all levels are stable, a level L is chosen and the gray level of every cell is propagated towards the son cells in level L-1. The process is repeated at all levels down to fovea level. The image will be divided into $4^{(M-L)}$ classes, M being the top level of the hierarchical structure. The process applied achieves two effects, one is to smooth the most homogeneous regions in the scene and the other is to enhance contrast at the edges of the regions, as can shown in Fig. 11.

7.2 Detection of Regions of Interest

The segmentation technique described in the above section divides the images into a controllable number of classes or segments. However, the classes may or may not correspond to objects or the classes could even be a set of unconnected regions. This is why the last step associated with class generation is not used to detect areas of interest. However, the links structure obtained in the segmentation phase will be useful as a source of object roots starting at cells located in the upper levels of the structure. In this sense, the attention mechanism will involve detecting potential roots. For this purpose, a set of simple criteria will be applied to discard or confirm whether a cell may or may not be an object root. The checking process is done in a top-down manner on the cells pertaining to the upper levels of the links structure, i.e. those above the waist. Since the number of cells in these levels is quite small, the checking criteria are simple and the cells descending from a cell labeled as a root are not subject to testing, the computational load of the overall checking process is very low.

The checking criteria applied to every cell are:

- Area: All cells whose area is below a certain threshold will be discarded.
- Contrast: All cells whose gray level is very similar to the level of their neighbor cells are also disregarded.
- Compactness: All cells whose area to associated bounding box area ratio is under a certain compactness threshold are discarded.

Since all criteria imply assignations of thresholds, they must be controlled by the gaze control module and, if lower threshold values are assigned, adding also more restrictive discarding criteria must also be added [29], with a higher complexity, such as a connectivity criterion. Thus, only the cells passing the above-mentioned tests will be used as potential roots and the associated bounding boxes will determine the coordinates of new fixation points, i.e. the new foveae.

The images in Fig. 12 are an example of the attention step detecting the areas of interest as well as processing the object within the fovea.

- Fig. 12a shows the uniresolution image and the two rings ($m = 2$) associated with the fixation and fovea dimensions chosen in this case.
- Fig. 12b is the segmented retinal image, where the detected areas of interest, bounding boxes and centroids, as well as the corners of the object within the fovea are marked.
- Fig. 12c presents the levels from fovea up to level 5, i.e. waist+3, of the segmented retinal polygon, indicating the root labeled cells in levels 4 and 5.

8 Conclusions

In this chapter, we have summarized some of the work performed by our group, incorporating multiresolution techniques into hardware subsystems

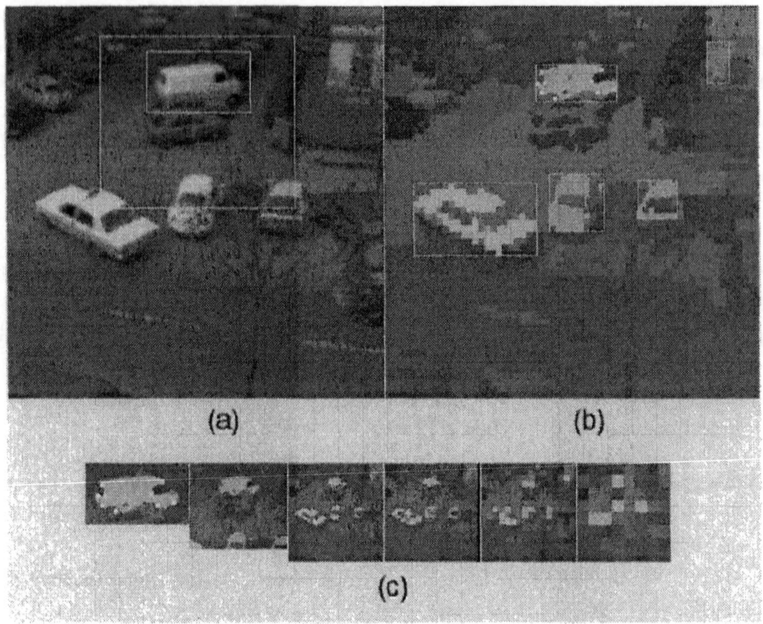

Fig. 12. a) Uniresolution image and resolution rings. b) Detected areas of interest on the retinal segmented image. c) Corners detection in fovea and roots detection in levels 4 and 5.

and software algorithms applied to systems operating in dynamic environments. As the systems front end, we use a multiresolution imager with adaptive characteristics to perform some of the recommended strategies for active vision: control of some camera parameters and foveal sensing.

The pyramid implementation is one of the most relevant imager functions. A similar software function would be computationally intensive, obtained with delay after receiving the base image, affecting processor activity in image processing and degrading system response. Besides, the bandwidth required to transmit full resolution images would be higher than required with the imager. In recent works for stereo applications, using a 32 I/O acquisition board, we have found that a VGA stereo imager with CMOS sensors can be implemented in a single 10K50E FPGA using 90% of its resources, with imager power consumption below 750 mW at 25 fps. However, for some applications, interface cables are a limiting constraint that, considering the low bandwidth required to transmit multiresolution data, we expect to overcome with a new imager design. Current work on the integration of hardware and software functions is oriented to segmenting motion, doing background extractions and robustly updating at the hardware level, using intermediate resolution pyramid levels and adapting algorithms used at applications level for edge detection and object motion.

Concerning the image processing execution times, as was to be expected, they are related to the number of cells in the hierarchical data associated with a fixation, i.e. the sizes of fovea and ROIs at the upper pyramid levels. Thus, working on a 900 MHz PC, scenes of 256×256 pixels processed with m=2 have processing times ranging from 50ms to 100ms for foveae sized 4Kpixels to 10Kpixels, while scenes of 512×512 pixels processed with m=3 and foveae from 4Kpixels up to 30Kpixels require up to 300ms.

Considering the processing delays of the fathers-sons linking algorithms, we are now implementing this algorithm in silicon, and expect to process hierarchical data structures in real-time [30].

Acknowledgements

This work has been partially funded by the Spanish Ministry of Science and Technology, MCYT and FEDER funds, under project TIC2001 - 1758.

References

1. Marr D (1982) Vision, MIT Press.
2. Swain MJ, Stricker M (1993) (eds.) Promising Directions in Active Vision. Int. Journal of Computer Vision 11:109-126.
3. Aloimonos JY, Weiss I, Bandyopadhyay A (1987) Active Vision. Int. Journal of Computer Vision 8:333-356.
4. Bajcsy R (1998) Active Perception. IEEE Proceedings, 76:996-1006.
5. Gremban KD, Ikeuchi K (1994) Planning Multiple Observations for Object Recognition. Int. Journal of Computer Vision 12:137-172.
6. Ahuja N, Abbot AL (1993) Active Stereo: Integrating Disparity, Vergence, Focus, Aperture and Calibration for Surface Estimation. IEEE Transactions on Pattern Analysis And Machine Intelligence 15:1007-1029.
7. Brunnstrm K, Eklundh J, Uhlin T (1996) Active Fixation for Scene Exploration. Intnl. Journal of Computer Vision 17:137-162.
8. Van der Spiegel J, Kreider G, Claeys C, Debusschere I, Sandini G, Dario P, Fantini F, Belluti P and Soncini G, (1989) A foveated retina-like sensor using CCD technology, In Mead C and Ismail M, (Eds.), Analog VLSI implementation of Neural Systems, Kluwer, 189-211.
9. Camacho P, Arrebola F, Sandoval F (1996) Shifted Fovea Multiresolution Geometries. IEEE Intnl. Conf. on Image Processing, ICIP-96, 1:307-310.
10. Santos-Victor J, Sandini G, Curotto F, Garibaldi S (1995) Divergent Stereo in Autonomous Navigation: From Bees to Robots. Int. Journal of Computer Vision 14:159-177.
11. Yamamoto H, Yeshurun Y, Levine M (1996) An Active Foveated Vision System: Attentional Mechanisms and Scan Path Convergence Measures. Computer Vision and Image Understanding 63:50-65.
12. Weiman CF, Chaikin G (1979) Logarithmic Spiral Grids for Image Processing. Computer Vision and Graphics Image Processing 11:197-226.

13. Kathman A, Johnson E (1992) Binary Optics: new diffractive elements for the designer tool kit. Photonic Spectra 9:125-132.
14. Camacho P, Arrebola F, Sandoval F (1998) Multiresolution Sensors with Adaptive Structure. 24th IEEE Intnl. Conf. Industrial Electronics, IECON'98, 2:1230-1235.
15. Rojer AS, Schwartz EL (1990) Design considerations for space-variant visual sensor with complex-logartihmic geometry. 10th Intnl. Conf. on Pattern Recognition 2:278-285.
16. Scott P, Bandera C (1990) Hierarchical Multiresolution Data Structures and Algorithms for Foveal Vision Systems. IEEE Intnl. Conf on System, Man and Cybernetics, 832-834.
17. Mead C (1989) Adaptive retina. In Analog VLSI implementation of Neural Systems, C. Mead and M. Ismail (eds.), Kluwer Acad. Publishers, N.Y. Chap. 8: 189-210.
18. Pardo F, Dierickx B, Scheffer D (1998) Space-Variant Non-orthogonal Structure CMOS Image Sensor Design. IEEE Journal of Solid-State Circuits 33:842-849.
19. Tistarelli M, Sandini G (1993) On the Advantages of Polar and Log-Polar Mapping for Direct Estimation of Time-to-impact from Optical Flow. IEEE Trans. PAMI, 15:401-410.
20. Capurro C, Panerai F, Sandini G (1997) Dynamic Vergence Using Log-Polar Images. Int. Journal of Computer Vision 24:79-94.
21. Bederson B, Wallace RS, Schwartz E (1994) A miniature pan-tilt actuator: the spherical pointing motor. IEEE Transactions on Robotics and Automation, 10a: 298-308.
22. Arrebola F, Urdiales C, Camacho P, Sandoval F (1998) Vision System Based on Shifted Fovea Multiresolution Retinotopologies. 24th IEEE Intnl. Conf. Industrial Electronics IECON'98, 3:1357-1361.
23. Camacho P, Arrebola F, Sandoval F (1997) Adaptive Fovea Structures for Space-variant Sensors, In A.del Bimbo (ed.) Image Analysis and Processing. Springer 1:422-429.
24. Jolion JM, Rosenfeld A, A Pyramid Framework for Early Vision, Kluwer Acad. Publishers, The Netherlands, 1994.
25. Arrebola F, Camacho P, Bandera A, Sandoval F (1999) Corner Detection and Curve Representation by Circular Histograms of Contour Chain Code. Electronics Letters 35: 1065-1067.
26. Bandera A, Urdiales C, Arrebola F, Sandoval F (1999) 2D Object Recognition Based on Curvature Functions Obtained from Local Histograms of the Contour Chain Code. Pattern Recognition Letters 20:49-55.
27. Burt PJ, Hong TH, Rosenfeld A (1981) Segmentation and Estimation of Image Region Properties Through Cooperative Hierarchical Computation IEEE Transactions on Systems, Man and Cybernetics 11:802:809.
28. Arrebola F (1998), Sistema Multirresolucin Basado en Imgenes Multirresolucin de Fvea Desplazable, (in Spanish) Ph. D. thesis, Malaga University.
29. Arrebola F, Camacho P, Sandoval F (1997) Generalization of Shifted Fovea Multiresolution Geometries Applied to Object Detection. In A. del Bimbo (ed.) Image Analysis and Processing. Springer 2:477-484.
30. Coslado F, Camacho P, Gonzlez M, Sandoval F (2001) Hardware Implementation of a Node Linking Segmentation Algorithm, Proc. XVI Conf. on Design of Circuits and Integrated Systems, DCIS2001, 654-659.

A Computer Vision Based Human-Robot Interface

José M. Buenaposada and Luis Baumela

Universidad Politécnica de Madrid, Departamento de Inteligencia Artificial,
Campus de Montegancedo s/n, 28660 Madrid, Spain
jmbuena@dia.fi.upm.es,lbaumela@fi.upm.es

Abstract. This chapter focuses on the real-time location and tracking of human faces in video sequences. The tracking is based on the cooperation of two low-level trackers based on colour and template information. The colour-based tracker is fast and robust, but it can only compute the 2D location of the face on the image. The template-based tracker, although it is more sensitive to environmental variations and more time consuming, it can compute the position and orientation of a human face in 3D space. As a result of the co-ordination of these two trackers, it emerges a robust real-time tracker that accurately computes face position and orientation in varying environmental conditions.

1 Introduction

The advent of Intelligent Service Robots has broadened the range of potential robot users from the traditional computer expert in an industrial setting to other professionals like firemen, doctors, police and even disabled or elderly people. These new users require a human-robot interface that allows the operator to interact with the robot in a "human-like" way, that is, by speaking (natural language understanding) and by visual observation of the user's face and gestures. This usage of different physical communication channels is what today we call "multimodal interaction." Building multimodal interfaces for human-robot interaction is a topic of intense research within the robotics community. Some researchers concentrate on natural language understanding [1], others have proposed vision-based interfaces that allow people to instruct robots via arm gestures [2,3], or by the integration of voice and vision [4,5].

The face is an important source of information for social interaction among humans, so it is also one of the most expressive elements of the human body, in terms of human-robot interaction. That is why facial expressions generation [6,7] and face location, tracking and expression analysis [8] are topics of interest for the creation of realistic humanoid robots. Face gestures have also been successfully used to control rehabilitation robots [9] and wheelchairs for handicapped users [10,11].

The development of a robust face tracker is a basic prerequisite for performing successful facial expression analysis. In this chapter we will focus on the real-time tracking and pose estimation of human heads in video sequences of head-and-shoulder images. The tracking is based on the co-operation of two trackers. A low-level colour-based tracker permits the location on the image of clusters of pixels of colour similar to the human skin. The second is a model-based tracker. It computes in real-time the position and orientation of a previously viewed planar patch. It performs reasonably well even for objects which, as a human face, are not perfectly planar.

The colour-based tracker is fast and robust to some extent to illumination changes, but it can only compute the 2D location of a cluster of coloured pixels on the image. On the other hand, the model-based tracker is able to decide whether a cluster of pixels is a known face. In that case it can compute the position and orientation of that face in 3D space. In our system both trackers co-operate in order to show a robust performance. The colour-based tracker is used as an initial estimate and as a recovery process when the more accurate and computer demanding model-based tracker cannot cope with face motion or with adverse conditions as face occlusion.

In the following sections we will describe the theoretical foundations of both trackers and test their performance with sequences of images taken in our laboratory.

2 Colour-based Face Tracking

Skin colour is the most frequently used feature for face detection and tracking [12]. The primary problem in automatic skin detection is colour constancy. The colour of an image pixel depends not only on the imaged object colour, but also on the lighting geometry, illuminant colour and camera response. This means that the RGB (Red, Green and Blue) values of a patch of skin can be very different depending on the camera used to capture the image, the colour and intensity of the illumination, the relative orientation between camera, skin surface and light source, or the existence of shadows or highlights in the image. Colour constancy algorithms model these effects and try to obtain colour invariants that facilitate the identification of a given colour under varying environmental conditions. For example [13], if the scene light intensity is scaled by a factor s, then each perceived pixel colour becomes $[sR, sG, sB]$. The rg-normalisation algorithm provides a colour constancy solution which is independent of the illuminant intensity:

$$[sR, sG, sB] \mapsto [\frac{sR}{s(R+G+B)}, \frac{sG}{s(R+G+B)}].$$

On the other hand, a change in illuminant colour can be modelled as a scaling α, β and γ in the R, G and B image colour channels. In this case the previous normalisation fails. The Grey World (GW) algorithm [13] provides a constancy solution independent of the illuminant colour by dividing each colour channel by its average value:

$$[\alpha R, \beta G, \gamma B] \mapsto [\frac{\alpha R}{\frac{\alpha}{n}\sum_i R}, \frac{\alpha G}{\frac{\beta}{n}\sum_i G}, \frac{\alpha B}{\frac{\gamma}{n}\sum_i B}].$$

The most widely used colour constancy algorithm in face tracking is rg-normalisation. As the skin colour distribution in the rg-normalised chromaticity space is Gaussian, a Bayesian classifier can be used to track fast and reliably a moving face [14]. Various improvements to this constancy model have been proposed in order to deal with small changes in the colour of the illumination. These consist on tracking the motion of the skin colour cluster in rg-space by using stochastic prediction models [15] or by clustering [16]. These algorithms have a problem in common: they cannot deal with sudden changes in lighting colour. Other algorithms that also work in rg-normalised space deal with sudden changes in lighting by matching the colour distributions by a shift in illuminant colour [17]. Unfortunately this solution is unfeasible for real-time tracking as it incurs in a substantial computational overhead.

Other colour constancy algorithms are more difficult to use in the analysis of a real-time image sequence, either because they need too much information, or because they are computationally too complex [18,19].

In this chapter we introduce a colour constancy algorithm that can be used for real-time colour-based image segmentation. The algorithm is based on GW and exploits the redundancy of the image sequence in order to compute the relative change in illumination between the images of the sequence. As shown in the experiments conducted, these constancy algorithms are clearly more robust to big sudden illuminant colour changes than the popular rg-normalised algorithm.

2.1 Grey World-based Colour Constancy

Colour constancy is the perceptual ability to assign the same colour to objects under different lighting conditions. The goal of any colour constancy algorithm is to transform the original $[RGB]$ values of the image into constant colour descriptors. In the case of Lambertian surfaces, the colour of an image pixel $I(ij)$ can be modelled by a lighting geometry component s_{ij}, which scales the $[rgb]$ surface reflectance of every pixel independently, and three colour illuminant components (α, β, γ), which scale respectively the red, green and blue colour channels of the image as a whole [13]. The lighting geometry component accounts for surface geometry and illuminant intensity variations, while the colour illuminant components account for variations in the illuminant colour. According to this model, two pixels $I(ij)$ and $I(kl)$ of an image would have the following $[RGB]$ values: $[s_{ij}\alpha r_{ij}, s_{ij}\beta g_{ij}, s_{ij}\gamma b_{ij}]$, $[s_{kl}\alpha r_{kl}, s_{kl}\beta g_{kl}, s_{kl}\gamma b_{kl}]$, where $[r_{ij}, g_{ij}, b_{ij}]$ and $[r_{kl}, g_{kl}, b_{kl}]$ represent surface reflectance; i.e. real object colour, independent of the illuminant.

The GW algorithm proposed by Buchsbaum [20] assumes that the average surface reflectance in an image with enough different surfaces is grey. So, the average reflected intensity corresponds to the illuminant colour, which can be used to compute the colour descriptors. This algorithm was refined in [21] by actually obtaining an average model of surface reflectance and proposing a procedure to compute the average image reflectance. On the basis of this, the colour normalisation proposed by GW consists on dividing each colour channel by its average value:

$$[s_{ij}\alpha r_{ij}, s_{ij}\beta g_{ij}, s_{ij}\gamma b_{ij}] \mapsto \left[\frac{\alpha s_{ij} r_{ij}}{\frac{\alpha}{n}\sum_I s_{ij} r_{ij}}, \frac{\beta s_{ij} g_{ij}}{\frac{\beta}{n}\sum_I s_{ij} g_{ij}}, \frac{\gamma s_{ij} b_{ij}}{\frac{\gamma}{n}\sum_I s_{ij} b_{ij}}\right] \tag{1}$$

Equation (1) is what we call *Basic GW Normalisation.* It is invariant to illuminant colour variations (α, β, γ), but it has one important drawback: it assumes that s_{ij} is constant. This means that:

- It does not account for all illuminant intensity variations (some of these variations will cancel with the colour coefficients). In the sequel we will present a normalisation procedure that will account for most of the illuminant intensity variations.
- It fails when the surface reflectances of the scene (r_{ij},g_{ij},b_{ij}) vary, i.e. when new objects appear or disappear in the scene. This means that GW is only valid for static scenes. In section 2.2 we will introduce a dynamic extension to the GW algorithm that solves this problem using the redundant information available in an image sequence.

Let us define the *image average geometrical reflectance*, $\bar{\mu}$, as

$$\bar{\mu} = [\mu_r, \mu_g, \mu_b] = \left[\frac{1}{n}\sum_{ij\in I} s_{ij} r_{ij}, \frac{1}{n}\sum_{ij\in I} s_{ij} g_{ij}, \frac{1}{n}\sum_{ij\in I} s_{ij} b_{ij}\right],$$

where n is the number of image pixels. It represents the average $[RGB]$ image values, once the colour illuminant component has been removed.

If we assume that the average geometrical reflectance is constant over the image sequence, then the following normalisation, called *Projective GW Normalisation*, removes the illuminant colour and intensity changes:

$$\begin{aligned}[s_{ij}\alpha r_{ij}, s_{ij}\beta g_{ij}, s_{ij}\gamma b_{ij}] &\mapsto \left[\frac{\alpha s_{ij} r_{ij}}{\alpha \mu_r}, \frac{\beta s_{ij} g_{ij}}{\beta \mu_g}, \frac{\gamma s_{ij} b_{ij}}{\gamma \mu_b}\right] \mapsto \\ &\mapsto \left[\frac{s_{ij} r_{ij}}{s_{ij}(\frac{r_{ij}}{\mu_r} + \frac{g_{ij}}{\mu_g} + \frac{b_{ij}}{\mu_b})}, \frac{s_{ij} g_{ij}}{s_{ij}(\frac{r_{ij}}{\mu_r} + \frac{g_{ij}}{\mu_g} + \frac{b_{ij}}{\mu_b})}\right]\end{aligned} \tag{2}$$

This normalisation is still valid just for static scenes. In section 2.2 we will present a dynamic extension to GW in order to overcome this problem.

2.2 Face Tracking using Dynamic Grey World

Here we present a colour-based face tracking algorithm. First we will briefly describe how to track a coloured patch using simple statistics, afterwards the Dynamic GW (DGW) algorithm is presented.

Face Segmentation and Tracking using a Skin Colour Model

Given a sequence of colour images, building a face tracker is straight forward if we have a reliable model of the image colour distributions. Let I_{rgb} be the $[RGB]$ channels of image I, and let $p(I_{rgb}|skin)$ and $p(I_{rgb}|back)$ be the conditional colour probability density functions (pfds) of the skin and background respectively (we assume that background is anything that is not skin). Using the Bayes formula, the probability that a pixel with colour I_{rgb} be *skin*, $P(skin|I_{rgb})$, can be computed as follows:

$$P(skin|I_{rgb}) = \frac{p(I_{rgb}|skin)P_s}{p(I_{rgb}|skin)P_s + p(I_{rgb}|back)P_b},$$

where P_s and P_b are the *a priori* probabilities of *skin* and *background.* The transformation $\mathcal{T}(I_{rgb}) = 255 \times P(skin|I_{rgb})$ returns an image whose grey values represent the probability of being skin (see Fig. 1). Face tracking on this image can be performed with a mode seeking algorithm, like [22], by computing the position and orientation of the face colour cluster in each frame [23].

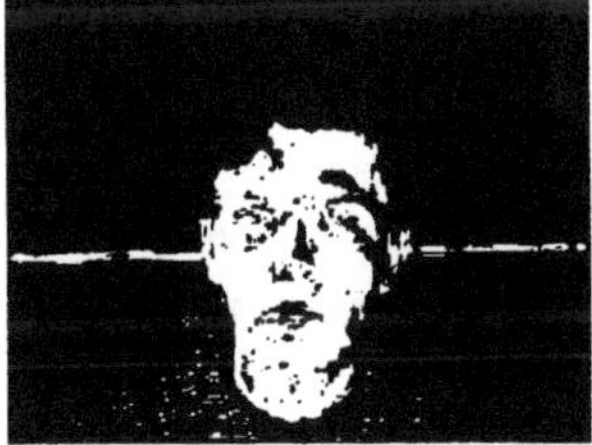

Fig. 1. Face segmentation based on skin colour. On the left is shown the colour image, on the right the probability image.

The problem now is to make the previous statistical model robust to variations in the scene illumination. This will be achieved by working in one of the normalised colour spaces presented in the previous section.

Various authors have indicated different preferences for modelling the colour distributions. In [15] Gaussian mixture models are used, whereas in [24] and [25] pure histogram-based representations are chosen. In our experiments we found that pure histogram-based models are faster and represent more accurately the skin colour distribution than continuous ones, if enough samples are available. On the other hand, continuous models are adequate when the number of samples is low.

The Projective GW Normalisation defines a two-dimensional colour space, $I_{\hat{r}\hat{g}}$, in which the colour distributions can be modelled with the skin, $h_s(I_{\hat{r}\hat{g}})$, and background, $h_b(I_{\hat{r}\hat{g}})$, colour histograms. Consequently, in our implementation we model the posterior probability as:

$$P(skin|I_{rgb}) = \frac{h_s(I_{\hat{r}\hat{g}})}{h_s(I_{\hat{r}\hat{g}}) + h_b(I_{\hat{r}\hat{g}})}.$$

The Dynamic Grey World Algorithm

The main problem of GW is that it was conceived for static images; i.e. it fails when there is a big change in the image average geometrical reflectance. In this section we propose a dynamic extension to GW (DGW) which will detect this situation and update the GW model.

In the following we assume that there exists a partition of the image sequence into a set of image sub-sequences such that the image average geometrical reflectance is constant over each sub-sequence; i.e. the basic GW algorithm can be used as a colour constancy criterion over each sub-sequence. We will use the first image of each sub-sequence as a *reference image*. The other images of the sub-sequence will be segmented using the colour descriptors of the reference image.

Let I^r_{rgb}, I^t_{rgb} and I^{t-1}_{rgb} be respectively the reference image, the present and the previous image, $F^r_{\hat{r}\hat{g}\hat{b}}$ be the face pixels in GW space, $\bar{\mu}^{I^t}_{rgb}$ be the average value for each colour channel in I^t_{rgb}, $\bar{\mu}^{F^r}_{\hat{r}\hat{g}\hat{b}}$ and $\bar{\mu}^{F^t}_{\hat{r}\hat{g}\hat{b}}$ be the average GW descriptors for the face pixels in the reference and present image respectively, and $\mathcal{E}$ be the statistical distribution of the GW space colour descriptors for the reference image.

The dynamic extension to the basic GW is based on the fact that when a change in the average geometrical reflectance ($\bar{\mu}$) is detected, the GW colour descriptors for the present image $I^t_{\hat{r}\hat{g}\hat{b}}$ can still be computed with the the average pixel values of the previous image, $\bar{\mu}^{I^{t-1}}_{rgb}$. In this situation we segment the present image with the the average pixel values of the previous one, and let the present image be the reference image.

The problem now is how to detect a change of sub-sequence. We do this just by searching for a change in the average geometrical reflectance. This cannot be accomplished on the basis of analysing $\bar{\mu}^{I}_{rgb}$, as it also changes with the illuminant colour. We solve this problem by monitoring the average GW descriptors of the face pixels. As they are invariant to illuminant colour changes, a change in these descriptors is necessarily caused by a change in average geometrical reflectance.

Based on these ideas in Fig. 2 we propose the Dynamic-GW (DGW) algorithm.

```
Initialisation
  /*Initialise the reference image model using motion
  segmentation and a precalculated colour model*/
  [ℰ, \bar{\mu}^{F^r}_{\hat{r}\hat{g}\hat{b}}] = InitTracking();
While (true)   /* tracker main loop */
  \bar{\mu}^{I^t}_{rgb} = Mean(I^t_{rgb}); /* image mean rgb values */
  I^t_{\hat{r}\hat{g}\hat{b}} = I^t_{rgb} / \bar{\mu}^{I^t}_{rgb}   /* Proj GW normalisation */
  F^t_{\hat{r}\hat{g}\hat{b}} = ProbabilisticSegment(I^t_{\hat{r}\hat{g}\hat{b}}, ℰ); /* segment img */
  \bar{\mu}^{F^t}_{\hat{r}\hat{g}\hat{b}} = ComputeAvgFaceGW(F^t_{\hat{r}\hat{g}\hat{b}}); /* face avg GW descriptors */
  If ||\bar{\mu}^{F^r}_{\hat{r}\hat{g}\hat{b}} - \bar{\mu}^{F^t}_{\hat{r}\hat{g}\hat{b}}|| > Δ then /* change of subsequence */
    I^t_{\hat{r}\hat{g}\hat{b}} = I^t_{rgb} / \bar{\mu}^{I^{t-1}}_{rgb}   /* Proj GW normalise with previous mean */
    F^t_{\hat{r}\hat{g}\hat{b}} = ProbabilisticSegment(I^t_{\hat{r}\hat{g}\hat{b}}, ℰ); /* segment image */
    I^r_{\hat{r}\hat{g}\hat{b}} = I^t_{\hat{r}\hat{g}\hat{b}} /* update reference image */
    \bar{\mu}^{F^r}_{\hat{r}\hat{g}\hat{b}} = ComputeAvgFaceGW(F^t_{\hat{r}\hat{g}\hat{b}}); /* face GW descriptors */
    [ℰ] = ColourDistrib(F^t_{\hat{r}\hat{g}\hat{b}}); /* ref. colour distrib */
  end /* if */
end /* while */
```

Fig. 2. Dymanic Grey World Algorithm

3 Model-based Face Tracking

In this section we will present a model-based tracking method to estimate in real-time the position and orientation of a previously viewed planar patch. We will show that this planar tracking procedure can be used to track a human face.

Tracking planar patches is a subject of interest in computer vision, with applications in mobile robot navigation [26], augmented reality [27], face tracking [28], or the generation of super-resolution images [29]. Traditional approaches to tracking are based on finding correspondences in successive images. This can be achieved by computing optical flow [30] or by matching a sparse collection of features [31]. In flow-based methods, a velocity vector is computed for each pixel, while in feature-based methods, image features, such as points and lines are matched across all frames in the sequence. Feature-based methods minimise an error measure based on geometrical constraints between a few corresponding features, while direct methods minimise an error measure based on direct image information collected from all pixels in the image, such as image brightness.

The tracking method presented in this section belongs to the first group of methods. It is based on minimising the sum-of-squared differences (SSD) between a selected set of pixels obtained from a previously stored image of the tracked patch (image template) and the current image of it. It extends

Hager's SSD tracker [32] by introducing a projective motion model, a method to compute the position and orientation of the tracked patch and a procedure to select the set of pixels used in tracking.

3.1 SSD Plane Tracking

Let P be the image of planar object. Assuming no changes in the scene illumination, the following constancy equation holds:

$$I(\bar{x}, t_0) = I(f(\bar{x}, \bar{\mu}), t_n) \forall \bar{x} \in P, \tag{3}$$

where $I(\bar{x}, t_0)$ is the template image of P and $I(f(\bar{x}, \bar{\mu}), t_n)$ is the rectified image at time t_n, with motion model $f(\bar{x}, \bar{\mu})$ and motion parameters $\bar{\mu}$.

The motion parameter vector $\bar{\mu}$ can be estimated from (3) by minimising the difference between the template and the rectified image:

$$\min_{\bar{\mu}} \left(\sum_{\bar{x} \in P} (I(f(\bar{x}, \bar{\mu}), t_n) - I(\bar{x}, t_0))^2 \right). \tag{4}$$

This minimisation problem can be solved linearly by computing $\bar{\mu}$ incrementally while tracking. We can achieve this by making a Taylor series expansion of (4) about ($\bar{\mu}$, t_n) and computing the increment, $\delta\mu$, between two time instants [32]:

$$\delta\bar{\mu} = -(M^\top M)^{-1} M^\top [I(\bar{x}, \bar{\mu}_n) - I(\bar{x}, \bar{\mu}_0)] \tag{5}$$

where M is the Jacobian matrix of the image and dependence of I on t has been dropped for convenience.

While tracking, matrix M must be re-calculated in each frame, as it depends on $\bar{\mu}$. This is computationally expensive, as M is of dimension $N \times n$, being N is the number of template pixels and n the number of motion parameters. In the sequel we will factor M in order to simplify this computation.

M can be written as

$$M(\bar{\mu}) = \begin{pmatrix} \nabla_x I(\bar{x}_1, \bar{\mu}_0)^\top f_x(\bar{x}_1, \bar{\mu})^{-1} f_\mu(\bar{x}_1, \bar{\mu}) \\ \nabla_x I(\bar{x}_2, \bar{\mu}_0)^\top f_x(\bar{x}_2, \bar{\mu})^{-1} f_\mu(\bar{x}_2, \bar{\mu}) \\ \vdots \\ \nabla_x I(\bar{x}_N, \bar{\mu}_0)^\top f_x(\bar{x}_N, \bar{\mu})^{-1} f_\mu(\bar{x}_N, \bar{\mu}) \end{pmatrix}, \tag{6}$$

where $\nabla_x I$ is the template image gradient, f_x is the derivative of the motion model with respect to the pixel coordinates and f_μ is the derivative of the motion model with respect to the motion parameters.

Depending on the motion model, M may be factored into the product of two matrices,

$$M(\bar{\mu}) = \begin{pmatrix} \nabla_x I(\bar{x}_1, \bar{\mu}_0)^\top \Gamma(\bar{x}_1) \\ \nabla_x I(\bar{x}_2, \bar{\mu}_0)^\top \Gamma(\bar{x}_2) \\ \vdots \\ \nabla_x I(\bar{x}_N, \bar{\mu}_0)^\top \Gamma(\bar{x}_N) \end{pmatrix} \Sigma(\bar{\mu}) = M_0 \Sigma(\bar{\mu}) \tag{7}$$

a constant matrix M_0 of dimension $N \times n$ and a matrix Σ of dimension $n \times n$, that depends on $\bar{\mu}$. As M_0 can be precomputed, this factorisation reduces the on line computation to the inversion of matrix Σ:

$$\delta\bar{\mu} = -\Sigma^{-1}(M_0^\top M_0)^{-1} M_0^\top [I(\bar{x}, \bar{\mu}_n) - I(\bar{x}, \bar{\mu}_0)]. \tag{8}$$

Matrix M_0 is the Jacobian of the template image. It is our *a priori* knowledge about target structure, that is, how the grey level value of each pixel changes as the object moves. It represents the information provided by each template pixel to the tracking process. Note that we cannot track any object, as in order to solve (7), a non singular $M_0^\top M_0$ matrix is needed.

3.2 Projective Model for Plane Tracking

Here we introduce a projective model of target motion. In order to do this, we need to obtain the Jacobian matrix decomposition that arises from this model.

Let $\bar{x} = (u, v)^\top$ and $\bar{x}_h = (r, s, t)^\top$ be respectively the Cartesian and Projective coordinates of an image pixel. They are related by:

$$\bar{x}_h = \begin{pmatrix} r \\ s \\ t \end{pmatrix} \rightarrow \bar{x} = \begin{pmatrix} r/t \\ s/t \end{pmatrix} = \begin{pmatrix} u \\ v \end{pmatrix}; \; t \neq 0. \tag{9}$$

The function f that describes the motion of a planar region is then a 2D projective linear transformation,

$$f(\bar{x}_h, \bar{\mu}) = H\bar{x}_h = \begin{pmatrix} a & d & g \\ b & e & h \\ c & f & 1 \end{pmatrix} \begin{pmatrix} r \\ s \\ t \end{pmatrix}, \tag{10}$$

where the motion parameters are $\bar{\mu} = (a, b, c, d, e, f, g)^\top$.

The Jacobian matrix decomposition of this motion model can be expressed in terms of the elements of (6):

$$\nabla_{x_h} I(\bar{x}_h, \bar{\mu}_0)^\top = \left(\frac{\partial I}{\partial u}, \frac{\partial I}{\partial v}, -\left(u\frac{\partial I}{\partial u} + v\frac{\partial I}{\partial v} \right) \right) \tag{11}$$

$$f_x(\bar{x}_h, \bar{\mu})^{-1} = H^{-1} \tag{12}$$

$$f_\mu(\bar{x}_h, \bar{\mu}) = \begin{pmatrix} r\,0\,0\,s\,0\,0\,t\,0 \\ 0\,r\,0\,0\,s\,0\,0\,t \\ 0\,0\,r\,0\,0\,s\,0\,0 \end{pmatrix} \tag{13}$$

Introducing (11), (12) and (13) into (6) M can be factored according to (7):

$$f_x(\bar{x}, \bar{\mu})^{-1} f_\mu(\bar{x}, \bar{\mu}) =$$

$$H^{-1}\left(rI_{3\times 3} \mid sI_{3\times 3} \mid \frac{tI_{2\times 2}}{0_{1\times 2}}\right) =$$
$$\left(rH^{-1} \mid sH^{-1} \mid tH_{\bar{12}}^{-1}\right) = \Gamma(\bar{x}_h)\Sigma(\bar{\mu}),$$

where $H_{\bar{12}}^{-1}$ is the matrix composed with the first two columns of H^{-1}, $I_{q\times q}$ is the $q \times q$ identity matrix and

$$\Gamma(\bar{x}_h) = \left(rI_{3\times 3} \,|\, sI_{3\times 3} \,|\, tI_{3\times 3}\right), \quad \Sigma(\bar{\mu}) = \begin{pmatrix} H^{-1} & 0 & 0 \\ 0 & H^{-1} & 0 \\ 0 & 0 & H_{\bar{12}}^{-1} \end{pmatrix}.$$

With this factorisation we can projectively track a planar patch with the computational cost of inverting an 8×8 matrix on each frame.

3.3 3D Pose Estimation

The tracking model presented previously computes the homography H_0^n between the present image and the stored template. In this section we are going to show that it is possible to estimate the pose of the tracked patch from H_0^n, if the camera acquiring the image sequence is calibrated.

So far we have only computed 2D information. In order to have 3D information we have to compute two more homographies: one from P to the image plane at t_0, H_W^0, and the other from P to the image plane at t_n, H_W^n (see Fig. 3).

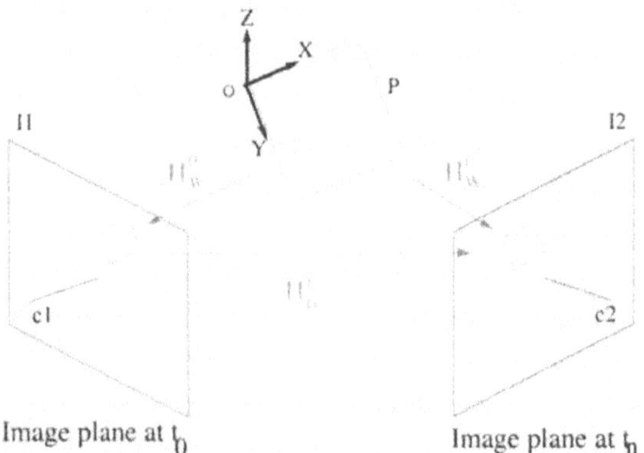

Fig. 3. Projective transformations involved in 3D plane tracking.

In order to simplify these equations we choose the scene coordinate system to have the X and Y axis on the plane P and the Z axis perpendicular to it (see Fig. 3).

Homography H_W^0 can be computed off line using the projection of, at least, four known points on P. Let $(X_P, Y_P)^\top$ be the Cartesian coordinates of a known point in P and let $(x_0, y_0)^\top$ be Cartesian coordinates of the projection of $(X_P, Y_P)^\top$ onto I at t_0 (i.e. at the template image). H_W^0 can be computed from:

$$\begin{pmatrix} x_0 \\ y_0 \\ 1 \end{pmatrix} = H_W^0 \begin{pmatrix} X_P \\ Y_P \\ 1 \end{pmatrix}. \tag{14}$$

On the other hand, the projection of point $(X_P, Y_P)^\top$ onto I at time instant t_n is given by

$$\begin{pmatrix} x_n \\ y_n \\ 1 \end{pmatrix} = \lambda K \, [R \,|\, t] \begin{pmatrix} X_P \\ Y_P \\ 0 \\ 1 \end{pmatrix}, \tag{15}$$

where R and t are respectively the orientation and the position of P in the camera coordinate system, λ is a scale factor and K is the camera instrinsics matrix.

Introducing in (15) the fact that all points of P have coordinate $Z = 0$, H_W^n can be written as:

$$\begin{pmatrix} x_n \\ y_n \\ 1 \end{pmatrix} = \lambda K \, [\bar{r}_1 \, \bar{r}_2 \, t] \begin{pmatrix} X_P \\ Y_P \\ 1 \end{pmatrix} = H_W^n \begin{pmatrix} X_P \\ Y_P \\ 1 \end{pmatrix} \tag{16}$$

where $\bar{r}_i$ is the ith column of matrix R.

Now, from (14) and (16)

$$\begin{pmatrix} x_n \\ y_n \\ 1 \end{pmatrix} = \underbrace{H_W^n (H_W^0)^{-1}}_{H_0^n} \begin{pmatrix} x_0 \\ y_0 \\ 1 \end{pmatrix} = \underbrace{\lambda K \, [\bar{r}_1 \, \bar{r}_2 \, t] (H_W^0)^{-1}}_{H_0^n} \begin{pmatrix} x_0 \\ y_0 \\ 1 \end{pmatrix} \tag{17}$$

From which we obtain the relation between the homography computed in the previous section, H_0^n, and the pose of P. So, if the intrinsics K and the homographies H_W^0 and H_0^n are known, we can compute H^* [27],

$$H^* = K^{-1} H_0^n H_W^0 = \lambda [\bar{r}_1 \, \bar{r}_2 \, t] \tag{18}$$

The translation is obtained directly from the third column of H^* but in order to obtain the rotation matrix we still have to impose some constraints:

- $||\bar{r}_1|| = ||\bar{r}_2|| = 1$, as R is a rotation matrix. In this way we get $\widehat{r}_1$ and $\widehat{r}_2$.
- $\bar{r}_3 \perp \bar{r}_1$ and $r_3 \perp \bar{r}_2$, from where we get $\widehat{r}_3$.

3.4 Template Pixel Selection

Only areas of high image contrast provide information about template motion (see Fig. 4, only the white pixels on the right image provide information for tracking). If in (8) we used all template pixels, most of the computational effort would be devoted non-informative pixels.

Fig. 4. Images of a template (left) and of $I(\bar{x}, \bar{\mu}_n) - I(\bar{x}, \bar{\mu}_0)$ (right), where parameter $\bar{\mu}$ represents a horizontal displacement.

In this section we will further improve the tracking procedure presented in the previous section by reducing the number of template pixels used for solving equation (8). This improvement comes not only from having a smaller matrix M_0, but mainly from diminishing the number of pixels warped to compute $I(\bar{x}, \bar{\mu}_n)$.

The Jacobian matrix M of image I can be expressed as:

$$M = (I_{\mu_1}, I_{\mu_2}, \cdots, I_{\mu_n}), \tag{19}$$

where $I_{\mu_i} = \frac{\partial I(\bar{x}, \bar{\mu})}{\partial \mu_i}$ is a column vector with an entry for every pixel in I. It represents the changes in image brightness induced by motion μ_i (see Fig. 5). Thus, M relates variations in motion parameters to variations in brightness values. Note that (8) works in the opposite direction, i.e. it uses M to compute motion from observed changes in brightness values.

Let us call $I_{\bar{\mu}}^{\top}(\bar{x})$ the row in M corresponding to image pixel $I(\bar{x})$. Each row entry is the derivative of image pixel $I(\bar{x})$ with respect to a model parameter μ_i ($\forall i = 1 \ldots n$). Intuitively, a pixel with a small $||I_{\bar{\mu}}(\bar{x})||$ provides almost no information for solving (8). So, a good pixel for tracking is one with a large $||I_{\bar{\mu}}(\bar{x})||$. Given two image pixels $I(\bar{x}_1)$ and $I(\bar{x}_2)$, one of them is redundant if $I_{\bar{\mu}}(\bar{x}_1) \approx I_{\bar{\mu}}(\bar{x}_2)$. So, a good set of pixels for tracking is one such that $M^{\top}M$ is not singular.

Selecting the "best" set of m pixels is a combinatorial search problem, as all $\binom{m}{N}$ sets of pixels should be considered in order to select the most informative one. In the context of image registration, Dellaert selects m pixels randomly from the top 20% of pixels with highest $||I_{\bar{\mu}}(\bar{x})||$ [33]. In our experiments we have found that the best set of pixels for tracking is the one with highest $||I_{\bar{\mu}}(\bar{x})||$, lowest redundancy and most even distribution on the

Fig. 5. Jacobian matrix for a translation (x, y), rotation (θ) and scale (s) motion model. In reading direction each image represents respectively I_x, I_y, I_θ, I_s.

image. In the sequel we will present a procedure to select a set of pixels with high $||I_{\bar{\mu}}(\bar{x})||$ and low redundancy.

If we consider each row vector $I_{\bar{\mu}}(\bar{x})$ as a point in n-dimensional space, then the points in the convex hull of this cloud are those with highest $||I_{\bar{\mu}}(\bar{x})||$ and lowest redundancy. Let us call this set of points the *Jacobian cloud.* Computing the convex hull of a Jacobian cloud with thousands of points in an 8-dimensional space (the projective motion model has 8 parameters) can be time consuming. On the other hand, as can be seen in Fig. 6 (right), the distribution of points for this model is highly correlated, with two space directions representing 99.96% of the total variance in the cloud. So, a good approximation to the convex hull of the cloud would be to compute the convex hull of its projection onto the two main directions (see Fig. 6, left).

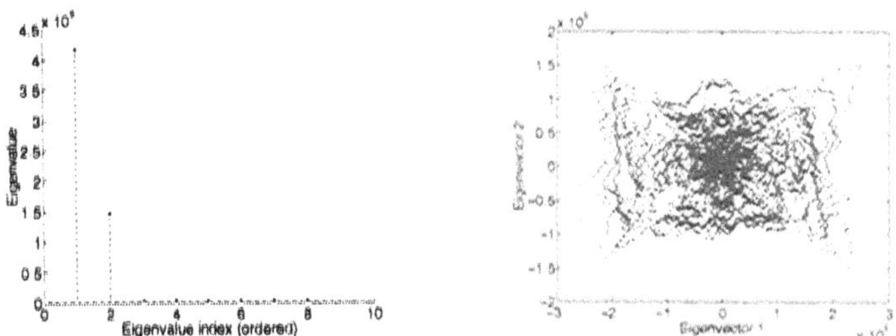

Fig. 6. Eigenvalues of the Jacobian cloud's covariance matrix (left) and view of the projection of the Jacobian cloud onto the two principal directions (right).

If we choose the points from the outer convex hulls (like peeling off an orange) then only the pixels in the strongest edges of the image would be selected. In order to achieve a more even spatial distribution of the selected pixels we choose all pixels of a randomly selected set of convex hulls from the outer 30% of them (see Fig. 7).

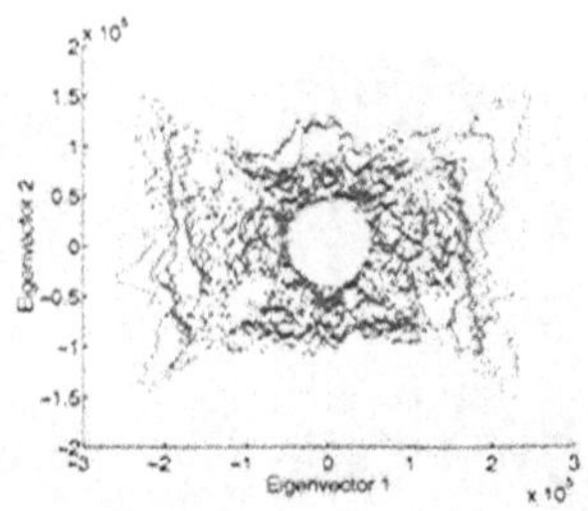

Fig. 7. Points in the outer 30% convex hulls in projected space.

4 Experiments

In our experiments we used a VL500 Sony colour digital camera at 320×240 resolution, iris open, no gain, no gamma correction. The system runs on an AMD K7 at 750MH under GNU/Linux. Images were taken with regular roof fluorescent lights and variations in illumination colour were obtained using a controlled tungsten light.

In the first experiment we validate the hypothesis on which the DGW algorithms rests, namely: variations in the average geometrical reflectance can be detected, and the reference image of each sub-sequence can be segmented. We acquired a sequence of 200 images with a green object appearing at one point and illuminant geometrical variations taking place at a different moment. The result of this experiment is shown in Fig. 8. Four images of the sequence are shown, each in one column of the figure. Each of them represents respectively the first image of the sequence (image 1), a change in the illuminant (roof lights turned off) (image 26), and the appearance and disappearance of an object (images 88 and 139). In this experiment the system detects three sub-sequences (1 to 87, 88 to 138, and 139 to 200). This is clearly visible in the plot at the bottom of Fig. 8. In image 26 the roof fluorescent lights are turned off. This geometrical illumination variation can be perceived again in the face GW descriptors plot. In this case the segmentation is good. This is an example of "worst case" test. In similar situations with stronger variations in the illuminant geometry, the system may not be able to segment the image and eventually may loose the target. Images 88 and 139 show the first segmented image in the two last sub-sequences, that coincide with the appearance and disappearance of an object in the image. Here we can see how the system detects a change of sub-sequence and correctly segments the images.

The goal of the next experiment is to check that the dynamic extension to GW is necessary; i.e. we want to check what would happen if we segment the previous sequence with the Basic GW Normalisation without the dynamic extension. In Fig. 9 the same sequence as in Fig. 8 is used and the same images are shown. We can clearly perceive that without the dynamic extension, the initial colour model is invalid when a change in the image average

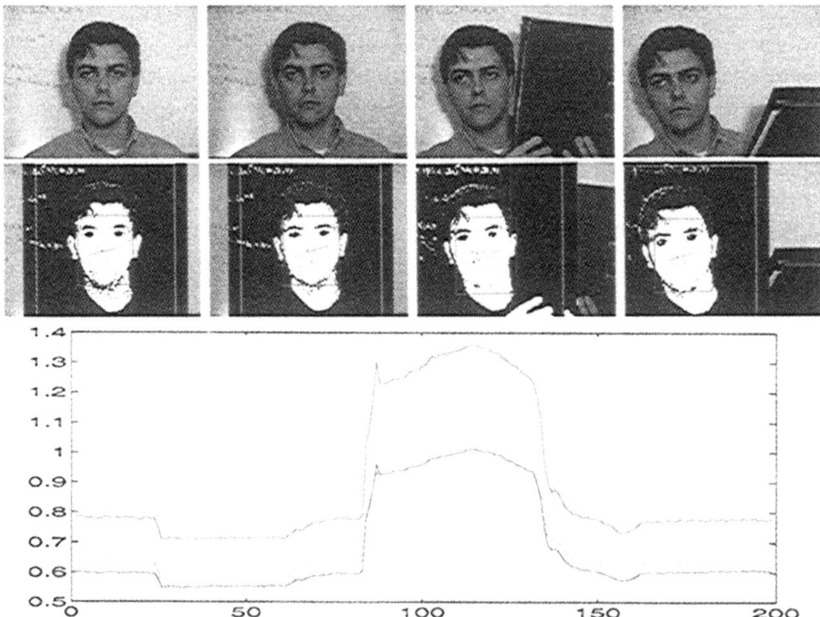

Fig. 8. Hypothesis validation experiment. On the first row four images of a sequence are shown. Their segmentation with the DGW algorithm using the Projective GW normalisation are presented on the second row. The average r,g and b face GW descriptors (in red, green and blue colour respectively) are shown on the third row.

geometrical reflectance (caused by the appearance of an object) takes place. The initial model gradually becomes valid again as the object disappears (see last column).

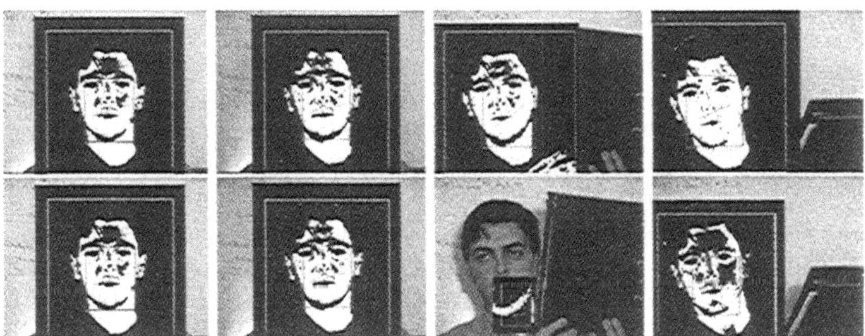

Fig. 9. DGW algorithm versus Basic GW without dynamic extension. Basic GW Normalisation with DGW algorithm results are shown in first row. Basic GW without dynamic extension is shown in the second row.

In the following experiment we compare the performance of the DGW algorithm with the rg-normalised colour constancy algorithm. We use a sequence with a set of images with "difficult" background (i.e. brownish books and shelves to distract the segmentation). Four frames of the sequence are shown in each column of Fig. 10, representing each one of them following situations: initial image, tungsten frontal light turned off, tungsten frontal light partially turned on, green object is introduced. Raw images are shown in the first row, rg-normalised results in the second row, and DGW segmentation results with Projective GW Normalisation are shown in the third row. Visual inspection of the results on the second and third row show a clear success of the DGW compared to the rg-normalisation when the illuminant colour abruptly changes (second column).

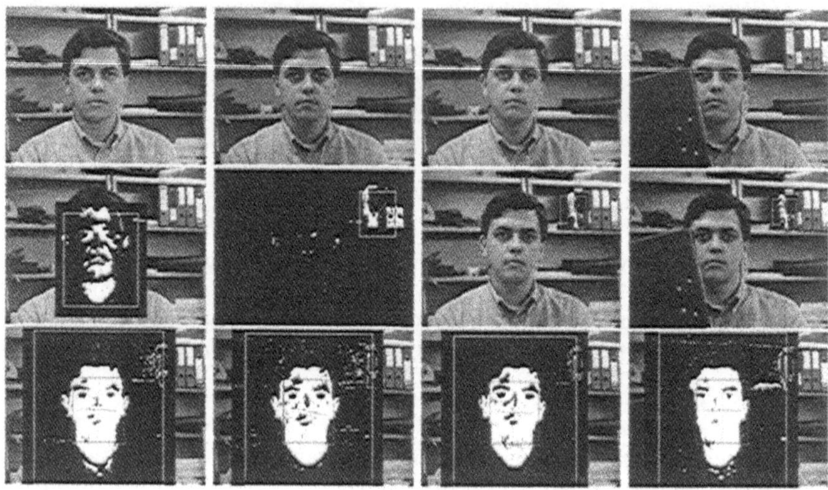

Fig. 10. Comparison of DGW and RG-normalisation colour constancy for face tracking.

For the pixel selection experiments we track a template of 149×104 pixels, shown on the left in Fig. 4. In the first experiment we study the gain in throughput achieved by tracking the template using only 407 pixels, instead of the 15.496 pixels of the full template. As can be seen in Fig. 11 (left), using pixel selection the system runs one order of magnitude faster.

Next we compare the performance of the pixel selection procedure presented in section 3.4 with Dellaert's method and with full frame tracking. The plots of the RMS tracking residual for full frame tracking and for a tracker using 407 pixels selected with the two methods discussed in this chapter are shown in Fig. 11 (right).

In the next experiment (see Fig. 12) we study the evolution of the average frame tracking residual as the number of pixels used in tracking increases.

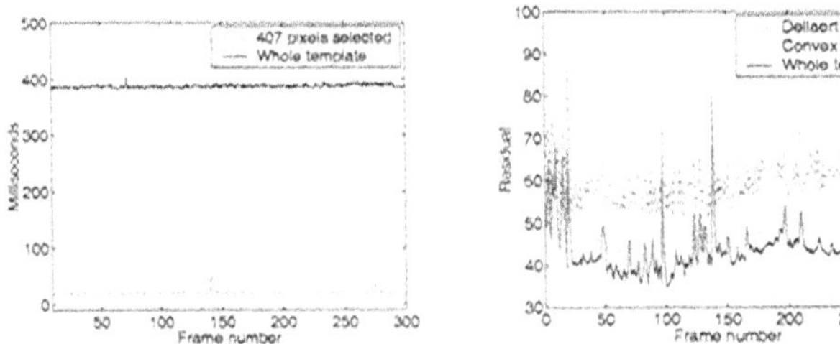

Fig. 11. The gain in system throughput achieved by using pixel selection is shown on the left plot. The tracking residual for different pixel selection procedures is shown on the right plot.

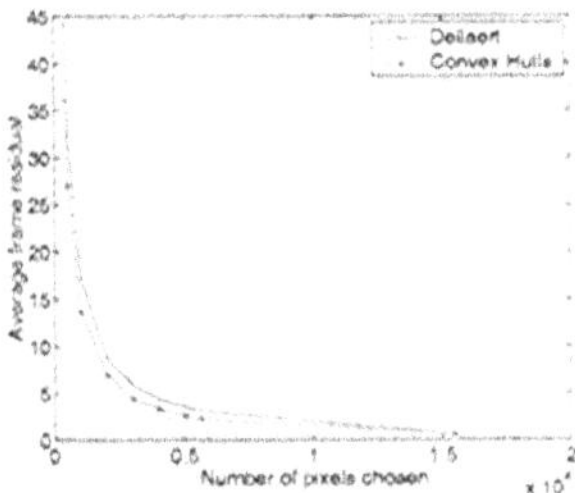

Fig. 12. Evolution of the average tracking residual for different numbers of selected pixels and different selection procedures.

These results show that by adequately selecting the pixels used in tracking, the amount of computation per frame can be reduced in one order of magnitude. The penalty that we pay for this improvement in processing time is an increase of about 20% in the tracking residual.

Finally, we are going to validate the 3D plane tracking algorithm with three more experiments: in the first one the target is the template used for the pixel selection experiments (see Fig. 13) in the second sequence we track a book hardcover (see Fig. 14) and in the third sequence we show that the algorithm can also track to some extent non planar objects, for example the human face (see Fig. 15). We have validated the model-based tracker by overlaying the coordinate axes of the tracked planar patch over the image, in this way we can get an indirect perception of the accuracy of pose estimation. As can be seen in the results presented in Figs. 8, 9, and 10, in all cases the axes over the image are coherent with the plane motion, except when the SSD homography estimation has less precision, as in the case of the head motion up and down.

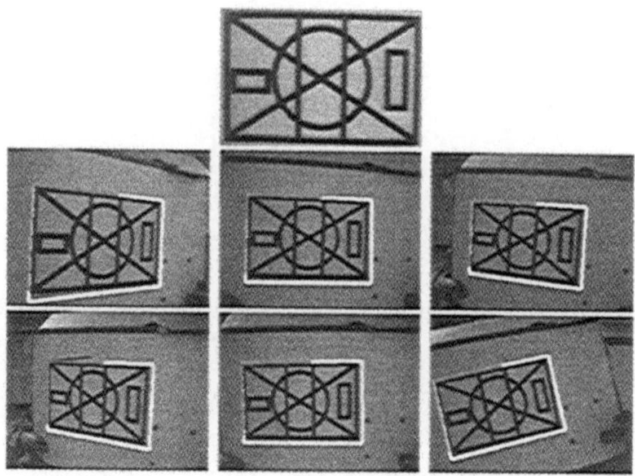

Fig. 13. In the first row: template image. Second and third rows, frames 15, 85, 160, 200, 235, 280 and 330 of a total of 400 in the sequence. The red rectangle is the position estimated from the model-based tracker. The 3D pose estimation can be perceived from the plane axes: Z axis in white, Y axis in blue and X axis in green.

Fig. 14. In the first row: template image. Second and third rows, frames 5, 35, 84, 115, 170, 200 and 250 of 350

5 Conclusions

In this chapter we have presented a system to track in real-time the position and orientation of human faces in video sequences. It is based on the co-operation of a low level colour-based tracker and a model-based tracker.

The colour-based tracker is based on the GW algorithm. We have presented a dynamic extension of the GW algorithm and a Projective GW Nor-

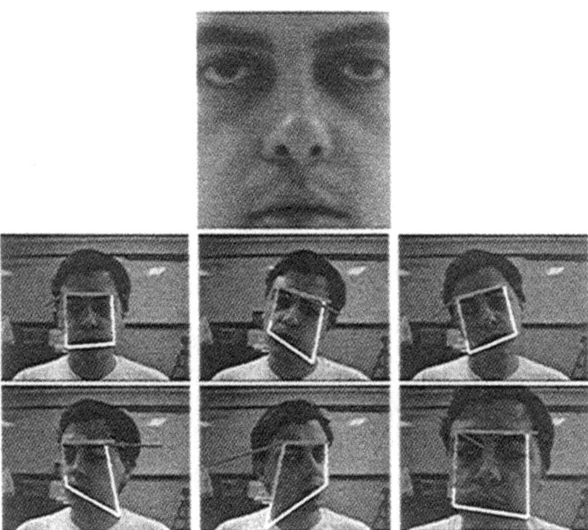

Fig. 15. First row: template image. Second and third rows, frames 37, 99, 130, 193, 233 and 289 of a total of 350 in a face moving sequence.

malisation procedure. These extensions of GW were designed to make it work in real-time with sequences of images with varying environmental conditions. In the experiments conducted, the DGW algorithm performed better than other normalisation algorithms when sudden changes in the illuminant colour take place.

The least favourable case for the algorithm occurs when strong changes in the illuminant geometry take place. In this case, the average geometrical reflectance is not constant and consequently the normalisation procedure is not successful. Also, in general, colour-based tracking has difficulty with similarly coloured distractions in the background. More research is needed in order to design normalisation procedures invariant to these changes. Another way to make this low level tracker more robust to environmental changes is introducing information which is not affected by these changes. We are now studying a low level tracker that combines colour information with motion.

In spite of these limitations, colour-based trackers are good as an initial estimate or follow-up verification of face location in the image plane or as a recovery process when more accurate and computer demanding trackers cannot cope with face motion.

The model-based tracker is based on minimising the sum-of-squared differences (SSD) between the image of a planar patch and a previously stored image of it. We have introduced a linear model for projectively tracking the planar patch, a method to compute the patch position and orientation in 3D space, and a procedure to speed up tracking by selecting only a special set of pixels from the tracked template. The pixel selection procedure introduced increases in one order of magnitude the speed at which frames are processed.

Being able to track a planar patch using a small set of pixels is important not only because of the increase in processing speed, but also because in this way we will be able to track regions of arbitrary shape. In the present implementation the system works at 24 frames per second for the full template tracker using Intel's IPL warping routines[1]. The results shown in the experimental section were obtained using a software not fully optimised. We are in the process of writing MMX-optimised routines for warping a selected set of pixels.

The spatial distribution of selected pixels on the image is an interesting line of research. We think that the performance of the tracker can be improved by evenly distributing the selected pixels in the image. This issue is specially important if we want to consider tracking with partial template occlusions.

The tracking system presented in this chapter could be used as a vision module in a multimodal interface for controlling a robot, like in [9–11], or as the substrate on which a facial expression analysis program can be constructed.

Acknowledgements

The authors gratefully acknowledge the Spanish *Comisión Interministerial de Ciencia y Tecnología* (CICyT) for funding this research under contract number TIC1999-1021. José Miguel Buenaposada was also funded by a FPU grant from the Spanish Ministry of Education.

References

1. H. Asoh, S. Hayamizu, I. Hara, Y. Motomura, S. Akao and T. Matsui. "Socially Embedded Learning of Office Conversant robot jijo-2," in *Proc. Int. Joint Conference on Artificial Intelligence*, 1997.
2. S. Waldherr, R. Romero, S. Thrun. "A Gesture-based Interface for Human-Robot Interaction," *Autonomous Robots*, 9, pp. 151-173. 2000.
3. R. Cipolla and N.J. Hollinghurst. "Human-Robot Interface by Pointing with Uncalibrated Stereo Vision," *Image and Vision Computing*, 14(3), pp. 171-178. 1996.
4. D. Perzanowski, A.C. Schultz, W. Adams, E. Marsh, M. Bugajska. "Building a Multimodal Human-Robot Interface," *IEEE Intelligent Systems*, pp. 16-21, Jan/Feb 2001.
5. Y. Yoshitomi, S-I. Kim, T. Kawao ans T. Kitazoe. "Effect of Sensor Fusion for Recognition of Emotional States Using Voice, Face Image and Thermal Image of Face," in *Proc. IEEE Int. Workshop on Robot and Human Interactive Communication*, Paris, France. 2001.

[1] `http://developer.intel.com/software/products/perflib/ipl/`

6. L. Cañamero and J. Fredslund. "I Show You How I Like You–Can You Read it in My Face," *IEEE Trans. on Systems, Man and Cybernetics-A*, 31(5), pp. 454-459. 2001.
7. H. Kobayashi, Y. Ichikawa and T. Tsuji. "Face Robot – Toward Realtime-Rich Facial Expressions," in *Proc. IEEE Int. Workshop on Robot and Human Interactive Communication*, Paris, France. 2001.
8. B. Scassellati, "Theory of Mind for a Humanoid Robot," *Autonomous Robots*, 12, pp. 13-24. 2002.
9. W-K. Song D-J. Kim, J-S. Kim and Z. Bien. "Visual Servoing for a User's Mouth with Effective Intention Reading in a Wheelchair-based Robotic Arm," in *Procceedings fo the IEEE Int. Conference on Robotics and Automation*, pp. 3662-3667. Seoul, Korea. 2001.
10. M. Mazo et al. "An Integral System for Assisted Mobility," *IEEE Robotics and Automation Magazine*, vol 8, no 1, pp. 46-56. March 2001.
11. Y. Matsumoto, T. Ino, T. Ogasawara. "Fast image-based tracking by selective pixel integration," in *Proc. IEEE Int. Workshop on Robot and Human Interactive Communication.* Paris, France. 2001.
12. K. Toyama. Prolegomena for robust face tracking. MSR-TR-98-65. Microsoft Research, November 1998.
13. G.D. Finlayson, B. Shiele and J.L. Crowley. Comprehensive colour normalization. *Proc. European Conf. on Computer Vison (ECCV). Vol. I*, 475–490, Freiburg, Germany. 1998.
14. J. Yang, W. Lu, A. Waibel. Skin-color modeling and adaptation. *Proc. Third Asian Conference on Computer Vision Vol. II*, 142-147. 1998.
15. Y. Raja, S.J. McKenna, S. Gong. Colour model selection and adaptation in dynamic scenes. *Proc. European Conference on Computer Vision. Vol. I*, 460–474. 1998.
16. Y. Wu, Q. Liu and T.S. Huang. Robust real-time hand localization by self-organizing color segmentation. *Proceedings RATFG'99*, 161–166. 1999.
17. D. Berwick and S.W. Lee. A chromaticity space for specularity-, illumination color- and illumination pose invariant 3-d object recognition. *Proc. of the Int. Conf. on Computer Vision.* Bombay, India. 1998.
18. M. Störring, H.J. Andersen and E. Granum. Estimation of the illuminant colour from human skin colour. *Proc. of the Int. Conference on Automatic Face and Gesture Recognition* (FG'00), 64–69, Grenoble. France. 2000.
19. M. D'Zmura and P. Lennie. Mechanisms of colour constancy. *Journal of the Optical Society of America A*, 3: 1662–1672, 1986.
20. G. Buchsbaum. A spatial processor model for object colour perception. *Journal of the Fanklin Institute*, 310: 1-26, 1980.
21. R. Gershon, A.D. Jepson and J.K. Tsotsos. From [R,G,B] to surface reflectance: Computing color constant descriptors in images. *Proc. Int. Joint Conf. on Artificial Intelligence*, 755–758, 1987.
22. Y. Cheng. Mean shift, mode seeking and clustering. *IEEE Trans. on Pattern Analysis and Machine Intelligence*, 17: 790-799, 1995.
23. G. Bradski. Computer Vision face tracking for use in a perceptual user interface. *Proc. of Workshop on applications of Computer Vision, WACV'98*, 214–219, 1998.
24. J. L. Crowley and J. Schwerdt. Robust tracking and compression for video communication. *Proc. of the Int. Workshop on Recognition, Analysis and Tracking of Faces and Gestures in Real-Time* (RATFG'99), 2–9, Corfu. Greece. 1999.

25. M. Soriano, B. Martinkauppi, S. Huovinen, M. Laaksonen. Skin detection in video under changing illumination conditions. *Proc. of the Int. Conference on Automatic Face and Gesture Recognition* (FG'00), 839–842, Grenoble. France. 2000.
26. F. Lerasle V. Ayala, J.B. Hayet and M. Devy, Visual localization of a mobile robot in indoor environments using planar landmarks. *Proceedings Intelligent Robots and Systems, 2000.* IEEE, 2000, pp. 275–280.
27. G. Simon, A. Fitzgibbon, and A. Zisserman, Markerless tracking using planar structures in the scene, *Proc. Int. Symposium on Augmented Reality*, October 2000.
28. M. J. Black and Y. Yacoob. "Recognizing facial expressions in image sequences using local parameterized models of image motion," *Int. Journal of Computer Vision*, vol. 25, no. 1, pp. 23–48, 1997.
29. C. Thorpe F. Dellaert and S. Thrun. "Super-resolved texture tracking of planar surface patches," in *Proceedings Intelligent Robots and Systems.* IEEE, 1998, pp. 197–203.
30. M Irani and P. Anandan. "All about direct methods," in *Vision Algorithms: Theory and practice*, W. Triggs, A. Zisserman, and R. Szeliski, Eds. Springer-Verlag, 1999.
31. P. H. S. Torr and A. Zisserman. "Feature based methods for structure and motion estimation," in *Vision Algorithms: Theory and practice*, W. Triggs, A. Zisserman, and R. Szeliski, Eds. Springer-Verlag, 1999, pp. 278–295.
32. Gregory D. Hager and Peter N. Belhumeur. "Efficient region tracking with parametric models of geometry and illumination," *IEEE Transactions on Pattern Analisys and Machine Intelligence*, vol. 20, no. 10, pp. 1025–1039, 1998.
33. F. Dellaert and R. Collins, "Fast image-based tracking by selective pixel integration," in *ICCV99 Workshop on frame-rate applications*, 1999.

Subject Index

Contributors

T.C. Ahn
Intelligent Information Control & System Lab
School of Electrical & Electronic Engineering
Won-Kwang University
344-2 Shinyong-Dong, Iksan
Chon-Buk, 570-749
Korea

José R. Álvarez
Departamento de Inteligencia Artificial
Facultad de Ciencias y ETSI Informática, UNED
28040 Madrid
Spain

Marcelo H. Ang Jr.
Department of Mechanical Engineering
National University of Singapore
Singapore 119260

Fabián Arrebola
Departamento de Tecnología Electrónica
ETSI Telecomunicación
Universidad de Málaga
Málaga, 29071
Spain

Marco Balsi
Dipartimento di Ingegneria Elettronica
Università "La Sapienza"
Rome, 00184
Italy

Antonio Bandera
Dpto. Tecnología Electrónica
ETSI Telecomunicación
Universidad de Málaga
Málaga, 29071
Spain

Luis Baumela
Departamento de Inteligencia Artificial
Universidad Politécnica de Madrid
Campus de Montegancedo
28660 Madrid
Spain

M. Borkowski
Department of Electrical and Computer Engineering
University of Manitoba
Winnipeg, Manitoba R3T 5V6
Canada

José Miguel Buenaposada
Departamento de Inteligencia Artificial
Universidad Politécnica de Madrid
Campus de Montegancedo
28660 Madrid
Spain

Victor Callaghan
Department of Computer Sciences
University of Essex
Wivenhoe Park
Colchester CO4 3SQ
England, UK

Pelegrín Camacho
Departamento de Tecnología Electrónica
ETSI Telecomunicación
Universidad de Málaga
Málaga, 29071
Spain

Graham Clarke
Department of Computer Sciences
University of Essex
Wivenhoe Park
Colchester CO4 3SQ
England, UK

Martin Colley
Department of Computer Sciences
University of Essex
Wivenhoe Park
Colchester CO4 3SQ
England, UK

V. Degtyaryov
Department of Electrical and Computer Engineering
University of Manitoba
Winnipeg, Manitoba R3T 5V6
Canada

Ana E. Delgado
Departamento de Inteligencia Artificial
Facultad de Ciencias y ETSI Informática, UNED
28040 Madrid
Spain

J. M. Dolan
The Robotics Institute
Carnegie Mellon University
Pittsburgh, PA 15213
USA

Hakan Duman
Department of Computer Sciences
University of Essex
Wivenhoe Park
Colchester CO4 3SQ
England, UK

Meng Joo Er
School of Electrical and Electronic Engineering
Nanyang Technological University
Singapore 639798

Yang Gao
School of Electrical and Electronic Engineering
Nanyang Technological University
Singapore 639798

Hani Hagras
Department of Computer Sciences
University of Essex
Wivenhoe Park
Colchester CO4 3SQ
England, UK

A. Howard
Telerobotic Research & Applications Group
NASA Jet Propulsion Laboratory
Caltech, Pasadena, CA 91109
USA

T. Huntsberger
Mechanical and Robotic Technologies
NASA Jet Propulsion Laboratory
Caltech, Pasadena, CA 91109
USA

Tong-Heng Lee
Department of Electrical and Computer Engineering
National University of Singapore
Singapore 117576

Javier de Lope
Department of Artificial Intelligence
Faculty of Computer Science
Universidad Politécnica de Madrid
Campus de Montegancedo
28660 Madrid
Spain

Darío Maravall
Department of Artificial Intelligence
Faculty of Computer Science
Universidad Politécnica de Madrid
Campus de Montegancedo
28660 Madrid
Spain

José Mira
Departamento de Inteligencia Artificial
Facultad de Ciencias y ETSI
Informática, UNED
28040 Madrid
Spain

M. Oussalah
City University, CSR
10 Northampton Square, EC1V
0HB, London
UK

Félix de la Paz
Departamento de Inteligencia Artificial
Facultad de Ciencias y ETSI
Informática, UNED
28040 Madrid
Spain

Eduardo Pérez
Departamento de Tecnología Electrónica
ETSI Telecomunicación
Universidad de Málaga
Málaga, 29071
Spain

J.F. Peters
Department of Electrical and
Computer Engineering
University of Manitoba
Winnipeg, Manitoba R3T 5V6
Canada

Alberto Poncela
Departamento deTecnología Electrónica
ETSI Telecomunicación
Universidad de Málaga
Málaga, 29071
Spain

S. Ramanna
Department of Electrical and
Computer Engineering
University of Manitoba
Winnipeg, Manitoba R3T 5V6
Canada

Bernd Rössler
Faculty of Technology
University of Bielefeld
Bielefeld 33501
Germany

Alessandro Saffiotti
Center for Applied Autonomous Sensor Systems, Orebro University
S-70182 Orebro
Sweden

Francisco Sandoval
Departamento de Tecnología Electrónica
ETSI Telecomunicación
Universidad de Málaga
Málaga, 29071
Spain

Ya Lei Sun
School of Electrical and Electronic Engineering
Nanyang Technological University
Singapore 639798

A. Trebi-Ollennu
Mechanical and Robotic Technologies
NASA Jet Propulsion Laboratory
Caltech, Pasadena, CA 91109
USA

E. Tunstel
Robotic Vehicles Group
NASA Jet Propulsion Laboratory
Caltech, Pasadena, CA 91109
USA

Cristina Urdiales
Departamento de Tecnología Electrónica
ETSI Telecomunicación
Universidad de Málaga
Málaga, 29071
Spain

Prahlad Vadakkepat
Department of Electrical and Computer Engineering
National University of Singapore
Singapore 117576

Xavier Vilasís-Cardona
Departament d'Electrònica
Enginyeria i Arquitectura La Salle
Universitat Ramon Llull
Barcelona
Spain

Zbigniew Wasik
Center for Applied Autonomous Sensor Systems, Orebro University
S-70182 Orebro
Sweden

Anthony Wong
Department of Mechanical Engineering
National University of Singapore
Singapore 119260

Liu Xin
Department of Electrical and Computer Engineering
National University of Singapore
Singapore 117576

Jianwei Zhang
Faculty of Technology
University of Bielefeld
Bielefeld 33501
Germany

GPSR Compliance
The European Union's (EU) General Product Safety Regulation (GPSR) is a set of rules that requires consumer products to be safe and our obligations to ensure this.

If you have any concerns about our products, you can contact us on

ProductSafety@springernature.com

In case Publisher is established outside the EU, the EU authorized representative is:

Springer Nature Customer Service Center GmbH
Europaplatz 3
69115 Heidelberg, Germany

www.ingramcontent.com/pod-product-compliance
Ingram Content Group UK Ltd.
Pitfield, Milton Keynes, MK11 3LW, UK
UKHW021900190726
13853UKWH00003B/1358

* 9 7 8 3 6 6 2 0 0 2 9 7 1 *